MARINE POLLUTION AND PHYSIOLOGY: Recent Advances

THE BELLE W. BARUCH LIBRARY IN MARINE SCIENCE

1 ESTUARINE MICROBIAL ECOLOGY
2 SYMBIOSIS IN THE SEA
3 PHYSIOLOGICAL ECOLOGY OF MARINE ORGANISMS
4 BIOLOGICAL RHYTHMS IN THE MARINE ENVIRONMENT
5 THE MECHANISMS OF MINERALIZATION IN THE INVERTEBRATES AND PLANTS
6 ECOLOGY OF MARINE BENTHOS
7 ESTUARINE TRANSPORT PROCESSES
8 MARSH-ESTUARINE SYSTEMS SIMULATION
9 REPRODUCTIVE ECOLOGY OF MARINE INVERTEBRATES
10 ADVANCED CONCEPTS IN OCEAN MEASUREMENTS FOR MARINE BIOLOGY
11 MARINE BENTHIC DYNAMICS
12 PROCESSES IN MARINE REMOTE SENSING

THE BELLE W. BARUCH LIBRARY IN MARINE SCIENCE NUMBER 13

Marine Pollution and Physiology: Recent Advances

Edited by F. John Vernberg, Frederick P. Thurberg, Anthony Calabrese, and Winona Vernberg

Published for the Belle W. Baruch Institute for Marine Biology and Coastal Research by the

UNIVERSITY OF SOUTH CAROLINA PRESS

Published in Columbia, South Carolina, by the
University of South Carolina Press

First Edition

Manufactured in the United States of America

Library of Congress Cataloging in Publication Data
Main entry under title:

Marine pollution and physiology, recent advances.

(Belle W. Baruch library in marine science; no. 13)
"Papers . . . presented at a symposium entitled Physiology and pollution of marine organisms, held at Mystic, Connecticut on November 1–3, 1983. The symposium was sponsored jointly by the Belle W. Baruch Institute for Marine Biology and Coastal Research and the School of Public Health, both at the University of South Carolina, and the Northeast Fisheries Center of the National Marine Fisheries Services, NOAA."
Includes index.
1. Marine fauna—Physiology—Congresses. 2. Marine pollution—Physiological effect—Congresses. I. Vernberg, Winona B., 1924– . II. Belle W. Baruch Institute for Marine Biology and Coastal Research. III. University of South Carolina. School of Public Health. IV. Northeast Fisheries Center (U.S.) V. Series.
QL121.M284 1985 591.2'4 84-27870
ISBN 0-87249-446-2

PARTICIPANTS

CONTRIBUTORS

MARIUS BROUWER, Duke University Marine Laboratory, Marine Biomedical Center, Beaufort, NC 28516.

DONALD W. BROWN, Northwest and Alaska Fisheries Center, Auke Bay Laboratory, National Marine Fisheries Service, National Oceanic and Atmospheric Administration, P.O. Box 210155, Auke Bay, AK 99821. Present address: Northwest and Alaska Fisheries Center, National Marine Fisheries Service, National Oceanic and Atmospheric Administration, 2725 Montlake Blvd. East, Seattle, WA 98112.

L.E. BURRIDGE, Fisheries and Environmental Sciences, Department of Fisheries and Oceans, Biological Station, St. Andrews, N.B. E0G 2X0, CANADA.

A. CALABRESE, National Marines Fisheries Service, National Oceanic and Atmospheric Administration, Northeast Fisheries Center, Milford Laboratory, Milford, CT 06460.

ANGELA CANTELMO-CRISTINI, Ramapo College of New Jersey, Mahwah, NJ 07430.

D. LOREN CHEATHAM, deceased.

PHILIP J. CONKLIN, Department of Biology, University of West Florida, Pensacola, FL 32514.

J.D. COSTLOW, Duke Marine Laboratory, Beaufort, NC 28516.

DANIEL G. DOUGHTIE, Department of Biology, University of West Florida, Pensacola, FL 32514.

DAVID W. ENGEL, National Marine Fisheries Service, National Oceanic and Atmospheric Administration, Southeast Fisheries Center, Beaufort Laboratory, Beaufort, NC 28516.

A. FARMANFARMAIAN, Physiology - Nelson Biological Laboratory, Rutgers University, Piscataway, NJ 08854.

LIU FAYI, Institute of Oceanology, Academia Sinica, Qingdao, Peoples' Republic of China.

R.B. FORWARD, JR., Duke Marine Laboratory, Beaufort, NC 28516.

JUDITH C. FOSTER, Bowdoin College Marine Research Laboratory, Bowdoin College Hydrocarbon Research Center, Brunswick, ME 04011.

M.A. FREADMAN, John B. Pierce Foundation, Yale University, 290 Congress Avenue, New Haven, CT 06519. Present address: Biology Department, Woods Hole Oceanographic Institution, Woods Hole, MA 02543.

D.B. GANT, Oregon State University, Marine Science Center, Newport, OR 97365.

STEPHEN G. GEORGE, N.E.R.C., Institute of Marine Biochemistry, St. Fitticks Road, Aberdeen, Scotland.

EDWARD S. GILFILLAN, Bowdoin College Marine Research Laboratory, Bowdoin College Hydrocarbon Research Center, Brunswick, ME 04011.

LINDA GONZALEZ, Bowdoin College Marine Research Laboratory, Bowdoin College Hydrocarbon Research Center, Brunswick, ME 04011.

E. GOULD, National Oceanic and Atmospheric Administration, National Marine Fisheries Service, Northeast Fisheries Center, Milford Laboratory, Milford, CT 06460.

R.A. GREIG, National Oceanic and Atmospheric Administration, National Marine Fisheries Service, Northeast Fisheries Center, Milford Laboratory, Milford, CT 06460.

RUTH GUTJAHR-GOBELL, Edgerton Research Laboratory, New England Aquarium, Central Wharf, Boston, MA 02110.

SHERRY A. HANSON, Bowdoin College Marine Research Laboratory, Bowdoin College Hydrocarbon Research Center, Brunswick, ME 04011.

BRUCE R. HARGREAVES, Department of Biology, Building #31, Lehigh University, Bethlehem, PA 18015.

K. HAYA, Fisheries and Environmental Sciences, Department of Fisheries and Oceans, Biological Station, St. Andrews, N.B. E0G 2X0, Canada.

FRANK E. HOSPOD, Ramapo College of New Jersey, Mahwah, NJ 07430.

D. MICHAEL JOHNS, Edgerton Research Laboratory, New England Aquarium, Central Wharf, Boston, MA 02110.

J.P. KAKAREKA, Edgerton Research Laboratory, New England Aquarium, Central Wharf, Boston, MA 02110.

SID KORN, Northwest and Alaska Fisheries Center, Auke Bay Laboratory, National Marine Fisheries Service, National Oceanic and Atmospheric Administration, P.O. Box 210155, Auke Bay, AK 99821.

ROBERT J. LAZELL, Ramapo College of New Jersey, Mahwah, NJ 07430.

B.C. MARKS, National Oceanic and Atmospheric Administration, National Marine Fisheries Service, Northeast Fisheries Center, Milford Laboratory, Milford, CT 06460.

CHARLES L. MCKENNEY, JR., U.S. Environmental Protection Agency, Environmental Research Laboratory, Gulf Breeze, FL 32561.

E.U. MEYHOFER, Northeastern University, Marine Science Laboratory, Nahant, MA 01908.

WILLIAM METCALF, Northwest and Alaska Fisheries Center, Auke Bay Laboratory, National Marine Fisheries Service,

National Oceanic and Atmospheric Administration, P.O. Box 210155, Auke Bay, AK 99821.

DOUGLAS P. MIDDAUGH, U.S. Environmental Protection Agency, Gulf Breeze Environmental Research Laboratory, Sabine Island, Gulf Breeze, FL 32561.

M.P. MORSE, Northeastern University, Marine Science Laboratory, Nahant, MA 01908.

JERRY M. NEFF, Battelle New England Marine Research Laboratory, 397 Washington St., Duxbury, MA 02332.

DAVID S. PAGE, Bowdoin College Marine Research Laboratory, Bowdoin College Hydrocarbon Research Center, Brunswick, ME 04011.

J.B. PEARCE, Sandy Hook Laboratory, Northeast Fisheries Center, National Marine Fisheries Service, National Oceanic and Atmospheric Administration, U.S. Department of Commerce, Highlands, NJ 07732.

ERIN PENDERGAST, Bowdoin College Marine Research Laboratory, Bowdoin College Hydrocarbon Research Center, Brunswick, ME 04011.

B.A. PENNEY, Edgerton Research Laboratory, New England Aquarium, Central Wharf, Boston, MA 02110.

K. RANGA RAO, Department of Biology, University of West Florida, Pensacola, FL 32514.

M. RENNA, Department of Zoology and Physiology, Rutgers University, Newark, NJ 07102.

STANLEY D. RICE, Northwest and Alaska Fisheries Center, Auke Bay Laboratory, National Marine Fisheries Service, National Oceanic and Atmospheric Administration, P.O. Box 210155, Auke Bay, AK 99821.

W.E. ROBINSON, Edgerton Research Laboratory, New England Aquarium, Central Wharf, Boston, MA 02110.

D. RUSANOWSKY, National Oceanic and Atmospheric Administration, National Marine Fisheries Service, Northeast

Fisheries Center, Milford Laboratory, Milford, CT 06460.

T.D. SABOURIN, BATTELLE, Columbus Laboratories, 505 King Ave., Columbus, OH 43201.

TOMMY I. SAMMONS, S.C. Department Health and Environmental Control, 2600 Bull Street, Columbia, SC 29202.

JACK P. SCHWARTZ, Institute of Marine Science, University of Alaska, Fairbanks, AK 99701. Present address: Auke Bay Fisheries Laboratory, P.O. Box 210155, Auke Bay, AK 99821.

GEOFFREY I. SCOTT, School of Public Health, University of South Carolina, Columbia, SC 29208.

J.R. SHARP, Department of Biology, Southeast Missouri State University, Cape Girardeau, MO 63701.

CARL J. SINDERMANN, Sandy Hook Laboratory, Northeast Fisheries Center, National Marine Fisheries Service, National Oceanic and Atmospheric Administration, U.S. Department of Commerce, Highlands, NJ 07732.

ROY L. SMITH, Department of Biology, Building #31, Lehigh University, Bethlehem, PA 18015.

R. SOCCI, Physiology - Nelson Biological Laboratory, Rutgers University, Piscataway, NJ 08854.

WILLIAM B. STICKLE, JR., Department of Zoology and Physiology, Louisiana State University, Baton Rouge, LA 70803.

JOANNE S. STOLEN, National Marine Fisheries Service, Northeast Fisheries Center, Sandy Hook Laboratory, Highlands, NJ 07732.

PETER THOMAS, The University of Texas at Austin, Port Aransas Marine Laboratory, Port Aransas, TX 78373.

ROBERT E. THOMAS, Department of Biological Science, Chico State University, Chico, CA 95926.

FREDERICK P. THURBERG, National Marine Fisheries Service, National Oceanic and Atmospheric Administration, Northeast Fisheries Center, Milford Laboratory, Milford, CT 06460.

S. VAIDYA, Department of Zoology and Physiology, Rutgers University, Newark, NJ 07102.

DONNA VALLAS, Bowdoin College Marine Research Laboratory, Bowdoin College Hydrocarbon Research Center, Brunswick, ME 04011.

L. VAN EECKHAUTE, Aquatic Industries Ltd., St. Andrews, N.B. EOG 2X0, Canada.

ALDO VIARENGO, Instituto di Fisiologia Generale, Universita de Genova, Genova, Italia.

CATHERINE VILLARS, Department of Zoology and Physiology, Louisiana State University, Baton Rouge, LA 70803.

B.A. WAIWOOD, Fisheries and Environmental Sciences, Department of Fisheries and Oceans, Biological Station, St. Andrews, N.B. EOG 2X0, Canada.

L.J. WEBER, Oregon State University, Marine Science Center, Newport, OR 97365.

J.S. WEIS, Department of Zoology and Physiology, Rutgers University, Newark, NJ 07102.

P. WEIS, Department of Anatomy, UMDNJ-NJ Medical School, Newark, NJ 07103.

J.E. WILSON, Duke University Marine Laboratory, Beaufort, NC 28516.

CAROLYN A. YEVICH, Environmental Research Laboratory, South Ferry Road, Narragansett, RI 02882.

PAUL P. YEVICH, Environmental Research Laboratory, South Ferry Road, Narragansett, RI 02882.

PREFACE

The papers in this volume were presented at a symposium entitled Physiology and Pollution of Marine Organisms held at Mystic, Connecticut on November 1-3, 1983. The symposium was sponsored jointly by the Belle W. Baruch Institute for Marine Biology and Coastal Research and the School of Public Health, both at the University of South Carolina, and the Northeast Fisheries Center of the National Marine Fisheries Service, NOAA.

This is the sixth symposium organized around the theme of physiological effects of pollutants. Since the first symposium was held in 1973, there has been an explosive increase in the body of knowledge dealing with the physiological response of marine animals to various types of pollutants. Research emphasis has dramatically shifted during this period of time. Initially greater attention was given to acute lethal effects, but there has been an evolution from physiological studies on intact organisms to tissues to cells to molecules and from lethal to sublethal responses. The papers included in the present volume represent current research directed toward the important question of physiological response to pollutant stress.

We are indebted to the authors for their cooperation, to the reviewers of all manuscripts, and especially to the staff of the University of South Carolina for their assistance in many ways.

CONTENTS

HEAVY METALS

PHYSIOLOGICAL EFFECTS OF MARINE POLLUTANT STRESS—A MANAGER'S PERSPECTIVE

J.B. Pearce

Beginning in the late 1960's, increased attention was paid to the problem of marine pollution. Initially, the principal research endeavors tended to concentrate on issues such as ocean dumping of sewage sludge, contaminated dredging spoils, and industrial wastes. When marine scientists recognized that certain contaminants, such as the pesticide DDT, were affecting several forms of marine life, including the ospreys, research was devoted to gaining an understanding of the effects of organochlorines and other synthetic organic substances. Some of the early research included how PCBs might affect phytoplankton productivity (O'Conner *et al.*, 1978); some individuals felt that there was a possibility that much of the phytoplankton of the world's seas would be affected and thus upset the balance of the oxygen and carbon dioxide cycles.

In these early research studies, both laboratory and field observations and experiments were conducted. It should be noted, however, that many laboratory studies of pollutants were never effectively "bridged" to field observations, or syntheses developed based on the two approaches. It has been extremely difficult for so-called environmental managers to relate the two rather different types of studies.

In recent years, scientists have begun to conduct more complex experimental studies in the field. A range of experimental protocols has been used to demonstrate how materials such as sewage sludge, toxic metals, petroleum hydrocarbons, and other organic contaminants might perturb the behavior of organisms in field situations (Olla *et al.*, 1980; Olla *et al.*, 1983). Other scientists have attempted to measure physiological and biochemical change that result from exposure to a range of gross or specific contaminants.

In the mid 1970's, there was increased international discussion devoted to environmental studies concerned with the relative efficacy of biological effects techniques (McIntyre & Pearce, 1980). That is, where scientists were attempting to determine *sources*, *fates*, and *effects* of contaminants, it was held important to measure how animals might respond from a physiological or biochemical point of view in a monitoring situation. This meant that there was a necessity for performing field biological effects studies in estuarine, coastal, and shelf waters to establish benchmarks for how individual organisms would respond to a range of contaminants under actual field conditions and over time; the basis for status and trends.

Major issues in relation to biological effects studies culminated in a workshop sponsored by the International Council for the Exploration of the Sea (ICES). This meeting, held at the Duke University Marine Laboratory, resulted in a significant publication (McIntyre & Pearce, 1980) which documented a number of approaches that might be used to study effects of contaminants in the field and, further, to monitor trends in space and time as pollution loading either increased or decreased in particular geographic regions.

At about the same time that marine scientists were attempting to assess accurately the impacts of contamination and pollution on living marine resources, many environmental issues were being taken to court or to hearings for review, evaluation, and decisions. In some instances it was possible to show in a court hearing how contaminants, for example a range of toxic metals, might affect the well-being of living marine resources based on laboratory studies. In at least one court case, the judiciary ruled against a company that was discharging trace metals into a domestic sewage system which drained wastes into Raritan Bay. In other cases, however, the court ruled against complainant agencies and environmentalists, and for the discharger or dumper. These court decisions were based on the legislation which regulated ocean discharges or dumping; legislation tended to use undefined phrases such as *unreasonable degradation* or *assimilative capacity*. Moreover, scientists interested in the effects of pollution on living marine resources have not, to date, related the results of controlled laboratory and field effects experiments to field measurements of contaminant levels, or contaminant effects which were observed to occur in relation to populations or communities.

Very recently, increasing attention has been given to the matter of risk or hazard assessments. In their simplest forms, hazard assessments suggest a range of environmental information or data arrayed in such a way that a judgement can be made as to whether a particular activity has had an impact on a specific resource, including human populations. Hazard or risk assessments have ranged from the well-known statistical calculations and predictions concerned with cigarette smoking to attempting to document what the total effects might be of Kepone discharge into the James River, Virginia. In many instances it has been possible to show that a contaminant or activity has a definite and specific effect on a resource. Once an effect is known, an evaluation of the relative worth of the resource to society can be made based on two disparate matters: 1) what are the benefits associated with an activity such as discharge of a chemical contaminant into a waterway, and 2) what is the relative economic worth of the resources that might be affected. It has been emphasized repeatedly that, in terms of living marine resources, one has to make projections as to their worth in the future. A particular resource, for example clams, might yield an initial landed revenue of \$100,000. However, by the time the resource is processed, sold, and consumed, it might be worth \$500,000. Beyond this, one can project that if the resource continues to reproduce itself, recruit effectively, and grow to be harvested in future years, then the resource may be worth millions of dollars when projected through future decades. Additionally, the material which will be or is being discharged might so compromise the environment that a resource would be worthless in future decades.

Many organizations are now looking to develop hazard assessments based on scientific data, including the results of laboratory and field observations (Kimerle *et al*., 1978; Gentile *et al*., 1983). Once such assessments have been developed, they can play an important role in future environmental decisions. Obviously, future laboratory and field environmental studies must be developed in a way that resulting data can be used effectively in risk assessments. Physical and chemical oceanographers must work closely with biologists to determine the sources and the fates of contaminants. Only when it is known how contaminants become distributed and sequestered in the physical environment, can one begin to make accurate assessments in regard to their effects on the living marine resources.

Other disciplines are important as well. In the past, physiological and biochemical data generally were not treated with sophisticated, state-of-the-art statistics. Increasingly, the results of environmental studies, whether done in the laboratory or field, must be accomplished such that resulting data can be placed into complex matrices or arrays and thus used in a total environmental assessment. Once the data have been developed in the form of a risk or hazard assessment, the resulting information must be presented or "packaged" in a way that it can be easily understood by non-scientists. As already noted, several principal environmental cases have come up for judicial review and decision-making; judges and juries usually are not oceanographers or marine scientists! Many environmental managers also cannot be considered scientists or experts in the truest sense. Yet they often must take a wide range of information and make important decisions that they are <u>not</u> fully able to understand themselves, at least in the context of how complex data have been developed in the past.

In order to advance the management of marine habitat quality, the National Marine Fisheries Service has developed a series of Habitat Conservation strategies (Federal Register [NOAA], 1983) which include the implementation of Regional Action Plans (RAP). In the northeast, personnel of the NMFS Regional Office and the Northeast Fisheries Center are working together to identify principal environmental issues and problems that involve living marine resources (LMRs). Where these problems have been delineated, the Regional Action Plan (RAP) concentrates on developing assessments of principal habitat areas along the east coast. These areas are designated as Water Management Units (WMUs) and are characterized by having certain similarities in regard to bathymetric ranges, sediment types, hydrographic conditions, and biota (Fig. 1). For instance, there is a coastal mid-Atlantic WMU as well as a shelf and slope WMU. Other WMUs include a unique hydrographic regime characterized by Georges Bank and its overlying waters, and further to the north the coastal Gulf of Maine and Central Gulf of Maine WMUs.

NMRS has begun the process of developing regional assessments for these WMUs. Recently a description of the Middle Atlantic shelf/slope waters that includes the

In recent years, scientists have begun to conduct more complex experimental studies in the field. A range of so-called 106-mile disposal site (Pearce <u>et al</u>., 1983) was completed. This will provide environmental managers with access to standard data on living marine resources, hydrographic conditions (Ingham et al., 1982), and quality of habitat for this important area; undoubtedly an area which will soon begin receiving a range of wastes: possibly sewage sludge, and certainly additional industrial and toxic wastes. This site characterization will also be used to evaluate petroleum exploration and production on the outer portions of the shelf and the slope. More brief regional assessments for several key east coast and Scandinavian estuaries have recently been published by the Internal Council for the Exploration of the Seas (ICES) (Pearce, 1983).

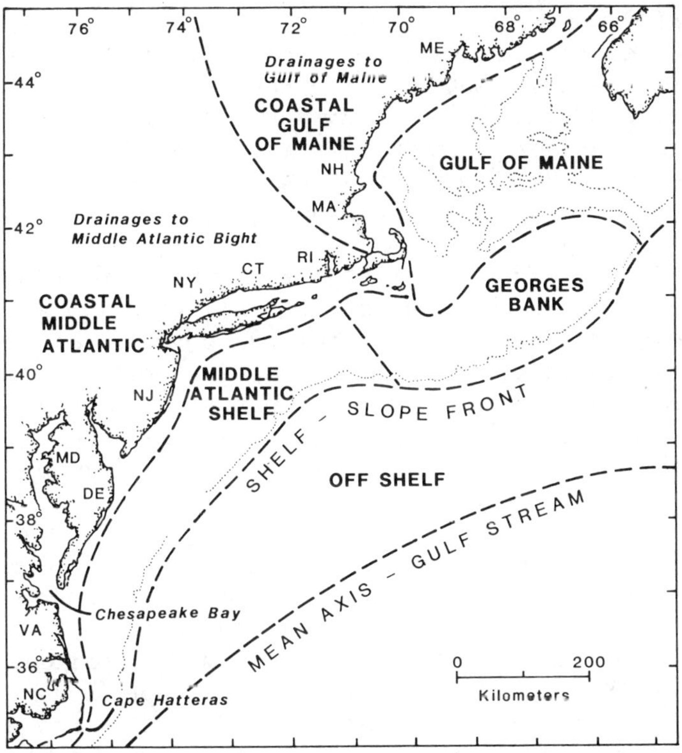

Fig. 1. Northeast Regional Action Plan (RAP), Water Management Units (WMU). This chart shows the boundaries of the principal WMUs in the Northeast Region. The WMUs are being described in detail, with information on the various categories of marine life and the status of their habitat given in sufficient detail to generate hazard assessments (see Fig. 3).

Such regional assessments identify the resources at risk. It remains for environmental and marine scientists to develop the risk assessments based on laboratory and field experiments and observations within the various WMUs. Obviously, certain of this information will have a generic application. For example, data on possible effects of the various components of sewage sludge would be applicable to a range of living marine resources, regardless of the geographic area that might be involved. Likewise, the effects of individual toxic wastes, as measured in the laboratory or field, will have application to fauna in the Gulf of Maine as well as off of Delaware Bay or even the Cornish Coast of the United Kingdom.

Such assessments will be paramount in future multiple-use activities in relation to estuaries, coastal zones, and shelf habitats (Fig. 1). There is no doubt that because of the nature of modern society, marine waters will increasingly be used not only for recreational and commercial fishing, but also for purposes of transportation, industrialization, domestic and recreational sitings, and waste disposal (Pearce, 1983). In the past, the results of these various activities have often resulted in conflicts between individuals having different interests in several economic elements. For instance, New York City and northern New Jersey communities have traditionally disposed of sewage sludge in nearshore waters. There is sufficient evidence to suggest that this has resulted in stresses in these habitats and, in many cases, the contaminants known to be associated with sludge and other wastes have often resulted in a buildup in body burdens of contaminants in fish species consumed by man (Belton *et al.*, 1982). Likewise, local communities have an interest in developing harbor areas to increase national and international surface transportation (Pacheco, 1984). Boston, New York, Delaware Bay, and Chesapeake Bay communities all have an interest in increasing the size and depth of harbor areas to accommodate ever larger vessels to move energy-related materials as well as other commodities. This requires the dredging of harbors to increase channel depths and turning and berthing areas. Increasingly, the public and many environmental groups are concerned about where the more contaminated dredging spoils will be placed. In the New York Bight area, should these spoils continue to be dumped at the so-called dredged material disposal area or should they be taken further offshore? Alternatively, should the materials continue to be dumped where they presently are, but a requirement be made that they be covered or capped with clean materials? Such questions can only be answered through the development of regional assessments and the concomitant production of hazard or risk assessments, which allow managers to make a proper evaluation of the situations.

At the same time as society is looking to the oceans to receive additional wastes and be used for transportation, other communities see the coastal zone as a place to be developed for recreational purposes. It is said that some 22 to 30 million people live within a one or two hour drive from the New Jersey shore. It often seems as if they have all arrived at the scene at the same time! Increasingly, people have more leisure time and wish to relax at the sea front while swimming, surfing or fishing. Again, this is another use of the coastal environment which is compromised by industrialization, waste disposal, and the numerous activities that accompany such developments. If mankind is to use the estuaries, coastal zones, and shelf areas in a multiplicity of ways, we must have a basis for rational decision-making. Thus, we come back to the matter of multiple-uses of the coastal zone and how these are best expedited; this must be done without compromising the long-term fishery production or biological yields of such habitats.

Perhaps even more important than the actual impact of contaminants on the fish stocks in some areas, is the matter of the public's perception that seafoods are contaminated and unfit to eat. Such matters must enter into risk assessments in a multiple-use mode.

It is now up to groups such as experimentalists, physiologists, biochemists, geneticists, and behaviorists to develop information in a way that will be useful to the so-called environmental managers. Increasingly, legislators, managers, and the public find fault with scientists in general because they have not provided answers to every single problem that seems to require a scientific solution. This has not been because of the lack of research and monitoring accomplished to date; to a great degree the position has resulted from a failure to use wisely and disseminate the data that we presently have.

To apply scientific information usefully in environmental decision-making will require that all scientists, field- or laboratory-oriented, pay more attention to how their data are collected, analyzed, manipulated, and used in behalf of environmental decision-making. The first steps leading to such decision-making have been taken by the United Nations' Group of Experts on Scientific Aspects of

Marine Pollution (GESAMP), which developed the general protocols necessary to effective research and long-term monitoring (Fig. 2). While many of the steps outlined by GESAMP seem obvious now, the sequence in which they should be conducted has rarely been followed. More recently the United States Environmental Protection Agency (US EPA) has been developing specific procedures to be followed in developing risk or hazard assessments, especially in relation to ocean disposal of a variety of wastes (Fig. 3). The challenge for the coming decade will be to develop laboratory research and field experimental data so that they can be used collectively to provide highly quantitative assessments useful in environmental decision-making. If scientists cannot, or will not, do this, then legislators, regulators, and managers are almost certain to turn to other parties or institutions for information on which to base their decisions.

1. Identification of contaminant or physically degraded "hot spots" using chemistry and biological effects monitoring.
2. Identification of trends; spatial and temporal.
3. Quantification of Effects:
   - 3.1 Sublethal effects,
   - 3.2 mortality, and
   - 3.3 effects on stocks.
4. Risk assessments (see Fig. 3).
5. Analyses of causation (field and laboratory experiments).
6. Management of fisheries habitats:
   - 6.1 Regulations,
   - 6.2 legislation, and
   - 6.3 fisheries management plans leading to:
     - 6.3.1 Maintenance of yields or increased yields, and
     - 6.3.2 Habitat enhancements or mitigation.

Fig. 2. The steps that have been agreed upon as leading to effective marine habitat management.

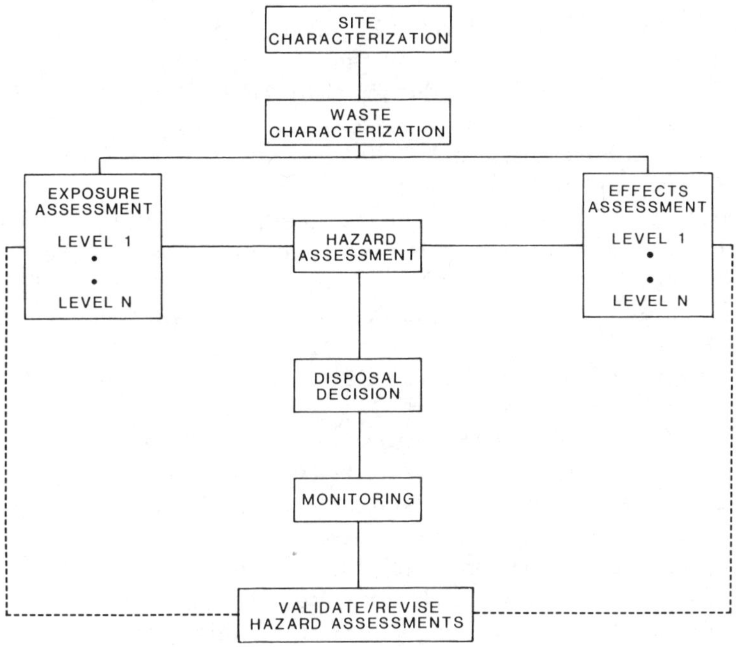

Fig. 3. Strategy for Ocean Disposal Hazard Assessments. From Bierman *et al*. (1982), Research Strategy for Ocean Disposal: Conceptual Framework and Case Study, US EPA, Narragansett, RI (Presented at the Workshop on Hazard Assessment for Complex Effluents, Cody, Wyoming, Aug. 22-27, 1982).

LITERATURE CITED

Belton, T.J., B.E. Ruppel, K. Lockwood, and M. Boriek. 1983. PCBs in Selected Finfish with Limited Chlordane Data (1981-1982). NJ Dept. of Environmental Protection, Office of Science & Research.

Gentile, J.H., V.J. Bierman, Jr., J.F. Paul, H.A. Walker, and D.C. Miller. 1983. A Hazard Assessment Research Strategy for Ocean Disposal: Concepts and Case Studies. *In*: IODS Special Symposium on "Ocean Waste Management: Policy and Strategies." University of Rhode Island, US EPA, Environmental Research Laboratory, Narragansett, RI.

Ingham, M.C. (ed.). 1982. Summary of the Physical Oceanographic Processes and Features Pertinent to Pollution Distribution in the Coastal and Offshore Waters of the Northeastern United States, Virginia to Maine. NOAA Tech. Memo. NMFS-F/NEC17. 166 pp.

Kimerle, R.A., W.E. Gledhill, and G.J. Levinskas. 1978. Environmental safety assessment of new materials. pp. 132-146. In: Estimating the Hazard of Chemical Substances to Aquatic Life, ASTM STP 657. J. Cairns, K.L Dickson, and A.W. Maki (eds.). American Society for Testing and Materials.

McIntyre, A.D. and J.B. Pearce (eds.). 1980. Biological Effects of Marine Pollution and Problems of Monitoring. Rapports et Proces-Verbaux des Reunions. Vol. 179.

National Oceanic and Atmospheric Administration. Habitat Conservation; Policy for National Marine Fisheries Service (NMFS). Federal Register, Vol. 48, No. 228. Dept. of Commerce.

O'Connor, H.B., Jr., C.F. Wurster, C.D. Powers, D.C. Biggs, and R.G. Rowland. 1978. Polychlorinated biphenyls may alter marine trophic pathways by reducing phytoplankton size and production. Science. 201: 737-739.

Olla, B.L., W.H. Pearson, and A.L. Studholme. 1980. Applicability of behavioral measures in environmental stress assessment. Rapp. P.-v. Reun. Cons. Int. Explor. Mer 179: 162-173.

Olla, B.L., A.J. Beejda, and W.H. Pearson. 1981. Effects of oiled sediment on the burrowing behavior of hard clam, *Mercenaria mercenaria*. Mar. Environ. Res. 9: 183-193.

Pacheco, A.L. (ed.). 1984. Raritan Bay - Its Multiple Uses and Abuses. Proceedings of the Walford Memorial Convocation, Sept. 30, 1983, Sandy Hook, NJ.

Pearce, J.B. 1983. Ocean waste disposal: Its demand on marine habitats. pp. 21-29. In: Improving Multiple Use of Coastal and Marine Resources. Proc. of Symp. Jointly Sponsored by the Internat. Assoc. of Fish and Wildl. Agencies and the Amer. Fish. Soc. J.W. Reintjes (ed.). Hilton Head, SC.

Pearce, J.B., D.C. Miller, and C. Berman (eds.). 1983. Northeast Monitoring Program - 106-Mile Site Characterization Update. NOAA Tech. Momo. NMFS-F/NEC26.

KEYNOTE ADDRESS:
NOTES OF A POLLUTION WATCHER

Carl J. Sindermann

SUMMARY

A close observer of the burgeoning research on physiological effects of pollutants soon begins to feel like a sorcerer's apprentice--as more and more data from constantly expanding specialty areas appear, demanding interpretation and integration. Looking at the field of pollution physiology as a whole, probably the best that can be done to provide an overview is to select for discussion some of the research areas which seem most active at the moment, and also those which offer expansion prospects for the future.

A review of the published volumes resulting from this symposium series suggests an increase in the sophistication of our analytical methodology and improvement in attempts to associate physiology and biochemistry. A molecular biological approach to understanding pollutant effects at cellular levels has been combined with physiological studies of all life history stages, and with measured changes in ecosystems, forming a continuum of research effort from molecular to community levels.

Efforts are being made to associate physiological/biochemical changes with morphological changes following exposure to pollutants. Histopathology should be a required concomitant of any physiological study, just as the chemical determination of tissue and environmental levels of pollutants should be. Furthermore, morphological changes at cellular and subcellular levels should be examined in relation to biochemistry at those levels. The detailed work of Moore and his associates with lysosomal responses to pollutants

(Moore, 1976, 1979; Moore *et al.*, 1978) is an excellent example of what is needed at the cytology/cytochemistry interface.

Among areas which could benefit from increased research attention is that of quantitation of mortality caused by pollution. Extensive studies of sublethal effect do not always include a component which would permit determination of the ultimate fate of stressed animals. A related issue is the examination of pollution effects on reproductive potential, where again an actual determination of the depressive effects of pollution on population size is not often made. As part of an examination of how physiological effects may be translated into population effects, the growing literature and increasing understanding of the role of mixed function oxygenases and metal-binding proteins in mitigating some possible toxic effects, and the counterpart--synergisms which may increase toxic effects--should be emphasized.

Although a further listing of potential expansion areas in pollution research could be a week-long exercise, it does seem that elucidation of pollution effects on the internal defenses of vertebrate and invertebrate animals will be a high priority. Availability of new and precise methodology in immunology, particularly as it relates to fish, should result in (and is already resulting in) a closer examination of pollutant stress, immune responses, and disease due to facultative pathogens.

Studies of physiological effects of pollutants can provide, as part of a continuum of research data, information vital to risk assessment for populations in areas affected by pollution.

INTRODUCTION

Examination of the literature on physiological effects of pollutants which has accumulated during the past decade discloses that knowledge of sublethal effects of pollutants on fish and shellfish has increased substantively. Research, of course, exposes more questions than it solves, but new information has been acquired, and insights have been achieved (even faint traces of new concepts); some of which are still debated--such as assimilative capacity and bioaccumulation.

To bring some order to this consideration of recent advances and research needs for the future, insofar as physiological effects of pollutants are concerned, I would like to propose five theses and then elaborate on each one:

1. The survival potential of marine animals in degraded habitats may have been underestimated.
2. Quantitation of population impacts, acute and subacute, of pollution is weak and must be strengthened (included would be effects of reproductive suppression and larval mortality in particular).
3. Physiological/biochemical effects of pollutants is an exciting research area, but it could profit from a more intimate and continuous relationship with histopathology, and from more complete analysis of environmental and tissue levels of contaminants.
4. The description, measurement, and management of stress is essential to any understanding of pollutant effects.
5. The relationships of immune responses and pollution are complex, but critical to survival of the animal; hence better understanding is important.

Each of these theses could be the subject of prolonged discussion, and some experts might disagree with or reject one or more of them. The following summaries expose the problems, but cannot, of course, explore each one fully.

SURVIVAL IN DEGRADED HABITATS

The survival potential of marine animals in degraded habitats has been underestimated or, conversely, mortality due to pollution has been overestimated. Capuzzo (1981) stated the case well (but less confrontationally) by pointing out that not all responses to contaminants are disruptive, and that a number of adaptive strategies exist which can mediate pollution impacts. The subject of abnormality and death in degraded habitats is always of popular interest, and contains elements of reality, but survival in degraded habitats is also a reality, so mitigating aspects must not be ignored. We should feel some small obligation to look at both sides of the marine pollution mirror--the dark side of disability and death in degraded habitats, and the shiny side of survival in those same habitats. The dark side (Fig. 1) could include:

1. Effects on local population abundance -- These effects would be the decline and disappearance of certain sensitive species or changes in entire communities in areas of severe impact.
2. Effects on early life history stages -- These effects are easier to demonstrate than those on

abundance, although they are certainly associated with changes in abundance. They would include mutagenic and cytologic effects on embryos, (as demonstrated by Longwell, 1976, 1977, 1978; Longwell & Huges, 1980) and counterpart studies of gross abnormalities in fish larvae as a consequence of exposure to contaminants (Rosenthal & Stelzer, 1970; Rosenthal, 1971; Westernhagen *et al.*, 1975; Rosenthal & Alderdice, 1976). The entire process of reproduction can be affected by pollutants, beginning with modification of hormonal control of oogenesis and vitellogenesis (*e.g.*, through increased catabolism of estrogen by cytochrome oxidases) and extending through effects on early growth of embryos and larvae, as affected by enzyme alterations.

3. *Pollution-associated diseases* -- This area of effects, which is always a favorite for news media as well as pollution watchers, includes disease conditions such as fin erosion, skeletal anomalies, and ulcers (which have reasonable statistical associations with pollution); and neoplasms, some of which have increasing statistical association with pollutants.
4. *Modification of intracellular enzyme activity* -- This would result in disturbed metabolism which would be reflected at higher levels of organization. An excellent example is seen in the sublethal effects of metal exposure on cellular enzymes (Sprague, 1971; Calabrese *et al.*, 1977; Gould, 1977).

Turning the mirror to its shiny side, we have many examples of the physiological/biochemical responses of marine animals which enhance their survival potential in moderately degraded waters (Fig. 2). Aquatic organisms, populations, and ecosystems are equipped with a wide variety of homeostatic or resistance mechanisms that permit survival in the presence of pollutants. Individuals can tolerate levels of contaminants that are within physiological limits (Bryan & Hummerstone, 1971; Bryan, 1976), and there is some suggestion that this tolerance may increase with continued exposure to sublethal doses of the contaminant (in some instances it may decrease, however). Furthermore, individual animals may respond to organic contaminants by sequestering and isolating them, or by induction of enzymes which metabolize the foreign chemicals. Detoxification of organic chemical pollutants through a number of metabolic pathways can be effective, although there are instances in which the metabolized compound (*e.g.*, benzo[a]pyrene) can be more toxic or carcinogenic than the compound itself.

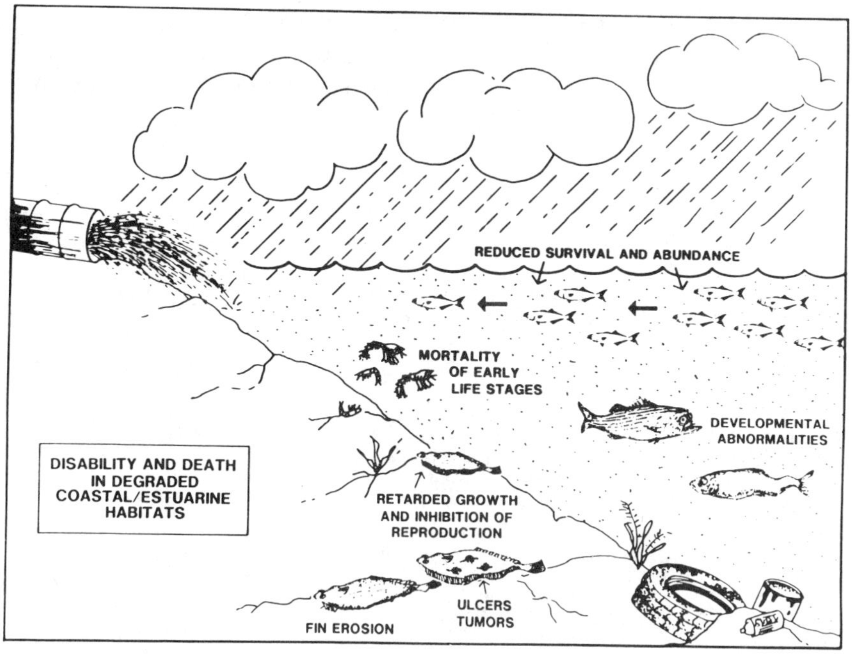

Fig. 1. Some examples of the effects of contaminants on marine organisms.

Judging from the numbers of papers presented at recent pollution meetings, two biochemical research areas have received intense study--cytochrome P-450s (hepatic microsomal mixed function oxygenases) and the metallothioneins (metal-binding proteins first recognized in fish in 1975). Both play an obviously important role in reducing toxicity of contaminants and enhancing survival in contaminated habitats. Cytochrome P-450s are clearly important in metabolizing organic contaminants including pre-carcinogens and carcinogens; toxic effects of polynuclear aromatic hydrocarbons (PAHs) are considered to be consequences of metabolic activation. Cytochrome P-450s may be important in metabolizing hormones such as estradiol in fish and molting hormone in crabs; conversely, estradiol may suppress induction of P-450 oxidases.

The metallothioneins constitute "heavy metal traps" (Piscator, 1964; Noël-Lambot *et al.*, 1980). The traps consist of binding heavy metals (such as cadmium and mercury) to low molecular weight proteins with high cysteine content (metallothioneins). The presence of heavy metals induces biosynthesis of metallothioneins in tissues; metallothioneins appear when the high molecular weight soluble proteins reach saturation with cadmium or certain other metals. The bound metal is toxicologically inert, enabling tolerance to high contamination levels in chronic exposures. Metals may also be detoxified by accumulation in insoluble form in lysosomes.

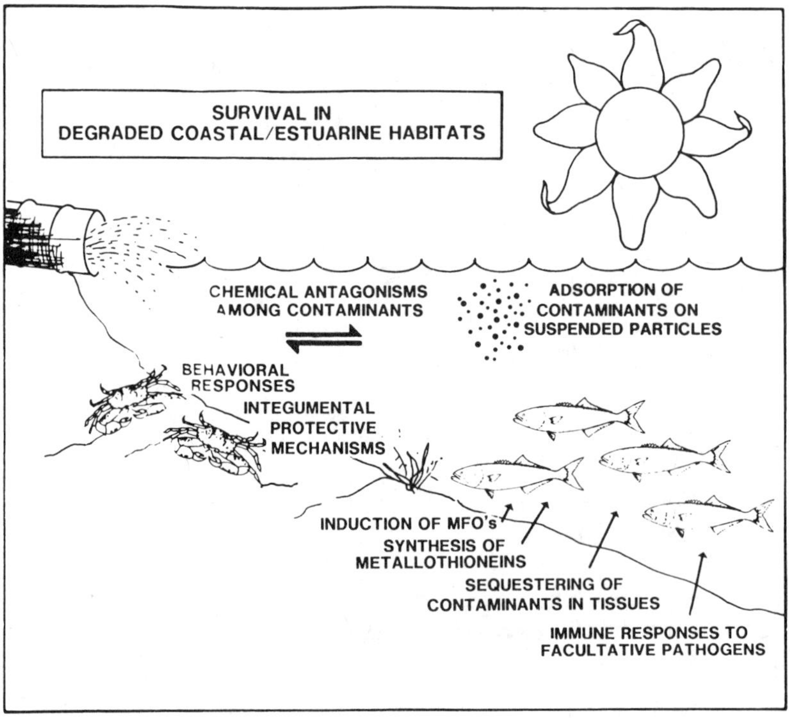

Fig. 2. Some mechanisms which enhance survival of marine organisms in degraded habitats.

Induction of other kinds of physiological changes which increase resistance in fish to several organochlorine and petroleum compounds has also been demonstration (Payne & Penrose, 1975; Watkins & Yarbrough, 1975). In one study, mechanisms of resistance to two pesticides included a membrane barrier which reduced uptake of the contaminants and a brain barrier in the form of insensitivity at the target site (Yarbrough & Wells, 1971). Mucus production is still another important mechanism of resistance to contaminants.

Other animals (*Nereis*) from metal-contaminated areas were found to be less permeable to metals in experimental studies--even to certain metals not in excess in that environment. Thus, the worms would be more likely to survive in heavily contaminated zones. This may be a membrane function of epidermal cells, where metal is absorbed and bound. (Another adaptive factor may be more effective excretion.)

A number of separate lines of investigation converge to illustrate strategies employed by fish and invertebrates for survival in degraded habitats. "Scope for growth" is a measure of the relative energy budget (energy available for somatic growth and gamete production) of animals in polluted zones as compared to unpolluted zones. Examination of this concept demonstrated that, at a critical level of contamination, animals were living at an energy deficit. Part of the survival strategy for species such as mussels was to reduce reproductive output during periods of high contamination, thereby conserving energy for growth (Bayne *et al*., 1978). Population replacement during such periods would depend on recruitment from populations outside the deficit zone. Resistance strategies impose a burden on the animal in that they are energy requiring, placing the organism at a physiological disadvantage compared to other members of the species living in non-polluted zones.

This kind of perspective on pollution effects suggests that damage occurs as a consequence of exposure to pollutants, but that important mechanisms for survival in modified habitats may mitigate part of the damage, although at energetic expense (Fig. 3). What emerges is, almost predictably, a limited system of checks and balances. Physiological mechanisms which compensate for environmental chemical imbalances function up to a point; beyond such limits the phenomena of toxic effects, collapse, and death appear. Sequestering, detoxification, and excretion of toxic chemicals are forms of protection which may be overloaded or overwhelmed (Fig. 4).

QUANTITATIVE EFFECTS ON POPULATIONS

Quantitation of acute and subacute effects of pollutants on marine populations is weak and must be strengthened, particularly those aspects concerned with reproductive suppression and mortality of early life stages. Examining the pluses and the minuses--the potentially harmful effects of pollutants on fish populations (as disclosed by experimentation) and the possible mitigating factors (many of them also indicated by experimentation), the evidence at hand suggests damage to fish populations in areas of highest contamination, but little direct indication of widespread effects on fish stocks. Baseline data on possible population fluctuations in the absence of pollution is, however, still inadequate enough to make any conclusion tenuous.

DISABILITY AND DEATH IN DEGRADED ESTUARINE/COASTAL HABITATS	SURVIVAL IN DEGRADED ESTUARINE/COASTAL HABITATS
REDUCED SURVIVAL AND ABUNDANCE RETARDED GROWTH INHIBITION OF REPRODUCTION MORTALITY OF EARLY LIFE STAGES DEVELOPMENTAL ABNORMALITIES POLLUTION-RELATED PATHOLOGY	EXTERNAL MITIGATION: DILUTION CHEMICAL ANTAGONISMS ADSORPTION OF CONTAMINANTS ON SUSPENDED PARTICLES BEHAVIORAL RESPONSES INTEGUMENTAL PROTECTIVE MECHANISMS INTERNAL MITIGATION: INDUCTION OF MFO's SYNTHESIS OF METALLOTHIONEINS SEQUESTERING OF CONTAMINANTS IN TISSUES IMMUNE RESPONSE TO PATHOGENS

Fig. 3. A balance sheet for disability and death versus survival in degraded habitats.

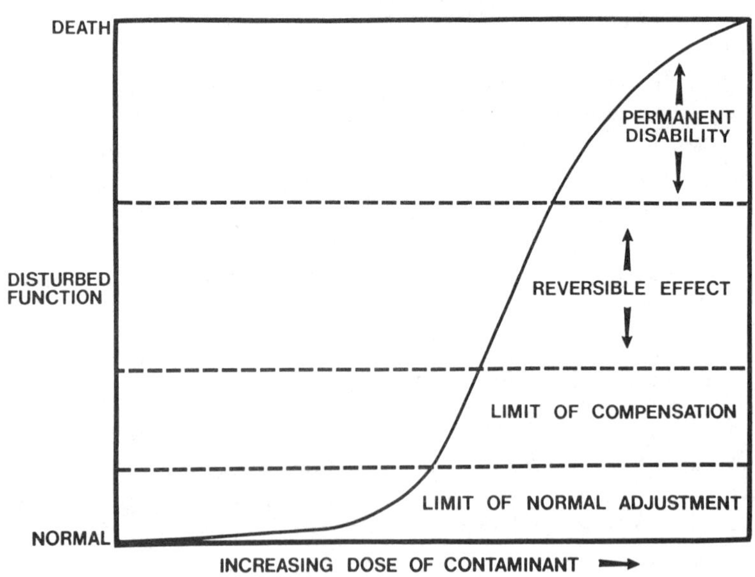

Fig. 4. General life zones in the presence of a varying environmental factor (modified from Wilson, 1980).

Hazard assessment (defined as "an attempt to quantify effects of particular acts or inputs," in this case pollution) for resource species in any geographic area must take into account all the complex variables, natural and man-induced, which may affect survival. The assessment may concern risk to the resources, as measured by acute and sub-acute effects, and it may be conducted on a biological or economic basis. Hazard assessment is a statistical examination; it may be defined in terms of societal values (as "acceptable" or "unacceptable" to society--*e.g.*, 5 ppm of mercury in tissues) or in terms of survival or non-survival of resource animals.

INTEGRATION OF PHYSIOLOGY AND MORPHOLOGY

Physiological/biochemical effects of pollutants are an exciting research area, but it could profit from a more intimate relationship with histopathology and with more consistent analyses of environmental and tissue contaminant levels.

"Integration" is a key operational term here, particularly an integration of histopathology and biochemistry. An excellent example of this is the continuing series of papers by Moore and his associates (Moore, 1976, 1979; Moore *et al.*, 1978) on the effects of pollutants on lysosomal membrane fragility. When this approach is tied to detailed analyses of tissue and environmental contaminant levels, the combination is a winner. Moore, working with invertebrates, has explored the relationship of stressors (including pollutants) on reduced lysosomal stability, using a cytochemical staining method for enzyme localization and intensity. Quantification of lysosomal changes has a clear potential as an index of pollution, and correlates well with scope for growth measurements.

Sublethal effects of pollutants can be expressed as physiological and biochemical alterations, behavioral modifications, genetic changes, and pathological changes. Pollution-associated pathology (diseases and abnormalities) can provide excellent clues about the extent of habitat degradation and the degree of risk for fish populations in any specified polluted environment. Pathology, in its broadest definition, includes everything from chromosomal damage in the unfertilized egg, to ultrastructural changes in gill epithelium, to tumors, infectious diseases, and skeletal anomalies. Among the most serious pathological effects, from the viewpoint of individual survival and effect on population size, are

- Mutagenic effects on eggs, embryos, and larvae, producing chromosomal damage;
- Environmentally-induced abnormalities, principally skeletal anomalies and somatic cell changes which result in tumors;
- Latent infections, particularly those of bacterial and viral etiology, provoked into active infections;
- Diseases caused by facultative pathogens, including fin erosion and ulcers in fish.

Clear and confirmed relationships between abnormalities and pollution still require additional documentation, but statistical evidence is accumulating rapidly.

Emphasis on these morphological effects is, of course, a reflection of one point of view--that of a pathologist. Most of the participants in this symposium are biochemists or physiologists. The biochemist would see all of the morphological changes as expressions of altered cellular metabolism, which produce changes at molecular and subcellular levels, mediated by hormonal and enzyme activity. The behaviorist and the physiologist would of course have quite different, but still correct, perceptions of pollution effects.

All this integration of physiology and morphology must be performed on a chemical stage--known tissue and environmental contaminant levels and the exchange rates of these contaminants across membranes, the sequestering rates, and the metabolic rates involved in degrading pollutants.

POLLUTION AS A STRESSOR

<u>The description, measurement, and management of stress is essential to any understanding of pollutant effects.</u>

Stress is a significant but elusive concept in biology; and like any concept, it can be misunderstood, misinterpreted, or abused. In its original sense, as developed by Selye (1950, 1952, 1955), stress represents the consequences of all the mechanisms whereby an organism attempts to maintain equilibrium in the face of environmental change. Selye's original description of stress emphasized physiological responses of the organism in three well-known phases: alarm, resistance, and exhaustion. The alarm phase includes immediate or short-term behavioral, biochemical, or physiological responses to non-optimum changes in the environment; the resistance or adaptation phase includes longer-term biochemical/physiological responses which improve likelihood of survival in the non-optimal environment; and the exhaustion phase includes failure of critical biochemical functions leading to physiological and morphological disorders and death.

An important, but often overlooked, aspect of stress is that it has two components: the non-specific stress responses which collectively make up the so-called "general adaptation syndrome" (Fig. 5), and the specific or localized stress responses superimposed as a consequence of injury or infection of a particular tissue or organ (direct physiological and/or morphological responses) (Fig. 6). These, as was pointed out by Selye (1955), may modify the non-specific reactions to the same stressor.

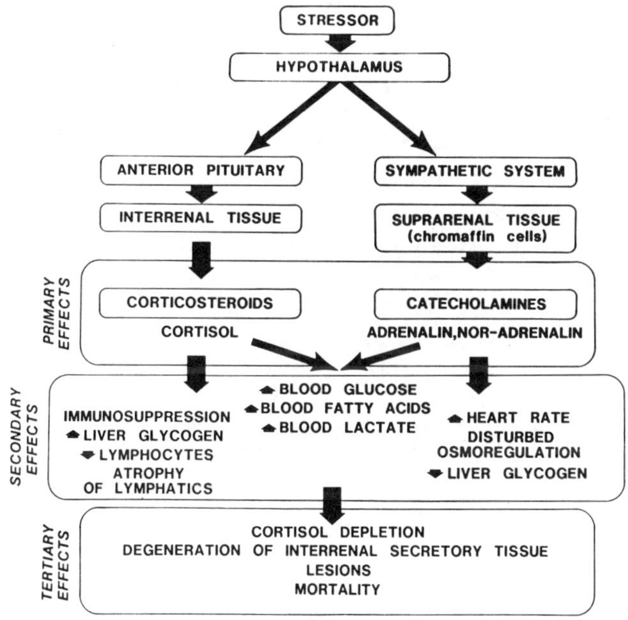

Fig. 5. Pathways of the stress response. Modified from Mazeaud et al. (1978) and Peters et al. (1980a).

The literature on stress in fish is abundant--to the point where two books have been written on the subject (Wedemeyer et al., 1976; Pickering, 1981). The effects of stress in eels have been the subject of a remarkable series of recent papers by Peters and her associates (Peters et al., 1980a,b,c; Peters, 1982). Multiple morphological and physiological indicators were useful in diagnosis of stress in the experimental systems used in those studies.

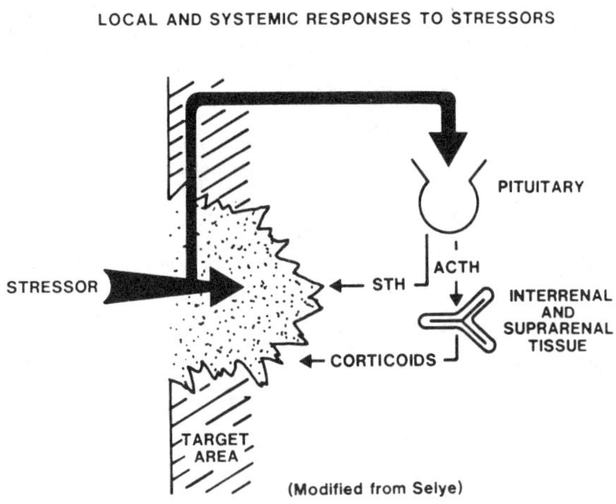

Fig. 6. Pathways of specific stress responses (after Selye, 1955).

Although most studies of the physiological and morphological consequences of stress have emphasized vertebrates, there are indications that counterpart phenomena may exist in invertebrates as well. Among Mollusca, several publications have provided evidence for a generalized response syndrome. Jeffries (1972) working with clams, Mercenaria mercenaria, found a shortened life span, mass mortalities, reduction in total amino acid and carbohydrate content, and shell invasion by Polydora in stressed clams. He considered amino acid ratios, particularly the taurine/glycine ratio in gill and mantle tissues, as the best indicators of stress. Ratios above 3 indicated stress, with ratios above 5 an indication of acute stress. High taurine ratios were also observed by Feng et al. (1970) in oysters infected with the trematode Bucephalus. In Jeffries' study, exposure of clams to petroleum resulted in black tar-like concretions in amebocytes and kidney tissues. The concretions collected in the renal sac and plugged the kidney tubules.

Other indices of stress in molluscs have been proposed --particularly reduced "scope for growth" (Bayne, 1975; Bayne et al., 1979) and the oxygen:nitrogen ratio (Corner & Cowry, 1968; Bayne & Scullard, 1977). Working with mussels, Bayne and associates observed a decline in condition and a depression of gametogenic activity under stress conditions. Other studies by Moore (1976, 1977) related degree of cell damage due to destabilization of lysosomal membranes to degrees of stress.

In bivalve molluscs, the signs of stress may include mantle recession, pale digestive gland, regression of digestive tubule epithelium, hemocyte infiltration of tissues, edema, lag in gametogenesis, and increased ceroid (brown bodies)--all or most of which constitute a stress syndrome (Fig. 7).

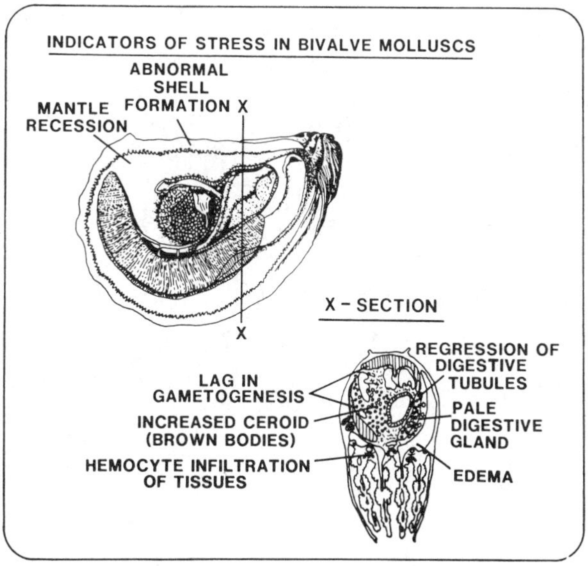

Fig. 7. Some responses to stress in bivalve molluscs.

In the larger Crustacea, signs of stress include black gills, abdominal muscle opacity, molt retardation, exoskeletal overgrowth with filamentous bacteria and protozoan epibionts, frequent occurrence of shell disease, disoriented or inappropriate behavior, presence in the tissues of gram positive bacteria, and clotting of the hemolymph (as a response to gram-negative endotoxin)--again constituting a stress syndrome (Fig. 8).

IMMUNE RESPONSES AND POLLUTION

Although immunosuppression is often cited as one of the important responses of fish to pollutant exposure, there are aspects of internal resistance which tend to favor survival of fish in polluted habitats. Robohm and Nitkowski (1974) found that experimental exposure of cunner, *Tautogolabrus adspersus*, to cadmium did not reduce humoral antibody production, but did reduce the bactericidal capabilities of phagocytes. However, other studies of fish have indicated suppression of humoral responses, and studies of contaminant-exposed oysters by Cheng (1983) disclosed suppression of phagocytosis by mercury, but enhancement by copper, iron, chromium, and tin.

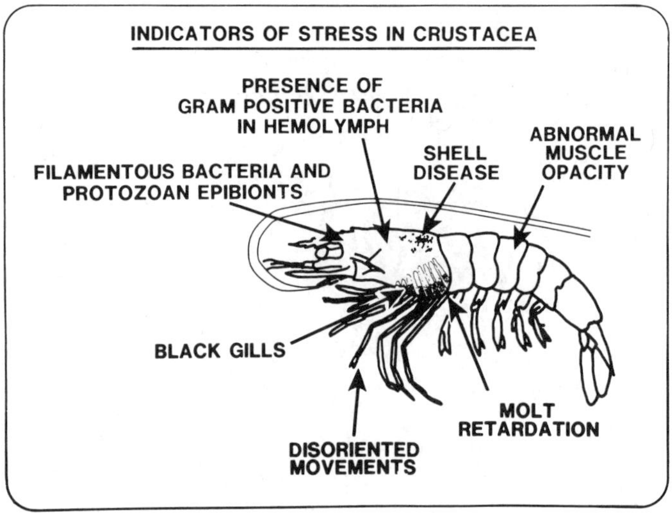

Fig. 8. Some responses to stress to crustaceans.

Robohm *et al.* (1979), in a survey of antibodies to a wide spectrum of bacteria in summer flounder, *Paralichthys dentatus*, from polluted and unpolluted habitats, found significantly higher antibody levels and a greater diversity of antibodies in samples from polluted waters (the New York Bight). Weakfish (*Cynoscion regalis*) also exhibited increased titers to many bacteria. The greatest proportion of increased titers was against *Vibrio* spp., although prominent titers against other fish pathogens such as *Aeromonas salmonicida* and *Haemophilus piscium* were seen.

In a related experimental study with summer flounder, Robohm and Sparrow (1982) reported a greater proportion of high antibody responders in fish taken from polluted vs. unpolluted sites. The fish pathogens, *Vibrio anguillarium* and *Aeromonas salmonicida*, elicited particularly strong responses. The authors suggested that fish surviving in polluted areas may be genetically selected as high antibody responders.

Other recent studies have demonstrated that many organic contaminants (Dioxin, PCBs, etc.) are immunosuppressive, and that some metals may selectively suppress parts of the immune system (*e.g.*, lead suppresses B and T lymphocytes and macrophages). More research is clearly indicated.

FUTURE RESEARCH DIRECTIONS

No certified pollution watcher would ever end a discussion voluntarily without offering some predictions and pipe dreams--some deliberatly fuzzy thoughts about the future. A few which emerge from the preceding considerations are

- Molecular biology will be important in understanding pollutant effects at cellular and subcellular levels. We have already learned much through cytochrome P-450 studies, and research on metallothioneins; both research areas have expanded exponentially in the past 5 years. Much more work on interactions of genes, enzymes, and hormones will come. Genetic manipulation to confront certain pollution problems seems feasible, once genetic sequencing has been done.
- Tumor development in marine animals in the presence of pollutants offers an opportunity to understand some of the basic mechanisms of carcinogenesis. Liver tumors seem to be the best prospects at present.
- Monoclonal antibodies can be a tool to revolutionize immunotoxicology, and immunology may prove to be a most sensitive test for toxic effects of pollutants. According to Koller (1983), altered immune capacity can be detected after exposure to 1 ppm methylmercury, 5 ppm pentachlorophenol, or 10 ppm PCB.
- Pollution effects on the entire reproductive process--hormonal changes, vitellogenesis, fecundity, larval survival--are receiving some attention, but deserve more.
- The search for consistent, sensitive, quantifiable biological indicators of pollution will continue, as it should. In 1979, a workshop was held at Beaufort, NC, under auspices of the International Council for the Exploration of the Sea, to assess the current status of research in this field. Optimistic statements were made about the immediate or near-term availability of methods for use of such indicators--including physiological/biochemical ones--in effects monitoring programs. Reality is invariably less than the initial optimistic projections, but several nations, particularly the United States, have developed ocean pollution monitoring programs with biological effects

included as one of several methods to assess contaminants. The physiological/biochemical contribution to such programs must be described as still in a pilot phase. Major problems so far have been variability unrelated to pollution, difficulties inherent in conducting laboratory work at sea, and lack of sensitivity of some of the criteria explored.

It seems clear from even a brief review of the published literature of the past decade that remarkable progress has been made in understanding the physiological effects of pollutants on marine animals. It seems equally clear that much remains to be done, with contributions required from many disciplines and specialties.

LITERATURE CITED

Bayne, B.L. 1975. Aspects of physiological condition in *Mytilus edulis* L. with special reference to the effects of oxygen tension and salinity. pp. 213-238. *In*: Proc. 9th European Marine Biol. Symp. H. Barnes (ed.). Aberdeen Univ. Press.

Bayne, B.L., D.L. Holland, M.N. Moore, D.M. Lowe, and J. Widdows. 1978. Further studies on the effects of stress in the adult on the eggs of *Mytilus edulis*. J. Mar. Biol. Assoc. UK 58: 825-841.

Bayne, B.L., M.N. Moore, J. Widdows, D.R. Livingstone, and P. Salkeld. 1979. Measurement of the response of individuals to environmental stress and pollution: Studies with bivalve molluscs. Phil. Trans. Roy. Soc. London B. 286: 563-581.

Bayne, B.L. and C. Scullard. 1977. Rates of nitrogen excretion by species of *Mytilus* (Bivalvia: Mollusca). J. Mar. Biol. Assoc. UK 57: 355-369.

Bryan, G.W. 1976. Some aspects of heavy metal tolerance in aquatic organisms. pp. 7-34. *In*: Effects of Pollutants on Aquatic Organisms. A. Lockwood (ed.). Cambridge Univ. Press.

Bryan, G.W. and L.G. Hummerstone. 1971. Adaptation of the polychaete *Nereis diversicolor* to estuarine sediments containing high concentrations of heavy metals. J. Mar. Biol. Assoc. UK 51: 845-863.

Calabrese, A., F.P. Thurberg, and E. Gould. 1977. Effects of cadmium, mercury, and silver on marine animals. Mar. Fish. Rev. 39(4): 5-11.

Capuzzo, J. 1981. Predicting pollution effects in the marine environment. Oceanus 24(1): 25-33.

Cheng, T.C. 1983. Effect of heavy metal ions on phagocytic activity of oyster hemocytes. Proc. (Abstracts) 2nd Internat. Symp. on Responses of Marine Organisms to Pollutants. Woods Hole, MA.

Corner, E.D.S. and C.B. Cowry. 1968. Biochemical studies on the production of marine zooplankton. Biol. Rev. 43: 393-426.

Feng, S.Y., E.A. Khairallah, and W.J. Canzonier. 1970. Hemolymph free amino acids and related nitrogenous compounds of *Crassostrea virginica* infected with *Bucephalus* sp. and *Minchinia nelsoni*. Comp. Biochem. Physiol. 34: 547-556.

Gould, E. 1977. Alteration of enzymes in winter flounder, *Pseudopleuronectes americanus*, exposed to sublethal amounts of cadmium chloride. pp. 209-223. *In*: Physiological Responses of Marine Biota to Pollutants. F.J. Vernberg, A. Calabrese, F.P. Thurberg, and W.B. Vernberg (eds.). Academic Press, New York.

Jeffries, H.P. 1972. A stress syndrome in the hard clam, *Mercenaria mercenaria*. J. Invert. Pathol. 20: 242-251.

Koller, L. 1983. Immuno-modulation by environmental pollutants. Proc. (Abstracts) 2nd Internat. Symp. on Responses of Marine Organisms to Pollutants. Woods Hole, MA.

Longwell, A.C. 1976. Chromosome mutagenesis in developing mackerel eggs samples from the New York Bight. NOAA Tech. Memo No. ERL MESA-7. 61 pp.

Longwell, A.C. 1977. Genetic effects. pp. 42-47. *In*: The *Argo Merchant* Oil Spill and the Fishery Resources of Nantucket Shoals and Georges Bank: A Summary of Assessment Activities and Preliminary Results. National Marine Fisheries Service, Northeast Fisheries Center, Narragansett Lab. Rept. No. 77-10.

Longwell, A.C. 1978. Field and laboratory measurements of stress response at the chromosome and cell levels in planktonic fish eggs and the oil problem. pp. 116-125. *In*: In the Wake of the *Argo Merchant*. Univ. Rhode Island, Kingston.

Longwell, A.C. and J.B. Hughes. 1980. Cytologic, cytogenetic and developmental state of Atlantic mackerel eggs from sea surface waters of the New York Bight, and prospects for biological effects monitoring with ichthyoplankton. Rapp. P.-v. Reun. Cons. Int. Explor. Mer 179: 275-291.

Mazeaud, M.M., F. Mazeaud, and E.M. Donaldson. 1978. Primary and secondary effects of stress in fish: Some new data with a general review. Trans. Am. Fish. Soc. 106: 201-212.

Moore, M.N. 1976. Cytochemical demonstration of latency of lysosomal hydrolyses in digestive cells of the common mussel, Mytilus edulis, and changes induces by thermal stress. Cell. Tissues Res. 175: 279-287.

Moore, M.N. 1977. Lysosomal responses to environmental chemicals in marine invertebrates. pp. 143-154. In: Pollutant Effects on Marine Organisms. C.S. Giam (ed.). D.C. Heath, Lexington, MA.

Moore, M.N. 1979. Cellular responses to polycyclic aromatic hydrocarbons and phenobarbital in Mytilus edulis. Mar. Envir. Res. 2: 255-263.

Moore, M.N., D.M. Lowe, and P.E.M. Fieth. 1978. Lysosomal responses to experimentally injected anthracene in the digestive cells of Mytilus edulis. Mar. Biol. 48: 297-302.

Noël-Lambot, F., J.M. Bouquegneau, and A. Disteche. 1980. Some mechanisms promoting or limiting bioaccumulation in marine organisms. ICES, CM 1980/E:39. 25 pp.

Payne, J.F. and W.R. Penrose. 1975. Induction of aryl hydrocarbon (benzo[a]pyrene) hydroxylase in fish by petroleum. Bull. Environ. Contam. Toxicol. 14: 112-116.

Peters, G. 1982. The effect of stress on the stomach of the European eel, Anguilla anguilla L. J. Fish. Biol. 21: 497-512.

Peters, G., H. Delvanthal, and H. Klinger. 1980a. Physiological and morphological effects of social stress in the eel (Anguilla anguilla L.). Arch. Fish. Wiss. 30: 157-180.

Peters, G., H. Delventhal, and H. Klinger. 1980b. Stress diagnosis for fish in intensive culture systems. FAO/EIFAC Symposium on New Developments in the Utilization of Heated Effluents and of Recirculation Systems for Intensive Aquaculture, Paper No. EIFAC/80/Symp.: E/37. 12 pp.

Peters, G., H. Delventhal, and H. Klinger. 1980c. Diagnosis of stress in the European eel. ICES, Special Meeting on Diseases of Commercially Important Marine Fish and Shellfish. Paper No. 27. 20 pp.

Pickering, A. (ed.). 1981. Stress and Fish. Academic Press, London. 365 pp.

Piscator, M. 1964. Cd in the kidneys of normal beings and the isolation of metallothionein from liver of rabbits exposed to Cd. Nord. Hyg. Tidskr. 45: 76-82.

Robohm, R.A. and M.F. Nitkowski. 1974. Physiological response of the cunner, *Tautogolabrus adspersus*, to cadmium. IV. Effects on the immune system. US Dept. Comm., NOAA Tech. Rept. NMFS SSRF-681, pp. 15-20.

Robohm, R.A. and D.S. Sparrow. 1981. Evidence for genetic selection of high antibody responders in summer flounder (*Paralichthys dentatus*) from polluted areas. pp. 273-278. *In*: Proc. Internat. Sympos. of Fish Biologics Serodiognostics and Vaccines. Leetown, WV, S. Karger, Basel.

Robohm, R.A., C. Brown, and R.A. Murchelano. 1979. Comparison of antibodies in marine fish from clean and polluted waters in the New York Bight: Relative levels against 36 bacteria. App. Env. Microbiol. 38: 248-257.

Rosenthal, H. 1971. Wirkungen von "Rotschlamm" von Embryonen und Larven des Herings *Clupea harengus*. Helgol. wiss. Meeresunters. 22: 366-376.

Rosenthal, H. and D.F. Alderdice. 1976. Sublethal effects on environmental stressors, natural and pollutional, on marine fish eggs and larvae. J. Fish. Res. Board Can. 33: 2048-2065.

Rosenthal, H. and R. Stelzer. 1970. Wirkungen von 2,4- und 2,5 Dinitrophenol auf die Embryonalentwicklung des Herings *Clupea harengus*. Mar. Biol. 5: 325-336.

Selye, H. 1950. Stress and the general adaptation syndrome. Br. Med. J. 1: 1383-1392.

Selye, H. 1952. The story of the adaptation syndrome. Acta, Inc., Montreal, Que. 255 pp.

Selye, H. 1955. Stress and disease. Science 122: 625-631.

Sprague, J.B. 1971. Measurement of pollutant toxicity to fish. III. Sublethal effects and "safe" concentrations. Water Res. 5: 245-266.

Watkins, J. and J.D. Yarbrough. 1975. Aldrin and Dieldrin uptake in insecticide-resistant and susceptible mosquito fish (*Gambusia affinis*). Bull. Envir. Contam. Tox. 14: 731-737.

Wedemeyer, G.A., F.P. Meyer, and L. Smith. 1976. Environmental stress and fish diseases. T.F.H. Publications. Neptune, NJ. 200 pp.

Westernhagen, H. von, V. Dethlefsen, and H. Rosenthal. 1975. Combined effects of cadmium and salinity on development and survival of garpike eggs. Helgol. wiss. Meeresunters. 27: 268-282.

Wilson, K.W. 1980. Monitoring dose-response relationship. ICES, Rap. P.-v. Reum. Cons. Int. Explor. Mer 179: 333-338.

Yarbrough, J.D. and M.R. Wells. 1971. Vertebrate insecticide resistance: The *in vitro* endrin effect on succinic dehydrogenase activity on endrin-resistant and susceptible mosquitofish. Bull Envir. Contam. Toxicol. 6: 171-177.

SWIMMING/LOCOMOTOR CAPACITY OF HUDSON RIVER STRIPED BASS

M.A. Freadman, F.P. Thurberg
and A. Calabrese

SUMMARY

The life history of active fishes, such as striped bass, is characterized by significant periods of swimming activity. The process of maneuvering in water relates to the swimming capacity of a fish and is essential for events such as migration, escape from predators, and food capture. Thus, reduced swimming capacity in active fish species could severely limit their survival, distribution, and abundance.

The present study was undertaken to investigate the swimming performance of striped bass from impacted and non-impacted waters. We also measured the swimming performance of striped bass exposed to a heavy metal prevalent in the Hudson River ecosystem.

Striped bass were collected from two Hudson River sites (River Mile 1 and River Mile 39), from eastern Long Island Sound (Greenport-Orient Point area), and from the Edenton National Hatchery (NC). The indices of swimming performance measured were critical swimming speed and burst swimming speed. These experiments were performed using a water tunnel treadmill. The results indicate both groups of fishes from the Hudson River had significantly lower swimming capacity compared to animals collected from eastern Long Island Sound and the Edenton Hatchery.

A further aspect of our study dealt with exposures of striped bass to lead (20, 100, and 200 ppb) for 60 d in the laboratory using a proportional diluter apparatus. Only at the highest exposure concentration were we able to detect significantly lower critical swimming speeds compared with control animals.

We measured PCB content of various tissues of striped bass collected from the Hudson River as preliminary work for a future study. PCBs were present in the ppm range and the highest quantities were found in the "red" swimming musculature. All routine swimming activity in striped bass is powered by this type of musculature. Organochlorines (such as PCBs) are sequestered in lipids, which are the predominate energy store in fish "red" muscle. PCBs may alter lipid profiles in the "red" swimming musculature.

INTRODUCTION

The distribution and abundance of organisms in marine ecosystems is influenced by many environmental parameters. In recent years, it has become increasingly evident that man-induced pollution is one of these parameters. Areas such as the Hudson-Raritan Estuary and western Long Island Sound receive substantial quantities of chemical constituents from natural phenomena (weathering processes of soils and bedrock) and from discharge of wastes from domestic and industrial sources (Gross, 1970; Sherk, 1972; Mueller *et al.*, 1976, 1982; Biejer & Jernelov, 1979). There is also substantial evidence of disruption of metal balance in the New York Bight and adjacent areas (O'Conner & Stanford, 1979; Mueller *et al.*, 1982). The Hudson-Raritan Estuary and western Long Island Sound have long histories of contaminant loading, yet the extent to which living resources (including finfish) have been affected, remains speculative.

The present study was undertaken to characterize the swimming performance of striped bass (*Morone saxatilis* Walbaum) from two areas of the Hudson River. A further aspect of our study measured swimming performance of striped bass exposed to lead (a heavy metal present in high concentrations in the Hudson) for 60 d in the laboratory using a proportional diluter apparatus. We also obtained some measurements on PCB content of various tissues of striped bass collected from the Hudson River.

MATERIALS AND METHODS

Striped bass (*Morone saxatilis* Walbaum) were collected with a beach seine, trap net, or by hook and line from waters in the vicinity of Hudson River Mile 1, Hudson River Mile 39 (Grassy Island), and eastern Long Island Sound (Greenport-Orient Point area) during late fall 1981 and during the summer and fall 1982. Animals were transported in aerated containers to the National Marine Fisheries

Service Laboratory, Milford, CT and held in plastic tanks which contained water collected from their environments. These animals were used for experimentation within 4-5 d of collection. Animals from the Edenton Hatchery were transported to the Milford Laboratory, Milford, CT in hatchery truck containers during late fall 1981. These animals were held in kiloliter tanks continuously supplied with sea water (23-27°/oo) over a 6-month period. Collections sites are shown in Figure 1.

Swimming/locomotor capacity was measured as critical swimming speed (= maximum sustainable swimming speed = U_{crit}) and burst swimming speed (Beamish, 1978). Critical swimming speed is the maximum speed achieved prior to fatigue in a water tunnel as swimming speed is increased in increments and held for a specific period of time. Critical swimming speeds of striped bass collected from the Hudson River and eastern Long Island Sound were measured in a Brett-type recirculating water tunnel (Brett, 1964) within 1 wk after animal collection. The stainless steel water tunnel has an acrylic swimming chamber (16.2 cm bore) with honeycomb grids upstream and downstream. Flow was microturbulent with a rectilinear profile. Flow velocity was measured with an Annubar Flowmeter which was calibrated with a movable pitot tube and an impeller knotmeter (Marine Advisers, Inc., Model B-7C). Further details of the swimming apparatus are available in an earlier publication (Freadman, 1979). For determination of critical swimming speeds, animals were lightly anaesthetized (tricaine methanesulphonate, 0.05 g/L), placed in the water tunnel for 16-20 h acclimation and training period, after which flow was increased in 6 cm/sec increments until the animal was exhausted. The test period at each velocity was 30 min. The time to exhaustion at the highest swimming velocity was recorded and the critical swimming speed calculated by adjustment of the last velocity increment in proportion to the length of time which the animal maintained station after that velocity was reached (Brett, 1964). Oxygen content of water in the tunnel was maintained above 85% saturation by periodic flushing with aerated water. Animals were tested individually in the water tunnel, which contained water collected from the same site at which a particular animal was collected. Two water tunnels were frequently used.

Burst swimming refers to very rapid swimming movements of high speed and short duration for periods of less than 20 sec. For determinations of burst swimming speed, individual animals were acclimated to the water tunnel as described

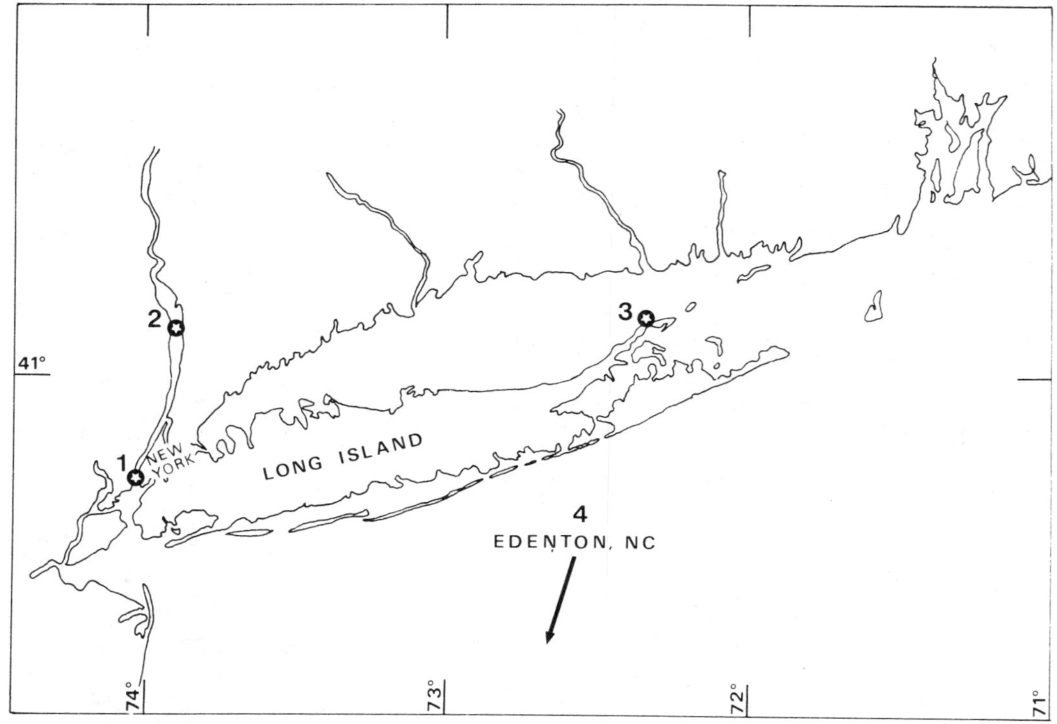

Fig. 1. Collection sites of striped bass from the Hudson River (Mile 1 and Mile 39, Site 1 and 2 respectively). Site 3 was in eastern Long Island Sound. A fourth collection site was the Edenton National Hatchery (Edenton, North Carolina); these striped bass were fingerlings and were subsequently raised at NMFS, Milford, CT.

above. Flow in the water tunnel was increased in 6-8 cm/sec steps and held at a particular velocity for only 45 sec. Burst speed was noted as the highest velocity an animal could maintain with rapid accelerations and a sprint; this occurred just before fatigue.

Striped bass originally collected from the Edenton National Hatchery (and subsequently reared at NMFS, Milford, CT) were used for a lead exposure study. Animals were exposed to 20, 100, or 200 ppb lead, as the nitrate ($Pb[NO_3]_2$), for 60 d in 80-L glass aquaria filled to 60 L with 23-27$^o/_{oo}$ sea water. A proportional diluter apparatus (Mount & Brungs, 1967) controlled intermittent delivery and dilution of toxicant-laced water and control water. The diluter provided a flow of about 1 L to each tank every 3 min (about 480 L/tank/day) and a 90% turnover time of 7 h (Sprague, 1969). Metal concentrations refer to calculated concentrations of the metal ion in solution; this does not include background levels of 0.5-1.5 ppb.

At the end of the 60-d exposure period, animals from each exposure regime (control, 20, 100, 200 ppb [$Pb(NO_3)_2$]), were individually tested for critical swimming speed using the same procedures described above.

Samples of gill tissue (not including the arch), "red" swimming muscle, and "white" swimming muscle were excised from striped bass freshly collected from the two Hudson River sites. Tissues were dissected using stainless steel instruments and stored in glass vials (with precleaned foil-lined lids) at -20°C prior to analyses. These tissues were analyzed for PCBs by a saponification procedure (Stout & Beezhold, 1979) followed by gas liquid chromotography using a packed column of 3% OV17 on gas chrom WHP (100/120 mesh) (Greig <u>et</u> <u>al</u>., 1983). Arochlor® 1254 was used as a standard.

The animals used in this study averaged 18.0 cm in length (range 15.5-21.0 cm) at the time of measurements of critical swimming speeds, burst swimming speeds, and at the end of the exposure study. Ambient water temperatures and laboratory experimental temperatures ranged from 14-17°C.

RESULTS

Maximum Sustainable Swimming Speeds (U_{crit})

Values for maximum sustainable swimming speeds of animals collected from four different areas are shown in Figure 2. Data were subjected to statistical treatment using ANOVA to determine if significant differences were measured. The Newman-Keuls multiple range test was employed to determine between which means significant differences exist. The 0.05 level of significance was used. The results show that both groups of fish from the Hudson River had significantly lower maximum sustainable swimming speeds than animals collected from eastern Long Island Sound or the Edenton Hatchery.

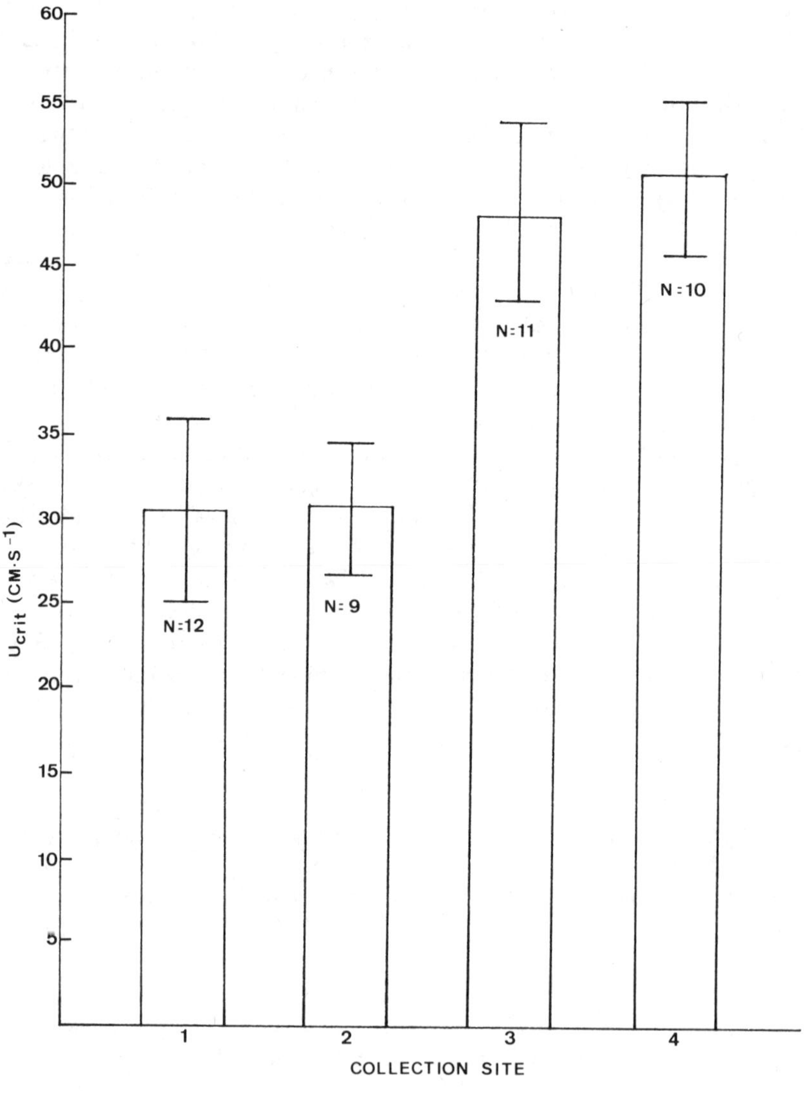

Fig. 2. Critical swimming speeds (cm/sec, ± 1 SD) of striped bass from the four collection sites, sites identified in legend of Figure 1.

Burst Swimming Speed

Figure 3 shows the results of burst swimming speed determinations for animals collected from the four areas. Using the same statistical techniques as described above, we again found that both groups of fish from the Hudson showed significantly lower burst swimming speeds compared to fish collected from eastern Long Island Sound or the Edenton Hatchery.

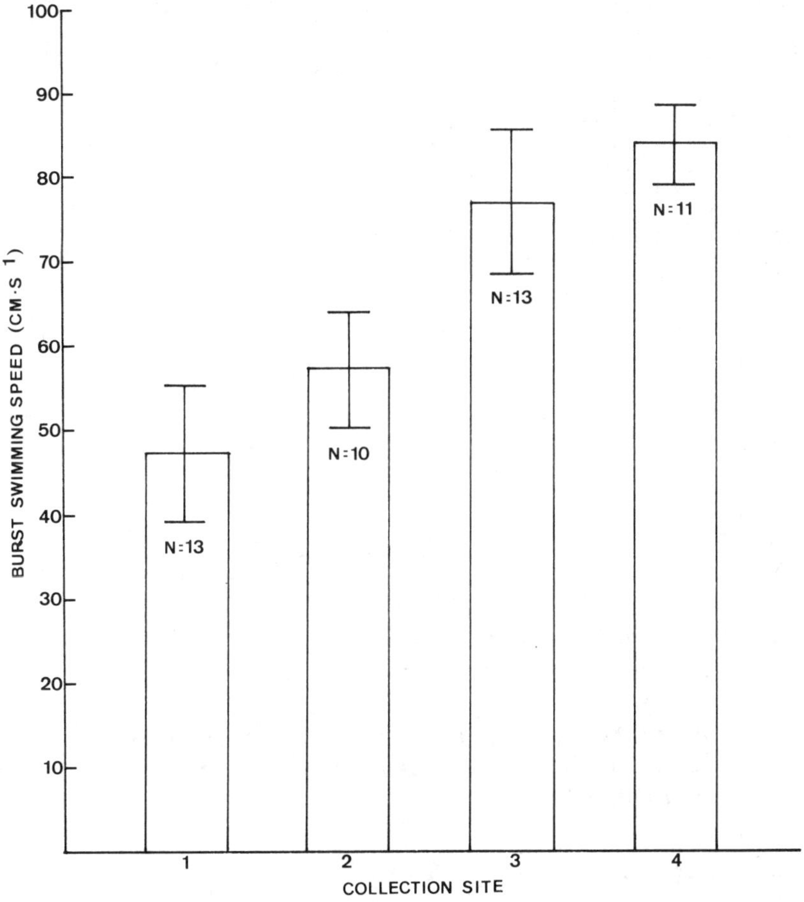

Fig. 3. Burst swimming speeds (cm/sec, ± 1 SD) of striped bass from the four collection sites, sites identified in legend of Figure 1.

Exposure to Lead

Striped bass exposed to 0, 20, 100, or 200 ppb lead for 60 d were tested for maximum sustainable swimming speeds. The results of the exposures are shown in Figure 4. Only at the highest exposure concentration (200 ppb) were we able to detect significantly lower critical swimming speeds.

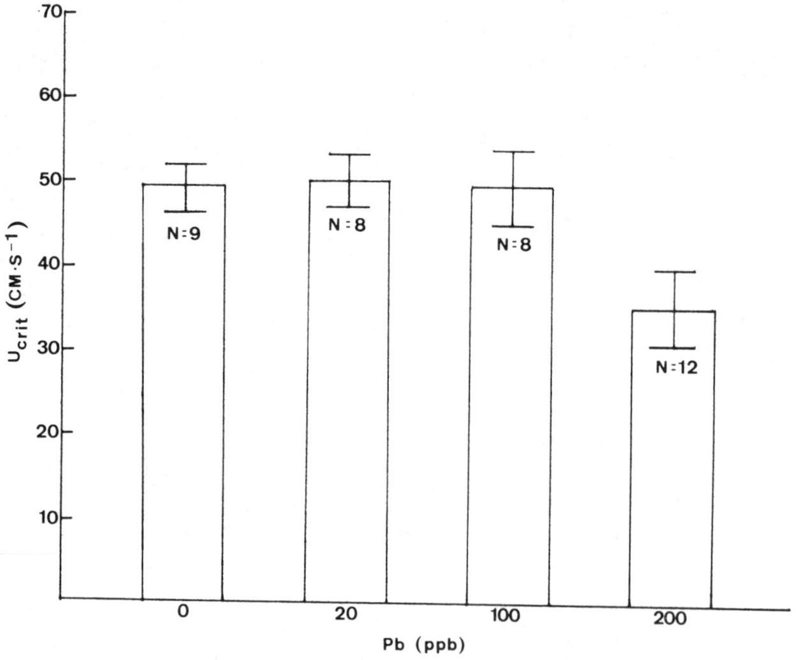

Fig. 4. Critical swimming speeds (cm/sec, ± 1 SD) of striped bass (collected from the Edenton National and subsequently reared at NMFS, Milford, CT) exposed to 0, 20, 100, or 200 ppb Pb (Pb $[NO_3]_2$) for 60 d in the laboratory using a proportionaldiluter apparatus.

PCB Concentrations

As a preliminary study for possible future work, we determined the concentration of PCBs in striped bass collected from the two Hudson River areas. The tissues analyzed were "red" swimming musculature, "white" swimming musculature, and gills (not including the arch). The highest concentrations, in the parts per million range, were found in the "red" swimming musculature (Table 1). "Red" swimming musculature and gill tissue were pooled from two or more animals for analyses, while the values in Table 1 for "white" swimming musculature represent individual animals.

Table 1. PCBs (ppm, wet wt) of selected tissues of striped bass collected from Hudson River Mile 1 and Mile 39.

	Mile 1	Mile 39
"RED" Swimming Musculature	1.80	0.55
	2.00	0.40
	0.76	0.40
	0.95	0.50
"WHITE" Swimming Musculature	0.16	0.06
	0.10	0.05
	0.44	0.05
	0.13	0.05
	0.19	
	0.09	
	0.32	
	0.38	
Gills	0.14	0.08
	0.28	0.08
	0.20	
	0.08	

DISCUSSION

The life history of active fishes is characterized by significant periods of swimming activity. Maneuvering in water reflects swimming performance and capacity which is essential for migration, food capture, and escape from predators. Reduced swimming capacity of active fish species could limit their survival, distribution, and abundance.

The present study was designed to characterize the swimming capacity and performance of striped bass from two Hudson River areas, from an eastern Long Island Sound area, and from the Edenton National Hatchery (North Carolina). The Hudson River areas represent "impacted" areas while the other two areas are indicative of much cleaner waters. The indices of swimming performance measured were critical swimming speed (maximum sustainable swimming speed) and burst swimming speed; these are levels of performance used by fish biologists to define swimming capacity (Fry, 1971; Beamish, 1978). The experiments represent tests performed to discern biological responses (locomotor capacity of fish) to contaminants.

Striped bass from both Hudson River areas exhibited significantly reduced critical swimming speeds and burst swimming speeds compared to animals collected from eastern Long Island Sound and the national hatchery. Bass from eastern Long Island Sound and the hatchery show critical swimming speeds very similar (on a body length/second comparison) to animals collected from non-impacted waters of Rhode Island and Massachusetts in an earlier study of striped bass swimming energetics by Freadman (1979).

PCBs are a recognized problem in the Hudson River (Mueller *et al*., 1982), yet specific effects on Hudson River biota are not well documented. Gadbois and Maney (1983) measured PCBs in Hudson River striped bass "fillets" and reported values similar to our determinations. We were particularly interested in PCB concentrations in the "red" swimming musculature. We found PCB concentrations in the parts per million range in the "red" swimming musculature of animals collected from Hudson River Mile 1 and Mile 39. This finding gains significance when viewed through the specific role of "red" swimming musculature of striped bass. The functional roles of "red" and "white" swimming musculature of striped bass show a clear division of labor. Sustained swimming at slow, cruise, and high velocities (up to and including maximum sustainable velocities) is accomplished by "red" muscle alone (Freadman, 1979). "White" muscle is utilized only at burst velocities normally sustained for short periods of time (less than 20 sec). The precise effects of PCBs on metabolism and energetics of "red" muscle are not known, but should prove to be an interesting area for future studies. Organochlorines (such as PCBs) are sequestered in lipids, which are the predominate energy store in fish "red" muscle. PCBs may alter lipid mobilization from the swimming musculature.

Under adverse environmental conditions, a population of fishes could move to a more favorable microenvironment. One of the limitations on predictive ability using fishes is precise information on the recent habitat of animals captured from the field. Available information about the movements of populations of striped bass in the Hudson River and Long Island Sound is gradually increasing (Young, 1981). Telemetry studies, coupled with water quality measurements, conducted over the course of a year would be helpful in determining the population-environment dynamics of striped bass.

Other environmental factors such as temperature at the extremes of thermal limits of a species, reduced water oxygen content, and contaminants (e.g., copper) have been reported to reduce the swimming stamina of a variety of fishes (Davis et al., 1963; Kutty & Saunders, 1973; Beamish, 1978; Waidwood & Beamish, 1978; Wedemeyer & McLeay, 1981). This study has demonstrated reduced swimming capacity of striped bass from the Hudson River compared to animals collected from cleaner waters. The specific cause per se is not known, yet appears to be related to water quality. Exposures to high concentrations of lead via the water column reduce the swimming capacity of striped bass as well. Our findings suggest that contaminant effects are detectable as measured by known, significant aspects (swimming/locomotor capacity) of the life history of striped bass, and that water quality can compromise the swimming capacity of striped bass.

ACKNOWLEDGMENTS

We are grateful to Mr. Byron Young and his staff (New York State Division of Environmental Conservation, Stony Brook, NY) for their unfailing cooperation in collection of animals and discussions of fish populations in the Hudson River and Long Island Sound. Mr. Guy Deblasio (New York, NY) greatly aided our collection of animals at Hudson River Mile 1. Mr. Richard Greig (NMFS, Milford, CT) performed the tissue analyses. We also thank two anonymous reviewers for their helpful comments. Elaine M. Lynch kindly typed the original manuscript.

The research was supported in part, by a contract to the National Marine Fisheries Service from NOAA Office of Marine Pollution Assessment and by Woods Hole Oceanographic Institution. This is Woods Hole Oceanographic Contribution No. 5558.

LITERATURE CITED

Beamish, F.W.H. 1978. Swimming capacity. pp. 101-187. In: Fish Physiology, Vol. VII. W.S. Hoar and D.J. Randall (eds.). Academic Press, New York.

Biejer, K. and A. Jernelov. 1979. Sources, transport and transportation of metals in the environment. pp. 47-

64. *In*: Handbook of the Toxicology of Metals. L. Fribert, G. Nordberg, and V. Vouk (eds.). Elsevier, New York.

Brett, J.R. 1964. The respiratory metabolism and swimming performance of young sockeye salmon. J. Fish. Res. Board Can. 21: 1183-1226.

Davis, G.E., J. Foster, C.E. Warren, and P. Doudoroff. 1963. The influence of oxygen concentration on the swimming performance of juvenile Pacific salmon at various temperatures. Trans. Am. Fish. Soc. 92: 111-124.

Freadman, M.A. 1979. Swimming energetics of striped bass (*Morone saxatilis*) and bluefish (*Pomatomus saltatrix*): Gill ventilation and swimming metabolism. J. Exp. Biol. 83: 217-230.

Fry, F.E.J. 1971. The effects of environmental factors on the physiology of fish. pp. 1-98. *In*: Fish Physiology, Vol. 6. S. Har and D.J. Randall (eds.). Academic Press, New York.

Gadbois, D.F. and R.S. Maney. 1983. Survey of polychlorinated biphenyls in selected finfish species from United States coastal waters. Fish. Bull. 81: 389-396.

Greig, R.A., S. Schurman, J. Pereira, and P. Naples. 1983. Metals and PCB concentrations in windowpane flounder from Long Island Sound. Bull. Environ. Con. Toxicol. 31: 257-262.

Gross, M.G. 1970. Preliminary analyses of urban wastes, New York metropolitan regions. Marine Sciences Research Center Tech. Rept. 7, SUNY, Stony Brook.

Kutty, M.N. and R.L. Saunders. 1973. Swimming performance of young Atlantic salmon (*Salmo salar*) as affected by reduced ambient oxygen concentration. J. Fish Res. Board Can. 30: 223-227.

Mount, D.I. and W.A. Brungs. 1967. A simplified dosing apparatus for fish toxicology studies. Water Res. 1: 21-29.

Mueller, J.A., T.A. Gerrish, and M.C. Casey. 1982. Contaminant inputs to the Hudson-Raritan Estuary, NOAA Tech. Mem. OMPA-21. 192 pp.

Mueller, J.A., J.S. Jeris, A.R. Anderson, and C.F. Hughes. 1976. Contaminant inputs to the New York Bight, NOAA Tech. Mem. ERL MESA-6. 34 pp.

O'Connor, J.S. and H.M. Stanford. 1979. Chemical pollutants of the New York Bight: Priorities for research, U.S. Dept. of Commerce, NOAA, ERL. 217 pp.

Sherk, J.A. 1972. Current status of the knowledge of the biological effects of suspended and deposited sediments in Chesapeake Bay. Chesapeake Sci. 13: Supplement; pp. 5137-5144.

Sprague, J.B. 1969. Measurement of pollutant toxicity to fish. I. Bioassay method for acute toxicity. Water Res. 3: 793-821.

Stout, V.F. and F.L. Beezhold. 1979. Analysis of chlorinated hydrocarbon pollutants: A simplified extraction and clean-up procedure for fishery product. Fish. Bull. 76: 880-886.

Waidwood, K.G. and F.W.H. Beamish. 1978. Effects of copper pH and hardness on the critical swimming performance of rainbow trout (*Salmo gairdneri*) (Richardson). Water Res. 12: 611-619.

Wedemeyer, G.A. and D.J. McLeay. 1981. Methods for determining the tolerance of fishes to environmental stressors. pp. 247-275. *In*: Stress and Fish. A.D. Pickering (ed.). Academic Press, New York.

Young, B.H. 1981. A study of striped bass in the marine district of New York III. Annual Report to NOAA. 76 pp.

AN *IN SITU* STUDY ON THE ADENYLATE ENERGY CHARGE OF *CORBICULA FLUMINEA* IN A FRESHWATER SYSTEM

Angela Cantelmo-Cristini, Frank E. Hospod, and Robert J. Lazell

SUMMARY

To gain clearer understanding of the responses of organisms to environmental levels of stress, the AEC, lipid, and glycogen content were monitored in the bivalve *Corbicula fluminea* while caged in the Raritan River, New Jersey. The AEC, lipid, and glycogen content were measured in one natural population and in clams caged at four different sites in the river for six months. Two pairs of cages were located above the combined outfall of an organic chemical products plant and the Somerset-Raritan Valley Sewage Authority and two pairs of cages were below the outfall.

There were no significant differences between the AEC values of clams from the natural population and those caged at the control station. This indicates that caging this species under the conditions of this study does not affect the AEC. The AEC values from the two control stations were not significantly different and indicate animals in optimal environmental conditions. These data suggest that transplanting a group of *C. fluminea* within Raritan River does not alter the AEC. Caged clams below the outfall exhibited AEC values that were significantly lower than the values from the control stations. The differences in AEC were observed within two weeks of the initiation of the study. All of these data suggest that AEC can be a sensitive indicator of stress in this species.

The glycogen concentrations indicate a distinct seasonal pattern for all the animals except those caged just below the outfall at station 3. All the clams at stations 1, 2, and 4 and the natural population exhibit a period of depressed glycogen concentration during the time that gametes are developing for the release of embryos into the environment in October. This is followed by a dramatic increase in glycogen from October to December. The glycogen values of the animals at station 3 decline throughout the entire study. There are no apparent seasonal changes in the lipid content of *C. fluminea* from July through December. Caged clams at station 3 exhibit significantly higher lipid values than those from all the other sampling sites. All of these data suggest that environmental levels of stressors have quantifiable effects on the physiology of this species.

INTRODUCTION

Many aquatic ecosystems in the United States serve as a depository for a myriad of waste materials. Toxic and non-toxic wastes from our society have had a heavy impact on several rivers in the northeast, particularly in New Jersey. The biota in such systems are continuously exposed to varying concentrations of mixtures of organic chemicals. Classically, field studies have established that degradation of the environment by chemical inputs results in alterations of community structure (Carpenter, 1924; Klein, 1957; Jeffries, 1962; Dean & Haskin, 1964; Pearce *et al*., 1976). Measurements of changes in species diversity and numbers of organisms are critical, but these give no information on how toxicants present in the system affect individuals in the population. Many laboratory studies have gone beyond the standard bioassay and have examined the sub-lethal effects of various chemicals on certain species of invertebrates (Kotch *et al*., 1972; Thurberg *et al*., 1973; Cairns *et al*., 1976; Calabrese *et al*., 1977; Rao *et al*., 1979; Cantelmo *et al*., 1981; Capuzzo & Lancaster, 1981). The data from such experiments are providing valuable insights as to the physiological, metabolic, and biochemical responses of an organism to a prolonged chemical insult. However, most of these studies are performed in the laboratory, usually with one model toxicant at a time, under constant environmental conditions. Such situations do not allow for many of the fluctuations that are present in the natural system. In addition, most studies do not take into account naturally occurring alterations in the physiology of an organism that occur with change in season.

To gain a more clear understanding of the effects of an environmental perturbation, studies should be performed to monitor sensitive physiological parameters on organisms _in situ_. These results would help to explain the response of organisms to "environmental levels' of stress. Few studies have examined the physiology of invertebrates at frequent intervals over a prolonged period of time in the field. Phelps _et al_. (1981) have studied several physiological parameters in the blue mussel, _Mytilus edulis_, along a gradient of pollution stress in Narragansett Bay, Rhode Island. Gould (1981) has measured enzyme activity of the sea scallop, _Placopecten magellanicus_, in the offshore waters of the New England and Mid-Atlantic states. Dickson and Giesy (1981) have studied the seasonal variations of AEC in naturally occurring populations of _Procambarus acutus acutus_, _Anodonta imbecillus_, and _Corbicula fluminea_. The present study monitored the changes in Adenylate Energy Charge (AEC), lipid, and glycogen content in the tissues of _Corbicula fluminea_ caged in the Raritan River, New Jersey.

Adenylate Energy Charge is defined as:

$$\frac{\text{ATP} + \frac{1}{2}\,\text{ADP}}{\text{ATP} + \text{ADP} + \text{AMP}}$$

(Atkinson & Walton, 1967). The values can theoretically range between 0 and 1. Values close to 1 indicate that a large amount of metabolic energy is available for use by the cells of an organism. Lower values suggest that less energy is available (Atkinson, 1977; Knowles, 1977). For example, studies have shown that AEC values range from 0.8 to 0.95 for molluscs grown in unstressed conditions (Beis & Newsholme, 1975; Wijsman _et al_., 1976; Ivanovici, 1980; Giesy & Dickson, 1981). Several workers have reported that AEC is an accurate indicator of naturally stressful conditions for a variety of organisms (Ball & Atkinson, 1975; Falkowski, 1977; Walker-Simmons & Atkinson, 1977; Rainer _et al_., 1979; Ivanovici, 1980). Conditions such as a change in temperature, salinity, or amount of food all tend to lower the AEC. Giesy _et al_. (1981) have shown that the AEC of the crayfish, _Procambarus pubescens_, and the grass shrimp, _Palaemonetes paludosus_, was lowered by exposure to cadmium. In another study, Giesy and his co-workers (in press) have shown that the AEC of _C. fluminea_ is altered upon exposure to cadmium in the laboratory. This species was chosen for the present study because of its general hardiness and high natural densities in the Raritan River. In addition, the fact that _C. fluminea_ is a filter feeder suggested that it could be caged for prolonged periods of time without adversely affecting its physiology.

MATERIALS AND METHODS

Establishment of Caged Populations

Four stations were selected along the Raritan River for establishment of caged populations of *C. fluminea* in 1982. Station 1 was established above Duke Island County Park. This area was selected as a control site for translocation of clams because of its high water quality. Station 2 was established at the confluence of the Raritan and Millstone Rivers, approximately one mile above the combined outfall of an organic chemical products plant and the Somerset-Raritan Valley Sewerage Authority. This site was selected because of its large natural population of *C. fluminea*. Clams from this station were used to stock all the cages in the study. Station 3 was established approximately one hundred feet below the combined outfall, and served as the primary test site. Station 4 was established approximately ½ mile below the combined outfall. This station allowed for the assessment of changes in AEC values in animals from an environment that is not directly exposed to the chemical input into the river.

The cages were constructed of wood with an internal dimension of approximately 1 m x 1 m x 25 cm; the top, bottom and two sides were screened with hardware cloth to allow flow of water through the boxes. Two cages were implanted in the river at each of the four stations in sites that had been matched for flow rate and depth. All of the cages were filled with sediment from station 2 and stocked with approximately 200 clams in the 0.8-0.2 cm size range collected from station 2. Two hundred animals were selected in order not to overcrowd the *C. fluminea* in the cages, while allowing for natural mortality during the course of the study. Nine clams could then be sampled biweekly without disturbance to the environment within the cages. Clams from the cages and from the natural population at station 2 were sampled biweekly for the duration of the study.

Environmental Parameters

The water temperature was measured in °C at all stations on all sampling dates. Water samples were obtained for the determination of alkalinity using the method outlined in *Standard Methods* (1980). Approximately 1 L of water was filtered through Whatman G/FC filters at each of the stations throughout the study. The filters were treated according to the method outlined by Strickland and Parsons (1965) for the determination of the Particulate Organic Carbon (POC).

Adenylate Energy Charge

Tissue samples from *C. fluminea* were collected following the methodology of Ivanovici (1981) with the following modifications. Three clams from each cage (6 per station) were removed from the substrate and kept submerged in water. Each clam was removed from the water, its valves opened, and its foot muscle clipped off and placed in a polyethylene bag. The bag and its contents were compressed between aluminum blocks stored in liquid nitrogen, freezing the tissue to at least -80°C within a few seconds. The tissue bag was then stored in a Dewar flask containing liquid nitrogen for the duration of the time in the field. The entire process of removing the animal from the water, dissecting, and freezing the tissue was always accomplished within 60 sec. This is well within the time limit of 90 sec reported by Ivanovici (1981). Upon return to the laboratory the samples were stored in a freezer at -80°C. Before bimonthly sampling commenced, a comparison was made between the AEC values of tissues stored in liquid nitrogen in the laboratory and tissues stored in a freezer at -80°C. No differences were observed between the two treatments, therefore the tissues were stored in the freezer for ease of handling. The tissues were extracted and analyzed for ATP, ADP, and AMP using a semi-micro spectrophotometric assay based on the method of Ivanovici (1981). These data were analyzed using a two-way ANOVA.

Lipid and Glycogen

Three clams were sampled from each station and each entire animal was placed in a polyethylene bag and frozen in liquid nitrogen. Upon return to the laboratory the samples were stored in a freezer at -80°C. The carcass of each clam was homogenized and extracted in a chloroform-methanol mixture according to the method of Barnes and Blackstock (1973). The analysis of the lipid content was also performed following their methodology. Tissue from which lipid was previously extracted was used for the glycogen determination. The analysis was performed following the method of Seifter *et al*. (1950) with the modifications of Helbert and Brown (1955). Glycogen concentrations were calculated from linear regressions derived from glucose standards. Both lipid and glycogen values are expressed as per cent dry weight of the tissue. These data were analyzed using a two-way ANOVA using an arcsine transformation.

RESULTS AND DISCUSSION

Environmental Parameters

The temperature profiles for all the stations are presented in Table 1. The river shows the expected seasonal temperature shifts, a high of 25°C in July and a low of 5°C in November. The data also indicate that there are no differences between the stations for any one sampling date. Examination of the alkalinity values also indicate that there are no significant differences between the stations ($p>0.05$).

Table 1. Temperatures in °C for sampling stations in the Raritan River, New Jersey 7/30 - 11/18, 1982.

Date	Station 1	Station 2	Station 3	Station 4
7/30	25	25	24.8	25
8/13	24.2	22.7	24	25
8/27	21	22.2	23	23
9/09	19	19.2	22	23
9/23	18.2	18.3	18.3	18.2
10/07	19	19.5	20.8	22
10/20	13	11.7	13.4	15.1
11/03	15.6	15	16	17
11/17	5	5.5	7	8

The POC concentrations at each station are presented in Table 2. There is a great deal of variability at all the stations; for example, values at station 4 range from 45 to 1729 mg C/L. The variability does not appear to be correlated with changes in season, although samples were not collected for the entire year. The POC values at station 3 are not significantly different from those at station 4 ($p>0.05$). However, the POC values at stations 3 and 4 are significantly higher than the values at stations 1 and 2 (3 & 4 vs. 1: $p<0.05$; 3 & 4 vs. 2: $p<0.01$). The method that was employed to measure the POC does not allow for the distinction between living and dead particles or between bacteria and algae. Therefore, the increase of POC at stations 3 and 4 does not necessarily imply that there is a greater supply of potential food at these stations. Rather, the observed increases are consistent with the sewage input at station 3 and its load of particulate matter.

Table 2. Particulate organic carbon in mg C/L for sampling stations in the Raritan River, New Jersey 7/30 - 11/17, 1982.

Date	Station 1	Station 2	Station 3	Station 4
8/27	656	425	823	1260
9/09	633	528	786	459
9/30	1051	533	1553	890
10/14	360	788	916	844
10/20	304	239	776	933
11/03	827	403	1802	1626
11/17	485	492	1322	1729

Adenylate Energy Charge

Examination of the seasonal profiles of all adenylates (AMP, ATP, ADP) indicates that the AEC is much less variable than the individual nucleotide concentrations. Giesy and Dickson (1981) also reported similar results in their study on AEC of *C. fluminea* in the Savannah River in Georgia. Ivanovici (1980) showed that the levels of variability between individual *Pyrazus ebeninus* were always lower for the AEC when compared with other nucleotide variables. Atkinson (1977) has suggested that organisms adjust their metabolic rates in order to keep their AEC constant. In addition, Giesy and Dickson (in press) remind us that AEC is a pure number that is not normalized to any factor, such as tissue weight. It is a ratio, and two of the terms (ATP and ADP) appear in both the numerator and denominator; this would further reduce the variability (Sokal & Rohlf, 1969). Ivanovici (1980) feels that this low variability is an advantage because a smaller sample size may be employed to assess the response of a population to stress. In light of the above information, we will restrict our discussion primarily to the AEC values for *C. fluminea*.

A comparison of the AEC values from the natural population and the clams caged at station 2 is presented in Table 3. There are no significant differences (p>0.05) between the two groups; both show similar seasonal changes. These results suggest that caging these animals for a prolonged period of time does not affect the AEC.

The AEC values of *C. fluminea* caged at the four stations are presented in Table 3. The data clearly show that the values at station 1 are not significantly different from the values at station 2 (p>0.05). The values at the pristine control station and at the source station range from 0.80 to 0.89 from July through November. Values in this range are indicative of clams in good to optimal conditions (Ching & Kronstad, 1972; Ball & Atkinson, 1975). These values are consistent with or higher than the AEC values determined by Giesy and Dickson (1981) for the same species in the Savannah River. The AEC stays fairly constant regardless of the temperature shift in the river and suggests that moving a group of animals per se does not have any long-term effect on the AEC.

Table 3. The AEC ± SEM of *C. fluminea* from the natural population and stations 1, 2, 3, and 4, 1982, n = 6.

DATE	STATION 1	STATION 2	NAT. POP.	STATION 3	STATION 4
7/30/82	0.810 ± 0.005	0.800 ± 0.017	0.820 ± 0.007	0.730 ± 0.013	0.810 ± 0.009
8/13/82	0.850 ± 0.006	0.840 ± 0.006	0.800 ± 0.018	0.770 ± 0.010	0.800 ± 0.010
8/27/82	0.800 ± 0.006	0.800 ± 0.009	0.840 ± 0.006	0.820 ± 0.010	0.820 ± 0.007
9/09/82	0.820 ± 0.010	0.830 ± 0.009	0.840 ± 0.005	0.730 ± 0.013	0.810 ± 0.011
9/23/82	0.820 ± 0.008	0.820 ± 0.012	0.790 ± 0.010	0.760 ± 0.022	0.790 ± 0.014
10/07/82	0.810 ± 0.010	0.810 ± 0.010	0.800 ± 0.010	0.760 ± 0.010	0.800 ± 0.010
10/20/82	0.840 ± 0.010	0.850 ± 0.004	0.830 ± 0.010	0.770 ± 0.030	0.820 ± 0.020
11/03/82	0.860 ± 0.010	0.850 ± 0.004	0.840 ± 0.010	0.850 ± 0.010	0.830 ± 0.010
11/17/82	0.810 ± 0.010	0.890 ± 0.010	0.880 ± 0.010	0.730 ± 0.202	0.790 ± 0.010

Examination of the AEC values of the clams at station 3 indicates that the group caged just below the combined outfall had AEC values significantly below (p<0.001) those at stations 1 and 2 for all but two sampling dates (Table 3). The values ranged from 0.73 to 0.77 for almost all of the study. These data indicate that the clams caged at this station were in a stressed condition; factors in the environment have measureably altered the energy state of these animals.

Comparison of the AEC values at station 4 to those at the other stations also indicates a reduction of the AEC values (Table 3). The AEC values of clams caged approximately ½ mile below the discharge are significantly lower (p<0.001) than the values at stations 1 and 2 for the 9/9,

10/20, 11/3, and 11/17 sample dates. These AEC values are not as low as those values measured for the animals caged at station 3. However, the data suggest that the adenylate system is very sensitive to any alteration in the environment.

The sewage input and the concomitant elevated POC values at stations 3 and 4 suggest that the animals caged at these stations are exposed to a greater amount of non-food materials than the animals at stations 1 and 2 (Table 2). The effects of nutrition on the AEC of multicellular organisms have not been clearly defined. Giesy and Dickson (1981) have reported an increase in the AEC of *C. fluminea* brought from the field and maintained in the laboratory at a higher temperature with nutritional augmentation. However, in another study they were unable to show that starvation lowered the AEC in two species of crayfish (Dickson & Giesy, 1981). Perhaps the increased organic load at stations 3 and 4 contributes to the total environmental stress on *C. fluminea*. However, there are no significant differences in the POC between stations 3 and 4, yet clams at station 3 have AEC values that are lower than the values at station 4. This suggests that the AEC appears sensitive enough to respond to other factors present in the environment that affect the energy state of this organism, such as chlorine or other chemicals present in the effluent at station 3.

In 1981, we performed a preliminary study on the natural population and two groups of *C. fluminea* caged at stations 2 and 3. A comparison of the data from the two years is presented in Table 4. The AEC values from the animals in the 1981 study are significantly lower ($p<0.05$) than the values from animals in the 1982 study at both stations. The AEC values at station 3 in 1981 never reached 0.7; this suggests that these animals were constantly in a very stressed condition (Chapman *et al.*, 1971; Montague & Dawes, 1974; Beis & Newsholme, 1975; Atkinson, 1977; Holmsen & Robkin, 1977; Walker-Simmons & Atkinson, 1977). The disparity in the AEC values can be understood if the differences in the environmental conditions between 1981 and 1982 are considered. In New Jersey, 1981 was a drought emergency year; the flow of the Raritan River was down to the minimum allowed by law (90 million gallons/day) during the time of the study. The reduction in flow would affect the amount of food available to *C. fluminea* and the dilution of the effluent at the outfall. In addition, heavy construction equipment in the river in August and September 1981 caused a considerable amount of siltation and turbidity at station 2. The lower

AEC values reflect the conditions in the environment. However, the relationship of AEC values between stations 1 and 2 is the same for both years; the clams below the effluent always had significantly lower values than the control station. These data again suggest the sensitivity of this biochemical parameter.

Table 4. The AEC ± SEM of C. fluminea from stations 2 and 3 in 1981 and 1982, n = 6.

1981 DATE	STATION 2	STATION 3	1982 DATE	STATION 2	STATION 3
8/03	0.690 ± 0.025	0.650 ± 0.030	7/30	0.800 ± 0.017	0.730 ± 0.013
8/19	0.690 ± 0.072	0.670 ± 0.040	8/13	0.840 ± 0.006	0.770 ± 0.010
-	-	-	8/27	0.800 ± 0.009	0.820 ± 0.010
9/09	0.800 ± 0.015	0.640 ± 0.060	9/09	0.830 ± 0.009	0.730 ± 0.013
-	-	-	9/23	0.820 ± 0.012	0.760 ± 0.022
10/07	0.740 ± 0.017	0.590 ± 0.030	10/07	0.810 ± 0.010	0.760 ± 0.010
-	-	-	10/20	0.850 ± 0.004	0.770 ± 0.030
-	-	-	11/03	0.850 ± 0.004	0.850 ± 0.010
11/12	0.820 ± 0.015	0.670 ± 0.013	11/17	0.890 ± 0.010	0.730 ± 0.020
12/18	0.690 ± 0.010	0.680 ± 0.010	-	-	-

Atkinson (1977) has suggested that AEC values in the high range indicate an increase in anabolic processes. Lower AEC values, conversely, would indicate an increase in catabolic processes. Such an increase in catabolism would be reflected in higher AMP values. Table 5 shows the AMP and the AEC values for the animals caged at station 3. The data reveal low AEC values accompanied by high AMP values and vice versa. This relationship is partially due to the fact that the AEC is a ratio that includes AMP in the denominator. However, such a relationship between AEC and AMP supports the idea that stressful conditions result in an increase in catabolic processes.

Table 5. The AEC ± SEM and the AMP ± SEM of C. fluminea from Station 3 in 1982, n = 6.

DATE	AEC	AMP
7/30	0.730 ± 0.013	0.235 ± 0.045
8/13	0.770 ± 0.010	0.145 ± 0.027
8/27	0.820 ± 0.010	0.156 ± 0.028
9/09	0.730 ± 0.013	0.218 ± 0.024
9/23	0.760 ± 0.022	0.208 ± 0.031
10/07	0.760 ± 0.010	0.210 ± 0.020
10/20	0.770 ± 0.030	0.220 ± 0.050
11/03	0.850 ± 0.010	0.100 ± 0.020
11/17	0.730 ± 0.020	0.220 ± 0.040

Lipid and Glycogen

The lipid and glycogen values are present in Figure 1. The lipid values at each of the stations remain fairly constant throughout the sampling period. There are no apparent seasonal changes from July through December. There are no significant differences between the lipid concentrations in the tissue of animals caged at stations 1, 2, and 4 and the natural population ($p>0.05$). Clams at station 3 exhibit significantly higher lipid values than those from all the sample sites ($p<0.001$). Examination of the glycogen values indicates that there is a distinct seasonal trend for the animals caged at stations 1, 2, and 4 and the natural population. Generally their values are lower from July through September and show a dramatic increase from October through December. This is in contrast with the seasonal pattern exhibited by the animals at station 3. The glycogen values of these clams decline throughout the entire study. They do not show any fall-winter increase; rather, the values continue to decrease through the December sample. The average glycogen values of these animals is significantly lower ($p<0.001$) than the values of all the other animals in this study.

The biochemical composition of molluscs is related to the availability of food as well as the reproductive cycle of the animal (Gabbott, 1983). Giesy and Dickson (1981) suggest that the population of *C. fluminea* in the Savannah River spawn in the spring and fall. Kraemer and Lott (1977) report that developing gametes were always present throughout the year in the gonads of *C. fluminea* from Arkansas, although the highest densities of eggs occurred in May. The preliminary histological examination made in the present study indicates that developing gametes are present in animals collected in July and October. Kraemer and Lott (1977) suggest that this species is hermaphroditic and capable of self-fertilization. In addition, their observations show that *C. fluminea* retain the developing embryos in the mantle cavity until they are well developed, indicating that the planktonic stage is very short. Very tiny juvenile *C. fluminea* were observed in the Raritan River during October, suggesting release of embryos in the fall. Since this sampling program started in July, no such evidence would be available for a possible spring spawning.

Several workers have reported that glycogen is utilized by the developing gonads. It may be used directly or converted into lipid and then incorporated into the developing

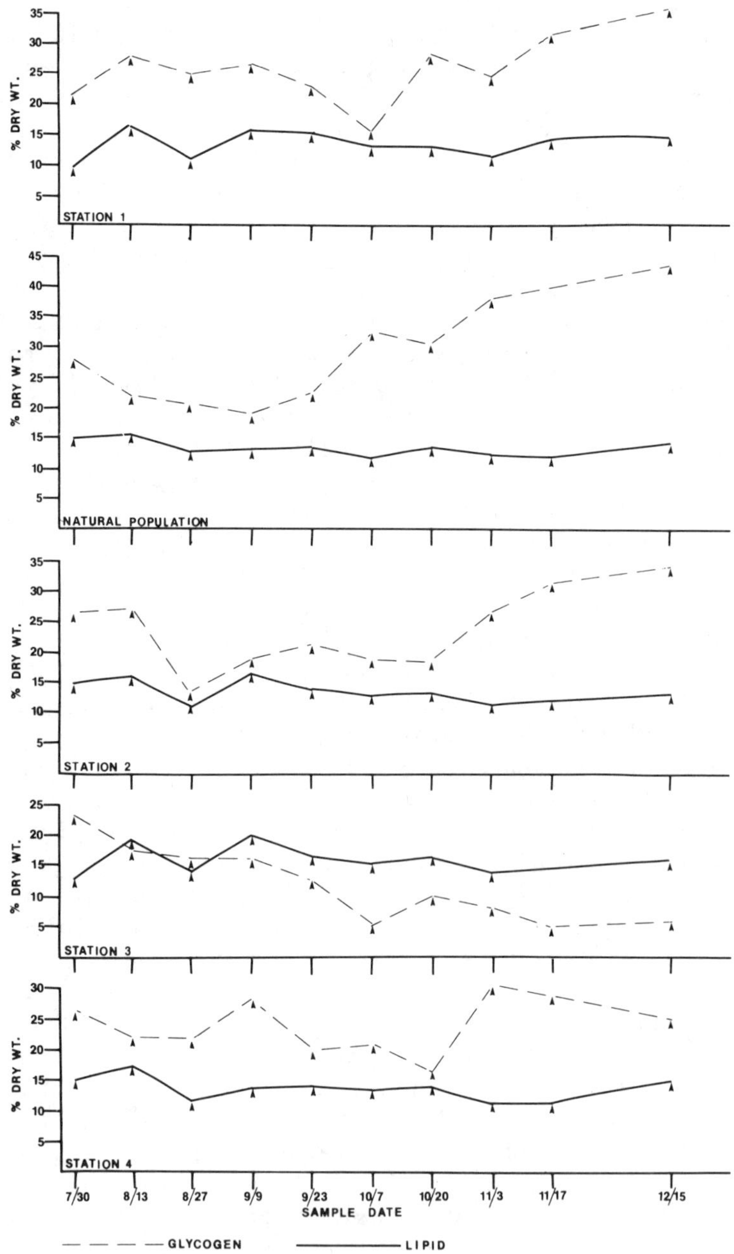

Fig. 1. The lipid and glycogen content of *C. fluminea* as % dry weight of tissue for stations 1, 2, 3, and 4 and the natural population, n = 3.

eggs (Williams, 1970; Lambert & Dehnel, 1974; Gabbott, 1975; Stickle, 1975; Pieters *et al.*, 1979). This appears to be consistent with the pattern of glycogen utilization found in *C. fluminea* in the present study. All the clams, except those caged at station 3, exhibit a period of stable, but depressed, glycogen levels during the time that gametes would be developing for release of embryos into the environment in October. This is followed by an increase in glycogen in their tissues. This increase may be an energy store for the winter or in preparation for gametogenesis for release of embryos in the spring.

The lipid concentration in tissues of *C. fluminea* does not exhibit any observable fluctuations associated with gamete production. There have been reports of similar seasonal profiles of lipid in other species of molluscs (Lambert & Dehnel, 1974; Pieters *et al.*, 1979; Zandee *et al.*, 1980). It is important to note that the AEC values were not affected by the normal reproductive cycle and the shifts in the biochemical composition of the tissues. As long as the animals are not under an environmental stress, the AEC remains relatively constant.

The pattern of lipid and glycogen levels and the AEC in animals caged at station 3 is different from the pattern observed in the other animals in this study. The clams caged below the outfall exhibited decreased stores of glycogen and consistently elevated lipid reserves, as well as AEC values indicative of catabolic processes. These data suggest that the animals are utilizing glycogen for metabolic processes, including gamete development, while not metabolizing their lipids to any great extent. Capuzzo and Lancaster (1981) have shown that exposure to petroleum hydrocarbons affects lipid storage and utilization in larval lobsters. Perhaps the organic chemicals present in the effluent at station 3 are affecting similar biochemical pathways in *C. fluminea*.

Responses to stress in vertebrates have been characterized by the general adaptation syndrome (Selye, 1976). According to Selye, all physiological systems of an organism initially respond to a stress, then a new level of resistance follows. This new level is characterized by an increase in catabolism, with a loss in energy available to the animal for processes such as growth and reproduction. If the stress continues and the organism cannot accommodate, it will eventually die. The results reported by Giesy *et al.* (in press) support this type of response to stress for *C. fluiminea*. Their study considered the effect of a laboratory exposure to cadmium on the AEC. The results of the present study indicate that stressors at levels found in the environment have significant and quantifiable effects on the physiology of this species. The AEC values and the glycogen concentrations obtained in clams caged at station 3 suggest that they are not accommodating to the stress. If the clams were able to survive the winter, it does not appear that they would be capable of gamete production in the spring.

The AEC has been shown to be a sensitive indicator of stress in *C. fluminea*. The effects of sub-optimal conditions in the environment can be detected over a long period of time by monitoring the changes in AEC in this species. Therefore, *C. fluminea* has the potential to be employed as an indicator organism for toxicological studies in freshwater systems. The data also indicate that the AEC is a more sensitive and better general indicator of stress than measuring lipid or glycogen concentrations. However, monitoring a combination of parameters including AEC, lipids, glycogen, protein, enzyme activity, and tissue histology will allow a more clear assessment of the effects of an environmental perturbation on organisms *in situ*. Studies should be conducted on other bivalves caged in the estuarine and marine environments in order to further assess the potential of this type of testing.

ACKNOWLEDGMENTS

This work was supported by the New Jersey Department of Environmental Protection, Office of Cancer and Toxic Substances Research. The authors are indebted to H. Garie who initially suggested the study. We also thank H. Garie and B. Ruppel for their many hours of field work; E. Saiff for the histological preparation, and H. Garie and L. Mantel for their assistance in preparation of the manuscript.

LITERATURE CITED

Atkinson, D.E. 1977. Cellular Energy Metabolism and its Regulation. Academic Press, New York. 293 pp.

Atkinson, D.E. and G.M. Walton. 1967. Adenosine triphosphate conservation in metabolic regulation: Rat liver citrate cleavage enzyme. J. Biol. Chem. 242: 3239-3241.

Ball, W.J., Jr. and D.E. Atkinson. 1975. Adenylate energy charge in *Saccharomyces cerevisiae* during starvation. J. Bacteriol. 121: 975-982.

Barnes, H. and J. Blackstock. 1973. Estimation of lipids in marine animals and tissues: Detailed investigation of the sulphophosphovanillin method for "total lipids." J. Exp. Mar. Biol. Ecol. 12: 102-118.

Beis, I. and E.A. Newsholme. 1975. The contents of adenine nucleotides, phosphagens and some glycolytic intermediates in resting muscles from vertebrates and invertebrates. Biochem. J. 152: 23-32.

Cairns, J., D.I. Messenger, and W.F. Calhoun. 1976. Invertebrate response to thermal shock following exposure to acutely sub-lethal concentrations of chemicals. Arch. Hydrobiol. 77: 164-175.

Calabrese, A., F.P. Thurberg, and E. Gould. 1977. Effects of cadmium, mercury, and silver on marine animals. Mar. Fish. Rev. 39: 5-11.

Cantelmo, A.C., R. Lazell, and L.H. Mantel. 1981. The effects of benzene on molting and limb regeneration in juvenile *Callinectes sapidus* Rathbun. Mar. Biol. Lett. 2: 333-343.

Capuzzo, J.M. and B.A. Lancaster. 1981. Physiological effects of South Louisiana crude oil on larvae of the American lobster (*Homarus americanus*). pp. 405-424. *In*: Biological Monitoring of Marine Pollutants. F.J. Vernberg, A. Calabrese, F. Thurberg, W.B. Vernberg (eds.). Academic Press, New York.

Carpenter, K.E. 1924. A study of the fauna of rivers polluted by lead mining in the Aberystwyth district of Cardiganshire. Ann. Appl. Biol. 11: 1-23.

Chapman, A.G., L. Fall, and D.E. Atkinson. 1971. Adenylate energy charge in *Escherichia coli* during growth and starvation. J. Bacteriol. 109: 1072-1086.

Ching, T.M. and W.E. Kronstad. 1972. Varietal differences in growth potential, adenylate energy level, and energy charge in wheat. Crop Sci. 12: 551-563.

Dean, D. and H.H. Haskin. 1964. Benthic repopulation of the Raritan River estuary following pollution abatement. Limnol. Oceanogr. 9: 551-563.

Dickson, G.W. and J.P. Giesy. 1981. Variation of phosphoadenylates and adeylate energy charge in crayfish (Decapoda: Astacidae) tail muscle due to habitat differences. Comp. Biochem. Physiol. 70A: 421-425.

Falkowski, P.G. 1977. The adenylate energy charge in marine phytoplankton: The effect of temperature on the physiological state of *Skeletonema costatum*. J. Exp. Mar. Biol. Ecol. 27: 37-45.

Gabbott, P.A. 1975. Storage cycles in marine bivalve molluscs: A hypothesis concerning the relationship between glycogen metabolism and gametogenesis. pp. 191-211. In: 9th Europ. Mar. Biol. Symp. H. Barnes (ed.). Aberdeen Univ. Press, Aberdeen.

Gabbott, P.A. 1983. Development and seasonal metabolic activities in marine molluscs. pp. 165-207. In: The Mollusca, Vol. 2. P.W. Hochachka (ed.). Academic Press, New York.

Giesy, J.P., S.R. Denzer, C.S. Duke, and G.W. Dickson. 1981. Phosphoadenylate concentrations and energy charge in two freshwater crustaceans: Responses to physical and chemical stressors. Verh. Internat. Verein. Limnol. 21: 205-220.

Giesy, J.P. and G.W. Dickson. 1981. The effect of season and location on phosphoadenylate concentrations and adenylate energy charge in two species of freshwater clams. Oecologia 49: 1-7.

Giesy, J.P. and G.W. Dickson. (in press). Changes in phosphoadenylate concentrations and adenylate energy chage as an integrating biochemical measure of stress in aquatic invertebrates. Environ. Toxicol. Chem.

Giesy, J.P., C.S. Duke, R.D. Bingham, G.W. Dickson and G.J. Leversee. (in press). Changes in phosphoadenylate concentrations and adenylate energy charge of the freshwater clam (*Corbicula fluminea*) when exposed to cadmium. Environ. Toxicol. Chem.

Gould, E. 1981. Monitoring sea scallops in the offshore waters of New England and the Mid-Atlantic states: Enzyme activity in phasic adductor muscle. pp. 377-388. In: Biological Monitoring of Marine Pollutants. F.J. Vernberg, A. Calabrese, F. Thurberg, W.B. Vernberg (eds.). Academic Press, New York.

Helbert, J.R. and K.D. Brown. 1955. Factors influencing quantitative determination of methylpentoses and ketohexoses with anthrone. Anal. Chem. 27: 1791-1796.

Holmsen, H. and L. Robkin. 1977. Hydrogen peroxide lowers ATP levels in platelets without altering adenylate energy charge and platelet function. J. Biol. Chem. 252: 1752-1757.

Ivanovici, A.M. 1980. The adenylate energy charge in the estuarine mollusc, *Pyrazus ebenius*. Laboratory studies of responses to salinity and temperature. Comp. Biochem. Physiol. 66A: 43-55.

Ivanovici, A.M. 1981. A method for extraction and analysis of adenine nucleotides for determination of adenylate energy charge in molluscan tissue. Common. Sci. & Ind. Res. Org., Div. Fish. Oceanog. Report # 118. Cronulla, Australia. 27 pp.

Jeffries, H.P. 1962. Environmental characteristics of Raritan Bay, a polluted estuary. Limnol. Oceanogr. 7: 21-31.

Klein, L. 1957. Aspects of River Pollution. Butterworths, London. 621 pp.

Knowles, C.J. 1977. Microbial metabolic regulations by adenine nucleotide pools. pp. 241-283. In: Microbial Energetics. B.A. Haddock and W.A. Hamilton (eds.). Cambridge University Press, Cambridge, England.

Koch, R.B., D. Desaiah, H.H. Yap, and L.K. Cutkomp. 1972. Polychlorinated biphenyls: Effects of long-term exposure on ATPase activity in fish, Pimephales promelas. Bull. Envir. Contam. Toxicol. 7: 87-92.

Kraemer, L.R. and S. Lott. 1977. Microscopic anatomy of the visceral mass of Corbicula (Bivalvia: Sphaeriaces). Bull. Amer. Malacol. Union Inc. 1977: 48-56.

Lambert, P. and P.A. Dehnel. 1974. Seasonal variations in biochemical composition during the reproductive cycle of the intertidal gastropod Thais lamellosa Gmelin (Gastropoda, Prosobranchia). Canadian J. Zool. 52(3): 305-318.

Montague, M.D. and E.A. Dawes. 1974. The survival of Peptococcus prevotii in relation to the adenylate energy charge. J. Gen. Microbiol. 80: 291-299.

Pearce, J.B., J.V. Carracciolo, M.B. Halsey, and L.H. Rogers. 1976. Temporal and spatial distribution of benthic macroinvertebrates in the New York Bight. Am. Soc. Limnol. Oceanogr. Spec. Symp. 2: 394-403.

Phelps, D.K., W. Galloway, F.P. Thurberg, E. Gould, and M.A. Dawson. 1981. Comparison of several physiological monitoring techniques as applied to the blue mussel, Mytilus edulis, along a gradient of pollutant stress in Narragansett Bay, RI. pp. 335-356. In: Biological Monitoring of Marine Pollutants. F.J. Vernberg, A. Calabrese, F. Thurberg, and W.B. Vernberg (eds.). Academic Press, New York.

Pieters, H., J.H. Kluytmans, W. Zurberg, and D.I. Zandee. 1979. The influence of seasonal changes on energy metabolism in Mytilus edulis (L.). I. Growth rate and biochemical composition in relation to environmental parameters and spawning. pp. 285-292. In: Cyclic Phenomena in Marine Plants and Animals. E. Naylor and R.G. Hartnoll (eds.). Pergamon Press, Oxford.

Rainer, S.F., A.M. Ivanovici, and V.A. Wadley. 1979. Effect of reduced salinity on adenylate energy charge in three estuarine molluscs. Mar. Biol. 54: 91-99.

Rao, K.R., F.R. Fox, P.J. Conklin, A.C. Cantelmo, and A.C. Brannon. 1979. Physiological and biochemical investigations of the toxicity of pentachlorophenol to crustaceans. pp. 307-339. In: Marine Pollution: Functional Responses. W.B. Vernberg, F.P. Thurberg, A. Calabrese, and F.J. Vernberg (eds.). Academic Press, New York.

Seifter, S., S. Dayton, B. Novic, and E. Muntwyler. 1950. The estimation of glycogen with the anthrone reagent. Arch. Biochem. 25: 191-200.

Selye, H. 1976. Stress in Health and Disease. Butterworth, Inc., Boston, Mass. 334 pp.

Sokal, R.R. and F.J. Rohlf. 1969. Biometry: The Principles and Practice of Statistics in Biological Research. W.H. Freeman and Co., San Francisco, CA. 776 pp.

Standard Methods for the Examination of Water and Wastewater. 15th Edition 1980. APHA, AWWA, WPCF, Washington, D.C. 403 pp.

Stickle, W.B. 1975. The reproductive physiology of the intertidal prosobranch *Thais lamellosa* (Gmelin). II. Seasonal changes in biochemical composition. Biol. Bull. (Woods Hole, MA) 148: 448-460.

Strickland, J.D.H. and R. Parsons. 1965. A manual of seawater analysis. Bulletin #125, Fish. Res. Board Can. pp. 137-140.

Thurberg, F.P., M.A. Dawson, and R. Collier. 1973. Effects of copper and cadmium on osmoregulation and oxygen consumption in two species of estuarine crabs. Mar. Biol. 23: 171-175.

Walker-Simmons, M. and D.E. Atkinson. 1977. Functional capacities and the adenylate energy charge in *Escherichia coli* under conditions of nutritional stress. J. Bacteriol. 130: 676-683.

Wijsman, T.C.M., A. Dezwaan, and R.H. Ebberink. 1976. Adenylate energy charge in *Mytilus edulis* L. during exposure to air. Biochem. Soc. Trans. 4: 442-443.

Williams, E.E. 1970. Seasonal variations in the biochemical composition of the edigle winkle *Littorina littoreas* (L.). Comp. Biochem. Physiol. 33: 655-661.

Zandee, D.I., J.H. Kluytmans, W. Zurburg, and H. Pieters. 1980. Seasonal variations in biochemical composition of *Mytilus edulis* with reference to energy metabolism and gametogenesis. Neth. J. Sea Res. 14: 1-29.

PLASMA CORTICOSTEROID AND GLUCOSE RESPONSES TO POLLUTANTS IN STRIPED MULLET: DIFFERENT EFFECTS OF NAPHTHALENE, BENZO[A]PYRENE AND CADMIUM EXPOSURE

Peter Thomas and Jerry M. Neff

SUMMARY

1. The plasma corticosteroid and glucose responses of striped mullet to naphthalene, benzo[a]pyrene, and cadmium exposure were compared.
2. A typical stress response, elevation of plasma cortisol accompanied by an increase in glucose concentrations, was seen only after naphthalene exposure (2 mg/L). Cortisol concentrations returned to control levels 12 h after the beginning of naphthalene exposure, whereas the hyperglycemia persisted for 10 d even though most of the original dose had disappeared from the exposure aquaria by this time.
3. Extremely low concentrations of benzo[a]pyrene (0.2-0.3 μg/L) caused a moderate acute elevation of plasma cortisol but this was not accompanied by changes in plasma glucose titers. A second rise in plasma cortisol was observed in mullet after 4 d exposure to benzo[a]pyrene. This pattern of plasma corticosteroid fluctuations during exposure to benzo[a]pyrene was confirmed in a second teleost species, Atlantic croaker.
4. Continuous exposure of mullet to 5 mg/L cadmium for 3 d did not significantly alter circulating cortisol and glucose levels. There was a dose-dependent hyperglycemia after 1-wk exposure to cadmium (2-25 mg/L) but this was not accompanied by an elevation of plasma cortisol.

5. The results indicate that mullet have different biochemical responses to the pollutants tested. Circulating cortisol titers and secondary metabolic disturbances such as hyperglycemia do not always show a good correlation. Possible mechanisms by which pollutants may influence plasma corticosteroid and glucose concentrations in teleosts are discussed.

INTRODUCTION

A wide variety of physiological and biochemical effects of sublethal pollutant exposure have been described in fish over the last decade (Waldichuk, 1979). Whereas many of these effects can be attributed to specific toxic actions of the pollutants themselves; others such as a pronounced hyperglycemia appear to be general responses to pollutants (Silbergeld, 1974; McLeay, 1977; DiMichele & Taylor, 1978; Thomas *et al.*, 1980, 1981a,b; Swift 1981). Certain pollutants also alter lipid metabolism and impair ionic and osmotic regulation (Holmberg *et al.*, 1972; Thomas *et al.*, 1980, 1981a,b; Englelhardt *et al.*, 1981). These pollutant effects probably represent secondary stress responses, since similar metabolic and osmoregulatory disturbances have been reported in fish exposed to a variety of physical stressors (Wedemeyer, 1972; Hattingh, 1976; Soivio & Oikari, 1976). It is thought that these secondary stress responses are mediated via elevations in circulating corticosteroids and catecholamines, termed primary stress responses (Mazeaud *et al.*, 1977; Leach & Taylor, 1980; Eddy, 1981).

Acute elevations of plasma corticosteroids preceeding the onset of secondary stress responses have been demonstrated in fish exposed to pollutants (DiMichele & Taylor, 1978; Thomas *et al.*, 1980, 1981a). A knowledge of the quantitative relationships between primary and secondary stress responses would permit an assessment of the major metabolic and osmoregulatory disturbances caused by a particular stressor from measurement of primary stress responses alone. Both plasma corticosteroid and glucose titers have been proposed as indicators of the degree of trauma in fish caused by various environmental stressors (Silbergeld, 1974; McLeay, 1977; Donaldson, 1981). However, there have been few attempts to determine whether there is a close correlation between these two blood parameters under a variety of conditions, including exposure to different pollutants.

In the present study the effects of three pollutants (naphthalene, benzo[a]pyrene, and cadmium) on the plasma cortisol and glucose concentrations in juvenile striped mullet (*Mugil cephalus*) were compared. Naphthalene was selected as the reference toxicant. Elevations in plasma cortisol and glucose have been reported in fish exposed to naphthalene (DiMichele & Taylor, 1978) as well as other hydrocarbons (Thomas *et al*., 1980, 1981a,b), which suggests that these pollutants elicit a general stress response in fish. The putative carcinogen benzo[a]pyrene (BaP) has an aqueous solubility of less than 1.0 μg/L and does not appear to be acutely toxic to fish at this concentration. The effects of BaP were investigated in order to determine whether extremely low hydrocarbon concentrations cause a corticosteroid stress response in fish. The results were compared to those obtained from Atlantic croaker (*Micropogonias undulatus*) to investigate if there are species differences in the biochemical responses to this pollutant. Schreck and Lorz (1978) could not demonstrate a significant corticosteroid stress response in salmon exposed to lethal doses of cadmium. However, blood corticosteroid titers were not monitored in salmon during the first 6 h of cadmium exposure (Schreck & Lorz, 1978), a period during which acute corticosteroid responses to pollutants have frequently been observed (Thomas *et al*., 1980, 1981a). Therefore, in the present study plasma cortisol levels were monitored during both acute and chronic exposure to cadmium.

MATERIALS AND METHODS

Juvenile striped mullet (*Mugil cephalus*), length 10-15 cm, were collected by seine or cast net near Port Aransas, Texas, USA. Atlantic croaker (*Micropogonias undulatus*), weight 20-40 g, were collected by trawl in the same area. The fish were acclimated in the laboratory for 2 wk in a 550-L recirculating aquaria maintained at 33°/$_{oo}$ S and 20 ± 0.5°C (lights on 0700 - 1900 hours). The fish were kept in artificial sea water (Instant Ocean artificial sea salts, Aquarium Systems Inc., Mentor, Ohio) and fed Tetramin algal flake (Tetrawerke, Germany) once a day at noon.

At the end of the acclimation period six fish were transferred to each exposure aquarium containing 40 L artificial sea water at 33°/$_{oo}$. Charcoal filters were removed from control and experimental aquaria 1 d before benzo[a]pyrene or naphthalene exposure and replaced in the tanks 3 d later. The fish were undisturbed for 16 h prior to the sampling period (between 10:00 and 11:00 am two weeks or longer after transfer to the experimental aquaria). The fish were removed with a dip net and bled by cardiac puncture. All the fish in each aquarium were bled within 2 min. The aquaria were positioned so that fish in other exposure tanks were not disturbed by the sampling procedure. This protocol prevented nonspecific activation of the interrenal glands. Blood was centrifuged at 1500 x g for 3 min to separate the plasma. Samples were analyzed for glucose immediately after collection and the remaining plasma was stored at -60°C until analyzed for cortisol.

Naphthalene exposure

Naphthalene was dissolved in ethanol (80 mg/L). One-ml aliquots of the stock solution were added to the exposure aquaria to give a final concentration of 2.0 mg/L. An equal volume of ethanol was added to the control aquaria at the same time. Mullet were exposed to a single dose of 2.0 mg/L naphthalene; a nominal concentration of 10 mg/L did not cause any mortalities. The naphthalene concentrations in the exposure aquaria were monitored using a spectrophotometric technique (Neff & Anderson, 1975). As much as 90% of the naphthalene had disappeared from the exposure tanks 1 d after dosing. However, certain mullet tissues bioaccumulate the hydrocarbon to a considerable extent and tissue burdens of naphthalene remain high for at least 2 wk after a single exposure (Thomas *et al.*, 1980).

Benzo[a]pyrene exposure

A sand dosing column was used to prepare saturated solutions of benzo[a]pyrene (Aldrich Gold Label 99% pure). Fifty mg of benzo[a]pyrene (BaP) was dissolved in 50 ml of ether, mixed with 1 kg course sand and evaporated to dryness under a stream of nitrogen. The charged sand was subsequently packed between layers of clean sand in a 7-cm diameter glass column and kept in place with glass wool plugs. Aluminum foil was wrapped around the column to exclude light. Water was pumped into the bottom of the column at a rate of 36 L/h with an Eheim filter pump. After 2-3 d operation, the concentration of BaP in the eluant stabilized (approx. 0.5 µg/L) and was constant thereafter for 6-8 wk.

For short-term exposures saturated stock solutions of BaP were prepared by recirculating 150 L of sea water through the dosing column for 48 h (final concentration approx. 1 μg/L). One day prior to BaP exposure, 20 L of water was removed from each exposure and control tank. Freshly prepared stock BaP solutions were diluted with artificial sea water to make final concentrations of 0.1, 0.2, and 0.3 μg BaP/L when 20-L volumes of the mixture were added to the exposure aquaria. At the same time, equal volumes of artificial sea water were added to the control aquaria (careful addition of water to the aquaria does not elicit a corticosteroid or hyperglycemic stress response in mullet). The aquaria were redosed with BaP every 2 d.

For long-term BaP exposures (greater than 168 h), water from the 150-L exposure tank was continuously recirculated through the dosing column. The concentration of BaP in the exposure tank reached 0.93 μg/L after 3 d and remained at approximately this level for the duration of the experiment. No mortalities occurred in either species during 6 wk exposure to this concentration of BaP. BaP concentrations in the water were determined fluorimetrically by the method of Giam *et al*. (1980). Eighty-to-ninety percent of the BaP was lost from the water within 2 d of dosing.

Mullet which had been exposed to BaP (0.9 μg/L) for 2 wk were challenged with a moderate stressor, exposure to a water-soluble fraction of a fuel oil (Thomas *et al*., 1980), to determine whether the corticosteroid stress response persisted in fish chronically exposed to BaP. Water-soluble fractions (WSF) of a No. 2 fuel oil were prepared by the method of Anderson *et al*. (1974) and added to the exposure tanks to give a final concentration of 20% WSF (Thomas *et al*., 1980).

Cadmium exposure

Cadmium chloride ($CdCl_2 \cdot 2H_2O$) was prepared as a concentrated stock solution (40 g Cd/ml) and added to the exposure aquaria to give final concentrations of 2, 5, and 25 mg Cd/L (highest dose killed 30-50% of animals after 96 h in 33°/oo sea water). The aquaria water was replaced every 3 d and redosed with cadmium. Cadmium concentrations in the water were monitored by atomic absorption spectrophotometry and declined 20% between doses.

Analytical methods

Cortisol was measured in 50 μl plasma samples by a radioimmunoassay technique (Thomas *et al*., 1980) using antiserum obtained from New England Nuclear Corp. Preliminary separation of corticosteroids was unnecessary since cortisol is the predominant corticosteroid in mullet and sciaenid plasma (Thomas and Wofford, 1984). The assay has a sensitivity of 10 pg and a mean within assay variance of 6.2%. Plasma glucose was analyzed with an automatic glucose analyzer (model 23A, Yellow Springs Instrument Co., Yellow Springs, Ohio). Plasma cholesterol and osmolality and hepatic glycogen were assayed by techniques described by Thomas _et al._ (1981b).

Data were analyzed by Duncan's multiple range test or Student's T test.

RESULTS

Exposure to naphthalene caused plasma cortisol levels in mullet to rise to 93 ng/ml, a 9-10 fold increase, within 1 h (Table 1). Circulating cortisol levels had declined to 28 ng/ml after 3 h and were not significantly different from control levels for the duration of the experiment. The increase in plasma cortisol concentrations was accompanied by a rise in blood glucose levels which reached 99 mg/dl, two times control levels, after 3 h exposure and remained elevated for the duration of the experiment (Table 1).

Plasma cortisol dynamics in mullet exposed to BaP are shown in Figure 1. Exposure to 0.1 μg BaP/L did not affect cortisol levels, whereas the plasma corticosteroid concentrations rose within one hour after exposure to 0.2-0.3 μg BaP/L. Plasma cortisol concentrations reached a peak at 2 h and thereafter declined to control levels by 5 h. A second rise in plasma cortisol concentrations occurred between 72 and 96 h of exposure, but by 168 h cortisol levels of mullet exposed to BaP were not significantly different from those of controls. Plasma glucose concentrations were also measured in these fish. None of the treatments or exposure periods caused a significant alteration in glucose levels (Table 2).

Responses of Atlantic croaker (_Micropogonias undulatus_) to BaP exposure were similar to those described in mullet (Table 2). The magnitude and the dynamics of the corticosteroid response to 0.2-0.3 μg BaP/L was similar in both species. As in mullet, there was no elevation of plasma glucose concentrations in _M. undulatus_ in response to BaP. Plasma cholesterol and osmolality and hepatic glycogen were also unaltered in _M. undulatus_ after acute and chronic exposure to BaP (range of mean osmolality values: controls 295-320 mOs/L, exposed 296-228 mOs/L; hepatic glycogen

Table 1. Fluctuations of plasma cortisol and glucose in *M. cephalus* after a single exposure to naphthalene (water concentration 2 mg/L).

	Duration of exposure						
	0h	1h	3h	12h	3d	6d	10d
plasma cortisol (ng/ml)	7.5 ± 1.5	92.5 ± 17.1*	28.0 ± 5.3*	5.3 ± 1.6	6.3 ± 3.7	10.1 ± 1.3	3.4 ± 0.5
plasma glucose (mg/dl)	53.2 ± 2.2	75.1 ± 6.8*	99.1 ± 7.4*	64.3 ± 3.5*	58.0 ± 1.8	65.9 ± 5.0*	67.0 ± 4.2*

* $p<0.05$ Duncan's. (N = 6)

Table 2. Fluctuations of plasma cortisol and glucose in *M. cephalus* and *M. undulatus* during continuous exposure to benzo[a]pyrene (nominal water concentration 0.2-0.3 μg/L).

	Duration of exposure (hours)							
	0	1	2	3	6	72	96	168
M. cephalus								
plasma cortisol (ng/ml)	4.0 ± 0.3	24.3 ± 4.1*	30.1 ± 6.9*	15.3 ± 0.4	4.0 ± 0.3	6.6 ± 0.5	32.7 ± 6.2*	8.1 ± 0.2
plasma glucose (mg/dl)	52.1 ± 1.7	54.1 ± 4.1	48.6 ± 5.6	49.2 ± 9.7	51.3 ± 1.3	46.0 ± 5.8	36.2 ± 9.7	47.2 ± 2.2
M. undulatus								
plasma cortisol (ng/ml)	10.8 ± 3.4	15.7 ± 6.2	---	36.5 ± 8.5*	17.8 ± 6.4	---	41.0 ± 10.2*	24.7 ± 12.4
plasma glucose (mg/dl)	44.0 ± 2.4	47.0 ± 3.0	---	46.6 ± 2.6	34.8 ± 2.6	---	43.3 ± 2.8	44.4 ± 2.8

*$p<0.05$ Duncan's. (N = 6)

values: controls 128-145 mg/g, exposed 92-149 mg/g, N = 4). Chronic exposure to BaP (0.9 μg/L) elevated plasma cortisol levels in croaker but not in mullet (croaker cortisol values [ng/ml]; controls 10.3 ± 2.8, 14-d BaP exposure 50.0 ± 9.2, 28-d BaP exposure 30.4 ± 23.6, p<0.05, N = 6: mullet cortisol values [ng/ml]: controls 5.7 ± 1.0, 10-d BaP exposure 8.0 ± 1.6, 22-d BaP exposure 4.7 ± 2.0, N = 6).

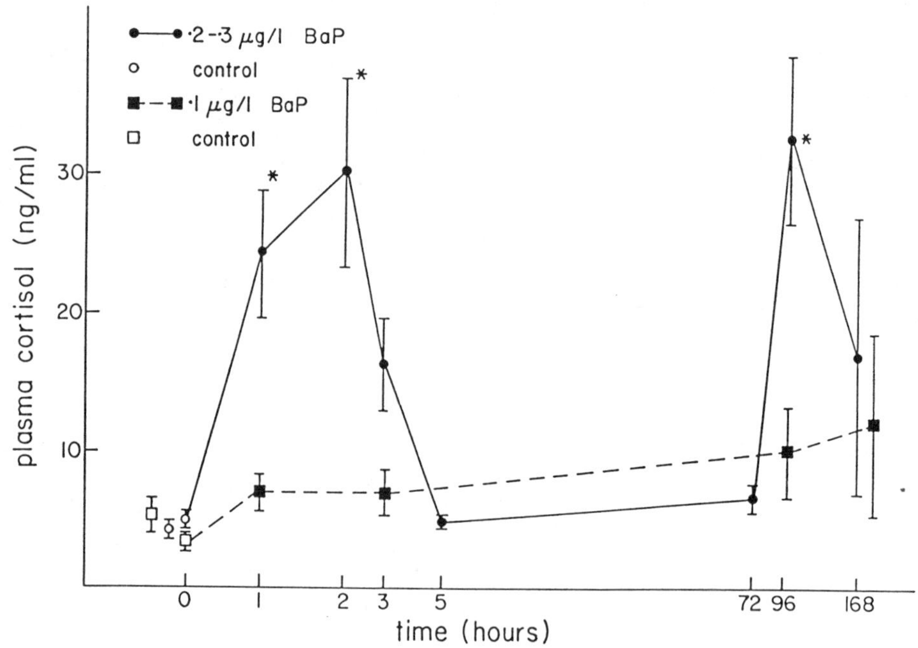

Fig. 1. Plasma cortisol fluctuations in M. cephalus exposed to 0.1, 0.2-0.3 μg BaP/L ($\bar{X}$ = 0.25 μg/L) (nominal concentration). Each point represents the mean (± SE) of at least 8 determinations. Asterisks denote mean values significantly different from controls. (p<0.05, Duncan's)

Mullet which had been exposed to BaP for 2 wk were subjected to a stressful procedure (exposure to a 20% water-soluble fraction [WSF] of a No. 2 fuel oil [Thomas et al., 1980]) to determine whether BaP caused a functional impairment of the hypothalamo-pituitary-interrenal axis. Plasma cortisol and glucose rose to 5 and 4 times control levels, respectively, after 1-3 h exposure to oil (cortisol: 1-h exposure to WSF, 115.2 ± 34.2 ng/ml; glucose: 3-h exposure to WSF, 167.6 ± 39.7 mg/dl; N = 6). Similar responses to oil were observed in mullet not pretreated with BaP (see Fig. 3 in discussion).

Plasma cortisol and glucose levels in mullet after 1-wk exposure to various cadmium concentrations are shown in Figure 2. None of the cadmium concentrations affected plasma corticosteroid levels, whereas a dose-dependent elevation of plasma glucose was observed. Glucose concentrations were twice control levels in mullet exposed to 25 mg Cd/L, a dose which resulted in 30% mortality after 1 wk exposure. Neither blood cortisol nor glucose concentrations were altered during acute exposure (up to 72 h) to 5 mg Cd/L (Table 3). Plasma glucose concentrations had increased by the seventh day of cadmium exposure (5 mg Cd/L), whereas plasma corticosteroid levels were unaltered (Fig. 2).

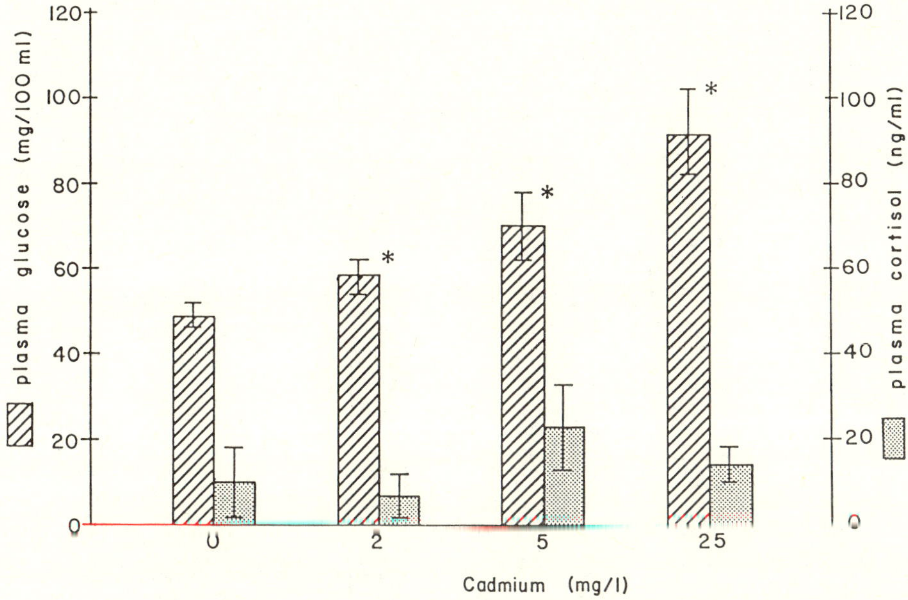

Fig. 2. Mean (± SE) plasma cortisol and glucose concentrations in _M. cephalus_ after 1-wk exposure to 0, 2, 5, and 25 mg Cd/L (N = 6). Asterisks denote mean values significantly different from controls. ($p<0.05$, Duncan's).

Table 3. Plasma glucose and cortisol fluctuations in *Mugil cephalus* during acute exposure to 5 mg Cd/L.

	Duration of exposure (hours)					
	0	1	3	12	24	72
plasma cortisol (ng/ml)	7.4 ± 2.8	4.3 ± 3.4	7.8 ± 5.6	16.4 ± 5.9	6.6 ± 3.5	2.6 ± 0.5
plasma glucose (mg/dl)	49.7 ± 1.3	44.4 ± 5.2	49.7 ± 4.6	48.0 ± 3.6	48.0 ± 2.1	49.8 ± 2.1

(N = 6)

DISCUSSION

Considerable controversy still exists over usage of the term stress, its definition and the classification of types of stress (for recent discussions see Henry, 1980; Vigas, 1980; Pickering, 1981). In the present discussion the term 'stressor' is used to describe the stimulus which evokes a 'stress response' in the organism. Although stress responses have been described as either emotional or physical, it has been shown that most mammalian responses to stressors have an emotional component which is associated with limbic system arousal (Mason 1975a,b). The evidence for an emotional or psychological component to stress responses in fish populations has recently been reviewed by Schreck (1981). It has been argued that emotional stress responses are not trivial, but are basic and widespread biological characteristics of vertebrates (Edholm, 1978). The significance of limbic system arousal and emotional stress responses in vertebrates can be appreciated if they are considered as part of a mechanism to rapidly mobilize the animal's defense mechanisms in response to signals of potentially dangerous situations. This specialized role of the limbic system occurred early in vertebrate phylogeny (Hilton, 1965) and it is easy to understand why such a system, which increases the chances of an animal's survival, would continue to evolve into a highly complex defense reaction to a wide variety of stimuli.

Plasma corticosteroid responses to pollutants

Benzo[a]pyrene, naphthalene and several other pollutants (Donaldson & Dye, 1975; Schreck & Lorz, 1978; Thomas *et al.*, 1980, 1981a), caused a rapid elevation of plasma cortisol concentrations. The threshold dose of BaP for activation of the hypothalamo-pituitary-interrenal axis in both mullet and croaker was between 0.1 and 0.25 μg/L, less than a tenth of the pollutant concentration necessary to cause a corticosteroid response to copper (Donaldson & Dye, 1975), chromium (Hill & Fromm, 1968), naphthalene (DiMichele & Taylor, 1978), and fluorene (Thomas *et al.*, 1981a). In contrast, 5 mg/L cadmium, a concentration twenty-five thousand times that of BaP, failed to elicit an acute corticosteroid stress response in mullet. Schreck and Lorz (1978) had previously shown that lethal doses of cadmium did not significantly increase plasma cortisol levels in salmon. These authors suggested that only stimuli which cause fright, discomfort, or pain elevate cortisol titers in fish.

Earlier, Mason (1975a,b) had suggested that the mammalian corticosteroid stress response to a wide variety of stimuli is not merely a general response to adverse stimuli as proposed in the General Adaptation Syndrome (Selye, 1950), but represents a specific response to a common element of all these stressors, namely that they cause the same emotional reaction. Thus, a deleterious environmental change would have to be detected by an individual in order to elicit a corticosteroid stress response. If the interrenal glands of teleosts are stimulated by a similar mechanism, then the present results would indicate that mullet are unable to detect cadmium or any of its toxic manifestations (described in later section on physiological stressors), but can detect BaP concentrations as low as 0.2 μg/L.

The acute minor elevation of plasma cortisol after exposure to BaP may be due solely to a fright reaction to the presence of this novel stimulus. Naphthalene caused a greater corticosteroid response possibly because it is noxious or causes physiological dysfunction at this concentration. This compound is toxic to fish, damaging sensory nerve endings and gill epithelia (DiMichele & Taylor, 1978) and causing osmoregulatory imbalance (Levitan & Taylor, 1979). A wide variety of pollutants, which includes metals (Hill & Fromm, 1968; Donaldson & Dye, 1975; Schreck & Lorz, 1978) and organic compounds (Grant & Mehrle, 1973; DiMichele & Taylor, 1978; Thomas *et al*., 1980, 1981a,b), have been shown to elicit a corticosteroid stress response in fish. However, the mechanisms by which pollutants activate the interrenal glands in fish are poorly understood, and the relative importance of physical (chemical) and emotional responses to pollutants is unknown. Therefore, it is unclear how cortisol production is stimulated by BaP and naphthalene, but not by cadmium. Although a narcotizing effect of cadmium has been reported in salmon (Schreck & Lorz, 1978), no such behavioral abnormalities were observed in cadmium-exposed mullet.

The fall in plasma cortisol concentrations 5 h after the beginning of BaP exposure is probably due to a delayed level-sensitive feedback mechanism (Yates & Maran, 1974). There is evidence that a corticosteroid feedback mechanism exists in teleosts (Fryer & Peter, 1977). However, feedback inhibition of cortisol release in vertebrates lasts only a few hours (Yates & Maran, 1974). The finding that plasma cortisol were at control values after 3-d exposure may indicate, therefore, that the hypothalamo-pituitary-interrenal axis of mullet is no longer activated by BaP and acclimation to the pollutant has occurred. An acclimation phase of the corticosteroid stress response to pollutants has been reported previously in teleosts (Hill & Fromm, 1968; Donaldson & Dye, 1975; Thomas et al., 1981a). However, there are insufficient data points in this study to determine whether acclimation to BaP occurs during the first 3 d of exposure.

Stressors can also stimulate cortisol secretion in mammals indirectly by causing changes in physiological parameters which exceed their normal limits (Vigas, 1980). Thus certain stimuli, called physiological stressors (Vigas, 1980), activate the hypothalamo-pituitary-adrenal axis by internal (e.g., humoral) pathways. Recent data suggest that ammonia and nitrite stimulate cortisol secretion in fish by causing physiological dysfunction, whereas merely the presence of these chemicals in the surrounding water was insufficient to induce a corticosteroid stress response (Tomasso et al., 1981). The delayed rise in plasma cortisol in mullet upon exposure to a low concentration of pentachlorophenol (PCP) appeared to be dependent on PCP accumulation in the tissues to a threshold level (Thomas et al., 1981b). The secondary moderate elevation of plasma cortisol after 4-d exposure to BaP may have also been caused by tissue accumulation of the hydrocarbon to a threshold level. Marked disturbances of carbohydrate and osmotic regulation were apparent after a few days exposure to pentachlorophenol. However, it is not known what toxic manifestations of BaP exposure are responsible for the secondary elevation of plasma cortisol since no disturbances of lipid, carbohydrate, and osmoregulatory function were apparent. The fact that disturbances of certain physiological parameters, such as carbohydrate metabolism, occur in cadmium-exposed mullet without significant activation of the hypothalamo-pituitary-interrenal axis indicates that no simple relationship exists between physiological dysfunction and stimulation of corticosteroid secretion in fish.

Exposure to BaP for more than a week elevated plasma cortisol concentrations in croaker but not in mullet; although the hypothalamo-pituitary-interrenal axis of mullet was still responsive to a stressful stimulus, oil. The decrease in corticosteroid levels in mullet may be the result of increased steroid metabolism, since hepatic cytochrome P-450 content in these fish had reached 150% control levels by this time (Woodin & Thomas, unpubl. data). The depression of plasma steroid concentrations in freshwater fish exposed to a polychlorinated biphenyl (Aroclor 1234) and industrial effluent has been attributed to such a mechanism (Sivarajah *et al.*, 1978; Hansson, 1981). However, direct evidence of enhanced hepatic clearance of circulating steroids in fish exposed to pollutants is lacking.

Pollutants can also influence plasma steroid levels by altering steroidogenesis (Colby, 1981). A direct inhibitory effect of *o,p*-DDD *in vitro* on ACTH stimulation of cortisol production by interrenal tissues has been demonstrated in *Sarotherodon aureus* (Ilan & Yaron, 1980). Cadmium has been shown to inhibit steroid hydroxylases in mammalian adrenals (Colby, 1981) and to alter androgen synthesis in salmonid testes (Freeman & Idler, 1975). Chronic exposure to 10 mg Cd/L did not elevate plasma corticosteroids, whereas they were elevated in mullet exposed to 1 mg/L (Thomas, unpubl. data), possibly reflecting impaired steroidogenesis at the higher concentration. However, there was no impairment of the corticosteroid stress response to a second stressor after chronic exposure to 10 mg Cd/L (Thomas, 1982), which indicates that the hypothalamo-pituitary-interrenal axis of these fish was still functionally intact.

Correlation between plasma cortisol concentration and glucose level

This study shows that the biochemical stress responses of mullet to benzo[a]pyrene, naphthalene, and cadmium differ. Only one of the pollutants, naphthalene, caused elevations of both plasma cortisol and glucose. Naphthalene has previously been shown to alter these parameters in *Fundulus heteroclitus* (DiMichele & Taylor, 1978). The dynamics of plasma corticosteroid and glucose fluctuations in mullet exposed to naphthalene were broadly similar to those observed in fish after exposure to oil and certain physical stressors. An acute rise in plasma cortisol occurred, followed by a hyperglycemic response which persisted for up to several days (Leach & Taylor, 1980; Thomas *et al.*, 1980). However, the hyperglycemia continued for 10 d after exposure to naphthalene, considerably longer than that observed after exposure to oil or physical stressors. It is possible that this persistent hyperglycemia partially reflects impaired insulin secretion, since naphthalene has been shown to exert a pancreotoxic effect in fish (DiMichele & Taylor, 1978).

It is difficult to distinguish the secondary stress responses mediated by corticosteroids from those controlled by catecholamines, as both of these parameters increase in response to stressors (Mazeaud & Mazeaud, 1981). Both adrenal catecholamines and corticosteroids have been implicated in the regulation of blood glucose levels in fish (see discussions by Leach & Taylor, 1980; Mazeaud & Mazeaud, 1981).

The cadmium study shows that plasma glucose levels can rise independently of corticosteroids in mullet (Fig. 3), although a close correlation between these two parameters has been found in previous pollutant studies. Cadmium can influence carbohydrate metabolism in mammals by a variety of mechanisms (Singhal & Merali, 1979), including stimulation of adrenal catecholamine secretion (Hart & Borowitz, 1974) and inhibition of pancreatic function, particularly insulin secretion (Ghafghazi & Mennear, 1975). The hyperglycemia induced by cadmium in fish appears to be at least partly due to impaired insulin secretion, since glucose intolerance was seen after feeding in mullet exposed for 1 wk to 10 mg Cd/L (Thomas, 1982).

The lack of a hyperglycemic response to benzo[a]pyrene was unexpected, since a dose of oil, which caused an elevation of plasma cortisol similar to that observed after exposure to BaP, significantly elevated plasma glucose (5% WSF, Fig. 3). Research by Leach and Taylor (1980) suggests that inhibition of cortisol synthesis by metapyrone does not affect the acute hyperglycemic response to a physical stress, although an elevation of plasma cortisol is necessary for chronic glucose elevation. Thus, it is possible that the hormones, presumably catecholamines, which are responsible for the acute hyperglycemia seen after exposure to a physical stressor are unaltered after benzo[a]pyrene exposure.

A marked hyperglycemia has been observed in fish exposed to various pollutants (Holmberg *et. al.*, 1972; Silbergeld, 1974; Hattingh, 1976; McLeay, 1977), but only a few studies have attempted to relate these metabolic disturbances to induction of primary stress responses (DiMichele & Taylor, 1978; Thomas *et al.*, 1980, 1981a,b). There was a good correlation between plasma cortisol and glucose concentrations in mullet after exposure to oil, fluorene, and pentacholorophenol (Fig. 3). However, the present study shows that these parameters are not always closely correlated and that the biochemical response of a teleost can vary with the pollutant under investigation. Consequently, measurement of a single stress parameter may be insufficient to accurately predict the metabolic disturbances associated with pollutant exposure. The present findings also emphasize the need for a better understanding of the mechanisms by which pollutants induce primary stress responses and secondary metabolic disturbances in fish.

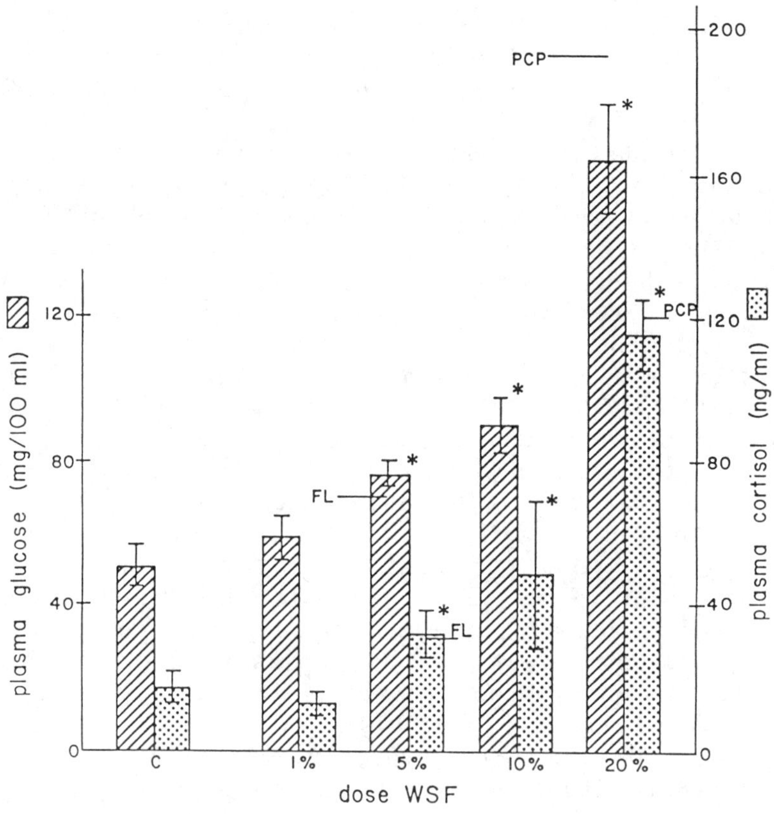

Fig. 3. Mean (± SE) plasma cortisol and glucose concentrations in *M. cephalus* 1 and 3 h, respectively, after exposure to 1%, 5%, 10% and 20% water-soluble fractions of a No. 2 fuel oil (adapted from Thomas *et al*. 1980, with permission). Asterisks denote means significantly different from controls. FL-cortisol and glucose concentrations in mullet after exposure to 100 μg/L fluorene (Thomas *et al*., 1981a). PCP-cortisol and glucose concentrations in mullet after exposure to 100 μg/L pentachlorophenol (Thomas *et al*., 1981b).

LITERATURE CITED

Anderson, J.W., J.M. Neff, B.A. Cox, H.E. Tatum, and G.M. Hightower. 1974. Characteristics of dispersions and water-soluble extracts of crude and refined oils and their toxicity to estuarine crustaceans and fish. Mar. Biol. 27: 75-88.

Colby, H.D. 1981. Chemical suppression of steroidogenesis. Environ. Health Persp. 38: 119-127.

DiMichele, L. and M.H. Taylor. 1978. Histopathological and physiological responses of *Fundulus heteroclitus* to naphthalene exposure. J. Fish. Res. Board Can. 35: 1060-1066.

Donaldson, E.M. 1981. The pituitary-interrenal axis as an indicator of stress in fish. pp. 11-47. *In*: Stress and Fish. A.D. Pickering (ed.). Academic Press, London.

Donaldson, E.M. and H.M. Dye. 1975. Corticosteroid concentrations in sockeye salmon (*Oncorhynchus nerka*) exposed to low concentrations of copper. J. Fish. Res. Board Can. 32: 533-539.

Eddy, F.B. 1981. Effects of stress on osmotic and ionic regulation in fish. pp. 77-102. *In*: Stress and Fish. A.D. Pickering (ed.). Academic Press, London.

Edholm, O.G. 1978. Introductory lecture - Emotion and stress. pp. 3-7. *In*: Environmental Endocrinology. I. Assenmacher and D.S. Farner (eds.). Springer-Verlag, Berlin.

Englelhardt, F.R., M.P. Wong, and M.E. Dvey. 1981. Hydromineral balance and gill morphology in rainbow trout, *Salmo gairdneri*, acclimated to fresh and sea water, as affected by petroleum exposure. Aquat. Toxicol. 1: 175-186.

Freeman, H.C. and D.R. Idler. 1975. The effect of polychlorinated biphenyl (PCB) on steroidogenesis and reproduction in the brook trout (*Salvelinus fontinalis*). Can. J. Biochem. 53: 666-670.

Fryer, J.N. and R.E. Peter. 1977. Hypothalamic control of ACTH secretion in goldfish. III. Hypothalamic cortisol implant studies. Gen. Comp. Endocr. 33: 215-225.

Ghafghazi, T. and J.H. Mennear. 1975. The inhibitory effect of cadmium on the secretory activity of the isolated perfused rat pancreas. Toxicol. Appl. Pharmacol. 31: 134-142.

Giam, C.S., D.A. Trujillo, S. Kira, and Y. Hrung. 1980. Simplified monitoring procedures for benzo[a]pyrene, hexachlorobenzene and pentachlorophenol in water. Bull. Environ. Contam. Toxicol. 25: 824-827.

Grant, B.F. and P.M. Mehrle. 1973. Endrin toxicosis in rainbow trout (Salmo gairdneri). J. Fish. Res. Board Can. 30: 31-40.

Hansson, T. 1981. Effects of treated municipal waste water on the hepatic metabolism of 4-androstene-3,1-dione in rainbow trout Salmo gairdneri Richardson. pp. 339. In: Stress in Fish. A.D. Pickering (ed.). Academic Press, London.

Hart, D.T. and J.L. Borowitz. 1974. Adrenal catecholamine release by divalent mercury and cadmium. Arch. Int. Pharmacodyn. 209: 94-99.

Hattingh, J. 1976. Blood sugar as an indicator of stress in the freshwater fish, Labeo capensis (Smith). J. Fish. Biol. 19: 191-195.

Henry, J.P. 1980. Present concept of stress theory. pp. 557-571. In: Catecholamines and Stress: Recent Advances. E. Usdin, R. Kvetnansky, and I.J. Kopin (eds.). Elsevier/North Holland, New York.

Hill, C.W. and P.O. Fromm. 1968. Response of the interrenal gland of rainbow trout (Salmo gairdneri) to stress. Gen. Comp. Endocr. 11: 69-77.

Hilton, S.M. 1965. Emotion. pp. 353-376. In: The Physiology of Human Survival. O.G. Edholm and A.L. Bacharach (eds.). Academic Press, London-New York.

Holmberg, B., S. Jensen, A. Larsson, K. Lewander, and M. Olsson. 1972. Metabolic effects of technical pentachlorophenol (PCP) on the eel Anguilla anguilla L. Comp. Biochem. Physiol. 43B: 171-183.

Ilan, Z. and Z. Yaron. 1980. Suppression by organochlorines of the response to adrenocorticotrophin of the interrenal tissue in Sarotherodon aureus (Teleostei). J. Endocr. 87: 185-193.

Leach, G.J. and M.H. Taylor. 1980. The role of cortisol in stress-induced metabolic changes in Fundulus heteroclitus. Gen. Comp. Endocr. 42: 219-227.

Levitan, W.M. and M.H. Taylor. 1979. Physiology of salinity-dependent naphthalene toxicity in Fundulus heteroclitus. J. Fish. Res. Board Can. 36: 615-620.

Mason, J.W. 1975a. A historical view of the stress field. Part I. J. Human Stress 1: 6-12.

Mason, J.W. 1975b. A historical view of the stress field. Part II. J. Human Stress 1: 22-36.

Mazeaud, M.M. and F. Mazeaud. 1981. Adrenergic responses to stress in fish. pp. 49-75. In: Stress and Fish. A.D. Pickering (ed.). Academic Press, London.

Mazeaud, M.M., F. Mazeaud, and E.M. Donaldson. 1977. Primary and secondary effects of stress in fish: Some new data with a general review. Trans. Am. Fish. Soc. 106: 201-212.

McLeay, D.J. 1977. Development of a blood sugar bioassay for rapidly measuring stressful levels of pulpmill effluent to salmonid fish. J. Fish. Res. Board Can. 34: 477-485.

Neff, J.M. and J.W. Anderson. 1975. An ultraviolet spectrophotometric method for the determination of naphthalene and alkyl naphthalenes in the tissues of oil contaminated marine animals. Bull. Environ. Contam. Toxicol. 14: 122-128.

Pickering, A.D. 1981. Introduction: The concept of biological stress. pp. 1-9. In: Stress and Fish. A.D. Pickering (ed.). Academic Press, London.

Schreck, C.B. 1981. Stress and compensation in teleostean fishes: Response to social and physical factors. pp. 295-321. In: Stress and Fish. A.D. Pickering (ed.). Academic Press, London.

Schreck, C.B. and H.W. Lorz. 1978. Stress response of coho salmon (Oncorhynchus kisutch) elicited by cadmium and copper and potential use of cortisol as an indicator of stress. J. Fish. Res. Board Can. 35: 1124-1129.

Selye, H. 1950. Stress and the general adaptation syndrome. Br. Med. J. 1: 1383-1392.

Silbergeld, E.K. 1974. Blood glucose: A sensitive indicator of environmental stress in fish. Bull. Environ. Contam. Toxicol. 11: 20-25.

Singhal, R.L. and Z. Merali. 1979. Biochemical toxicity of cadmium. pp. 61-112. In: Cadmium Toxicity. J.H. Mennear (ed.). Marcel Dekker Inc., New York.

Sivarajah, K., C.S. Franklin, and W.P. Williams. 1978. The effects of polychlorinated biphenyls on plasma steroid levels and hepatic microsomal enzymes in fish. J. Fish. Biol. 13: 401-409.

Soivio, A. and A. Oikari. 1976. Haematological effects of stress in a teleost, Esox lucius L. J. Fish. Biol. 8: 397-411.

Swift, D.J. 1981. Changes in selected blood component concentrations of rainbow trout (Salmo gairdneri Richardson) exposed to hypoxia or sublethal concentrations of phenol or ammonia. J. Fish. Biol. 19: 45-61.

Thomas, P. 1982. Effect of cadmium exposure on plasma cortisol levels and carbohydrate metabolism in mullet (Mugil cephalus). J. Endocr. Suppl. 94: 35.

Thomas, P., H.W. Wofford, and J.M. Neff. 1981a. Biochemical stress responses of striped mullet (Mugil cephalus L.) to fluorene analogs. Aquat. Toxicol. 1: 329-342.

Thomas, P., R.S. Carr, and J.M. Neff. 1981b. Biochemical stress responses of striped mullet Mugil cephalus and polychaete worms Neanthes virens to pentachorophenol. pp. 73-104. In: Biological Monitoring of Marine Pollutants. F.J. Vernberg, A. Calabrese, F.P. Thurberg, and W.B. Vernberg (eds.). Academic Press, New York.

Thomas, P., B.W. Woodin, and J.M. Neff. 1980. Biochemical responses of the striped mullet Mugil cephalus to oil exposure. I. Acute responses - Interrenal activation and secondary stress responses. Mar. Biol. 59: 141-149.

Thomas, P. and H.W. Wofford. 1984. High-performance liquid chromatography of corticosteroids in vertebrate plasma: Assay of cortisol in mullet and corticosterone in the rat. Comp. Biochem. Physiol. 78: 473-479.

Tomasso, J.R., K.B. Davis, and B.A. Simco. 1981. Plasma corticosteroid dynamics in channel catfish (Ictalurus punctatus) exposed to ammonia and nitrite. Can. J. Fish. Aquat. Sci. 38: 1106-1112.

Vigas, M. 1980. Contribution to the understanding of the stress concept. pp. 573-578. In: Catecholamines and Stress: Recent Advances. E. Usdin, R. Kvetnansky, and I.J. Kopin (eds.). Elsevier/North Holland, New York.

Waldichuk, M. 1979. Review of the problems. Phil. Trans. R. Soc. Lond. B286: 399-424.

Wedemeyer, G. 1972. Some physiological consequences of handling stress in the juvenile coho salmon (Oncorhynchus kisutch) and steelhead trout (Salmo gairdneri). J. Fish. Res. Board Can. 29: 1780-1783.

Yates, F.E. and J.W. Maran. 1974. Stimulation and inhibition of adrenocorticotropin release. pp. 367-404. In: Handbook of Physiology, Section 7, Vol. IV. E. Knobil and W.H. Sawyer (eds.). Waverly Press, Baltimore.

THE EULAMELLIBRANCH *MERCENARIA MERCENARIA* (L): A REVIEW AND CURRENT DATA ON METALS ACCUMULATION AND THE INTERNAL TRANSPORT OF CADMIUM

W.E. Robinson, M.P. Morse, B.A. Penney,
J.P. Kakareka, and E.U. Meyhöfer

SUMMARY

The content and concentration of various metals (e.g., Ag, Ba, Cd, Cu, Fe, Zn) were measured in individual tissues (adductor muscles, blood cells, blood plasma, digestive glands, gills, kidneys, mantle, pericardial glands, and remaining viscera) of the quahog, Mercenaria mercenaria. Similar analyses were conducted on tissues from clams exposed to elevated levels of either Cd or radiolabeled ^{109}Cd (0.1 µg/ml) in 0.45 µm filtered sea water (15°C, 30-32°/$_{oo}$). Although small in size, the kidneys were clearly the major organs that concentrated Ag, Ba, Cd, Cu, Fe, and Zn, and to some extent Mn and Ni. Following exposure to 100 ppb Cd for 15 d, a 14-fold increase in Cd concentration was detected in the kidneys, with a lesser, but appreciable, accumulation of Cd in the gills, digestive glands, mantle, and remaining viscera.

Different metals have different subcellular distributions within the kidneys. For example, Fe and Zn are primarily associated with kidney granules and other subcellular organelles which can be easily separated from kidney homogenates by centrifugation. Cd and Cu are predominately associated with a soluble metal-binding protein(s) with an apparent molecular weight of 14,500-16,500.

Data from chemical analyses and radiotracer studies have revealed a possible metal transport mechanism which has not been previously described. Blood plasma normally contains significant amounts of such metals as Ba, Cu, Fe, and Zn. Within a sample of whole blood, more total metal is contained in the blood plasma than the blood cells. Similarly, in quahogs exposed to ^{109}Cd in the surrounding sea water, more label (nCi per ml of whole blood) is measured in the blood plasma than the blood cells. Similar concentrations of ^{109}Cd can be found in the blood plasma and pericardial fluid, indicating that the label passes through the ultrafiltration pores of the pericardial glands. Therefore, although the importance of blood cells in metal transport cannot be discounted, metals in the blood plasma may have a direct pathway to the kidneys via the pericardial fluid.

INTRODUCTION

Most research on mechanisms of metal accumulation, transport, and elimination in marine bivalves has centered on species within the class Filibranchia, such as the blue mussel *Mytilus edulis*, the oysters *Ostrea edulis*, *Crassostrea virginica*, and *Crassostrea gigas*, and the scallop *Pecten maximus*. All of these species are epifaunal and, with the exception of *M. edulis*, are generally found in relatively clean sea water with low suspended-particulate loads. In contrast, the Eulamellibranchia has received little attention, even though this class contains approximately six times the number of species that comprise the Filibranchia. Furthermore, the majority of the approximately 6000 eulamellibranch species are infaunal and are found in a wider range of habitats than the filibranchs.

In light of this disproportionate emphasis on filibranchs, we chose to study metal accumulation and internal transport mechanisms in the infaunal eulamellibranch bivalve *Mercenaria mercenaria*. The "hard clam" or "quahog" is an ecologically prominent and commercially important species which ranges from the Gulf of St. Lawrence to Florida in the western Atlantic (Abbott, 1974). Much is known concerning the quahog's basic biology and physiology (see McHugh *et al.*, 1982). More importantly with regard to metal accumulation and transport, the morphology and function of the blood cells (amoebocytes) have been previously investigated (Zacks, 1955; Cheng & Foley, 1975; Moore & Eble, 1977; Moore & Gelder, 1983). However, little is known about metal transport and accumulation processes associated with the excretory organs of *M. mercenaria*. Although the glandular coelomoducts (referred to as "kidneys") and the pericardial glands in the dorsal wall of the pericardial cavity accumulate and excrete materials, the functional roles of these organs have not been described adequately.

The purpose of this paper is twofold. First, we will present data on metal accumulation and the internal transport of cadmium in *M. mercenaria*. Second, we will provide a review of the literature concerning metal accumulation in this species in relation to current knowledge of metal accumulation in other eulamellibranch bivalves.

MATERIALS AND METHODS

Quahogs were collected from various sites on Cape Cod, MA (*e.g.*, West Bay, Osterville; Cataumet Bay, Bourne) and maintained in running sea water at ambient temperature. Groups of animals were exposed to 0.1 μg Cd/ml (Fisher Scientific Cadmium Reference Standard, Cd metal in 0.1 N HNO_3) in a closed-system aquarium containing 20 L of 0.45 μm filtered sea water at 15°C and 30-32°/oo salinity. Length of exposure time (7-15 d) and number of quahogs exposed (n = 7-16) varied with experimental design. Control animals were maintained under identical conditions in an adjacent aquarium. Sea water was continually aerated and changed every 2-3 d to preclude the build-up of metabolites.

Metals were analyzed in individual tissues (adductor muscles, digestive glands, gills, kidneys, mantle, pericardial glands, and remaining viscera) from seven quahogs ($\bar{X}$ animal length = 76.4 mm, SD = 10.1 mm) exposed to 0.1 μg Cd/ml for 15 d and seven control animals ($\bar{X}$ animal length = 67.8 mm, SD = 6.4 mm) by d.c. Plasma Emission Spectrometry (DCP). Following lyophilization, tissues were digested in redistilled concentrated HNO_3, evaporated to dryness, and redissolved in 0.1 N HNO_3.

Pooled samples of kidneys (n = 16) from both control and Cd-exposed (15 d) quahogs were prepared for liquid chromatographic protein analysis according to the methodology of Roesijadi (1982). Tissues were homogenized with a Ten Broeck tissue grinder in an equal volume of cold 20 mM TRIS-HCl buffer (pH 8.6) containing 0.1 mM PMSF (phenylmethylsulfonylfluoride) and 275 mM NaCl. Homogenates were centrifuged at 20,200 x g for 10 min. Due to equipment limitations, the resulting supernatant was recentrifuged at 34,800 x g for 60 min, as opposed to the 80,000 x g, 90 min centrifugation employed by Roesijadi (1982). Residues from each centrifugation step and a subsample of the final supernatant were analyzed for metal content. A 5-ml subsample of the final supernatant was applied to a 2.5 by 88 cm Sephadex

G-75 liquid chromatography column eluted with 20 mM TRIS-HCl buffer (pH 8.6) containing 0.02% NaN_3, at a flow rate of 25 ml/h. Fractions (6 ml) were automatically collected, their absorbances read at 250 and 280 nm, and their metal content determined by DCP.

Individual quahogs were exposed to mixtures of ^{109}Cd (New England Nuclear, $^{109}CdCl_2$ in 0.1 N HNO_3) and cold carrier Cd, at a final concentration of 0.1 μg Cd/ml, for periods of 1 h to 10 d (15°C, 30-32°/$_{oo}$, n = 16; $\bar{X}$ animal length = 76.8 mm, SD = 4.1 mm). Animals were maintained in the same 2.5 L of continually aerated sea water for the entire exposure period, resulting in an average Cd depletion of 2 ng/ml per day. Following exposure, one valve was removed, and the mantle cavity was quickly rinsed with running water. Whole blood was collected, by vacuum (30 cm Hg), from slits made in the anterior and posterior adductor muscles. Pericardial fluid (approximately 0.2 ml) was removed by vacuum from a small slit made in the wall of the pericardial cavity. Tissues (adductor muscles, digestive glands, gills, kidneys, mantle, and pericardial glands) were dissected from the remaining viscera and rinsed in filtered sea water. Blood cells were separated from blood plasma by centrifugation (20 min, 2750 x g). Blood plasma and pericardial fluid were mixed directly with 10 ml Aquasol (New England Nuclear) liquid scintillation cocktail in scintillation vials. Following lyophilization, all other tissues were digested in a 1:4 mixture of concentrated $HClO_4$ and HNO_3, heated to drive off the HNO_3, and brought to volume with glass-distilled, deionized water (Super-Q system, Millipore Corp.). Subsamples (1 ml) of each tissue digest were added to 10 ml Aquasol. Samples were counted in a Beckman LS-100 liquid scintillation counter set at 0.2% counting efficiency. Results from triplicate 20 min counts were averaged and corrected for background, decay, and quench (external standard ratio method).

Kidneys were pooled from 30 adult quahogs and homogenized as described previously. Homogenates were layered on a 2.5 M sucrose bed and centrifuged for 30 min at 1500 x g. The resulting pellet of kidney granules was lyophilized, digested in concentration HNO_3, and analyzed for metals by DCP.

RESULTS

The kidneys are clearly the major organs that accumulate metal in *M. mercenaria*, containing the highest

Table 1. Mean concentration (± SD) of six metals in tissues of *Mercenaria mercenaria*. Control quahogs (n = 7) reflect background metal levels in tissue; Cd-exposed quahogs (n = 7) subjected to 0.1 μg Cd/ml in sea water for 15 d.

		Metal Concentration (μg/g dry wt)					
Tissue	Tissue dry Weight (g)	Ag	Ba	Cd	Cu	Fe	Zn
Control							
Adductors	0.56 (0.28)	<0.2	0.8 (1.2)	<0.3	2.3 (0.5)	27.3 (7.1)	34.9 (4.4)
Digestive gland	0.14 (0.08)	0.8 (0.4)	2.3 (4.1)	<0.1	33.6 (7.0)	114.2 (36.4)	90.8 (12.2)
Gills	0.14 (0.06)	<0.4	1.0 (0.8)	<0.3	10.8 (1.6)	105.1 (47.7)	86.7 (7.6)
Kidneys	0.06 (0.05)	7.6 (3.7)	3.4 (4.5)	21.0 (8.0)	357.4 (81.8)	637.9 (321.4)	420.8 (164.2)
Mantle	0.33 (0.11)	<0.2	3.2 (6.0)	<0.3	7.3 (2.2)	66.5 (26.2)	45.0 (5.7)
Viscera	2.00 (0.67)	<0.2	1.3 (1.7)	<0.4	6.6 (1.5)	61.6 (26.8)	43.2 (8.8)
Cd-exposed							
Adductors	0.75 (0.33)	<0.2	0.2 (0.1)	<0.2	1.8 (0.3)	32.5 (9.2)	35.1 (3.2)
Digestive gland	0.14 (0.06)	0.6 (0.2)	2.4 (1.8)	4.6 (1.5)	53.2 (23.2)	96.9 (16.9)	90.3 (8.3)
Gills	0.18 (0.06)	<0.2	1.0 (0.6)	9.4 (2.2)	10.8 (2.5)	195.2 (123.7)	85.0 (15.5)
Kidneys	0.07 (0.05)	4.8 (2.5)	28.2 (25.4)	288.4 (62.6)	337.1 (89.7)	3336.2 (1585.7)	966.9 (635.6)
Mantle	0.39 (0.13)	0.2 (0.1)	4.7 (2.6)	5.4 (1.6)	10.6 (5.2)	74.3 (22.6)	68.5 (41.9)
Viscera	2.64 (1.24)	0.2 (0.04)	1.4 (0.8)	3.2 (1.1)	6.8 (1.4)	55.8 (17.7)	43.0 (5.2)

concentrations of Ag, Ba, Cd, Cu, Fe, and Zn (Table 1). In addition to high metal concentrations, the kidneys contain similar total amounts (μg per organ) of Cu, Cd, and Ag as the remaining viscera, even though kidney weight is only 1/50 as great. The remaining viscera had relatively large amounts of Ag, Ba, Cd, Cu, Fe, and Zn, although the concentration of each of these elements (Table 1) is low because of the large mass of this tissue. Upon exposure to Cd (0.1 μg/ml; 15 d), there is a 14-fold increase in Cd concentration in the kidneys, and appreciable, but lower, accumulation of Cd in gills, digestive glands, mantle, and remaining viscera. Higher Ba and Fe concentrations were also found in the kidneys of Cd-exposed quahogs, although this is probably attributed to high inter-animal variability of Ba and Fe levels, rather than a possible interaction between Cd and Ba or Fe levels.

Quahogs exposed to ^{109}Cd rapidly accumulate the label in all organs examined (Fig. 1). Within 1 h, 20 ng/g (wet wt) of accumulated Cd can be measured in the kidneys and gills, and between 1 to 10 ng/g in all other tissues (adductor muscles, digestive glands, mantle, pericardial glands, and remaining viscera). Accumulation of ^{109}Cd in each organ over the 10-d exposure period follows a pattern similar to the allometric growth curve (Sokal & Rohlf, 1969), since the data is best fit to a linear regression line following logarithmic transformation of both ^{109}Cd concentration and exposure time. The data poorly fit both the exponential and polynomial curves. Accumulation by the kidneys (log [Cd] = 1.37 log time + 1.34; r^2 = 0.961) is much more rapid than for the remaining tissues, all of which accumulate ^{109}Cd at similar rates (*e.g.*, gills: log [Cd] = 0.650 log time + 1.23, r^2 = 0.923; adductor muscles: log [Cd] = 0.601 log time + 0.104, r^2 = 0.911). The concentration of ^{109}Cd in the kidneys exceeds the level of the surrounding sea water in approximately 3 h and is 300 times greater than the initial concentration of Cd in the sea water (0.1 μg/ml) after 10 d. No other organ concentrates ^{109}Cd to such an extent. Concentrations of ^{109}Cd in the gills, for example, only reach 4 to 8 times the initial Cd level, whereas those in the adductor muscles do not exceed the concentration of Cd in the sea water even after 10 d.

Since the kidneys of *M. mercenaria* are the primary organs that concentrate a variety of metals, the subcellular distribution of kidney metals was determined in both control and Cd-exposed groups of quahogs. Typical results for Ba, Cd, Cu, Fe, and Zn are presented in Table 2. Similar

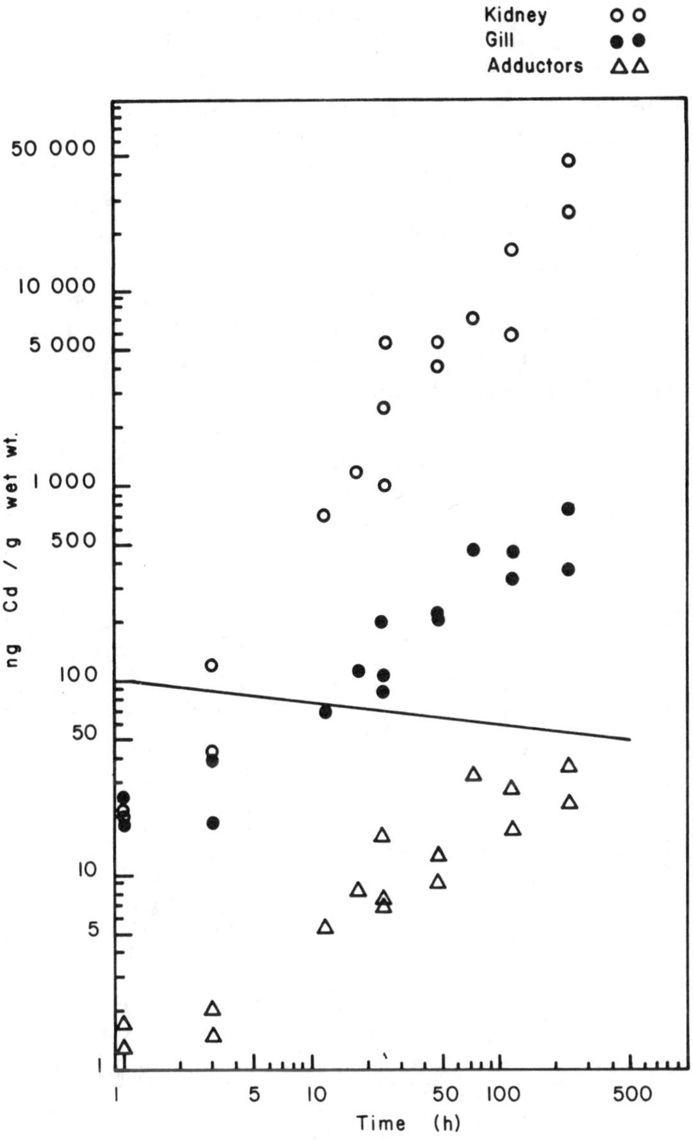

Fig. 1. Accumulation of cadmium in the kidneys, gills, and adductor muscles of Mercenaria mercenaria. Each symbol represents the concentration of Cd (^{109}Cd + carrier Cd) taken up by organs of each individual following continuous exposure of the quahog to 0.1 μg Cd/ml (1-h to 10-d exposure periods). Solid line represents the mean concentration of Cd in the surrounding sea water over a 10-d exposure period.

subcellular distributions of metals have been measured in seven additional pooled kidney samples (data not presented). Metals such as Ba, Fe, and Zn are primarily associated with the residue from the first centrifugation step (20,200 x g, 10 min). This pellet is expected to contain all the kidney granules, plus intact cells and nuclei. Cd and Cu, on the other hand, are mainly found within the supernatant following the final centrifugation step. This would indicate that Cd and Cu are either free, bound to soluble proteins, or associated with a cell organelle which is too light to rapidly settle out of suspension. Most of the Cd and Cu associated with the centrifugation residues is found in the 34,800 x g residue (8-16% of total kidney Cd or Cu).

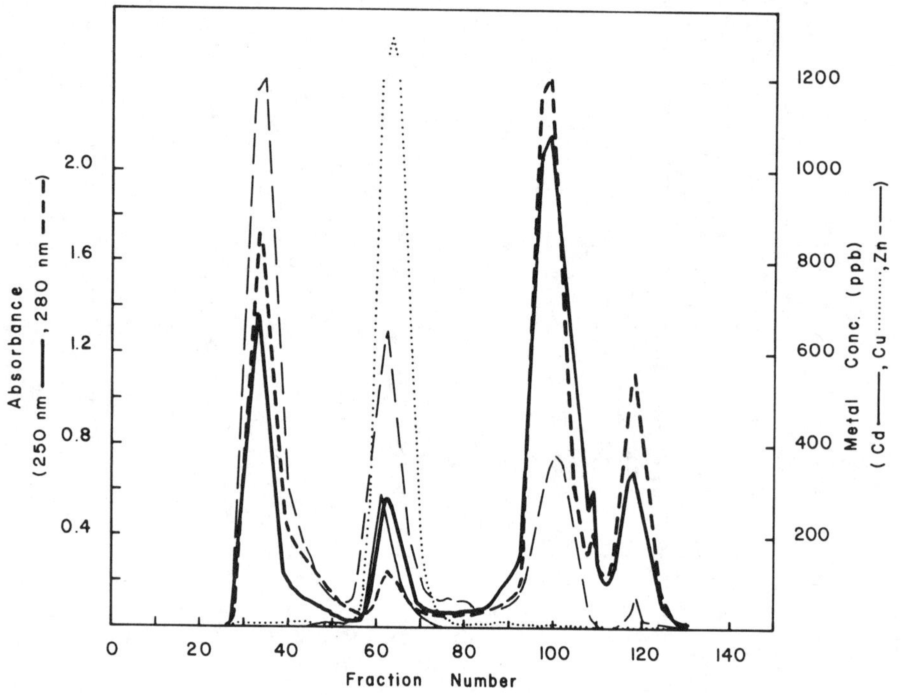

Fig. 2. Sephadex G-75 LC chromatogram of soluble proteins from pooled *Mercenaria mercenaria* kidneys (n = 16). Copper concentrations 2.5 times higher than graphed values.

Figure 2 is a typical LC chromatogram (Sephadex G-75) of the soluble proteins from pooled control kidneys. Fraction numbers 27-40 contain only high molecular weight (HMW) proteins (column void volume, MW >80,000), whereas fractions 93 and higher represent proteins with a molecular weight <3000 (column bed volume). The second low molecular weight (LMW) peak (fractions 113-126) may be indicative of interactions between organic compounds and the Sephadex G-75. For example, L-tryptophan (MW 204) elutes with this second peak, although vitamin B-12 (MW 3000) elutes with the bed volume (fractions 93-112). The small LMW peak between fractions 59-70 corresponds to a molecular weight of 14,500-16,500 and is referred to as the "metal-binding protein" fractions. Virtually all of the soluble Cd and Cu is associated with this metal-binding protein or proteins. Soluble Zn, on the other hand, is primarily associated with HMW proteins and only to a much lesser extent with LMW and metal-binding protein fractions. Neither Ba nor Fe are associated with the metal-binding protein fractions.

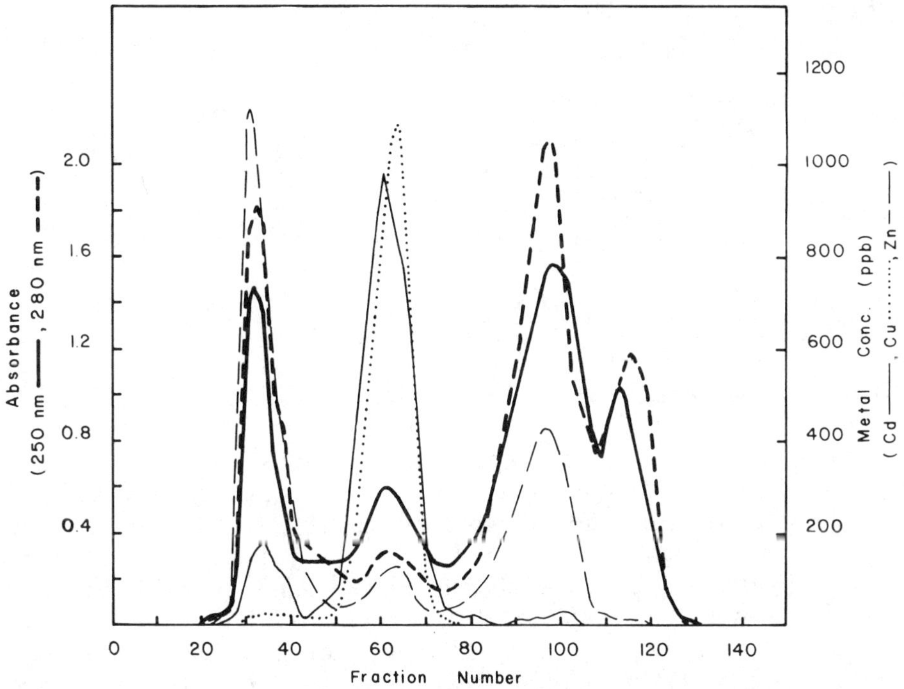

Fig. 3. Sephadex G-75 LC chromatogram of soluble proteins from pooled *Mercenaria mercenaria* kidneys (n = 16). Quahogs exposed to 0.1 μg Cd/ml for 15 d. Copper concentrations 2.5 times higher than graphed values.

Table 2. Distribution of selected elements (%) in tissue fractions of *Mercenaria mercenaria* kidney tissue. Residue = pelleted material following centrifugation (20,200 x g, 10 min; 34,800 x g, 60 min). MBP(s) = 14,500-16,500 MW metal binding protein(s). Percent recovery of metals following LC separation: Cd = 83-86%; Cu = 82-94%; Zn = 94-99.7%.

Group	Fraction	Ba	Cd	Cu	Fe	Zn
Control						
	Residue	100	13.5	12.8	98	72.1
	20,200 x g	93	3.6	4.3	91	57.5
	34,800 x g	7	9.9	8.5	7	14.6
	Soluble Fraction	0	86.5	87.2	2	27.9
	<3,000 MW Proteins	0	0.0	0.0	-	6.1
	MBP(s)	0	74.5	71.1	-	7.0
	>80,000 MW Proteins	0	0.0	0.0	-	12.8
	Unrecovered Metal	0	12.0	16.1	-	2.0
Exposed to 0.1 µg Cd/ml for 15 d						
	Residue	100	21.2	21.3	96	74.4
	20,200 x g	100	5.0	6.1	88	59.8
	34,800 x g	0	16.2	15.2	8	14.6
	Soluble Fraction	0	78.8	78.7	4	25.6
	<3,000 MW Proteins	0	1.8	0.2	-	9.1
	MBP(s)	0	55.9	71.5	-	2.4
	>80,000 MW Proteins	0	7.7	2.0	-	14.0
	Unrecovered Metal	0	13.4	5.0	-	0.1

For comparison, the LC chromatogram of the soluble kidney proteins from the Cd-exposed quahogs is presented in Figure 3. Although the UV-absorbance of the metal-binding protein fractions (53-72) are similar, the chromatogram of the Cd-exposed kidneys represents approximately 1/3 less mass of kidney tissue, due to differences in animal size (2.23 g wet wt for the Cd-exposed vs. 3.21 g wet wt for the control). As with the controls, the majority of the Cu and Cd was found associated with the metal-binding protein fractions. However, an appreciable amount of Cd was also found with the HMW proteins. As with the controls, Zn was associated primarily with the HMW fraction, and to a lesser degree with the metal-binding protein and LMW peaks. However, less Zn was found associated with the metal-binding proteins than in the controls, probably indicating that Zn was displaced from the metal-binding proteins by Cd during exposure.

Table 2 summarizes the percent distribution of Cd, Cu, and Zn in the soluble portion of kidney homogenates from both control and Cd-exposed quahogs. Percent metal recovery

Table 3. Concentration of selected elements in *Mercenaria mercenaria* kidney granules (mg/g dry wt).

Reference	Ba	Ca	Cd	Cr	Cu	Fe	Mg	Mn	Ni	P	Zn
This paper	0.486	80.7	0.029	0.011	0.373	89.5	13.6	31.0	0.035	89.4	14.7
Doyle *et al.*, 1978	0.075	NA	0.011	0.020	4	NA	NA	NA	0.050-0.070	NA	59.0
Rheinberger *et al.*, 1979	NA	89.4	0.008	NA	2.0	30.3	NA	85.6	NA	NA	52.9
Gold *et al.*, 1982	NA	NA	NA	NA	NA	NA	NA	NA	NA	167	NA

NA = Not Analyzed

Table 4. Concentrations of selected elements (μg/g of whole blood) in blood plasma and blood cells from three individuals (unexposed) and in a pooled sample of blood from five *Mercenaria mercenaria* exposed to 0.1 μg Cd/ml for 7 d.

Quahog #	Sample	Ag	Ba	Cd	Cu	Fe	Zn
1	Plasma	<0.001	0.070	<0.01	0.116	0.540	8.110
	Blood cells	<0.001	0.020	<0.01	0.006	0.108	0.241
2	Plasma	<0.001	0.057	<0.01	0.072	0.500	3.37
	Blood cells	<0.001	0.014	<0.01	0.005	0.070	0.081
3	Plasma	<0.003	0.118	<0.01	0.090	1.910	10.600
	Blood cells	<0.001	0.011	<0.01	0.007	0.156	0.386
4-8 (Pooled)	Whole blood	<0.005	0.070	<0.03	0.205	0.73	7.14
	Plasma	<0.005	0.062	<0.03	0.192	0.68	6.61
	Blood cells	<0.002	0.012	<0.03	0.015	0.11	0.14

after LC protein separation varied (Cd = 83-86%; Cu = 82-94%; Zn = 94-99.7%). In kidneys of control animals, 74% of the Cd, 71% of the Cu, and 7% of the Zn is associated with metal-binding proteins, compared to 56% of the Cd, 72% of the Cu, and 2% of the Zn in kidneys from Cd-exposed animals.

In addition to determining the subcellular metal distribution pattern, metal concentrations were measured in kidney granules prepared by differential centrifugation. Histological and ultrastructural evidence indicated that the kidneys of *M. mercenaria* contain both intracellular granules and numerous black granules (up to 150-200 μm dia.) that lie free in the kidney lumen. Samples containing a mixture of these granule types contain high concentrations of Ca, P, Fe, and to a lesser extent, Mg, Mn, and Zn, together with a broad array of trace elements (Table 3).

Although the data presented indicate that different metals have a different subcellular distribution within the kidneys of *M. mercenaria*, how the metals are transported to the quahog kidneys is not understood. Since the blood and open circulatory system are obviously involved, we have begun to investigate the importance of the blood cells vs. the blood plasma in the transport of metals.

A variety of metals (*e.g.*, Ba, Cu, Fe, Zn) were detected in whole blood samples, and in blood cells and blood plasma which were separated by centrifugation. Values for both blood cells and plasma are presented (Table 4) as μg of metal per gram of the initial whole blood samples (plasma + cells). Metal levels therefore can be directly compared in the plasma and blood cells. In all cases, more metal was detected in the blood plasma than in the blood cells that were suspended in the plasma sample. Although the plasma contains more *total* metal than the blood cells, the *concentration* of metal (on a per gram wet or dry wt basis) is much higher in the blood cells than for an equal weight of blood plasma. Cd and Ag, if present, were below detection limits, even in the pooled sample from five quahogs exposed to 0.1 μg Cd/ml for 7 d.

Cd accumulation was easily detected in the blood of quahogs exposed to ^{109}Cd (0.1 μg/ml) for only 1 h (Fig. 4). In contrast to other quahog tissues, blood cells and blood plasma concentrations reached an apparent asymptote after 24 h. Blood cells accumulated a 2-4 times higher concentration of ^{109}Cd than was present in the surrounding sea water, whereas the concentration of ^{109}Cd in the blood plasma remained below the sea water concentration. For the first 24 h of exposure, ^{109}Cd accumulation data were best fit to a linear regression following logarithmic transformation of both Cd concentration and exposure time (blood cell: log [Cd] = 0.740 log time + 1.35, r^2 = 0.887; blood plasma: log [Cd] = 0.695 log time + 0.596, r^2 = 0.901). As with detectable Ba, Cu, Fe, and Zn, the total amount of ^{109}Cd per volume of whole blood was greater in the plasma than in the blood cells.

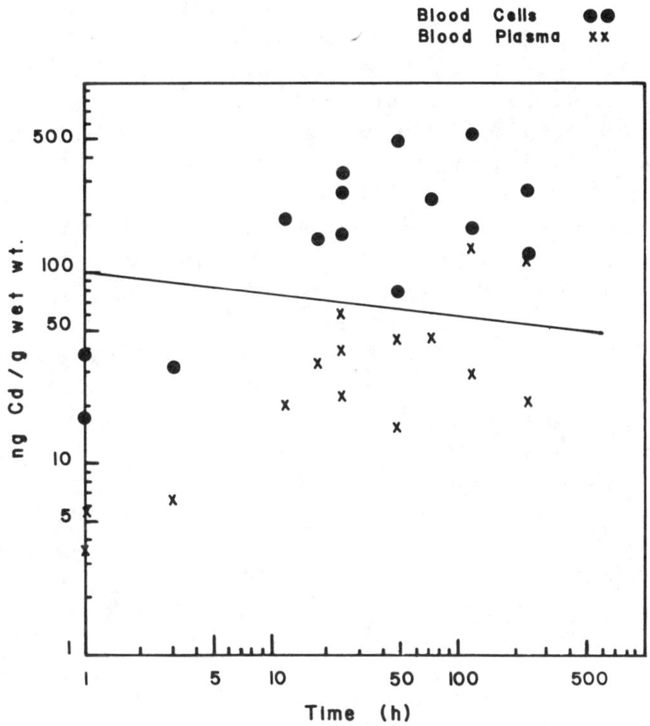

Fig. 4. Accumulation of cadmium in *Mercenaria mercenaria* blood cells and blood plasma (cell-free) following continuous exposure of quahogs to 0.1 μg Cd/ml (^{109}Cd + cold carrier Cd) for various periods of time (1 h to 10 d). Solid line represents mean concentration of Cd in the surrounding sea water over the 10-d exposure periods.

DISCUSSION

We have shown that a variety of metals accumulate primarily in the kidneys of *Mercenaria mercenaria*. This accumulation process, as indicated by our ^{109}Cd accumulation experiments, is rapid. No other tissues concentrate Cd to such an extent as the kidneys. Within the kidneys, the subcellular distribution of each metal varies. Some metals, such as Ag, Ba, Fe, and Zn, are principally associated with cellular constituents that are easily centrifuged out from kidney homogenates (*e.g.*, granules and nuclei, in addition to large pieces of cell fragments). Within this centrifugation residue, extracellular and intracellular kidney granules incorporate a variety of metals in trace amounts (Ag, Ba, Cd, Cr, Cu, Ni, Pb) along with their major constituents (Ca, P, Fe, Mg, Mn, Zn). Other metals are associated with soluble kidney proteins (*e.g.*, Cd, Cu and, to a lesser extent, Zn). Within the kidneys, the majority of Cd and Cu is associated with metal-binding protein(s) that are detectable not only in Cd-exposed quahogs, but in control animals as well.

The concentrations of various metals have been measured in a variety of eulamellibranchs collected from both relatively clean and obviously polluted sites: *Clinocardium nuttallii* (Caldwell & Buhler, 1983), *Hippopus hippopus* (Benson & Summons, 1981), *Macoma baltica* (Ray *et al*., 1981; Strong & Luoma, 1981), *Macoma nasuta* (Bradford & Luoma, 1980), *Mya arenaria* (Pringle *et al*., 1968; Bradford & Luoma, 1980; Caldwell & Buhler, 1983), *Protothaca staminea* (Hardy & Roesijadi, 1982), *Rangia cuneata* (Lytle & Lytle, 1982), *Spisula solidissima* (Pringle *et al*., 1968), *Tapes japonica* (Bradford & Luoma, 1980; Kumagai & Saeki, 1982), *Trichomya hirsuta* (Klumpp & Burdon-Jones, 1982), *Tridacna maxima* and *Tridacna derasa* (Benson & Summons, 1981), and *Venerupis decussata* (Boyden, 1977). Metal accumulation in *Mercenaria mercenaria* has been well documented for numerous metals (As, Cd, Co, Cr, Cu, Fe, Hg, Ni, Mn, Pb, Si, Zn) (Pringle *et al*., 1968; Raymont, 1972; Boyden, 1977; Larsen, 1979; Romeril, 1979; Behrens & Duedall, 1981a; Genest & Hatch, 1981). Concentrations of Cd, Cu, Fe, and Zn reported in this study are similar to concentrations measured by Larsen (1979) for Chesapeake Bay quahogs (Table 5), although Zn and Fe concentrations are generally lower than those reported for other locales.

In addition to environmental monitoring, metal accumulation has been followed in a few laboratory-exposed eulamellibranchs, such as *Cerastoderma edule* (Ni; Wilson, 1983), *Donax trunculus* (Mn; Mauri, 1980; Mauri & Orlando, 1982), *Macoma balthica* (Luoma, 1983), *Mulinia lateralis* (Cd; Jackim *et al*., 1977), *Mya arenaria* (Cd, Cu, Pb, Zn; Pringle *et al*., 1968; Ag; Thurberg *et al*., 1974; Cd; Jackim *et al*., 1977), and *Protothaca staminea* (Ni; Hardy & Roesijadi, 1982). In *Mercenaria mercenaria*, Carmichael *et al*. (1980) have

Table 5. Comparison of whole body metal concentrations in *Mercenaria mercenaria* collected from different locales.

Collection Sites	Cd	Cu	Fe	Zn	Reference
	μg metal/g dry wt				
Cape Cod, MA, USA	0.3 - 0.6	7.6 - 13.5	24.8 - 87.8	24.3 - 62.9	This study
Great South Bay, NY, USA	0.8 - 1.0	1 - 2	---	150 - 200	Behrens & Duedall, 1981a
Chesapeake Bay, USA	0.0 - 0.8	1.1 - 7.9	---	5 - 112	Larsen, 1979
Southampton Water, U.K.	---	~20 - 55	~75 - 275	~75 - 200	Romeril, 1979
Southampton Water, U.K.	0.9 - 2.4	15 - 40	---	~190	Boyden, 1977

Collection Sites	Cd	Cu	Fe	Zn	Reference
	μg metal/g wet wt				
Cape Cod, MA, USA	0.1 - 0.2	2.1 - 3.2	6.7 - 21.4	6.6 - 15.8	This study
Atlantic Coast, USA	0.1 - 0.7	1 - 16.5	9.0 - 83.0	11.5 - 40.2	Pringle *et al.*, 1968

described the subcellular distribution of ^{109}Cd, ^{54}Mn, and ^{65}Zn in the kidneys following a 5-d exposure to these radioisotopes. The accumulation of soluble Cu and Pb have been measured in *M. mercenaria* (Pringle *et al.*, 1968), as well as the uptake of Cd via the food chain (Kerfoot & Jacobs, 1976). Although Cd could be accumulated from sewage sludge, no uptake of Zn or Pb was found (Valiela *et al.*, 1974). In addition, neither Cd nor Hg accumulated in quahogs which were living in Cd- and Hg-contaminated sediments during a 100-d exposure period (Rubenstein *et al.*, 1983). These varied results point out the importance that metal speciation and organic binding may have on the metal bioavailability to marine organisms (Engel *et al.*, 1981; Langston, 1982; Luoma, 1983).

The process of metal elimination by *M. mercenaria* is undoubtedly related to the subcellular distribution of metals within the kidneys. Depuration studies, however, have shown that metals are eliminated by the quahog at very slow rates, regardless of the subcellular distribution of the particular metal investigated. Behrens and Duedall (1981b), for example, found no decrease in Cd, Cr, Cu, Pb, or Zn over a 50-d depuration period; Ni levels actually increased approximately 56%. Similarly, Pringle *et al.* (1968) measured no decline in Fe concentration in 84 d, although Zn and Cu levels dropped by 32% and 26%, respectively. Their data indicate that the depuration rate for Cu was two orders of magnitude lower for the quahog than for the soft-shell clam, *Mya arenaria*, although the initial concentration of Cu was 2 to 9 times less.

Metals contained in kidney granules are probably eliminated from the body via the urine with the discharge of the granule (Doyle *et al.*, 1978; Rheinberger *et al.*, 1979). However, Gold *et al.* (1982) have shown that larger quahogs have more kidney granules, and therefore more bound metals, than smaller, younger quahogs collected from the same population. This suggests that either the rate of granule production is greater for larger quahogs or that larger quahogs retain granules for longer periods of time. Boyden (1977) has shown that older quahogs have lower concentrations of Cd, Cu, and Fe than younger animals, although the body burden of these metals continues to increase with age. However, the concentrations of other metals which are generally found in higher concentrations in kidney granules (Ni, Pb, and Zn) remained constant over the size range of quahogs investigated.

Quahogs from Cape Cod waters contain an average of 0.81 mg of kidney granules per animal, compared with 7.0 mg/animal for Sabin's Point, Narragansett Bay, RI (Doyle *et al.*, 1978) and 0.11 to 7.45 mg/animal for waters off of New York (Gold *et al.*, 1982). The number of granules per animal varies tremendously between individuals and between populations. Rheinberger *et al.* (1979) found more and larger granules in quahogs from polluted sites. The production of granules may be rapid, as evidenced by the incorporation of significant amounts of ^{109}Cd, ^{54}Mn, and ^{65}Zn into granules within only 8 d from initial exposure to these isotopes (Carmichael *et al.*, 1980).

Granules are primarily composed of calcium phosphate (Doyle *et al.*, 1978) and contain additional high concentrations of Fe, Mn, and Zn. Organic carbon ranges from 2-4% (Rheinberger *et al.*, 1979). The concentration of Ca reported in the present study (80.7 mg/g) is similar to that measured by Rheinberger *et al.* (89.4 mg/g; 1979), although the concentrations of the other major constituents vary substantially (Table 3). For example, we measured 89.4 mg P/g, whereas Gold *et al.* (1982) measured twice as much (167 mg/g). This variability in component concentrations is even greater for the detectable trace elements, where differences as high as 10 times are found between quahog populations. Similar major constituents and variable concentrations of minor elements have been reported for kidney granules of other eulamellibranchs, such as *Donax trunculus* (Mauri, 1980; Mauri & Orlando, 1982), *Macrocallista nimbosa* (Tiffany *et al.*, 1980; Tiffany, 1982), and *Tridacna maxima* (Hignette, 1979).

Aside from the analysis of kidney granules, few studies have been made on the subcellular distribution of metals in eulamellibranch tissues. Metal-binding proteins have been investigated in *Macoma baltica* and *Rangia cuneata* (Bray et al., 1983) and a Cd-binding protein (MW 22,000) characterized for *Tapes philippinarum* (Ishiguro *et al.*, 1982). Roesijadi (1980) investigated the association of Cu, Zn, and Cd with soluble proteins from the gill, muscle, kidney, and remaining viscera of *Protothaca staminea* that were exposed to a range of Cu concentrations. A metallothionein-like Cu-binding protein (MW 14,000) and a Zn-binding protein (MW 10,500) were present in the gill and muscle, although both Cu and Zn were associated with the 10,500 dalton protein in the kidney and remaining viscera.

Carmichael *et al.* (1980) described a metallothionein-like protein as well as the subcellular distribution of accumulated radiolabels (^{109}Cd, ^{54}Mn, ^{65}Zn) in the kidney of *Mercenaria mercenaria*. Their results differ slightly from our findings. In this previous study, quahogs were exposed to 5.43 µg Cd/L for 5 d and then allowed to depurate for 3 d prior to sampling. Following centrifugation of kidney homogenates, soluble proteins were first dialyzed against distilled water for 24 h and then lyophilized prior to liquid chromatographic (LC) separation. In the present study, quahogs were exposed to 100 µg Cd/L for 15 d. LC separation of the soluble proteins was begun within 3 h of dissection. As a consequence, 87% of the Cd was found to be associated with the soluble fraction in the present study. Virtually all of the Cd in the soluble fraction was associated with a metal-binding protein of 14,500 to 16,500 daltons. In comparison, only 44% of the ^{109}Cd in Carmichael *et al.*'s (1980) study was associated with soluble protein; 83% of this Cd eluted with the column void volume and only 17% was associated with a metallothionein-like protein (MW 10,000 to 14,000). The lengthy dialysis and the lyophilization may account for the differences observed in subcellular distribution of Cd in the two studies. The molecular weights of the metal-binding proteins are fairly similar in both studies (14,500 to 16,500 vs. 10,000 to 14,000). However, since total Cd content was not measured by Carmichael *et al.* (1980), it is not known whether the distribution of the introduced ^{109}Cd was the same as the distribution of the entire pool of Cd in the kidney.

The subcellular distribution of metals in the kidneys of *Mercenaria mercenaria* must interrelate with the mechanism(s) of metal transport to the kidneys. Metal transport processes have not been studied in the quahog, nor is metal transport well understood in other bivalves (see reviews by Coombs & George, 1978; George, 1980, 1982; Fowler *et al.*, 1981). As with the filibranchs on which most work has been done, uptake is probably mediated by the epithelial tissues that are in direct contact with the external environment. Although the mantle epithelial cells may be capable of phagocytosis of metals such as particulate Fe and Hg (Fowler *et al.*, 1975), the usual route of entry is probably through the gill and mantle epidermis and into the blood system. Metals could then enter the kidney in one of two ways. Ultrastructural studies on the pericardial glands of *M. mercenaria* in our laboratories (Meyhöfer *et al.*, 1983) have revealed specialized cells called podocytes that are putative sites of ultrafiltration for the formation of primary urine (Kummel, 1977). Soluble metals or metals bound to small organic molecules in the blood plasma may pass through these podocytes, enter the pericardial cavity, and subsequently pass into the kidney lumen. Within the kidney lumen, these metals could either be eliminated through the renal pore or taken up by the kidney epithelial cells at their distal tips. This uptake process may simply be coincidental with the normal reabsorption of useful organic molecules. Alternatively or additionally, metals may enter the kidney epithelial cells basally, having been carried to the kidney blood spaces by the blood plasma and/or amoeboid blood cells.

Chemical analyses have shown that blood cells are capable of concentrating many metals (Ba, Cu, Fe, and Zn). Therefore, the possible role of blood cells in metal transport cannot be discounted. Chemical analyses and radioisotope studies indicate that large quantities of such metals as Cd are present in the blood plasma and are detectable in the pericardial fluid at similar levels. Therefore, Cd in the blood plasma appears to have a direct pathway to the kidney lumen. This potential route of metal transport is presently being examined in more detail in our laboratory.

Although metal accumulation in bivalves has received the most attention over the past decade, metal accumulation *per se* has no physiological meaning. Animals which accumulate metals may store them in a biologically unavailable form. Physiologically harmful effects may be totally unrelated to metal accumulation. Metal transport processes, however, probably mediate elevated body burdens of metals due to pollution stress as well as normal physiological metal turnover. Further investigations of metal transport processes should lead to a better understanding of these normal physiological processes of metal regulation.

ACKNOWLEDGMENTS

We sincerely thank D. J. Elwood for preparation of the kidney granules for chemical analysis, and P.J. Boyle, D.K. Ryan, and A.J. Barker for critical comments on the study. Research was supported by Department of Energy contract DE-AC02-77EV04580.

LITERATURE CITED

Abbott, R.T. 1974. American Seashells, 2nd ed. Van Nostrand Reinhold Co., New York. 663 pp.

Behrens, W.J. and I.W. Duedall. 1981a. Temporal variations of heavy metals in *Mercenaria mercenaria*. J. Cons. Int. Explor. Mer 39: 219-222.

Behrens, W.J. and I.W. Duedall. 1981b. The behaviour of heavy metals in transplanted hard clams, *Mercenaria mercenaria*. J. Cons. Int. Explor. Mer 39: 223-230.

Benson, A.A. and R.E. Summons. 1981. Arsenic accumulation in Great Barrier Reef invertebrates. Science 211: 482-483.

Boyden, C.R. 1977. Effect of size upon metal content of shellfish. J. Mar. Biol. Assoc. U.K. 57: 675-714.

Bradford, W.L. and S.N. Luoma. 1980. Some perspectives on heavy metal concentrations in shellfish and sediment in San Francisco Bay, California. pp. 501-532. *In*: Contaminants and Sediments, Vol. 2. R.A. Baker (ed.). Science Publishers, Inc., Ann Arbor, MI.

Bray, J.T., L.A. Webb, and F.J. Reilly, Jr. 1983. Multi-element analysis of metal-binding proteins in cytosol fractions. Sci. Total Environ. 28: 367-374.

Caldwell, R.S. and D.R. Buhler. 1983. Heavy metals in estuarine shellfish from Oregon. Arch. Environ. Contam. Toxicol. 12: 15-23.

Carmichael, N.G., K.S. Squibb, D.W. Engel, and B.A. Fowler. 1980. Metals in the molluscan kidney: Uptake and sub-cellular distribution of ^{109}Cd, ^{54}Mn and ^{65}Zn by the clam *Mercenaria mercenaria*. Comp. Biochem. Physiol. 65A: 203-206.

Cheng, T.C. and D. Foley. 1975. Hemolymph cells of the bivalve mollusc *Mercenaria mercenaria*: An electron microscopical study. J. Invert. Pathol. 26: 341-351.

Coombs, T.L. and S.G. George. 1978. Mechanisms of immobilization and detoxication of metals in marine organisms. pp. 179-187. *In*: Physiology and Behavior of Marine Organisms. D.S. McLusky and A.J. Berry (eds.). Pergamon Press, Oxford.

Doyle, L.J., N.J. Blake, C.C. Woo, and P. Yevich. 1978. Recent biogenic phosphorite: Concretions in mollusc kidneys. Science 199: 1431-1433.

Engel, D.W., W.G. Sunda, and B.A. Fowler. 1981. Factors affecting trace metal uptake and toxicity to estuarine organisms. I. Environmental parameters. pp. 127-144. *In*: Biological Monitoring of Marine Pollutants. F.J. Vernberg, A. Calabrese, F.P. Thurberg, and W.B. Vernberg (eds.). Academic Press, New York.

Fowler, B.A., D.A. Wolfe, and W.F. Hettler. 1975. Mercury and iron uptake by cytosomes in mantle epithelial cells of quahog clams Mercenaria mercenaria exposed to mercury. J. Fish. Res. Board Can. 32: 1767-1775.

Fowler, B.A., N.G. Carmichael, K.S. Squibb, and D.W. Engel. 1981. Factors affecting trace metal uptake and toxicity to estuarine organisms. II. Cellular mechanisms. pp. 145-163. In: Biological Monitoring of Marine Pollutants. F.J. Vernberg, A. Calabrese, F.P. Thurberg, and W.B. Vernberg (eds.). Academic Press, New York.

Genest, P.E. and W.I. Hatch. 1981. Heavy metals in Mercenaria mercenaria and sediments from the New Bedford Harbor region of Buzzards Bay, Massachusetts. Bull. Environ. Contam. Toxicol. 26: 124-130.

George, S.G. 1980. Correlation of metal accumulation in mussels with the mechanisms of uptake, metabolism and detoxification: A review. Thalassia Jugosl. 16: 347-365.

George, S.G. 1982. Subcellular accumulation and detoxication of metals in aquatic animals. pp. 3-52. In: Physiological Mechanisms of Marine Pollutant Toxicity. W.B. Vernberg, A. Calabrese, F.P. Thurberg, and F.J. Vernberg (eds.). Academic Press, New York.

Gold, K., G. Capriulo, and K. Keeling. 1982. Variability in the calcium phosphate concretion load in the kidney of Mercenaria mercenaria. Mar. Ecol. Prog. Ser. 10: 97-99.

Hardy, J.T. and G. Roesijadi. 1982. Bioaccumulation kinetics and organ distribution of nickel in the marine clam (Protothaca staminea). Bull. Environ. Contam. Toxicol. 28: 566-572.

Hignette, M. 1979. Composition des concretions minerales contenues dans les reins de 2 mollusques lammellibranches: Pinna nobilis L. et Tridacna maxima. C.R. Acad. Sci. Paris 289: 1069-1092.

Ishiguro, T., K. Kitajima, M. Chiba, R. Fujisawa, F Ichikawa, K. Kondo, and H. Takehana. 1982. Studies on Cd-binding proteins in short-necked clam. Bull. Jpn. Soc. Sci. Fish. 48: 793-798.

Jackim, E., G. Morrison, and R. Steele. 1977. Effects of environmental factors on radiocadmium uptake by four species of marine bivalves. Mar. Biol. 40: 303-308.

Kerfoot, W.B. and S.A. Jacobs. 1976. Cadmium accrual in combined wastewater treatment-aquaculture system. Environ. Sci. Technol. 10: 662-667.

Klumpp, D.W. and C. Burdon-Jones. 1982. Investigations of the potential of bivalve molluscs as indicators of heavy metal levels in tropical marine waters. Aust. J. Mar. Freshwater Res. 33: 285-300.

Kumagai, H. and K. Saeki. 1982. Heavy metal contents of short-neck clams *Tapes japonica* and the nearby mud. Bull. Jpn. Soc. Sci. Fish. 48: 837-841.

Kummel, G. 1977. Der gegenwärtige stand der forschung zur funktionsmorphologie exkretorischer systeme. Versuch einer vergleichenden darstellung. Verh. Dtsch. Zool. Ges. 1977: 154-174.

Langston, W.J. 1982. The distribution of mercury in British estuarine sediments and its availability to deposit-feeding bivalves. J. Mar. Biol. Assoc. U.K. 62: 667-684.

Larsen, P.F. 1979. The distribution of heavy metals in the hard clam, *Mercenaria mercenaria*, in the lower Chesapeake Bay region. Estuaries 2: 1-8.

Luoma, S.N. 1983. Bioavailability of trace metals to aquatic organisms - A review. Sci. Total Environ. 28: 1-22.

Lytle, T.F. and J.S. Lytle. 1982. Heavy metals in oysters and clams of St. Louis Bay, Mississippi. Bull. Environ. Contam. Toxicol. 29: 50-57.

Mauri, M. 1980. Formazione di nefroliti contenenti manganese in *Donax trunculus* L. (Bivalvia). Mem. Biol. Mar. Oceanogr. Suppl. 10: 235-240.

Mauri, M. and E. Orlando. 1982. Experimental study on renal concretions in the wedge shell *Donax trunculus* L. J. Exp. Mar. Biol. Ecol. 63: 47-57.

McHugh, J.L., M.W. Sumner, P.J. Flagg, D.W. Lipton, and W.J. Behrens. 1982. Annotated Bibliography of the Hard Clam (*Mercenaria mercenaria*). NOAA Technical Report, NMFS SSRF-756. 845 pp.

Meyhöfer, E., M.P. Morse, and W.E. Robinson. 1983. Functional aspects of the heart-kidney complex of *Mercenaria mercenaria* with reference to heavy metal pollution. Amer. Zool. 23: 1024.

Moore, C.A. and A.F. Eble. 1977. Cytochemical aspects of *Mercenaria mercenaria* hemocytes. Biol. Bull. 152: 105-119.

Moore, C.A. and S.R. Gelder. 1983. The role of the "blunt" granules in hemocytes of *Mercenaria mercenaria* following phagocytosis. J. Invert. Pathol. 41: 369-377.

Pringle, B.H., D.E. Hissong, E.L. Katz, and S.T. Mulawka. 1968. Trace metal accumulation by estuarine mollusks. J. Sanit. Eng. Div. (Proc. Am. Soc. Civil Eng.) 94 (No. SA3): 455-475.

Ray, D., D.W. McLeese, and M.R. Peterson. 1981. Accumulation of copper, zinc, cadmium and lead from two contaminated sediments by three marine invertebrates - A laboratory study. Bull. Environ. Contam. Toxicol. 26: 315-322.

Raymont, J.E.G. 1972. Some aspects of pollution in Southhampton water. Proc. R. Soc. Lond. 180B: 451-468.

Rheinberger, R., G.L. Hoffman, and P.P. Yevich. 1979. The kidney of the quahog (*Mercenaria mercenaria*) as a pollution indicator. pp. 119-129. *In*: Animals as Monitors of Environmental Pollutants. National Academy of Sciences, Washington, DC.

Roesijadi, G. 1980. Influence of copper on the clam *Protothaca staminea*: Effects on gills and occurrence of copper-binding proteins. Biol. Bull. 158: 233-247.

Roesijadi, G. 1982. Uptake and incorporation of mercury into mercury-binding proteins of gills of *Mytilus edulis* as a function of time. Mar. Biol. 66: 151-157.

Romeril, M.G. 1979. The occurrenc of copper, iron and zinc in the hardshell clam, *Mercenaria mercenaria*, and sediments of Southampton water. Est. Coast. Mar. Sci. 9: 423-434.

Rubinstein, N.I., E. Lores, and N.R. Gregory. 1983. Accumulation of PCB's, mercury and cadmium by *Nereis virens*, *Mercenaria mercenaria* and *Palaemonetes pugio* from contaminated harbor sediments. Aquat. Toxicol. 3: 249-260.

Sokal, R.R. and F.J. Rohlf. 1969. Biometry: The Principles and Practice of Statistics in Biological Research. W.H. Freeman, San Francisco. 776 pp.

Strong, C.R. and S.N. Luoma. 1981. Variations in the correlation of body size with concentrations of Cu and Ag in the bivalve *Macoma balthica*. Can. J. Fish. Aquat. Sci. 38: 1059-1064.

Thurberg, F.P., A. Calabrese, and M.A. Dawson. 1974. Effects of silver on oxygen consumption of bivalves at various salinities. pp. 67-78. *In*: Pollution and Physiology of Marine Organisms. F.J. Vernberg and W.B. Vernberg (eds.). Academic Press, New York.

Tiffany, W.J., III. 1982. Excretory concretions in the sunray Venus clam, *Macrocallista nimbosa* (Bivalvia: Veneridae). Veliger 25: 77-79.

Tiffany, W.J., III, W.H. Luer, and M.A. Watkins. 1980. Intracellular and intraluminal aspects of renal calculosis in the marine mollusc *Macrocallista nimbosa*. Invest. Urol. 18: 139-143.

Valiela, I., M.D. Banus, and J.M. Teal. 1974. Response of salt marsh bivalves to enrichment with metal-containing sewage sludge and retention of lead, zinc and cadmium by marsh sediments. Environ. Pollut. 7: 149-157.

Wilson, J.G. 1983. The uptake and accumulation of Ni by *Cerastoderma edule* and its effect on mortality, body condition and respiration rate. Mar. Environ. Res. 8: 129-148.

Zacks, S.I. 1955. The cytochemistry of the amoeboyctes and intestinal epithelium of *Venus mercenaria* (Lamellibranchiata) with remarks on a pigment resembling ceroid. Quart. J. Microscop. Soc. 96: 57-71.

RELATIONSHIP BETWEEN GLUCOSE-6-PHOSPHATE DEHYDROGENASE AND ASPARTATE AMINO TRANSFERASE ACTIVITIES, SCOPE FOR GROWTH AND BODY BURDEN OF AG, CD, CU, CR, PB AND ZN IN POPULATIONS OF *MYTILUS EDULIS* FROM A POLLUTED ESTUARY

E.S. Gilfillan, D.S. Page, D. Vallas, L. Gonzalez, E. Pendergast, J.C. Foster, and S.A. Hanson.

SUMMARY

1. An appropriate sampling design and statistical analysis scheme can help make up for the lack of experimental control in the field environment.
2. In this study, only a few of the metals determined were important in affecting enzyme activity in *Mytilus edulis*; those metals which were important in 1982 were not important in 1983.

INTRODUCTION

The value of using mussels (*Mytilus edulis*) as *in situ* indicators of heavy metal pollutant uptake has been well documented (Phillips, 1977, 1980; Goldberg *et al.*, 1978; Southgate *et al.*, 1983). The fate and effect of heavy metals on marine bivalves is a multivariate process. The ability of an organism to absorb a given metal is dependent on the environmental form of the metal. For example, sediment/chromium ion interactions had a significant effect on the response of *Mytilus edulis* to Cr as measured by filtration rate (Capuzzo & Sasner, 1977). The processes of absorption and intracellular fate of heavy metals in mussels has recently been reviewed (Brown, 1983; George & Viarengo,

1983). These processes involve uptake of bioavailable heavy metals followed by detoxification through the formation of metal complexes, often with protein. Chronic environmental exposure to heavy metals results in intracellular metabolic alterations beyond the direct effects of metal toxicity.

The measurement of various physiological parameters in marine animals as biochemical sublethal stress indicators has been discussed in a recent review (Livingstone, 1982). Ideally, a biochemical stress indicator should relate to a given stressor in a dose-response relationship; this has been demonstrated in a number of systems (Bayne *et al*., 1980). This approach breaks down, however, in the field situation where there is no overwhelmingly large input of a single stressor or pollutant. In such a situation, the relationship of a physiological measurement to the body burden of heavy metals will be complicated by antagonistic and synergistic interactions between the pollutants themselves (Phillips, 1980), seasonal variations within the animal population under study, size and sex (Latouche & Mix, 1982); as well as environmental factors such as salinity and temperature. Here, the signal to noise ratio will be large for any given pollutant/response relationship. For example, a field study showed that a number of sublethal stress indicators varied significantly in mussel populations transplanted along a generalized pollutant gradient in Narragansett Bay, RI (Phelps *et al*., 1981). Four metals (Cd, Pb, Ni, and Fe) were analyzed by Phelps *et al*., as well as petroleum hydrocarbons. It was found that only the tissue concentrations of Ni and hydrocarbons could be related to the stress indicators studied.

The work reported here originated as a damage assessment study relating to an oil spill that occurred in the Piscataqua River near Portsmouth, New Hampshire in 1979. It was found that physiological changes measured in mussels (*Mytilus edulis*) and clams (*Mya arenaria*) from sampling sites within the Piscataqua River estuary related more closely to tissue concentrations of heavy metals than to the tissue concentrations of aliphatic and aromatic hydrocarbons (Gilfillan & Page, unpubl. data).

The Piscataqua River estuary is the location of a major shipyard in Kittery, Maine as well as other smaller industries on the Maine and New Hampshire side of the river. The effects of chromium inputs from tannery waste on *Mytilus edulis* was studied in Great Bay, a part of this river system, by Capuzzo and Sasner (1977). They found that increased levels of Cr in sediments depressed filtration rates in *M. edulis*. Since 1977, major inputs of heavy metals from tanneries and other industrial sources into the Piscataqua have diminished as a result of federal and state water quality legislation and enforcement. During the period of the present study (1982-1983), there were no major, single sources of metal pollutants. There were a number of variable sources such as the shipyard (Pb, Cu, Cr, Zn, Cd, and possibly Ag), another industrial source (Cr, Al, and Cu), and US Route 1 and Interstate 95 (Pb). The purpose of this study was to determine the following:

1. Tissue concentrations of Cu, Cr, Pb, and Cd in whole body samples of *Mytilus edulis* taken from specific sites within the study area at intervals over a two-year period.
2. Specific activities of the enzymes, glucose-6-phosphate dehydrogenase (E.C.# 1.1.1.49) and aspartate amino transferase (E.C.# 2.6.1.1) in whole body subsamples; and scope for growth using subsamples from the same groups of animals sampled for (1) above.
3. Possible relationships between the physiological measurements and the tissue metal concentration as determined using multivariate statistical methods.

METHODS

Study Design

The Piscataqua estuary receives very little freshwater runoff; the tidal prism is large. As a result the salinity of the water at the sampling sites remains relatively stable, *i.e.*, 30-31$^{o}/_{oo}$; therefore salinity was not routinely monitored. Previous work had shown that there was no statistically significant interdependence between the body burden of petroleum hydrocarbons and aspartate amino transferase, glucose-6-phosphate activity, or scope for growth in *M. edulis* (Gilfillan & Page, unpubl. data). Those metals whose presence was suspected were determined as potential stressors.

The physiological responses chosen were the activity of glucose-6-phosphate dehydrogenase (G-6-P) and the activity of aspartate amino transferase (AAT). Aspartate aminotransferase is a key member of a family of enzymes that link protein metabolism with carbohydrate metabolism. Protein catabolism and, thus, transaminase activity can be enhanced in stressed animals (Gould, 1977). Glucose-6-phosphate dehydrogenase is also a key enzyme that regulates the flow of glucose-6-phosphate through the pentose phosphide pathway, in the process producing NADPH. Stressed animals undergo shifts in the demand for such precursors; the activity of the key enzymes involved are thus affected (Thurberg *et al.*, 1977); To gain information on a higher level of physiological organization, scope for growth was determined. The results of the 1982 sampling showed that G-6-P and scope for growth gave comparable results. Scope for growth measurements are very labor intensive, thus in 1983 G-6-P and AAT were determined and not scope for growth.

Field Sampling

Figure 1 shows a map of the study area with the sampling locations marked. On each sampling date (see Table 1) samples of *Mytilus edulis* were taken from those stations that were sampled. Mussels 45-60 mm in length were collected from near the top of the mussel band. Mussels were transported to the laboratory in an insulated cooler; once at the laboratory they were frozen at -18°C pending analysis. Animals were analyzed within 30 d of collection.

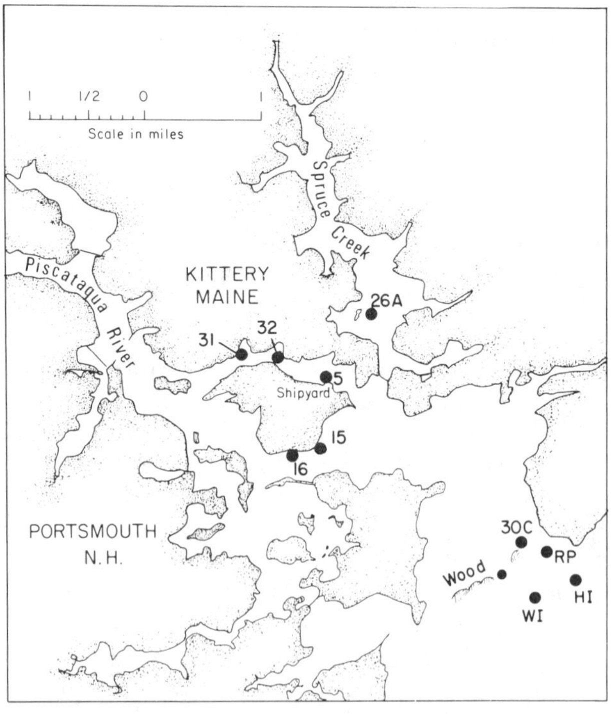

Fig. 1. Map of study area. Sampling locations are marked.

Table 1. Sampling matrix for Piscataqua mussel study.

Station	9/8/82	10/4/82	6/14/83	8/2/83	9/16/83
5	S*	S	S	S	S
15	S	S	S	S	S
16	S	S	S	S	S
26A	S	S	S	S	S
31	S	S	S	S	S
32	S	S	S	S	S
30C	S	S	S	S	S
HI			S	S	S
WI			S	S	S
RP			S	S	S
WOOD			S	S	S

*Sample taken at this location on the date.

Tissue Metals Analysis

Concentrations of Ag, Cd, Cu, Cr, Pb, and Zn were determined in the tissue samples using flame atomic absorption spectrometry on nitric acid digestates of freeze-dried whole animal tissue homogenates. Each field sample consisted of a least 50 mussels, which were subsampled on a random basis for chemical analysis and physiological measurements. Chemical analyses for each field sample were performed in triplicate using randomly selected subsamples of four individuals for each replicate. The average of each triplicate tissue analysis was used in the statistical data analysis.

Tissue Enzyme Analysis

Crude enzyme preparation: Whole soft-body tissue from four individuals was used for each assay. In 1982, the tissue was combined with three parts ice-cold distilled water (w/v) and ground thoroughly for 5 min in a Virtis homogenizer. In 1983, the tissue was combined with three parts (w/v) of ice-cold buffer solution consisting of 0.01 M Triethanolamine-HCl and 2 mM Ethylenediaminetetraacetic Acid (EDTA), ground thoroughly, then centrifuged at 11,000 g at 4°C. The supernatant was used as the crude enzyme extract. It was found that using a buffered system produced more reproducible results. The extracts were frozen for no longer than 2 wk prior to analysis (1982) or analyzed immediately

(1983). Protein content of the extract was determined by the Biuret test using bovine serum albumin as a standard.

Glucose-6-phosphate dehydrogenase (G-6-P) activity: The method used for the assay of glucose-6-phosphate activity is adapted from that of Gould (1977). Assays were performed at 340 nm using a double beam spectrophotometer (Beckman 24, 1982; Perkin Elmer Lambda 3, 1983) with the chamber temperature at 25°C, using 1-cm path length cuvettes. Absorbance was recorded on a strip chart recorder. Reaction rates were read beginning about 90-120 sec after the start of the reaction. The reaction rate was calculated using the linear portion of the curve.

For the G-6-P assay each cuvette contained 2.55 ml (2.45 ml, 1983) tris buffer, 0.1 M, pH 8; 0.1 ml nicotinamide adenine dinucleotide phosphate, 8 mg/ml; 0.1 ml d-glucose-6-phosphate disodium salt hydrate, 15 mM; and 0.15 ml $MgCl_2 \cdot 6H_2O$, 50 mM. In 1983, 0.1 ml maleimide, 8 mg/ml H_2O was also added to the reaction mixture. The reaction was started by addition of 0.1 ml of the crude enzyme extract. The unit of activity was 0.001 change in absorbance/min/mg protein. Each sample was run against a cuvette containing all reagents except the crude enzyme preparation. Each assay was performed in triplicate.

Aspartate amino transferase (AAT) activity: The assay procedure is adapted from Thurberg *et al*. (1977). These assays were performed using the same instruments and cuvettes as for G-6-P assays. Reaction rates were determined by measuring decreased absorbance over 7 min; the steepest linear portion of the curve was used to calculate the reaction rate. The unit of activity was 0.001 change in absorbance/min/mg protein (Gould, 1977).

The reaction mixture contained 2.60 ml phosphate (KH_2PO_4) buffer, 0.1 M pH 7.5, which also contained L-aspartic acid (0.25 M) (neutralized with KOH prior to addition); 0.1 ml reduced nicotinamide adenine dinucleotide, 9 mg/ml 0.1 M phosphate buffer; and 0.05 ml crude enzyme extract. The reaction was started by addition of 0.2 ml alpha-keto glutarate (0.15 M) neutralized with KOH. Each sample was run against a blank containing all reagents except the substrate, alpha-keto glutarate. Assays were performed in triplicate.

Scope for Growth Methods

For scope for growth measurements, mussels less than 37 mm in length were collected from the same locations as animals for enzyme and metal analysis. The experimental animals were transported to the laboratory in coolers; they were then acclimated in running sea water for 24 h.

Methods were adapted from Gilfillan *et al*. (1976). After acclimation, respiration rates were determined for five animals from each sampling location using a Gilson Differential Respirometer. Respiration rates were determined at the acclimation temperature, which was always close to the temperature observed in the field. After the respiration measurements, the animals were placed in individual, labeled, paper envelopes and frozen. Subsequently the animals were freeze dried and the dry tissue weighed (± 0.1 mg). Respiration rates were calculated as ul oxygen consumed/mg/h.

In order to determine filtration rates, one animal was placed in each of five 4-L beakers filled with glass-fiber filtered sea water to which sufficient *Dunalliella tertiolecta* and been added to equal the food concentration observed at the collecting site. Prior to placing the experimental animals into the beakers, two 25-ml samples were drawn and the number of food particles counted using a Coulter (model ZBI) particle counter. After three hours the animals were removed, two additional 25-ml samples were drawn, and the number of food particles determined as above. All experiments were determined at the acclimation temperature. Following the removal of the animals, the beakers were rinsed with glass-fiber filtered sea water, more *D. tertiolecta* was added and the animals replaced. After 24 h the animals were removed, placed in labeled paper envelopes and frozen. Feces produced by the animals were collected on tared, pre-combusted glass-fiber filters. Animals were freeze dried and weighed (± 0.1 mg). The volume swept clear of particles was calculated from the relation:

$$V = v(\ln Co - Ct)/t$$

where V = volume swept clear of particles in L/h; v = volume of the experimental container; Co = initial food concentration; Ct = final food concentration; and t = duration of the experiment in hours.

The filter with the feces was washed with distilled water to remove salt, dried, and weighed (± 0.01 mg). The filter was then ashed at 600°C and weighed. The organic fraction of the feces was determined as ash-free dry weight divided by total dry weight. A filter containing food was treated the same way to yield organic fraction of the food.

Assimilation efficiency of the animals was determined from the following relationship:

$$\text{Assimilation efficiency} = (A - B)/(1 - B)$$
$$A = \text{organic fraction of food}$$
$$B = \text{organic fraction of feces}$$

Values of respiration and filtration rate observed will vary depending on the size of the experimental animals. Before comparisons between animals at different stations could be made, it was necessary to reduce the data to some common measure. One way to overcome unavoidable differences in size between groups of experimental animals is to express results in terms of a 'standard' animal.

The work of Zeuthen (1947, 1953) and Hemmingson (1950, 1960) showed that many of the rate functions of nearly all marine poiklilotherms vary logarithmically with body weight. As a result, relations of this type are frequently used to calculate values in terms of standard animals (Ansell & Lander, 1967; Gilfillan et al., 1977). The relationship between the common logarithm of the animals dry weight and either the volume of oxygen consumed or the volume of water filtered was used to calculate either the volume of oxygen consumed or the volume of water filtered by a standard (300 mg) animal.

Once standard values for respiration and filtration are obtained for standard animals, it is possible to convert these values to carbon units and calculate scope for growth for the animals in question. A value of 0.85, typical for an animal metabolizing a mixture of fat and carbohydrate, has been assumed for the animal's respiratory quotient (RQ). Since bivalves are herbivores, it is possible that the RQ will be higher than 0.85. However, an assumed value of 0.85 will not introduce error because this is a comparative study. Assuming an RQ of 0.85, carbon respired in µg/h is calculated as 0.43 x µl O_2 consumed/h.

The amount of carbon consumed by the animals (gross carbon) was calculated as the filtration rate of the standard animal in L/h x 2000 mg. Two thousand milligrams is a good average concentration of carbon available to the animals during the summer months. Using a constant amount of available food facilitates comparison of data between sampling dates and introduces no error in a comparative study.

Net carbon was calculated as gross carbon x assimilation ratio. Scope for growth was calculated as net carbon - respired carbon.

Methods for Data Analysis

Data for enzyme activity and body burden of the metals were subjected to stepwise multiple linear regression (Box *et al.*, 1978). Activity of AAT, G-6-P, or scope for growth was used as the dependent variable; the corresponding body burden of metals was used as the independent variable. The regressions were carried out using an Apple IIe and the Regress II (Human Systems Dynamic Co.) software package. In carrying out the analyses the level of significance required to accept an independent variable into the regression was set at $p<0.05$.

Table 2. Results of analyses for body burden of selected metals and for G-6-P activity in *Mytilus edulis* collected from the Piscataqua estuary in 1982.

Station	Cu	Cr	Pb	Cd	Ag	Zn	G-6-P
9/8/82							
16	7.12*	2.67	3.56	1.89	0.05	64.05	44.4**
15	0.06	2.56	11.32	2.12	0.14	69.75	55.9
5	7.41	2.41	3.31	2.05	0.37	61.89	59.9
26A	5.94	4.98	6.57	2.56	0.14	70.00	62.4
30C	8.38	1.44	2.89	2.82	0.16	62.97	52.2
31	8.25	3.01	5.16	2.40	0.15	63.98	68.0
32	9.26	4.82	67.98	2.18	0.46	80.62	58.4
10/4/82							
16	7.69	2.03	2.80	1.83	0.19	44.99	48.5
14	9.12	1.90	4.40	1.35	0.19	64.55	41.8
5	6.69	3.69	4.50	1.58	0.34	60.37	47.4
26A	6.40	3.96	5.66	2.01	0	51.49	59.0
30C	6.06	.87	3.69	1.56	0	63.21	32.1
31	6.87	4.33	6.4	1.54	0.11	75.17	41.5
32	8.6	3.14	33.12	1.8	0	78.4	37.1

*Values for metal concentrations are ppm/dry weight; the mean of triplicate analyses.
**Values are relative activity units (0.001 change in absorbance/min/mg protein); mean of triplicate determinations.

Table 3. Results of analyses for body burden of selected metals and for AAT and G-6-P activity in *Mytilus edulis* collected from the Piscataqua estuary in 1983.

Station	Pb	Cu	Cd	Cr	Zn	Ag	G-6-P	AAT
6/14/83								
5	9.5*	10.1	4.5	6.2	122	1.1	42.3**	214**
15	10.8	9.1	6.4	7.2	150	0.81	52.8	260
16	6.6	9.0	5.6	6.5	125	1.1	56.3	168
26A	8.8	8.8	3.9	5.0	133	0.89	44.4	187
31	7.0	10.4	2.6	6.4	117	0.96	58.8	201
32	12.4	8.2	2.7	4.0	158	0.88	40.8	165
30C	4.6	7.8	2.7	4.7	117	0.67	65.8	337
HI	2.6	7.8	4.0	4.2	84	0.70	63.7	219
WI	3.8	6.7	3.7	5.8	85	0.76	50.3	211
RP	2.2	7.9	1.9	2.9	112	0.58	69.4	230
WOOD	6.0	9.9	5.3	10.2	88	0.78	42.7	267
8/2/83								
5	5.8	6.4	1.9	5.0	82	0.99	49.0	55.9
15	6.4	9.2	5.9	13.9	83	0.69	39.0	85.9
16	6.1	7.3	2.6	11.4	90	0.82	51.5	26.8
26A	5.8	7.3	2.3	7.3	85	0.82	48.0	41.0
31	8.4	6.2	4.0	8.3	94	0.83	51.8	50.0
32	8.0	6.8	3.8	5.3	113	0.54	50.3	58.7
30C	5.0	6.4	1.8	7.2	85	0.67	54.2	197
HI	2.5	6.0	2.2	4.0	84	0.51	60.6	128
WI	3.6	5.4	2.2	3.5	67	0.53	56.7	59.6
RP	3.3	5.9	3.0	6.8	92	0.68	58.1	71.3
Wood	3.8	8.1	3.2	4.8	105	0.71	60.4	61.6
9/16/83								
5	6.5	11.6	1.9	3.8	74	0.72	50.5	126
15	7.4	9.7	2.3	2.8	81	0.65	29.8	120
16	7.9	9.9	2.4	7.3	81	0.84	63.8	202
26A	7.8	7.0	1.8	5.3	85	0.86	43.3	152
31	9.2	7.4	2.1	5.6	100	0.52	51.9	146
32	11.4	8.6	2.1	5.8	97	0.67	54.6	124
30C	5.0	8.3	2.0	22.3	72	0.68	28.2	109
HI	5.5	6.7	1.7	2.4	97	0.54	36.1	163
WI	4.6	6.1	2.0	0.69	102	0.66	49.8	127
RP	5.7	6.6	2.8	7.0	63	0.61	61.4	192
WOOD	7.5	7.0	2.0	1.1	86	0.5	40.3	118

*Values are ppm/dry tissue weight; the mean of triplicate determinations.
**Values are relative activity units (0.001 change in absorbance/min/mg protein); the mean of triplicate determinations.

RESULTS

Results of assays for the activity of G-6-P carried out in 1982 as well as the results of metals analyses are shown in Table 2. These data show considerable variation in both enzyme activity and metal concentration both between and within each of the two sampling dates.

Results of assays for the activity of G-6-P and AAT carried out in 1983 as well as the results of metals analyses are shown in Table 3. As in 1982, there is considerable variation in both enzyme activity and metal concentration between sites at each sampling period.

Results of scope for growth measurements carried out in 1982 (Table 4) show similar variation.

Table 4. Elements of scope for growth expressed as μg carbon for *Mytilus edulis* collected from the Piscataqua estuary in 1982.

Station	Filtration[1] (l/h)	Assimilation Ratio	Respiration[1] (μg C/h)	Scope for Growth[1] (μg C/h)
9/8/82				
16	1.77	.85	138	2935
15	2.01	.84	163	3305
5	2.32	.90	103	3733
30C	2.54	.98	198	4874
26A	3.26	.75	157	4847
31	2.56	.98	97	4957
32	2.45	.99	192	4781
10/4/82				
16	1.11	.80	90	2228
15	1.58	.85	141	2616
5	1.36	.92	175	2432
30C	2.12	.84	116	3527
26A	1.65	.94	88	3060
31	1.74	.84	94	2904
32	2.11	.92	78	3832

[1]Values shown are for 300 mg 'standard' animal.

Results of Statistical Analysis

Analysis of data for 9/82 and 10/82 was performed by stepwise regression (Table 5). Increased body burden of Cd was found to significantly stimulate G-6-P activity. Both scope for growth and filtration were significantly increased by increased body burdens of Cd and Zn. Respiration was significantly increased by increased body burdens of Ag.

Table 5. Summary of results of stepwise regression of body burden of metals on G-6-P activity and on scope for growth in *Mytilus edulis* collected from the Piscataqua estuary in 1982.

Dependent Variable	Independent Variables Pb	Cu	Cd	Cr	Zn	Ag
G-6-P	ns*	ns	$p<0.001$	ns	ns	ns
Filtration	ns	ns	$p<0.002$	ns	$p<0.002$	ns
Assimilation	ns	ns	ns	ns	ns	ns
Respiration	ns	ns	ns	ns	ns	$p<0.002$
Scope for Growth	ns	ns	$p<0.001$	ns	$p<0.001$	ns

*No significant interdependence.

Table 6 shows the results of stepwise multiple regression on the 1983 metal and enzyme data. For all data taken together, G-6-P activity is significantly reduced by increased body burden of Cr. The activity of AAT is significantly stimulated by increased body burden of Cu.

When the data for each sampling period were analyzed separately the results were variable. For the June sampling, G-6-P activity was significantly increased by increased body burden of Zn and significantly decreased by increased body burden of Pb. The activity of AAT was significantly decreased by increased body burden of Ag. For the August sampling, increased body burden of Cr was found to significantly decrease G-6-P activity; no significant regression was found for AAT. For the September sampling, no significant regression was found for G-6-P; increased body burden of Cr significantly stimulates AAT activity.

Table 6. Results of stepwise regression of body burden of selected metals on AAT and G-6-P activity in *Mytilus edulis* collected from the Piscataqua estuary in 1983.

Dependent Variable	Independent Variables Cu	Cr	Pb	Cd	Ag	Zn
All data						
AAT	$p<0.006$	ns*	ns	ns	ns	ns
G-6-P	ns	$p<0.006$	ns	ns	ns	ns
6/14/83						
AAT	ns	ns	ns	ns	$p<0.048$	ns
G-6-P	ns	ns	$p<0.047$	ns	ns	$p<0.004$
8/2/83						
AAT	ns	ns	ns	ns	ns	ns
G-6-P	$p<0.01$	ns	ns	ns	ns	ns
9/16/83						
AAT	ns	$p<0.018$	ns	ns	ns	ns
G-6-P	ns	ns	ns	ns	ns	ns

*No significant interdependence.

DISCUSSION

In a field situation marine organisms are responding to a number of environmental factors, some of which may be anthropogenic. In a situation where a population of marine animals is exposed to a number of environmental factors, each of which can potentially affect the population in an adverse way, it is very difficult to sort out which of the factors actually are affecting the population adversely. In the field situation, it is almost certain that more than one environmental factor is operative; there is no possibility of investigating one factor at a time as in a controlled situation. As a result the investigator has to rely upon his sampling plan and data analysis to make up for the lack of a controlled environment.

The sampling design must be such that simultaneous measurements are made of important physiological processes in an organism and of the body burden of bioavailable stressors that are considered important. It is essential to include measurements of as many potential stressors as possible. Measurements should be carried out over a time series to gain information about the dynamics of stressor-organism interaction. The resulting data must then be analyzed using some form of multivariate analysis.

A very useful form of multivariate analysis is multiple correlation and regression, where physiological response is the dependent variable and the potential stressors are the independent variables. In a situation where a significant degree of interdependence is not expected for all independent variables, the use of stepwise multiple regression is preferred. This is because in simultaneous multiple regression, least squares coefficients are computed for all independent variables regardless of significance. As a result significant regressions can be obscured. In stepwise multiple regression, variables are included in the regression only if they have a significant effect on the dependent variable.

Regression techniques can be confounded if there is a significant independent variable (stressor) that is not included in the sampling plan. Also, if a variable which does have a significant effect on a physiological occurs over a small dynamic range (*i.e.*, all stations are similarly affected), then no significant regression will be observed. If the occurrence of one independent variable is correlated with the occurrence of another, then, if the first actually has a significant regression on the dependent variable, the chances are that the second may as well, even though no causal relationship may exist. In this study the mussels filtration was a major factor in scope for growth; when filtration was significantly increased by increased Cd, a similar relationship was found for scope for growth (Table 5). In this case, both regressions have a causal basis, but the effect of Cd on scope for growth is brought about by an effect on filtration. The same situation also occurs for Zn (Table 5).

It is also possible for spurious regressions to arise by chance. A possible example is the barely significant regression of G-6-P on Pb observed in the 6/14/83 sampling (Table 6). No other significant regression on Pb occurs in any of the other data. Lead levels on 6/14/83 are slightly higher than at other samplings, but not significantly. Thus, the observed regression may be real; at the same time it may be an artifact.

In the present study, a study plan and data analysis scheme based on the above principles allowed detection of physiological effects by a number of metals on mussels living in an estuary that is not grossly polluted. In 1982, significant stimulatory effects of Cd were found on G-6-P activity as well as on filtration rate and scope for growth (Table 5). A similar regression of Zn on filtration and scope for growth was also observed (Table 5). A significant stimulation of respiration by increased Ag was also observed. However, the effect of silver on respiration was not sufficiently great to significantly affect scope for growth.

In 1983, scope for growth was not determined. Significant stimulation of AAT activity by Cu was observed as was a significant reduction of G-6-P activity by Cr when the data set was analyzed as a whole (Table 6). When the data arising from each of the three samplings were analyzed separately, the results were not consistent (Table 6).

In comparing the results obtained for 1982 with those obtained for 1983, it is clear that where Cd, Zn, and Ag were important in 1982, only Cu and Cr are important in 1983. It is very likely that this difference reflects a change in inputs to the estuary. The inconsistency between the overall analysis and the analyses of each sampling period in 1983 suggests that the somewhat smaller dynamic range of independent variables within sampling periods could be the cause.

In comparing the results obtained for G-6-P and scope for growth in 1982, it is clear that scope for growth makes a somewhat better monitoring tool as it is affected by more stressors than is G-6-P activity. However, the increased sensitivity of scope for growth may be largely offset by the increased labor costs involved in scope for growth determinations.

Recently, Popham and D'Auria (1983) have suggested that Principal Component Analysis can be used to determine whether a sample of *Mytilus edulis* comes from a population polluted by heavy metals. Results of analyses of suspect mussels are compared statistically with a reference group and the probability of similarity determined. If the probability of similarity is less than 1% or 5%, the sample is declared polluted. This method may be useful to detect differences between populations of mussels having differing heavy metal body burdens. Having identified polluted areas, the biological effects of the pollution can be established using the methods discussed above.

The statistical analysis of data dealing with metals in marine organisms is complicated by the fact that many of these organisms have evolved mechanisms that greatly reduce the bioavailability of metals within the animal through the action of metal-complexing proteins and/or the formation of insoluble metal-containing granules (Brown, 1983; George & Viarengo, 1983). Therefore, the concentration of metals determined by analysis will consist of *both* the concentration of complexed metal *and* the concentration of available metal in the cytosol (*i.e.*, the biologically active concentration). Since the concentration of available metal in the cytosol represents an equilibrium concentration with the complexed metal, an increased body burden of metal would reflect an increased concentration in the cytosol. The problem with the statistical analysis is that, at present, it has only been possible to determine the total amount of metals in body tissues. As a result, one must use measured total-tissue concentrations to estimate the unknown, and biologically active, cytosol concentration. It is not known whether varying levels of chronic exposure in the field result in the same fraction of available/non-available metal for all levels of exposure. This introduces additional 'noise' into the statistical analysis. This is particularly true when comparing the effects of metals having different ligand binding characteristics. A major research goal should be to gain quantitative determinations of the parameters describing the partitioning of metals into bioavailable and non-bioavailable intracellular forms.

In summary, this investigation has shown that it is possible to sort out the active from the potential stressors affecting populations of *Mytilus edulis* in a moderately polluted estuary using an appropriate sampling design and statistical analysis. Significant effect of Cu, Cr, and Cd on the G-6-P activity, AAT, and scope for growth have been shown. These results demonstrate the importance of simultaneous chemical and biological observations as well as the utility of stepwise multiple regression in separating the effects of multiple stressors in the environment.

ACKNOWLEDGMENTS

The authors are grateful to the many Bowdoin students who worked on this study and to Mrs. Betty Pike who typed the manuscript.

LITERATURE CITED

Ansell, A.D. and K.F. Lander. 1967. Studies on the hard shell clam, *Venus mercenaria*, in British waters. III. Further observations on the seasonal biochemical cycle and on spawning. J. Appl. Ecol. 4: 425-435.

Bayne, B.L., J. Anderson, D. Engel, E. Gilfillan, D. Hoss, R. Lloyd, and F.P. Thurberg. 1980. Physiological techniques for measuring the biological effects of pollution in the sea. Rapp. P.-v. Reun. Cons. Int. Explor. Mer 179: 88-99.

Box, G.E.P., W.G. Hunter, and J. Stuart Hunter. 1978. Statistics for Experimenters. John Wiley & Sons, Inc., New York. 653 pp.

Brown, B.E. 1983. Metal containing granules - Potential biological indicators of pollution? Mar. Poll. Bull. 14: 369-372.

Capuzzo, J.M. and J.J. Sasner, Jr. 1977. The effect of chromium on filtration rates and metabolic activity of _Mytilus edulis_ L. and _Mya arenaria_ L. pp. 225-237. _In_: Physiological Responses of Marine Biota to Pollutants. F.J. Vernberg, A. Calabrese, F.P. Thurberg, and W.B. Vernberg (eds.). Academic Press, New York.

George, S.G. and A. Viarengo. 1983. A model for heavy metal homeostasis and detoxication in mussels. _In_: Pollution and Physiology of Marine Organisms. (This volume).

Gilfillan, E.S., D. Mayo, S. Hanson, D. Donovan, and L.C. Jiang. 1976. Reduction in carbon flux in _Mya arenaria_ caused by a spill of no. 6 fuel oil. Mar. Biol. 337: 115-123.

Gilfillan, E.S., D.W. Mayo, D.S. Page, D. Donovan, and S.A. Hanson. 1977. Effects of varying concentrations of hydrocarbons in sediments on carbon flux in _Mya arenaria_. pp. 299-314. _In_: Physiological Responses of Marine Biota to Pollutants. F.J. Vernberg, A. Calabrese, F.P. Thurberg and W.B. Vernberg (eds.). Academic Press, New York.

Goldberg, E.D., V.T. Bowen, J.W. Farrington, G. Harvey, J. Martin, P.L. Parker, R.W. Risebrough, E. Schneider, and E. Gamble. 1978. The mussel watch. Environ. Conservation 5: 101-125.

Gould, E. 1977. Alteration of enzymes in winter flounder, _Pseudopleuronectes americanus_, exposed to sublethal amounts of cadmium chloride. pp. 209-224. _In_: Physiological Responses on Marine Biota to Pollutants. F.J. Vernberg, A. Calabrese, F.P. Thurberg, and W.B. Vernberg (eds.). Academic Press, New York.

Hemmingson, A.M. 1950. The relationship of standard (basal) energy metabolism to total fresh weight in living organisms. Rep. Steno. Hosp. Copenhagen. 4: 7-58.

Hemmingson, A.M. 1960. Energy metabolism as related to total fresh weight in living organisms. Rep. Steno. Hosp. Copenhagen. 9: 7-110.

Latouche, Y.D. and M.C. Mix. 1982. The effects of depuration, size and sex on trace metal levels in bay mussels. Mar. Poll. Bull. 13: 27-29.

Livingstone, D.R. 1982. General biochemical indices of sublethal stress. Mar. Poll. Bull. 13: 261-263.

Phelps, D.K., W. Galloway, F.P. Thurberg, E. Gould, and M.A. Dawson. 1981. Comparison of several physiological monitoring techniques as applied to the blue mussel, *Mytilus edulis*, along a gradient of pollutant stress in Narragansett Bay, Rhode Island. pp. 335-355. *In*: Biological Monitoring of the Marine Pollutants. F.J. Vernberg, A. Calabrese, F.P. Thurberg, and W.B. Vernberg (eds.). Academic Press, New York.

Phillips, D.J.H. 1977. The use of biological indicator organisms to monitor trace metal pollution in marine environments - A review. Environ. Pollution 13: 281-317.

Phillips, D.J.H. 1980. Quantitative Aquatic Biological Indicators. Applied Science Publishers Ltd., London. pp. 292-325.

Popham, J.D. and J.M. D'Auria. 1983. Statistical approach for deciding if mussels (*Mytilus edulis*) have been collected from a water body polluted with trace metals. Env. Sci. Technol. 17: 576-582.

Southgate, T., D.J. Slinn, and J.F. Eastham. 1983. Mine-derived metal pollution in the Isle of Man. Mar. Poll. Bull. 14: 137-140.

Thurberg, F.P., A. Calabrese, E. Gould. R.A. Greig, M.A. Dawson, and R.K. Tucker. 1977. Response of the lobster, *Homarus americanus*, to sublethal levels of cadmium and mercury. pp. 185-197. *In*: Physiological Responses of Marine Biota to Pollutants. F.J. Vernberg, A Calabrese, F.P. Thurberg, and W.B. Vernberg (eds.). Academic Press, New York.

Zeuthen, F. 1947. Body size and metabolic rate in the animal kingdom with special regard to the marine microfauna. Comptes rendues des travaux du Laboratoire, Carlsberg. Sreri Chimique 26: 17-161.

Zeuthen, F. 1953. Oxygen uptake as related to body size in organisms. Quart. Rev. Biol. 28: 1-12.

A MODEL FOR HEAVY METAL HOMEOSTASIS AND DETOXICATION IN MUSSELS

Stephen G. George and Aldo Viarengo

SUMMARY

The current knowledge of the processes of uptake, metabolism and intracellular control of heavy metals in mussels is summarized, and a metabolic model is presented for cellular homeostasis which emphasizes the crucial role of lysosomes in control of cellular metal turnover and concentrations. This model appears to be generally applicable since it is also consistent with data from mammalian studies.

INTRODUCTION

Metal accumulation by lamellibranch bivalves such as *Mytilus* sp. has been extensively studied over the past few years since animals living in environments with elevated concentrations of certain trace metals show little regulation of uptake as indicated by high tissue concentrations and linear uptake kinetics (George, 1980). Consequently, they are now widely used as biological indicators of pollution in monitoring programs (Goldberg, 1980). This ability to accumulate high metal concentrations without toxic effects makes *Mytilus* a good organism to use for elucidating more generally applicable mechanisms of intracellular metal homeostasis and toxicology of heavy metals. The general concepts of homeostasis, essentiality of trace metals, and ligand competition between essential and pollutant metals were discussed in a previous symposium of this series (George, 1982) and, therefore will not be reiterated, even though they are crucial to our understanding of the metabolism of metals such as Cu, Cd, Hg, Pb, and Zn. Metal accumulation in Mytilus and other aquatic organisms have been reviewed by George (1980, 1982). Sufficient knowledge of these processes exists to allow formulation of a metabolic model which explains the observed subcellular and tissue heavy metal distributions, and which can be used as a basis for future experimental work in defining the capacity of these organisms to tolerate environmental contaminants.

UPTAKE OF HEAVY METALS

The exact mechanisms of heavy metal uptake by bivalves (or any other animals) are not known; however, several studies have been carried out towards this goal (Wrench, 1978; Carpene & George, 1980; Gutkneckt, 1981, 1983; Roesijadi, 1982; Simkiss, 1983; Simkiss & Mason, in press).

For soluble metal species, which are the predominant forms of Cu, Cd, Zn, etc., there is a growing body of evidence to suggest that in many marine organisms (including phytoplankton, crustaceans, and some bivalves) uptake is proportional to the metal ion activity in the water (Sunda & Guillard, 1976; Sunda et al., 1978; Engel et al., 1981; Sanders et al., 1983). These data would indicate that it is an inorganic species which is transported. Since the different metal species are in equilibrium with one another, the presence of organic ligands in the sea water will, therefore, merely alter this equilibrium and will affect the rate of transport by altering the bioavailability.

In Mytilus, the uptake of Cd by isolated gills has been characterized (Carpene & George, 1981). Uptake increased linearly with time (up to 30 min) and was directly proportional to the external Cd concentration (up to 0.5 mg Cd/L). It was also a concentrative process (after about 5 min incubation) with no energy, temperature, aerobic dependence, or competition by Cu, Hg, Pb, or Zn being demonstrable. We concluded that this was consistent with passive diffusion through the membrane, followed by intracellular complexation. The measured Cd flux through artificial black lipid membranes (Gutkneckt, 1983) is 100 times slower than the calculated permeability coefficient (10^{-6} cm/sec) observed in our isolated gill experiments, which would infer the presence of a transport protein. Since we did not observe saturation kinetics, there must be a large number of transport sites. Alternatively, a lipid carrier could be postulated. On the basis of olive oil-water partition coefficients, Simkiss (1983) postulated that it is the uncharged

$[CdCl_2]^\circ$ species which crosses the membrane by dissolution in the hydrophobic portion of the membrane lipid. This is thought unlikely on the basis of steric considerations and the strong interaction of Cd with the phosphate head groups of membrane lipids (Liao & Prestegrad, 1981; Gutkneckt, pers. comm.). However, transmembrane Cd transport might be effected by removal of Cd from the phospholipid on the inner face of the membrane by a strong intracellular complexing agent (such as thionein or glutathione). Further studies are required to resolve these uncertainties.

Hg acts differently, forming a planar complex which can rapidly cross a lipid bilayer membrane (permeability coefficient 10^{-2} cm/sec) by passive diffusion (Gutkneckt, 1981). However, experimental work with isolated gills of mussels (Roesijadi, 1982) showed linear uptake with time (up to 2 h), but dependence upon energy generating metabolic processes. This characteristic was also observed in a study of oyster gills by Wrench (1978), but the uptake (over 5 h) was exponential. One possible explanation of this Hg data could be the uptake of precipitates of Hg by a second process of uptake, that of pinocytosis. This method of uptake has been demonstrated for Fe in gills of *Mytilus* (George *et al*., 1976) and additional biochemical experiments showed that it was partially energy dependent (George *et al*., 1977). Pinocytosis of Hg has been demonstrated in clam mantle by Nakahara and Bevelander (1976). In the absence of carbonate, the inorganic speciation of Hg is largely chloride ion dependent and in sea water the equilibrium is strongly in favor of chlorocomplexes over the hydroxide. *In vivo*, however, Hg may readily complex with gill mucus, while the epithelium might never be exposed to $HgCl_2$. The data of Roesijadi (1981) for Hg uptake into *Mytilus* gills (redrawn in Fig. 1) show that after 24-h exposure, 73% of the Hg was present in the cell pellet fraction which contains the lysosomes and pinocytic vesicles. This proportion decreased with increased time of exposure. Although the first time point (24 h) is rather long, this may be indicative of pinocytic uptake of Hg.

Similarly, Pb has been identified in endocytic vesicles in mussel gills (Coombs, 1977; Schulz-Baldes, 1978; Marshall & Talbot, 1979; George, 1982), although the pinocytic event has not been observed. However, Pb easily precipitates as an insoluble carbonate from sea water and, if we examine the uptake data of Schulz-Baldes (1978) shown in Figure 2, two interpretations are possible. The author fitted a two compartment model: first, "I" an exponential phase "indicating a reversible exchange of lead" until 70 h, and second, "II" a linear phase "with irreversible uptake." This model could be equated with initial, reversible, non-specific binding of Pb to cellular ligands, followed by immobilization in lysosomes as observed microscopically (Coombs, 1977; Schulz-Baldes, 1978; Marshall & Talbot, 1979; George, 1982) or as insoluble precipates of lead carbonate (Marshall & Talbot, 1979; George, 1982). Alternatively, a series of exponential curves may be drawn for uptake in Mytilus (dashed lines in Fig. 3), which are very similar to those observed for phagocytosis in protozoa, where uptake takes place in bursts, separated by periods of membrane recovery and resynthesis.

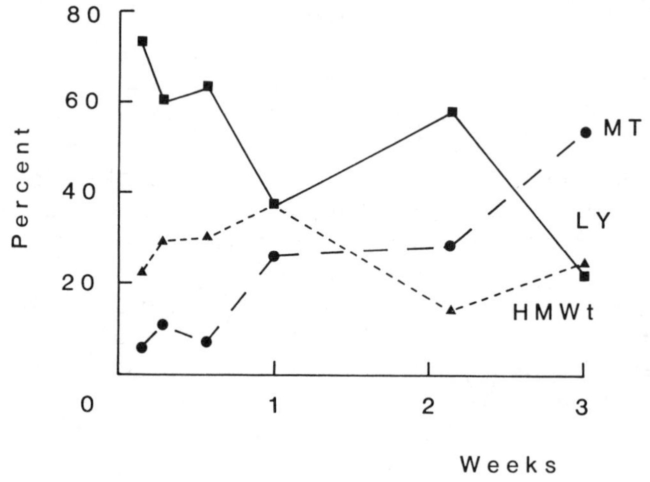

Fig. 1. Uptake of Hg into components of gills. Data calculated from Roesijadi (1982). Animals were continuously exposed to 5 μg Hg/L sea water. Percentages of total metal in the gills. LY = particulate fraction, HMWt = high molecular weight macromolecules, MT = metallothionein fraction.

Definitive experiments will be required to establish the exact mechanisms of heavy metal uptake, probably utilizing artificial and reconstituted membrane systems. Additionally, short-term uptake experiments must be repeated with gill tissues, and concomitant, investigation of the subcellular metal localization and binding should be carried out. Study of the subcellular partitioning should be sufficient to establish whether the metal has been transported into the cell across the membrane or by pinocytosis.

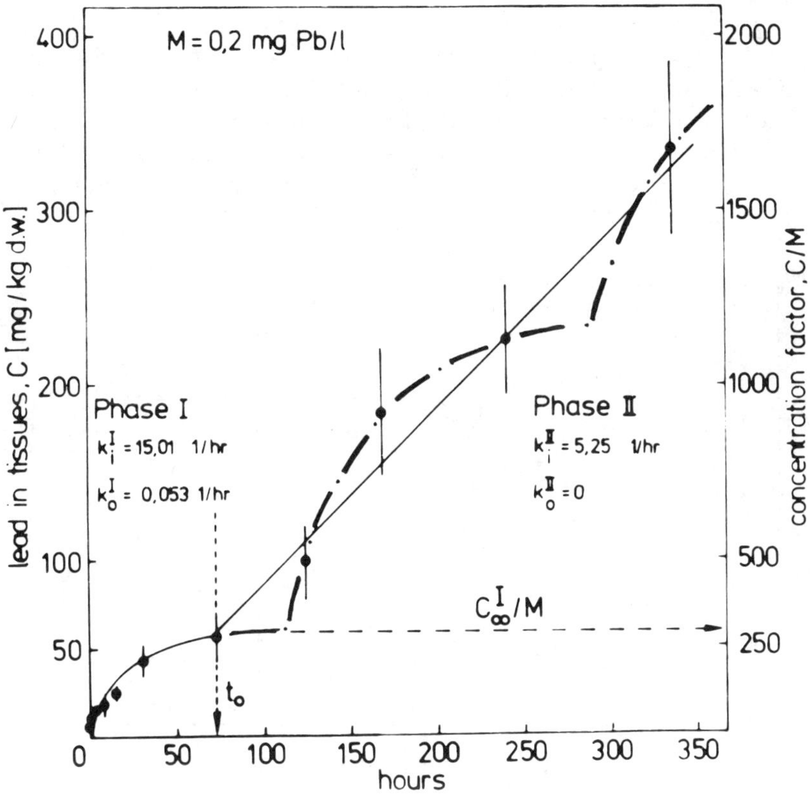

Fig. 2. Uptake of Pb into soft tissues from 0.2 mg Pb/L sea water. Reproduced from Schultz-Baldes (1978). We have inserted the dashed curve to indicate a possible series of cyclic events (pinocytosis).

SUB-CELLULAR METAL PARTITIONING

The only report thus far for metal distribution in the initial stages of metal uptake is that for Cd (Carpene & George, 1981). After 30-min exposure in Cd-containing sea water, subcellular fractionation showed that Cd was not bound to the exterior membrane, but bound to high molecular weight soluble macromolecules. Thus, Cd is clearly transported as a soluble species since it is not present in pinocytic vesicles.

Gel permeation chromatography is the most widely applied technique for fractionation of the soluble cellular components and separates them according to molecular size. In non metal-exposed mussels, or those exposed to heavy metals for a few hours, soluble Cd, Cu, Hg, and Zn are all predominately associated with macromolecules >60 K daltons mol. wt. In response to metal exposure, this distribution alters as the predominant metal-binding component is shifted to a lower molecular weight fraction with an apparent mol. wt. of about 15 K daltons (generally two peaks at 20 and 10 K daltons). The nature of this component has been the object of many recent biochemical studies on Mytilus.

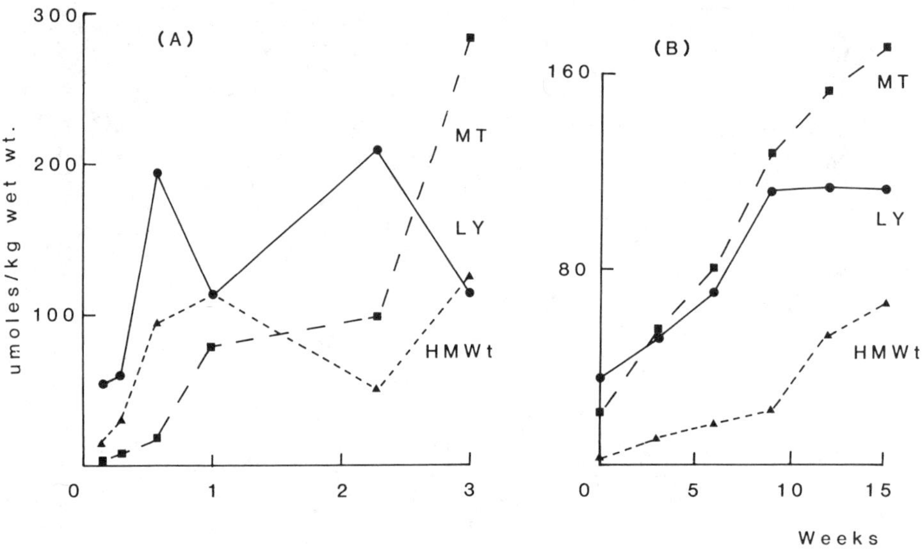

Fig. 3. Uptake of metals into tissue components of mussels. A) uptake of Hg into gills. Data calculated from Roesijadi (1982). Exposure to 5 µg Hg/L. B) Uptake of Cu into digestive gland. Exposure to 25 µg Cu/L. Data generously provided by F. Harrison. LY, HMWt, Mt as in Figure 1.

SOLUBLE METAL BINDING PROTEINS

In response to Cd exposure, mammalian cells synthesize a unique protein known as thionein (reviewed by Kagi & Nordberg, 1979). This protein, which binds Ag, Cd, Cu, Hg, and Zn very strongly, functions in pollutant metal detoxication and is also believed to function in the storage, intracellular regulation, and metabolism of the essential metals, Cu and Zn (Brady, 1982). Metallothioneins are a group of low molecular weight proteins (about 6800 daltons but with apparent mol. wts. of 10 K daltons by gel filtration) characterized by their high metal-binding capacities (up to 7 g-atoms/mol), high content of cysteine (30-35%), and virtual absence of aromatic amino acids (Table 1).

Table 1. Amino acid composition of group IIB metal-binding proteins from *Mytilus* sp.

The Hg and Ag proteins are not pure.

1. from Kagi & Nordberg (1979),
2. from Roesijadi & Hall (1981),
3. from Viarengo *et al*. (in press)

	RAT	Mussel			
	Cd-BP[1]	Cd-BP	Hg-BP[2]	Ag-BP	Cu-BP[3]
Amino Acid	mole %		mole %		
Cys	29	26	8	13	17
Asx	9	8	12	14	10
Thr	6	7	7	7	8
Ser	18	8	7	7	10
Glx	4	4	12	13	5
Pro	3	7	6	?	5
Gly	10	16	9	10	18
Ala	6	5	6	6	8
Val	2	5	5	4	4
Met	1		1		
Ileu		3	5	2	3
Leu		1	6	4	2
Tyr			2		
Phe			3	2	
Lys	12	8	7	7	6
His			1	1	1
Arg	2	2	3	3	3

Noel-Lambot *et al*. (1976) first reported the presence of a Cd-binding protein in Cd-exposed *Mytilus* with a molecular weight similar to mammallian metallothionein. This finding has been confirmed in many later studies (Talbot & Magee, 1978; Marshall & Talbot, 1979; Scholz, 1980; Carpene *et al*., 1983; Kohler & Riisgard, 1982). The protein has been purified and has very similar composition and properties to other metallothioneins (George *et al*., 1979; Frankenne *et al*., 1980; Table 1), and although it has not yet been sequenced, we feel justified in establishing the Cd-binding protein of *Mytilus* as a thionein. Unlike many other thioneins, the *Mytilus* protein appears to exist in monomeric and dimeric forms and at least five charge variants. From a comparative point of view, Cd-thionein purified from mussel shows a typical thionein-like composition with high cysteine and a low aromatic amino acid content. It does not, however, appear to contain a methionine residue and contains a lower serine and higher glycine content than vertebrate thioneins. It may therefore represent an evolutionary intermediate between the crustacean and yeast Cd-binding proteins. As in mammals, synthesis is also induced by exposure to Ag (George, unpubl. data), Cu (Viarengo *et al*., 1980; Viarengo *et al*., 1981), and Hg (Roesijadi & Hall, 1981). The amino acid composition of these proteins are compared in Table 1 and it may be noted that these have lower cysteine contents than the Cd-protein. However, most of these amino acid analyses are for preparations obtained by gel permeation chromatography and not purified further, consequently the differences are probably due to the presence of impurities.

Metal distributions between the soluble cytosolic components are altered by differences in the duration of metal exposure. Tests of animals with no pre-exposure to heavy metals have indicated that, following brief exposure, the high molecular weight macromolecules (HMWM) are the predominant metal-binding fraction in the cytosol (Carpene & George, 1981; Roesijadi, 1982), and there appears to be a lag period before thionein is synthesized. In animals where thionein has been pre-induced, there may be initial binding to thionein by metal-displacement reactions, as indicated by the results of Harrison *et al*. (1983). This past history of the animals may explain why thionein synthesis was observed as early as 24 h after onset of exposure to Cu (Viarengo *et al*., 1981) or Hg (Roesijadi, 1982) and after 4-d exposure to Cd (Viarengo, unpubl. data). The available data infers that there may be a longer lag period before Cd-thionein appears (see Scholz, 1980). During continuous exposure, metallothionein becomes the major metal-binding component for Hg (Roesijadi, 1982), Cd (Scholz, 1980), and Cu (Harrison *et al*., submitted) as shown in Figure 3.

Providing that precautions are taken to avoid protein degradation (use of low temperature, proteolysis inhibitors), very low molecular weight Cd-, Cu- and Zn-binding ligands are only observed in the kidneys of *Mytilus* (George & Pirie, 1980; George 1983a). In oysters, very low molecular weight ligands are the major Cd-, Cu- and Zn-binding components (Frazier & George, 1983).

PARTICULATE METALS

With increased exposure time, the heavy metals content of the particulate organelles increases (Fig. 3). In this respect, *Mytilus* provides a good model system for studying metal uptake by subcellular organelles, since (with the exception of certain disease states) a much higher proportion of metal is associated with these structures than in vertebrates. The occurrence of heavy metals in subcellular structures of aquatic animals has been reviewed by George (1982).

In a previous study of Fe accumulation in *Mytilus* by x-ray microprobe analysis, we found that the lysosomes contained a high concentration of Fe and detectable quantities of Zn were also present in the kidney (George *et al.*, 1976). Subsequent studies of mussels exposed to elevated concentrations of trace metals using the same analytical technique have shown that excesses of Ag (George, unpubl. data), Cd (George & Pirie, 1979; Marshall & Talbot, 1979), Pb (Coombs, 1977; Coombs & George, 1978; Schulz-Baldes, 1978; Marshall & Talbot, 1979), and Zn (George *et al.*, 1976; George & Pirie, 1980) also accumulate in the tertiary lysosomes of *Mytilus*. By histochemical methods, Cd (Janssen & Schultz, 1979), Cu (Viarengo *et al.*, 1981), Hg (Janssen & Schultz, 1979), and Zn (Lowe & Moore, 1979) have also been found in these structures. This lysosomal accumulation of metals is a general phenomenon in all animal phyla which has been reviewed previously by Sternlieb and Goldfisher (1976), Coombs and George (1978), and George (1982). The stages of the genesis of lysosomes are as follows: 1) primary lysosomes containing digestive enzymes at acid pH are pinched off from the golgi apparatus of the cell, 2) the primary lysosomes then engulf cellular macromolecules (proteins, polysaccharides, nucleic acids) (whereupon they are termed secondary lysosomes) and digest them into their monomeric units; and 3) the secondary lysosomes also engulf and digest organelles (mitochondria, membrane components, pinocytic vesicles) and, when aged are termed tertiary lysosomes, which primarily contain the indigestible remnants. Recently we have successfull purified and chemically characterized these tertiary lysosomes from *Mytilus* kidney (George *et al.*, 1982), they are made up of highly insoluble, peroxidized, lipoprotein pigments (lipofuschins) containing up to 10% metal (principally Fe, Zn, and Ca). *In vitro* metal-binding experiments have shown that they may play a dual role in the control of intracellular metal concentrations (George,

1983b). These studies have shown that (1) tertiary lysosomes can exert a thermodynamic control by equilibration with cytoplasmic metal. This binding is a non-specific, weak, exchangeable binding of metal, either present in the cytoplasm as the free ion (*e.g.*, Ca^{2+} or Mg^{2+}) or in the case of divalent heavy metals as weak amino acid or protein complexes; and (2) they can exert a kinetic control and act as a sink for excess metal by irreversible binding of metals caused by intramolecular cross-linkage. This results in the accumulation and detoxication of high concentrations of a variety of divalent metals. These two types of metal binding sites are represented diagramatically in Figure 4 (from George, 1983b). Studies with mussels exposed to Cu and Cd *in vivo* have shown that (like all cytosolic proteins) Cu- and Cd-thionein are taken up and can be degraded in the lysosomes (George, 1983a; Viarengo, in prep.), which leads to an accumulation of these metals in the tertiary lysosomes.

THE MODEL FOR CELLULAR HOMEOSTASIS

The components of the pathway for metal metabolism appear to be present in all tissues examined, and it is therefore valid to discuss a generalized model for intracellular metal homeostasis; although, as will be discussed later, the rates of metabolism and consequently the intracellular partitioning of metals will vary according to both the metal in question and the tissue. The metabolic model is summarized in Figure 5.

It is essential that excess free Cu^{++}, Zn^{++}, etc., ions are prevented from reacting with cytosolic enzymes and inhibiting their activities. While the metals appear to be non-specifically bound to high molecular weight proteins upon initial entry into the cell, within hours they become bound to thionein since its affinity for heavy metals is much higher (cf. stability constants of ~ 11-15 for metallothionein vs. ~ 3 for imidazole coordination in a protein such as albumin (Table 2). The binding of Zn probably requires synthesis of new thionein, while metals such as Cu, Cd, and Hg will be bound by displacement of Zn from existing Zn-thionein in the cell, and new synthesis will only occur when saturation occurs. This complexation will therefore reduce the reactive metal concentration in the cytosol to negligible amounts. During rapid growth the cellular Cu and Zn requirements are quite large and therefore a store of these essential trace metals must be maintained. Metabolically it is probably disadvantageous to store them as metallothionein

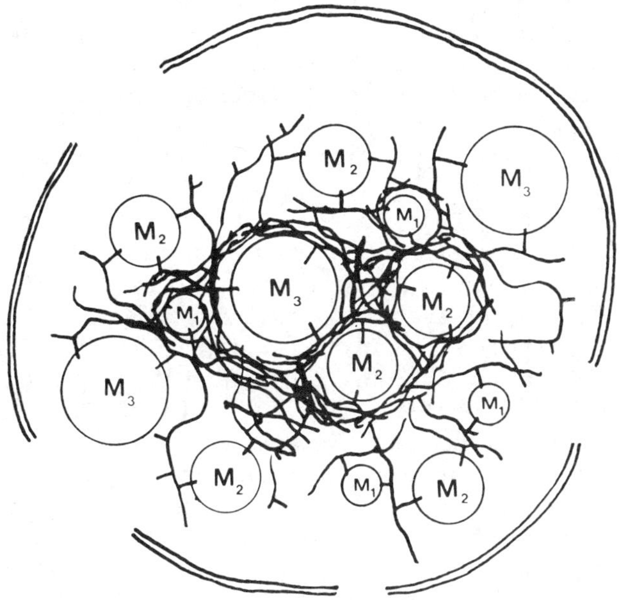

Fig. 4. Schematic representation of metal-binding sites in mussel kidney tertiary lysosomes. Reproduced from George (1983b). Metal-binding sites are distributed randomly. Different metals (M1, M2, etc.) can be bound if geometry is correct (position and distances of binding ligands). Metals in outer "flexible" regions can be exchanged, in other regions lipid peroxidation reactions increase intramolecular cross-linkages and lock metals within the structures.

Table 2. Metal-binding constants of some macromolecules.

Ligand	Metal			Comments
	Cd	Cu	Zn	
Glutathione		4.7*	5.1	Tripeptide, 1-5 mM in cell
Carboxpeptidase	10.8	10.6	10.5	Zn-metalloenzyme
Alk. phosphatase	-	-	11	Zn-metalloenzyme
Carbonic anhydrase			12	Zn-metalloenzyme
Serum Albumin	2.8	3.7	2.9	imidazole binding
3° lysosomes	4		3.5	approximate
Thionein	15		11	approximate

*believed to be inaccurate, expected to be higher.

since the cell has a high requirement for cysteine in glutathione synthesis (glutathione has a key role in detoxication of oxygen-induced organic peroxides and organic electrophiles and is present in mM concentrations in the cytosol [see *e.g.* Sies & Wendel, 1978]). However, since metallothionein is continually turned over by the lysosomes, another coordination site is available for storage of metals: this is the acidic lipoprotein matrix of the secondary and tertiary lysosomes. *In vitro* studies have shown that Zn is readily released from Zn-thionein (probably by the acidic pH) and binds to the lysosomal matrix, whereas Cd is not released from Cd-thionein in this way (George, 1983b), thus binding of Cd (also Cu and Hg) probably requires enzymatic hydrolysis of the protein as well. The lysosomes probably act as a store for Cu and Zn since the metal released from metallothionein during its degradation is only weakly bound (log. $K \sim 3.5$) and is freely exchangeable; therefore, it will also be available for coordination to newly synthesized thionein. Besides its role in detoxication, metallothionein may have a metabolic role in the transport of Cu and Zn to newly synthesized apoenzymes (reviewed by Brady, 1982) since several *in vitro* studies have demonstrated that activation of enzymes such as alkaline phosphatase, carbonic anhydrase and tyrosinase can be mediated by Zn-, Cu-thionein. If the lysosomes continued to act as a sink for more and more exchangeable metal, their efficiency as an intracellular buffer would be exceeded and metal would leak out by mass action effects. This is prevented by the intrinsic nature of the lysosomal matrix itself to form an insoluble peroxidized polymer. During this process, excess metal becomes trapped and therefore completely unavailable. Consequently, the tertiary lysosomes act as the most efficient type of metal buffer system, effectively an insoluble precipitate of the metal. This concept of the relative buffering efficiencies of differing types of metal coordination has been recently proposed by Williams (in press). The cell is therefore equipped with a very efficient mechanism for buffering (homeostasis) and detoxication of heavy metals. The final step in the metabolic pathway for detoxication of excess metals is by excretion of these tertiary lysosomes by exocytosis. It is interesting to note that since these metal-rich tertiary lysosomes are entirely resistant to degradation by enzymes of eucaryotes, they are probably very persistant in the environment and are only likely to be degraded by heterotrophic bacteria in the sediments.

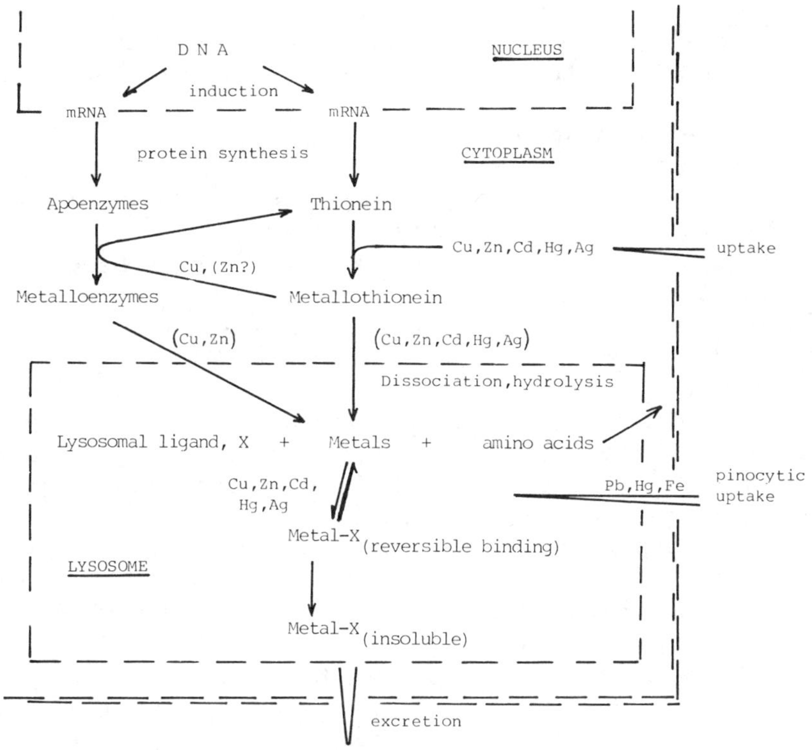

Fig. 5. Metabolic pathways for Group B metals in animal cells.

CONSEQUENCES OF THE MODEL ON METAL PARTITIONING AND TURNOVER

Thus, although this group of metals (Cd, Cu, Hg, and Zn) share the same metabolic pathway, when the intracellular distribution of these metals is investigated, differences are observed between metals and also between tissues. These are explained by differences in the rates of degradation of metallothionein according to its metal composition and differences in the rates of lysosomal exocytosis. *In vitro* studies of mammalian metallothioneins have shown that the pH at which 50% dissociation of metal occurs are about pH 4 for Zn-thionein, pH 3 for Cd-thionein, and pH<1 for Cu-thionein; Cd-thionein which has an ordered secondary structure is enzymically hydrolysed more slowly than Cu-thionein which has little secondary organization (see Kagi & Nordberg,

1979). Consequently, in mammals the *in vivo* turnover times vary, for Cu-thionein it was 12-17 h, for Zn-thionein 10-65 h, for Cd-thionein 70-110 h, and Hg-thionein about 110 h (Webb, 1979). It is interesting to note that the thioneins of the essential metals are turned over faster than those of the pollutant metals. Another relevant property is the tendency of Cu-thionein to oxidize and form an insoluble polymer which accumulates in the lysosomes. In our studies of *Mytilus*, the different distributions and turnover of various heavy metals indicates that these factors are operating *in vivo*. Thus, in the digestive glands of Cu-exposed mussels where exocytosis of residual bodies is rapid ($t\frac{1}{2}$ 5-6 d), Cu-thionein is taken up by the lysosomes where the pH is not low enough to remove the metal and the protein is not digested (and also tends to polymerize), thus the metal is not recycled back into the cytosol before the lysosomes are exocytosed (Viarengo *et al.*, 1981, in press). Consequently, there is a rapid excretion of Cu from the digestive gland cells.

On the other hand, Cd appears to be released from Cd-thionein in the secondary lysosomes (George, 1983a) and is either bound by the lysosomal matrix or rebound by newly synthesized cytoplasmic thionein. The small lysosomal component will be excreted ($t\frac{1}{2}$ 5-6 d, as for Cu); the major intracellular Cd component will be cytosolic Cd-thionein which will have a long half life. This has been confirmed experimentally (Viarengo *et al.*, in prep.). In the kidney, however, where the lysosomal half life is very much longer (ca. 60 d, George & Pirie, 1980), proportionately much more Cd is found in the lysosomes where it has been trapped by peroxidation (George & Pirie, 1979; George, 1983b).

Pb is not metabolized in the cell and does not bind to thionein, therefore any soluble Pb which enters the cell will partition into the lysosomes, which will also contain Pb taken up by pinocytosis. This difference in half lives of lysosomes in kidney and digestive gland probably explains why heavy metals such as Cd, Zn, and Pb reach such very high concentrations in the kidney. These tissue differences are summarized in Figure 6.

The proposed model appears to fit current data for metabolism, tissue partitioning, and dynamics of metals both in *Mytilus* and in mammals. We believe that, although it is a simplistic view of the complex situation which occurs in the environment (where multiple metal exposures and ligand competition occur), it is of general applicability and provides a start to a clarification of our understanding of the processes of metal accumulation and intracellular homeostasis.

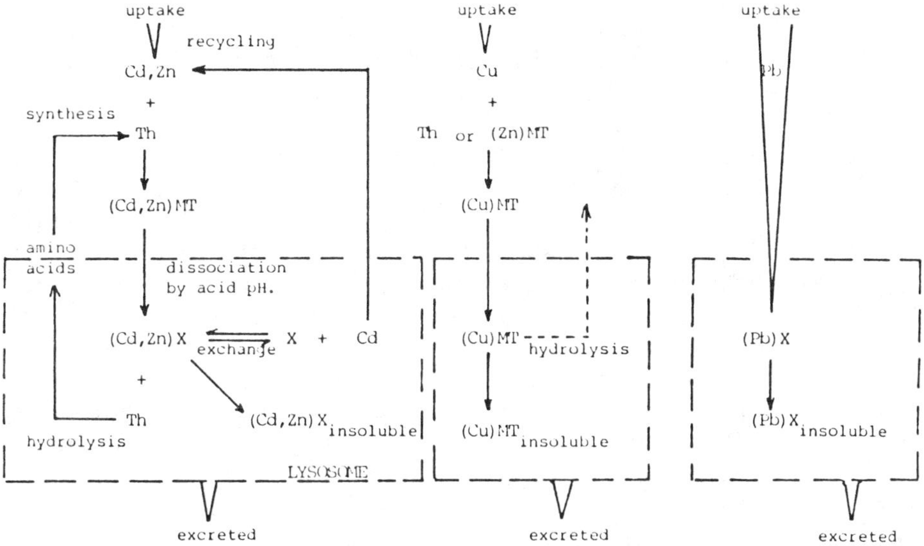

Fig. 6. Lysosomal control of tissue metal concentrations and subcellular distributions. There is rapid cycling of Cd and Zn, slow recycling of Cu (and probably Hg), and no recycling of Pb from lysosomes back into the cytosol. Digestive gland, $t_{1/2}$ lysosomes 4-5 d. Cd and Zn accumulate in metallothionein in cytosol, slow tissue loss, $t_{1/2}$ approx. 60 d; Cu and Pb rapidly lost from tissue in lysosomes, $t_{1/2}$ approx. 5 d. Kidney, $t_{1/2}$ lysosomes 40-50 d. Slow excretion, metals trapped in lysosomes where they accumulate in very high concentrations. $t_{1/2}$ Cd and Zn approx. 60 d, Pb 40 d loss via the lysosomes. $t_{1/2}$ Cu approx. 5 d.

ACKNOWLEDGMENTS

We gratefully acknowledge the work of others in this field whose data we have interpreted freely and hopefully, correctly. We are particularly grateful to Florence Harrison for her magnanimous gesture in providing us with the Cu data from a draft manuscript which is reproduced in Figure 3.

LITERATURE CITED

Brady, F.O. 1982. The physiological function of metallothionein. Trends Biochem. Sci. 7: 143-145.

Carpene, E., O. Cattani, K. Hakim, and G.P. Serrazanetti. 1983. Metallothionein from foot and posterior adductor muscle of *Mytilus galloprovinvialis*. Comp. Biochem. Physiol. 74C: 331-336.

Carpene, E. and S.G. George. 1980. Absorption of cadmium by gills of *Mytilus edulis* (L.). Mol. Physiol. I: 23-24.

Coombs, T.L. 1977. Uptake and storage mechanisms of heavy metals in marine organisms. Proc. Analyt. Div. Chem. Soc. 14: 219-222.

Coombs, T.L. and S.G. George. 1980. Mechanisms of immobilization and detoxication of metals in marine organisms. pp. 179-187. *In*: Proc. 12th Europ. Symp. Mar. Biol. D.S. McLusky and A.J. Berry (eds.). Pergamon Press, Oxford.

Engel, D.W., W.G. Sunda, and B.A. Fowler. 1981. Factors affecting trace metal uptake and toxicity to estuarine organisms. I. Environmental parameters. pp. 127-144. *In*: Biological Monitoring of Marine Pollutants. F.J. Vernberg, A. Calabrese, F.P. Thurberg, and W.B. Vernberg (eds.). Academic Press, New York.

Frankenne, F., F. Noel-Lambot, and A. Disteche. 1980. Isolation and characterization of metallothioneins from cadmium-loaded mussel *Mytilus edulis*. Comp. Biochem. Physiol. 66: 179-182.

Frazier, J.M. and S.G. George. 1983. Cadmium kinetics and oysters -- A comparative study of *Ostrea edulis* and *Crassostrea gigas*. Mar. Biol. 76: 55-61.

George, S.G. 1980. Correlation of metal accumulation in mussels with the mechanisms of uptake, metabolism, and detoxication. A review. Thalassia Jugosl. 16: 347-365.

George, S.G. 1982. Subcellular accumulation and detoxication of metals in aquatic animals. pp. 1-60. *In*: Physiological Mechanisms of Marine Pollutant Toxicity. W.B. Vernberg, A. Calabrese, F.P. Thurberg, and F.J. Vernberg (eds.). Academic Press, New York.

George, S.G. 1983a. Heavy metal detoxication in the mussel, *Mytilus edulis* - Composition of Cd-containing kidney granules (tertiary lysosomes). Comp. Biochem. Physiol. 76C: 53-58.

George, S.G. 1983b. Heavy metal detoxication in Mytilus kidney -- An in vitro study of Cd and Zn-binding to isolated tertiary lysosomes. Comp. Biochem. Physiol. 76C: 59-66.

George, S.G., E. Carpene, T.L. Coombs, J. Overnell, and A. Youngson. 1979. Characterization of cadmium-binding proteins from the common mussel, Mytilus edulis. Biochim. Biophys. Acta. 580: 225-233.

George, S.G., T.L. Coombs, and B.J.S. Pirie. 1982. Characterization of metal-containing granules from the kidney of the common mussel, Mytilis edulis. Biochim. Biophys. Acta 716: 61-71.

George, S.G. and B.J.S Pirie. 1979. Subcellular localization of cadmium in the kidneys of Mytilus edulis after cadmium exposure by electron probe x-ray microanalysis. Biochim. Biophys. Acta 580: 234-244.

George, S.G. and Pirie, B.J.S. 1980. Metabolism of zinc in the common mussel, Mytilus edulis. J. Mar. Biol. Ass. U.K. 60: 575-590.

George, S.G., B.J.S. Pirie, and T.L. Coombs. 1976. The kinetics of accumulation and excretion of ferric hydroxide in Mytilus edulis (L) and it's distribution in the tissues. J. Exp. Mar. Biol. Ecol. 39: 261-268.

George, S.G., B.J.S. Pirie and T.L. Coombs. 1977. Metabolic characteristics of endocytosis of ferritin by gills of a marine bivalve mollusc. Biochem. Soc. Trans. 5: 136-137.

Goldberg, E.D. (ed.). 1980. The International Mussel Watch. National Academy of Science, Washington, D.C. 248 pp.

Gutkneckt, J. 1981. Inorganic mercury (Hg^{2+}) transport through lipid bilayer membranes. J. Membrane Biol. 61: 61-66.

Gutkneckt, J. 1983. Cadmium and thallous ion permeabilities through lipid bilayer membranes. Biochim. Biophys. Acta. 735: 185-188.

Harrison, F.L., J.R. Lam, and R. Berger. 1983. Sublethal responses of Mytilus edulis to increased dissolved copper. Sci. Total Environ. 28: 141-158.

Janssen, H.H. and N. Schultz. 1979. Uptake and cellular distribution of cadmium in Mytilus edulis. Mar. Biol. 55: 133-141.

Kagi, J.H.R. and M. Nordberg. 1979. Metallothionein. Experientia Supplementum 34, Birkhauser, Basel.

Kohler, K. and H.U. Riisgard. 1982. Formation of metallothioneins in relation to accumulation of cadmium in the common mussel Mytilus edulis. Mar. Biol. 66: 53-58.

Liao, M.J. and J.H. Prestegard. 1981. Structural properties of a Ca^{2+}-phosphatidic acid complex, small angle x-ray scattering and calorimetric results. Biochim. Biophys. Acta 645: 149-156.

Lowe, D.M. and M.N. Moore. 1979. The cytochemical distribution of zinc (Zn II) and iron (Fe III) in the common mussel, *Mytilus edulis*, and their relationship with lysosomes. J. Mar. Biol. Assoc. U.K. 59: 851-856.

Marshall, A.T. and V. Talbot. 1979. Accumulation of cadmium and lead in the gills of *Mytilus edulis*: X-ray microanalysis and chemical analysis. Chem. Biol. Interactions 27: 111-23.

Nakahara, H. and G. Bevelander. 1976. Ingestion of particulate matter by the outer surface cells of the mollusc mantle. J. Morph. 122: 139-146.

Noël-Lambot, F. 1976. Distribution of cadmium, zinc and copper in the mussel, *Mytilus edulis*. Existence of cadmium-binding proteins similar to metallothioneins. Experientia 32: 324-326.

Roesijadi, G. 1982. Uptake and incorporation of mercury into mercury-binding proteins of gills of *Mytilus edulis* as a function of time. Mar. Biol. 66: 151-157.

Roesijadi, G. and R.E. Hall. 1981. Characterization of mercury-binding proteins from the gills of marine mussels exposed to mercury. Comp. Biochem. Physiol. 70C: 59-64.

Sanders, B.M. K.D. Jenkins, W.G. Sunda, and J.D. Costlow. 1983. Free cupric ion activity in sea water: Effects on metallothionein and growth in crab larvae. Science 222: 53-54.

Scholz, N. 1980. Accumulation, loss and molecular distribution of cadmium in *Mytilus edulis*. Helgol. wiss. Meeresunters. 33: 68-78.

Schulz-Baldes, M. 1978. Lead transport in the common mussel *Mytilus edulis*. pp. 211-218. *In*: Proc. 12th Europ. Symp. Mar. Biol. D.S. McLusky and A.J. Berry (eds.). Pergamon Press, Oxford.

Sies, M. and A. Wendel (eds.). 1978. Functions of Glutathione in Liver and Kidney. Springer-Verlag, Berlin. 262 pp.

Simkiss, K. 1983. Lipid solubility of heavy metals in saline solutions. J. Mar. Biol. Assoc. U.K. 63: 1-9.

Simkiss, K. and A.Z. Mason. (in press). Bioaccumulation of metals - A conceptual basis. Mar. Environ. Res.

Sternleib, I. and S. Goldfischer. 1976. Heavy metals and lysosomes. pp. 185-200. *In*: Lysosomes in Biology and

Pathology, Vol. 5. J.T. Dingle and R.T. Dean (eds.). American Elsevier Publishing Company, New York.

Sunda, W.G., D.W. Engel, and R.M. Thuotte. 1978. Effect of chemical speciation on toxicity of cadmium to grass shrimp, Palaemonetes pugio: Importance of free cadmium ion. Environ. Sci. Technol. 12: 409-413.

Sunda, W.G. and R.R.L. Guillard. 1976. The relationship between cupric ion activity and the toxicity of copper to phytoplankton. J. Mar. Res. 34: 511-529.

Talbot, V. and R.J. Magee. 1978. Naturally occurring heavy metal-binding proteins in invertebrates. Arch. Environ. Contam. Toxicol. 7: 73-81.

Viarengo, A., M. Pertica, G. Mancinelli, G. Zanicchi, and M. Orunesu. 1980. Rapid induction of copper-binding proteins in the gills of metal exposed mussels. Comp. Biochem. Physiol. 67: 215-218.

Viarengo, A., G. Zanicchi, M.N. Moore, and M. Orunesu. 1981. Accumulation and detoxication of copper by the mussel Mytilus galloprovincialis (L.): A study of the subcellular distribution in the digestive gland cells. Aquat. Toxicol. 1: 147-157.

Viarengo, G., M. Pertica, G. Mancinelli, G. Zanicchi, M. Orunesu, and M. Bouquegneau. (in press). Biochemical characterization of Cu-thionein isolated from the tissues of mussels exposed to the metal. Mol. physiol.

Webb, M. (ed.). 1979. The Chemistry, Biochemistry and Biology of Cadmium. Elsevier, Amsterdam. 465 pp.

Williams, R.J.P. (in press). Proceedings of the Dahlem Conference Workshop. Berlin, 1983.

Wrench, J.J. 1978. Biochemical correlates of dissolved mercury uptake by the oyster Ostrea edulis. Mar. Biol. 47: 79-86.

PURIFICATION OF VERY LOW MOLECULAR WEIGHT CU-COMPLEXES FROM THE EUROPEAN OYSTER

Liu Fayi and S.G. George

SUMMARY

The preparation and separation of six Cu-complexes with molecular weights less than 2 K daltons from Cu-contaminated (green-sick) oysters is described. The major Cu-complex which accounts for ca. 70% of the Cu was purified and the results indicate that it is not homarine- or amino acid-containing, but is a nitrogen heterocycle. The other Cu-complexes contain amino acids and may include Gly2, Cys-Gly, and glutathione complexes.

INTRODUCTION

Oysters can accumulate very high concentrations of Cu and Zn in their tissues without apparent toxicity (Thrower & Eustace, 1973; George *et al.*, 1978). Previous chromatographic studies of soluble tissue extracts have shown that these metals are associated with ligands of <2 K daltons molecular weight (Coombs, 1974; Frazier & George, 1983). The study of Coombs (1974) indicated that the metal-complexing ligands may be taurine, lysine, ATP, and homarine. Microscopical studies have shown that the metals are accumulated in the amoeboid blood cells (Boyce & Herdmann, 1897; Ruddell, 1971; George *et al.*, 1978), and the metal concentrations are up to 300 mM Cu and 1 M Zn (Pirie *et al.*, 1984). Apart from their obvious role in metal detoxication, the Cu-containing cells are thought to have a defensive role akin to mast cells or platelets in vertebrates (Ruddell, 1971; Ruddell *et al.*, 1978; George *et al.*, 1983). In the present paper we report the purification of the copper-binding components as the first stage in the investigation of the biochemistry of this detoxication and defense system.

MATERIALS AND METHODS

Adult specimens of _Ostrea edulis_ (ca. 65 mm shell length) were obtained from Restronguet Creek, Cornwall, England (a metal-contaminated environment), shipped to Aberdeen and maintained in running sea water for at least 3 wk before sacrifice. Batches of 50 oysters were shucked, rinsed in distilled water, drained for 5 min, and the soft tissues dissected from the adductor muscle, pooled and homogenized in 0.5 M NaCl 20 mM Tris pH 8 (1:1 vol. buffer to wet wt tissue). All successive operations were either carried out under N_2 or with N_2-gassed solutions. The homogenate was centrifuged successively at 3000 x g for 20 min and 100,000 x g for 60 min, and the blue supernatants (cytosol) pooled. Ethanol was added to the cytosol to 33% (v/v) and, after standing in ice for 60 min, the precipitate was removed by centrifugation at 3000 x g for 20 min. The ethanol was removed by cooling to -80°C and lyophilizing. The lyophilized material was then dissolved in distilled water, and the insoluble residue was removed by centrifugation at 100,000 x g for 60 min. The green supernatant was then fractionated by ultrafiltration through different pore-sized membranes using N_2 pressure. The membranes were obtained either from Amicon Ltd. or Chemlab Ltd., and had nominal cut off values of 30 K daltons, 5 K daltons, and 500 daltons.

The Cu-containing ligands in the 500-5000 dalton fraction were separated by ion exchange chromatography after transference to phosphate buffer. One ml 0.2 M phosphate buffer (pH 7.5) was added to 10 ml oyster fraction and after standing for 30 min the precipitated phosphates were removed by centrifugation. The supernatant was then diluted to 25 ml and applied at a flow rate of 40 ml/h to an 87 x 1.5 cm column of DE 52 cellulose (Whatman Ltd.) equilibrated with 10 mM phosphate buffer (pH 7.5). After application of the sample, the column was washed with 80 ml equilibration buffer and then developed with a linear salt gradient generated from 200 ml 0.01 M phosphate and 200 ml 0.6 M NaCl 0.01 M phosphate (pH 7.5). Fractions of 5 ml were collected. The fractions were analyzed for Cu and Zn and the absorbance at 260 and 320 nm measured. Four Cu-containing and one Zn-containing peak were eluted (designated peaks I - V).

Peak V was purified further. The combined fractions with metal concentrations greater than half peak height were concentrated by ultrafiltration on a 500 dalton cutoff membrane, diluted to 100 ml, and reconcentrated to 10 ml. The material was then rechromatographed on DE 52 using a gradient produced from 200 ml 0.1 M NaCl and 200 ml 0.4 M NaCl in 0.01 M phosphate buffer (pH 7.5). The peak fractions were pooled and lyophilized, dissolved in 3 ml H_2O, then desalted on Sephadex G25 (97 x 2.5 cm column) using distilled water as eluate.

Due to the high salt concentration in the <500 dalton ultrafiltrate, which prevented absorption onto DEAE cellulose, the material was lyophilized, dissolved in a minimal volume of water and partially desalted on a Sephadex G15 gel permeation column (1.6 x 95 cm). The Cu-containing fractions were pooled and subjected to DE 52 cellulose chromatography, as for the higher molecular weight material.

Samples were analyzed for amino acids (JEOL JL6 6AH amino acid analyzer) after hydrolysis in 6N HCl for 48 h and removal of HCl by rotary evaporation.

UV/VIS spectra were recorded on a Perkin Elmer Model 555 Spectrophotometer, and fluorescence spectra on an Aminco Bowman Model SPF 125 Spectrophotofluorimeter. Metals were analyzed by atomic absorption spectrophotometry as described previously (Frazier & George, 1983); inorganic phosphate and protein were analyzed as described by George _et al_. (1982).

RESULTS AND DISCUSSION

The oysters used in this work were collected from an extremely metalliferous estuary and contained 0.32 g Cu/kg wet tissue and 4.34 g Zn/kg wet tissue.

The results of the preliminary fractionation of the oyster tissues are given in Table 1. Seventy three percent of the Cu and 32% of the Zn was released into the soluble fraction by the homogenization procedure; this is due to incomplete tissue disruption rather than insoluble metal, since more vigorous procedures with a more dilute homogenate give a higher yield in the cytosolic fraction. Very little metal was precipitated by ethanol even though nearly 70% of the protein was removed. Ultrafiltration showed that practically all the Cu and Zn passed through 30 K and 5 K dalton membranes and were almost equally distributed between the fraction retained by a 500 dalton membrane and that passing through. (While a 5 K dalton membrane was used routinely, previous experiments had shown that practically all of the metal passed through a 2 K dalton membrane.)

Table 1. Preliminary fractionation of oyster tissues. 50 oysters, 268 g wet wt. tissue were homogenized and fractionated as detailed in "Materials and Methods."

Fraction	Cu		Zn	
	mg.	% recovery	mg.	% recovery
Homogenate	85	100	1164	100
Particulates	22	26	780	67
Cytosol	62	73	372	32
33% ethanol SN	56	66	338	29
Ultrafiltration				
30 K daltons	2.6	3	11.6	1
5-30 K daltons	0.1	<1	1.6	<1
0.5-5 K daltons	21	25	151	13
0.5 K daltons	20	24	163	14

The low molecular weight fractions contained such high concentrations of Zn that when they were subjected to ion exchange chromatography using a tris buffer system the Zn was distributed in two very broad peaks which overlapped most of the other components. It was found, however, that addition of phosphate buffer produced a white precipitate which contained about 65% of the Zn and less than 5% of the Cu. Therefore, precipitation with phosphate was utilized as a method of removing a large proportion of the Zn. Control experiments with sulphates of Cu and Zn showed that over 99% of the metals were precipitated by this procedure, which indicates that in the oyster all the Cu and at least 35% of the Zn is complexed and not present as the free ions.

Chromatography of the 500-5000 dalton and the <500 dalton fractions on DEAE DE-52 cellulose produced the profiles shown in Figure 1. In the 500-5000 dalton fraction, four Cu-containing peaks were eluted (Fig. 1A). The first peak (designated IA) contained 7% of the eluted Cu and was eluted with the non-bound material and is therefore neutral or cationic. Three peaks representing anionic components were eluted with the salt gradient and comprised 17, 2, and 74% of the eluted Cu, respectively. Zn was eluted in one major component (peak IV) separable from the Cu peaks and some minor peaks which were not resolved from the Cu-containing peaks.

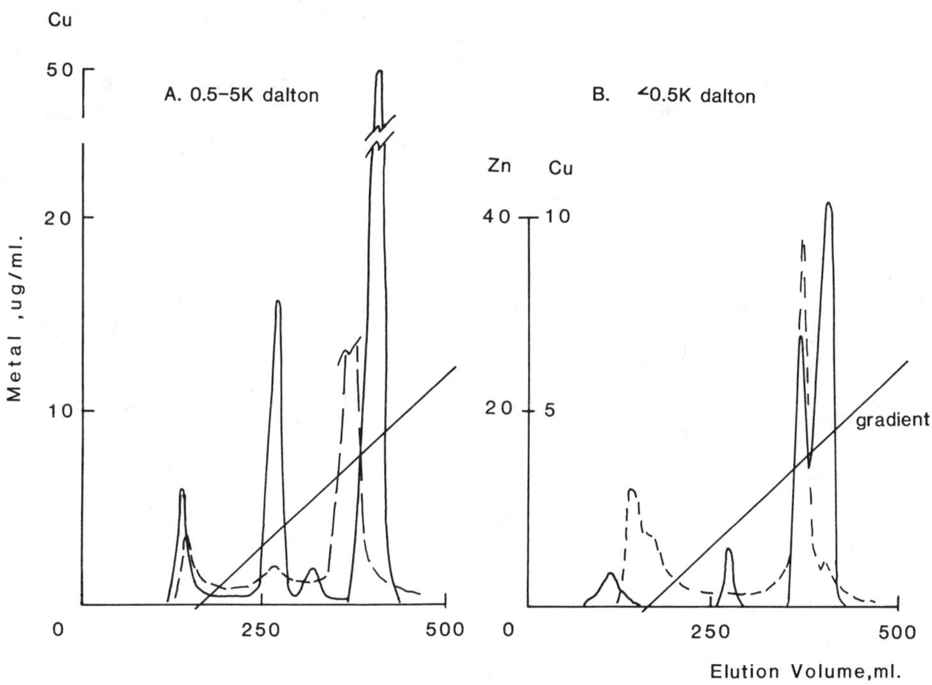

Fig. 1. DEAE chromatography proflies of very low molecular weight oyster material. (A) sample, 500 - 5000 dalton material. (B) sample, 500 dalton membrane filtrate material. Column 87 x 1.5 cm, equilibrated and washed with 10 mM phosphate pH 7.5, developed with linear salt gradient in 10 mM phosphate to 0.6 m NaCl. ——— Cu, ----- Zn.

The elution profile of the partially desalted <500 k dalton material chromatographed under the same conditions is shown in Figure 1B. There was a Cu-containing peak in the non-retained material which contained only about 3% of the eluted Cu. This peak was not coincident with that for the higher molecular weight material and is therefore designated peak IB; however, this difference in elution position may be due to minor differences in salt content of the applied samples. Three Cu-containing peaks were eluted by the salt gradient. The first, comprising 5% of the Cu, coincided with peak II of the material >500 daltons; the second corresponded with the major Zn peak (peak IV) in both samples; the third, which accounted for 60% of the Cu in this fraction, was coincident with the major Cu-component of the >500 dalton fraction. Since the nominal molecular weight cut off of the ultrafiltration membrane is about 500 daltons (i.e.,

95% retention value), material of about this molecular weight will be found in both the filtrate and retentate. In view of the correspondence in peak positions, we conclude that the Cu-binding components in peaks II and V have a molecular weight of about 500 daltons, and there are two higher molecular weight Cu-binding ligands (peaks IA & III) and two lower molecular weight Cu-binding ligands (peaks IB & IV). The relative partitioning of Cu in the cytosol between these components is calculated in Table 2. The major Cu-binding ligands were found in peak II (11%), IV (16%), and V (67%).

Table 2. Distribution of Cu between very low molecular weight binding ligands.

DEAE peak No.	IA	IB	II	III	IV	V	Approx. mol. wt.
Percentage of Cytosol Cu	3.5			1			> 500
			11			67	~ 500
		1.5			16		< 500
Zn present	+	+	?	-	+	-	

Samples of the peak fractions from DE 52 chromatography of the >500 dalton material were hydrolyzed and analyzed for amino acids. The results are presented in Table 3. Peak IB, which only contained 1.5% of the cytosolic Cu, contained all the common amino acids in at least equivalent concentrations to Cu and excess Pro, Gly, Ala, and Tau. Peak II (comprising 11% of the Cu) had a much simpler composition, containing only Gly, Cys, and Ala in amounts comparable with Cu. It is probable that this Cu-complex is the dipeptide Cys-Gly (a component of the γ-glutamyl cycle) and that Gly and Ala are also coordinated. Peak III, which has a very minor Cu component (1%), can probably be assigned to glutathione (Glu-Cys-Gly). Peak IV, which in the higher molecular weight material only contained Gly, is probably Zn $(Gly)_2$ or Zn (Gly-Gly). In either case, crystal structures indicate that the Zn atom is also coordinated to adjacent molecules (Freeman, 1967), which would explain why the material was retained by the ultrafiltration membrane even though the monomeric molecular weight is only about 215. On the other hand, Cu $(Gly)_2$ and Cu (Gly-Gly) do not tend to form these polymeric structures and would pass through the membrane. Therefore, we conclude that diglycine complexes of both Cu and Zn are probably present in these oysters. Peak V is essentially devoid of amino acids.

Table 3. Amino acid analyses of DEAE purified oyster metal-binding ligands. The four peak fractions from each of the Cu-containing maxima were pooled, hydrolysed and analyzed for amino acid content.

Peak No.	Metal : Amino acid ratios	Comments
IB	Cu 1 : Tau 30 : Pro 4 : Gly 4 : Ala 3	All other common amino approx. 1.
II	Cu 1 : Gly 2.8 : Cys 1.4 : Ala 1.2	Traces other (0.5).
III	Cu 1 : Gly 1.4 : Cys 1.9 : Glx 0.7	Trace Asx 0.3, probably Cys coordination.
IV	Zn 1 : Gly 2	
V	Cu 215 : Tau 2 : Gly 1 : Ala 1 : Pro 1.	Trace amino acids due to contaminants.

The nature and properties of this peak V material were investigated further. After reduction of the salt concentration, the material was rechromatographed on DE 52. The Cu was eluted as a single symmetrical peak which showed a UV absorption maximum at about 260 nm. This peak was completely resolved from a Zn peak (i.e., the tail of peak IV) and other UV-absorbing material. The Cu peak was lyophilized, dissolved in distilled water, and chromatographed on Sephadex G25. The elution profile (Fig. 2) showed only one Cu-containing peak, which corresponded with the UV absorbing peak. A peak of phosphate (previous chromatography buffer) was eluted later, although there was a small peak of phosphate which appeared under the first half of the Cu peak. The ratios of phosphate (Cu in the phosphate-containing fractions) varied from 0.01-0.06 : 1.

Spectroscopic analysis of the purified Cu-binding ligand gave absorption maxima at 257 and about 755 nm with a shoulder at 215 nm (Fig. 3) and a fluorescence excitation

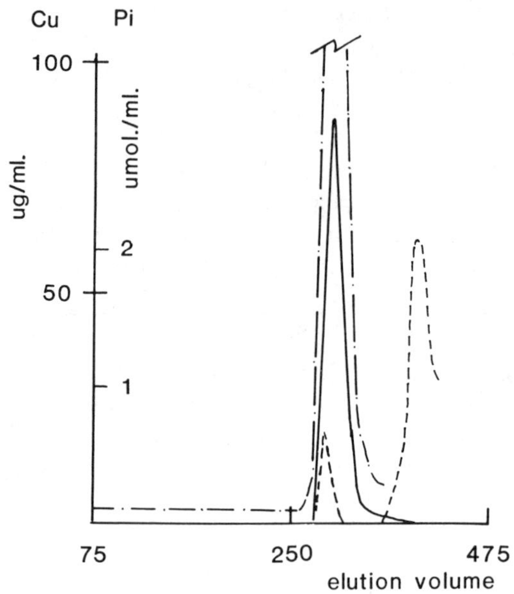

Fig. 2. Gel permeation chromatography of Cu-complex V. Sample from DE 52 cellulose chromatography (twice), lyophilized and dissolved in water. Column 95 x 1.6 cm Sephadex G15, eluant water. ——— Cu, ----- phosphate, — - — E260 nm.

max. at 329 nm with emission max. at 427 nm. X-ray analysis showed that the material contained some S (less than Cu) and a trace of P. From this data we conclude that this Cu-binding ligand is not homarine (absorption maximum 272 nm, Fig. 3) as suggested by Coombs (1974). In view of the very low phosphate it cannot be ATP, although the absorption maximum is similar (Fig. 3) and the fluorescence spectrum indicates a nitrogen heterocycle. Thin-layer chromatography of the Cu-ligand V on silica gel SIL-20 in 1 M LiCl showed that it co-migrated with ADP, ATP and their Cu-complexes, but not with NAD, NADP, homarine, or their Cu-complexes. When a sample of a Cu-ATP complex was chromatographed on DE 52 cellulose, the elution characteristics were also different. Unlike homarine or Cu-homarine, which have pH-independent UV absorption spectra, the UV spectrum of the oyster Cu-complex is pH dependent (Fig. 3). The absorption at 257 nm decreased very markedly as the pH was reduced. Titration showed that the pK of this effect was 4.7 (Fig. 4). It was demonstrated that this absorption maximum was due to Cu-complexation by addition of Cu to the Cu-free ligand (prepared by G15 chromatography at pH 3). When Cu is added to ATP an increase in 260 nm absorption is also produced; however, the pK for the reaction is at pH 5.6 and unlike the oyster material an absorption peak remains at an acid pH. NMR spectroscopy was performed on the Cu-free material and indicated aromaticity, however, we had insufficient material for characterization. We are currently using MS techniques to identify this ligand, where upon some indication of the function of this Cu-complex might be obtained, however, the molecule has so far proved intractible to FAB techniques and it is planned to use DCI.

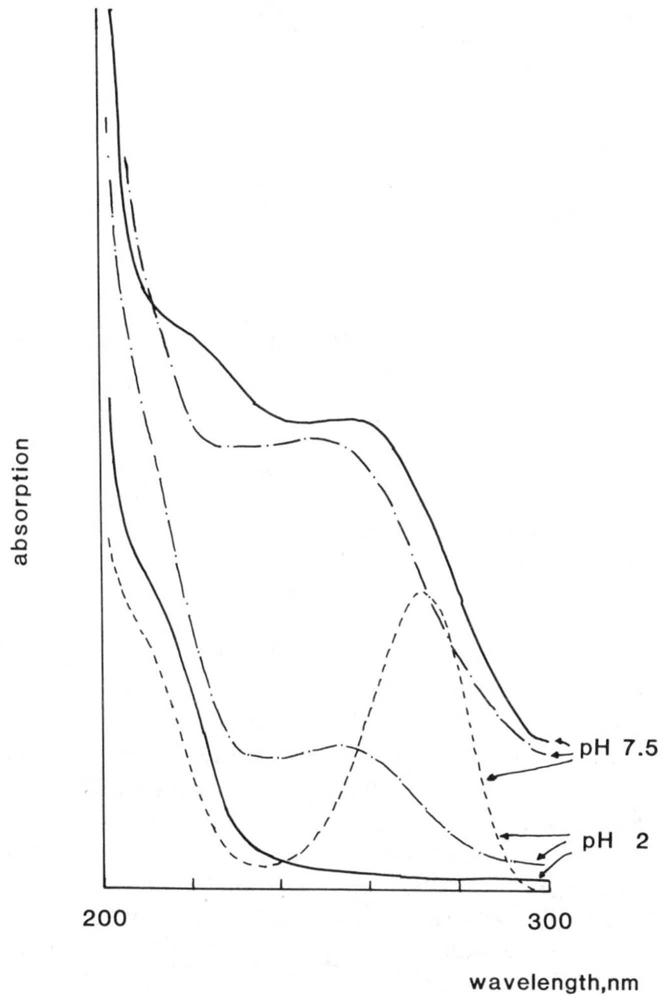

Fig. 3. UV absorption spectra of Cu-complexes. ——— oyster Cu-complex V, — - — Cu-ATP, — — Cu-homarine.

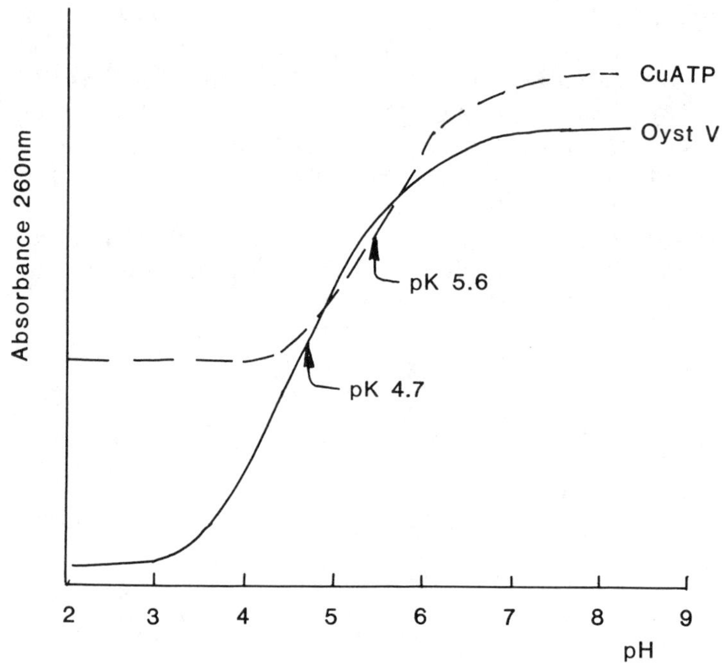

Fig. 4. pH profiles of UV absorption maxima of Cu-ATP and oyster V material.

LITERATURE CITED

Boyce, R, and W.G. Herdman. 1897. On a green leucocytosis in oysters. Proc. R. Soc. 62: 30-38.

Coombs, T.L. 1974. The nature of the zinc and copper complexes in the oyster, Ostrea edulis. Mar. Biol. 28: 1-10.

Frazier, J.M. and S.G. George. 1983. Cadmium kinetics in oysters - A comparative study of Crassostrea gigas and Ostrea edulis. Mar. Biol. 76: 55-61.

Freeman, H.C. 1967. Crystal structure of metal peptide complexes. pp. 257-424. In: Advanced Protein Chemistry, Vol. 22. C.B. Anfinsen, M.L. Anson, J.T. Edsall, and F.M. Richards (eds.). Academic Press, New York.

George, S.G., T.L. Coombs, and B.J.S. Pirie. 1982. Characterization of metal-containing granules from the kidney of the common mussel, Mytilus edulis. Biochim. Biophys. Acta. 716: 61-71.

George, S.G., B.J.S. Pirie, A.R. Cheyne, T.L. Coombs, and P.T. Grant. 1978. Detoxication of metals by marine bivalves: An ultrastructural study of the compartmentalization of copper and zinc in the oyster, *Ostrea edulis*. Mar. Biol. 45: 147-156.

George, S.G., B.J.S. Pirie, and J.M. Frazier. 1983. Effects of cadmium exposure on metal-containing amoebocytes of the oyster *Ostrea edulis*. Mar. Biol. 76: 63-66.

Pirie, B.J.S., S.G. George, D. Lytton, and J.D. Thomson. 1984. Metal containing blood cells of oysters: Ultrastructure, histochemistry and x-ray microanalysis. J. mar. Biol. Ass. U.K. 64: 115-123.

Ruddell, C.L. 1971. Elucidation of the nature and function of the granular oyster amoebocytes through histochemical studies of normal and traumatized oyster tissues. Histochemie. 26: 98-112.

Ruddell, C.L., T. Dunlap, R.K. Okazaki, and R. Munn. 1978. The effect of selected basic dyes on the blood cells, in particular the basophils, of the Pacific oyster, *Crassostrea gigas*. J. Invert. Pathol. 31: 313-323.

Thrower, S.J. and I.J. Eustace. 1973. Heavy metal accumulation in oysters grown in Tasmanian waters. Fd. Technol. Aust. Nov. : 546-553.

METAL-EXPOSED SEA SCALLOPS, *PLACOPECTEN MAGELLANICUS* (GMELIN): A COMPARISON OF THE EFFECTS AND UPTAKE OF CADMIUM AND COPPER

E. Gould, R.A. Greig,
D. Rusanowsky, and B.C. Marks

SUMMARY

1. The giant sea scallop (Placopecten magellanicus Gmelin) was more susceptible to toxic effects of copper (20 μg/L, 7 wk) in winter, when the animal was in the early stages of gametogenesis, than in spring, when scallops from the same population had fully differentiated gonads.
2. In scallops undergoing early gametogenesis, similar cadmium exposure provoked early gamete maturation, whereas copper inhibited it. In scallops with fully differentiated gonads, copper exposure induced gonad recession.
3. By the end of the 7-wk exposure, kidneys of Cu-exposed scallops had become pale yellow to transparent rather than the normal brown, and contained no more copper than kidneys of control animals, apparently having excreted the toxic overloads of copper. Consistently low kidney IDH indicated mitochondrial impairment.
4. The Cd-exposed scallops continued to accumulate that metal in the kidney, even during a 2-wk depuration period.
5. Kidney and digestive gland in metal-exposed scallops accumulated more than did the gill. The gonad also accumulated highly significant amounts of both metals and had not cleared after the 2-wk depuration period.

6. Muscle glycogen concentrations in Cd-exposed scallops indicated an earlier than normal transfer of energy reserves to the gonad, whereas muscle glycogen in Cu-exposed scallops indicated gamete resorption.

INTRODUCTION

Calcium-phosphate concretions in the kidney of bivalve molluscs serve as a major sequestering mechanism by which the animals adapt to the high concentrations of heavy metals often found in their tissues. These inorganic concretions, by which overloads of toxic metals are apparently sequestered and eventually eliminated, are thought to be inducible (*e.g.*, Carmichael & Fowler, 1981). They appear to serve a normal excretory function in molluscs undergoing reproductive, environmental, or pollutant-induced stress (Doyle *et al*., 1978). Another important physiological strategy for inactivating heavy metals utilizes metal-binding proteins, normally found in most tissues but primarily concentrated in the kidney and digestive gland (*e.g.*, Engel & Fowler, 1979a; Ridlington & Fowler, 1979; Roesijadi, 1981). George (1982) has written a comprehensive explication of these and other metal-detoxication mechanisms.

The kidney concretions have been investigated in a variety of marine bivalves with particular respect to cadmium and copper (*e.g.*, Coombs & George, 1978; Carmichael *et al*., 1979, 1980; Fowler *et al*., 1981). In 1980 George *et al*. isolated the metal-rich granules from the kidney of field-collected *Pecten maximus*; analysis showed that they contained only a small fraction of the kidney's cadmium burden, which was predominantly associated with another particulate fraction of lower density. Only 20% was found in the cytosol fraction containing the metal-binding proteins. Quite a different compartmentalization was produced experimentally in another pectinid, *Argopecten irradians*, after an acute, brief (700 µg/L, 5 d) cadmium exposure (Carmichael & Fowler, 1981). There was a massive induction of kidney concretions, which contained 60% of the total cadmium in the kidney, exclusively located in the surface layers; only 2% was found in the cytosol. Although the severe metal challenge was to a different scallop species from George's field collections, the two studies appear to support the observation of Engel and Fowler (1979b): the rate at which metals are taken up by tissue is important in determining the immediate mechanism of detoxication.

Two general categories of Cd-rich metal-storage granules have been reported for the barnacle *Balanus balanoides*

(Walker, 1977; Coombs & George, 1978): one type contained Zn and P, and the other, Cu and S. A similar pair of granule types was found in the marine prosobranch *Littorea littorina*, each with a different composition, structure, and cellular location (Mason & Nott, 1981), suggesting functional differences as well. Questions therefore arise whether toxic-metal cadmium and essential trace-metal copper are immobilized by different metabolic strategies, whether the two metals are stored in different areas of the molluscan kidney, whether they are cleared from different tissues at different rates and by different mechanisms, and what the physiological effects of each may be.

Our field-monitoring efforts for the past few years (Gould, 1983) have focused on the sea scallop, *Placopecten magellanicus*. Applying these questions to that species, therefore, we undertook an experimental series exposing *P. magellanicus* to sublethal levels of cadmium or copper. The series comprised three sequential steps: (1) In a preliminary exposure to low concentrations of the two metals, scallops were sampled at 3-wk intervals to discover the animals' relative metal tolerance (Spring 1982). (2) The principal exposure series (Winter 1983) was for a single exposure period, with two subsequent periods of clearing, and used scallops in the early stages of gametogenesis. Our purpose here was to determine the rates of uptake and clearance of the two metals in scallop tissues, and to learn something of the effects of copper and cadmium on tissue metabolism. Whole scallops from this exposure group were provided to C.A. and P.P. Yevich (USEPA Narragansett, RI) for histopathological examination (this volume). (3) The experimental design of the Winter 1983 exposure was repeated in the spring, using scallops with fully differentiated gonads. Our twofold purpose, arising from observations during the winter exposure, was to observe metal effects on the reproductive effort and to compare the metal tolerance of scallops in a highly active metabolic state (Spring 1983) to that of scallops in a metabolic "resting" stage (Winter 1983). Fixed gonad sections from scallops in the Spring 1983 exposure group were sent to R.J. Thompson (Marine Sciences Research Laboratory, Memorial University, St. John's, NFLD), who generously provided us with an assessment of gamete maturation in these scallops.

METHODS AND MATERIALS

Animal collection and exposure

Scallops were trawl-collected in late November 1982 from a site 66 m deep ESE of Block Island, off the Rhode Island coast. They were transported in chilled coolers to the Milford Laboratory, where they were acclimated for 5 d in sand-filtered water from Milford Harbor (10.5°C, 27°/$_{oo}$), and selected for a shell-height range of 95-112 mm. Digestive gland and kidney were taken from eight scallops at 0-time for trace-metal analysis. The animals were spawned out and had very small gonads of visually indeterminate sex. The scallops were apportioned among four control tanks and four exposure tanks in each of two dosing systems with flowing sea water, as described in Thurberg *et al*. (1977), after Mount and Brungs (1967). The animals were exposed for 7 wk to either 20 μg/L copper (as $CuSO_4 \cdot 5H_2O$) or to 20 μg/L cadmium (as $CdCl_2 \cdot 2\frac{1}{2}H_2O$). They were fed thrice weekly with an algal mixture comprising *Isochrysis galbana*, *Pavlova lutheri*, *Dunaliella euchlora*, *Tetraselmis maculata*, *Thalassiosira pseudonana*, and *Phaeodactylum tricornutum*. Algal density was equivalent to a packed-cell volume of 0.15 ml per scallop per feeding. Water temperature during the 7-wk exposure ranged from 7.3-14.4°C, and salinity 25.3-27.0°/$_{oo}$. After a sampling at the end of the 7-wk period (Jan 10-12), the remaining scallops were allowed to depurate for either 1 or 2 wk in flowing sea water to which no metal had been added. This study, which constitutes the bulk of this report, is referred to as "Winter 1983."

For the initial study to establish a metal tolerance range (March 1982), scallops from the same Block Island population were exposed to cadmium (10 μg/L) and sampled every 3 wk up to 15 wk. A smaller group of scallops were exposed to 20 μg/L copper and sampled at 6, 9, and 12 wk. Tissue handling and analytical procedures were the same as for the 7-wk exposure with clearance periods described below. This preliminary exercise is referred to as "Spring 1982."

The Winter 1983 study was repeated (March 1983) with scallops that had fully differentiated gonads. We were unable to obtain a Block Island collection and instead used scallops from a site SE of Montauk Point, Long Island. This study is referred to as "Spring 1983."

Tissue handling

After the 7-wk exposure, tissues were removed from scallops for biochemical analysis (adductor muscle, kidney) and for trace-metal analysis (kidney, digestive gland, gill, gonad). Gonad differentiation and weight (without crystalline style) were recorded. Tissue specimens for biochemical analysis were placed in small plastic pouches from which as much air as possible was excluded, tightly wrapped with masking tape, labeled, and stored at -80°C until testing. For trace-metal analysis, tissue samples were placed in acid-washed vials and stored at -20°C until testing. The remaining scallops were sampled after 1 and 2 wk depuration in flowing sea water, following the same procedure as above.

Biochemical analysis

Parameters used: Biochemical parameters in kidney tissue were 1) activities of glucose-6-phosphate dehydrogenase (E.C.1.1.1.49, G6PDH), a pentose shunt regulator providing metabolites for the biosynthesis of sterols and of nucleic and fatty acids; 2) malate dehydrogenase (E.C.1.1.1.37, MDH), a regulator of the cellular redox balance; and 3) isocitrate dehydrogenase (E.C.1.1.1.42, IDH), an enzyme of the citric acid cycle that produces reducing equivalents. Glycogen concentrations were measured in the phasic adductor muscle; together with lipids stored in the digestive gland, it has been shown to be the primary source of energy reserves for gametogenesis and gamete maturation in *P. magellanicus* (Robinson *et al.*, 1981), two European scallops, *Chlamys septemradiata* (Ansell, 1974) and *C. opercularis* (Taylor & Venn, 1979), and the common mussel, *Mytilus edulis* (Gabbott, 1975). Enzyme activities measured in muscle were 1) pyruvate kinase (E.C.2.7.1.40, PK), a unidirectional regulator of glucose breakdown for energy needs; 2) MDH; and 3) glutamate dehydrogenase (E.C.1.4.1.3, GDH), which is involved in both protein and carbohydrate metabolism and which, because low levels of ATP can induce its activity, may serve as an indicator of the cellular energy charge.

Tissue preparations: All tissue homogenates were prepared at 0°C in small glass homogenizers containing 25-μm glass powder to facilitate grinding. Adductor muscle preparations consisted of a 10% homogenate (1:9, w/v) in 20 mM Tris-HCl buffer pH 8.0 with 1 mM dithiothreitol (DTT), centrifuged for 1 h; the supernatant is referred to as "E^{10x}." Kidney tissue was prepared as a 20% homogenate (1:19, w/v) in glass-distilled deionized water (DI), centrifuged for 45 min ("E^{20x}"). All centrifugation was at 40,000 g and 4°C.

Enzyme reaction mixtures: Enzyme rate studies were followed spectrophotometrically at 340 nm and 25°C using NADP or reduced NAD, either directly or in a coupled assay. Reaction media were adjusted for each enzyme and each tissue to provide for maximal activity in a 3.00-ml final volume as specified below. All reagents and solutions were prepared with glass-distilled DI water. Protein was determined by the biuret method. Because only one metal concentration and only one period of metal exposure were used in this study, statistical analysis was limited to the Student's t-test. The Scheffé contrast, applied to the glycogen data, corroborated (99%) t-test results.

Adductor muscle: PK (based on Bücher & Pfleiderer, 1955) - 1.75 ml 0.1 M triethanolamine (TEA) buffer pH 7.5, 0.25 ml 0.1 M $MgCl_2$, 0.50 ml 0.45 M KCl, 0.10 ml 4.5 mM reduced nicotinamide adenine dinucleotide (NADH), 0.10 ml lactate dehydrogenase (Sigma #L-2625, diluted 1:2), 0.10 ml E^{10x}, and to start the reactions, 0.20 ml of a 1:1 mixture of 6.9 mM adenosine diphosphate (ADP) and 24 mM phosphoenolpyruvate in buffer.

GDH (based on Schmidt, 1974) - 1.50 ml 0.10 M phosphate buffer pH 7.5, 0.10 ml 3 mM NADH, 1.00 ml 0.21 M ammonium acetate, 0.10 ml 40 mM ADP, 0.10 ml E^{10x}, and to start the reaction, 0.20 ml 0.10 M α-ketoglutarate (K salt).

MDH (based on Grimm & Doherty, 1961) - 2.70 ml 0.1 M glycine-1 mM K_2 EDTA buffer pH 9.0, 0.10 ml 4.5 mM NADH, 0.10 ml 30 mM cis-oxaloacetate, and to start the reaction, 0.10 ml E^{100x} (E^{10x} diluted 1:9, v/v, with buffer).

Kidney: G6PDH - 2.30 ml 0.1 M TEA buffer pH 8.5, 0.10 ml 9.5 mM NAD phosphate (NADP), 0.30 ml 0.1 M $MgCl_2$, 0.10 ml E^{20x}, and to start the reaction, 0.20 ml 15 mM Na glucose-6-phosphate.

IDH - 0.90 ml 0.1 M Tris-HCl buffer pH 8.0, 0.10 ml 9.5 mM NADP, 0.10 ml 0.10 M $MgCl_2$ in 0.15 M NaCl, 1.50 ml 10 mM DL-isocitrate in 160 mM TEA buffer (Sigma #176-1), and to start the reaction, 0.40 ml E^{20x}.

MDH - same as for muscle, with 0.10 E^{20x} to start the reaction.

Muscle glycogen: The glycogen method was based primarily upon methods provided by the Sigma Chemical Company (St. Louis, MO 63178, USA), adapted for use with the scallop muscle preparations. It is described here in detail to assist those who may wish to use or adapt it further.

Duplicate tubes for each sample were prepared containing 0.50 ml E^{10x} muscle supernatant and 0.50 ml DI H_2O. To one tube of each pair was added 0.30 ml 50% trichloroacetic acid (1:1, v/v, 100% TCA:H_2O), the whole vortex-mixed, and 1.00 ml commercial amyloglycosidase (Sigma #A-7255, AGS: 10.0 mg in 100 ml 50 mM acetate buffer pH 4.5) added and the whole again mixed, capped with parafilm and set aside. This first group represents free simple sugars (FSS). A standard glucose curve (0-100 μg solubilized in the muscle extractant) was set up for each group of assays and treated similarly.

To the second tube of each pair, 1.0 ml AGS solution was added, parafilm-capped, mixed gently, and incubated with moderate shaking in a 55°C water bath for 2 h, after which time 0.30 ml 50% TCA was added and the whole vortex-mixed. This second group represents the sum of FSS plus the glucose liberated by glycogen hydrolysis.

All tubes were neutralized by careful addition of solid $NaHCO_3$ and centrifuged for 10 min at 4,200 g, room temperature (RT). Aliquots of the supernatants were used for a standard oxidase/peroxidase (PGO) test (Sigma Tech. Bull. 510). The contents of one capsule of commercial PGO enzymes (Sigma #510-6) were dissolved in 100 ml DI H_2O with gentle mixing; this should be prepared fresh each day of assay. To this solution was added 1.60 ml chromogen (50.0 mg o-dianisidine dihydrochloride in 20 ml DI H_2O), a solution that is stable for about 3 months under refrigeration, unless turbidity or color develops. For the colorimetric assay, 0.50 ml supernatant was added to 5.00 ml PGO-chromogen reagent in a colorimeter tube, allowed to stand for 45 min at RT, and read at 440 nm. The developed color is stable for 30 min. The differences between the hydrolyzed group and the FSS group, both expressed as μg glucose, represents muscle glycogen.

Chemical analysis

Sea water in the diluter systems was analyzed weekly for copper and cadmium by a modification of a procedure obtained from the Graduate School of Oceanography, University of Rhode Island: 0.40 ml 10% nitric acid (1:9, v/v, quartz-distilled HNO_3:DI H_2O) was added to 200 ml sea water, followed by 2.0 ml 200-mg/L cobalt ($CoCl_2$, AA std) and 2% ammonium pyrrolidine dithiocarbamate, both in DI H_2O. The sample was allowed to stand for ½ h, then put through a Biorad 0.2-μ Unipore membrane filter. The filter paper was placed in a plastic test tube to which 2.0 ml quartz-distilled HNO_3 and 8.0 ml DI H_2O were added, and then it was placed in a shaker bath at RT for 1 h. The solution was analyzed by atomic absorption employing a graphite furnace.

Tissues of the experimentally exposed scallops (kidney, digestive gland, gonad, gill) were analyzed for Cu and Cd according to the procedure described in Greig *et al.*, 1982.

RESULTS AND DISCUSSION

The three series of experimental exposures of *P. magellanicus* to the two metals are considered below chronologically, with emphasis on the winter study.

SPRING 1982

During the Spring 1982 exposure of scallops to copper (20 μg/L, 12 wk) or to cadmium (10 μg/L, 15 wk), all the animals remained in apparently good health for the duration of the metal exposure. There were no mortalities in either group of scallops, in spite of temperatures that rose (>18°C) into the range of incipient stress (Dickie, 1958; Dickie & Medcof, 1963). Figure 1 shows the gradual increase in cadmium concentrations in the kidney of the Cd-exposed scallops from that exposure study, with a concomitant small increase in copper as well. In the Cu-exposed scallops, copper accumulated in the kidney through 6 wk (Fig. 2), but at 9 wk half of the four kidney pools (2 animals per pool) were still accumulating copper and the other half had apparently eliminated most of the copper taken up by that organ. The 5700 datum at 9 wk probably was the result of a pool of one cleared kidney and one non-cleared kidney. At 12 wk, all three remaining kidney pools showed very low levels of remaining copper. If it is assumed that most of the copper (and cadmium) taken up by the kidney was associated with mineralized phosphate granules, then excretion of those granules would account both for the loss of copper and for the low cadmium concentration (inset, Fig. 2).

WINTER 1983

A considerably different picture was seen in the Winter 1983 study, although scallops were used from the same population as in Spring 1982. Of the 92 Cu-exposed scallops, 20 died during the exposure period. As a consequence, the metal exposures were terminated at 7 wk rather than the originally planned 9, and the longest depuration period was 2 wk rather than 4. During the depuration, 7 more of the Cu-exposed scallops died. All Cd-exposed scallops survived in apparent good health.

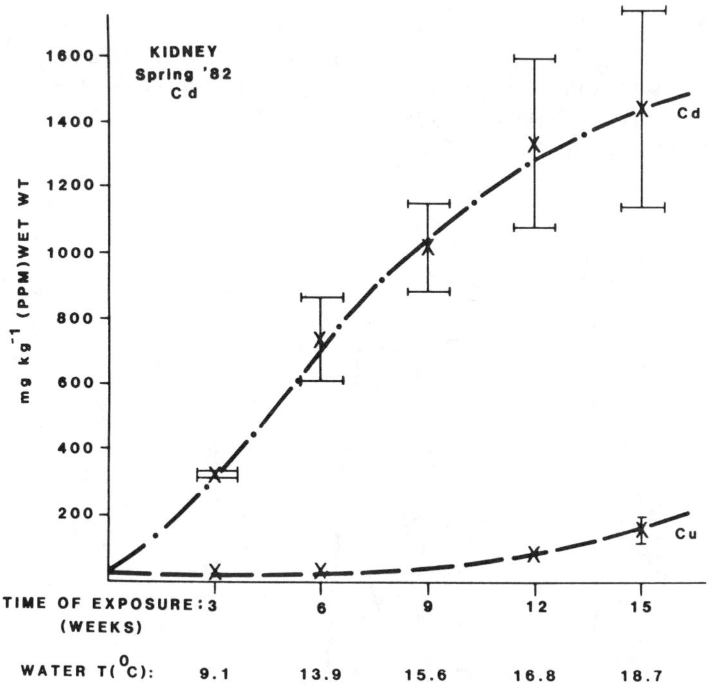

Fig. 1. Concentrations of cadmium and copper in kidney tissue of *P. magellanicus* exposed to 10 μg/L Cd, Spring 1982. Crosses indicate arithmetic mean and ticked vertical lines, standard error. Water temperatures are for the date of animal take-down.

Effects and uptake of kidney

Normal bivalve kidneys are brown, the intensity of the coloration depending on the concentration of brown mineralized concretions within the cells, as has been reported in *Mytilus edulis* (Pirie & George, 1979), *Pecten maximus* (George *et al*., 1980), and *Argopecten irradians* (Carmichael & Fowler, 1981). In this study, kidneys of both control and Cd-exposed scallops were dark brown (Fig. 3a), whereas those of the Cu-exposed scallops were almost transparent (Fig. 3b) or, at most, pale lemon-colored. Carmichael and Fowler (1981) observed a similar phenomenon in the kidneys of *A. irradians* that had been given a massive overload of cadmium; microscopy confirmed that almost all epithelial cells had extruded their concretions. If the brown color of normal *Placopecten* kidneys is similarly attributable to such mineralized granules, as seems likely, then the pale kidneys of the Cu-exposed scallops give evidence that the mineralized concretions had been excreted. Trace-metal analysis (Fig. 4) supports this inference: copper concentrations in the kidney of Cu-exposed animals were no different from those in controls, and cadmium concentrations were only a fraction of those in controls ($p<0.001$).

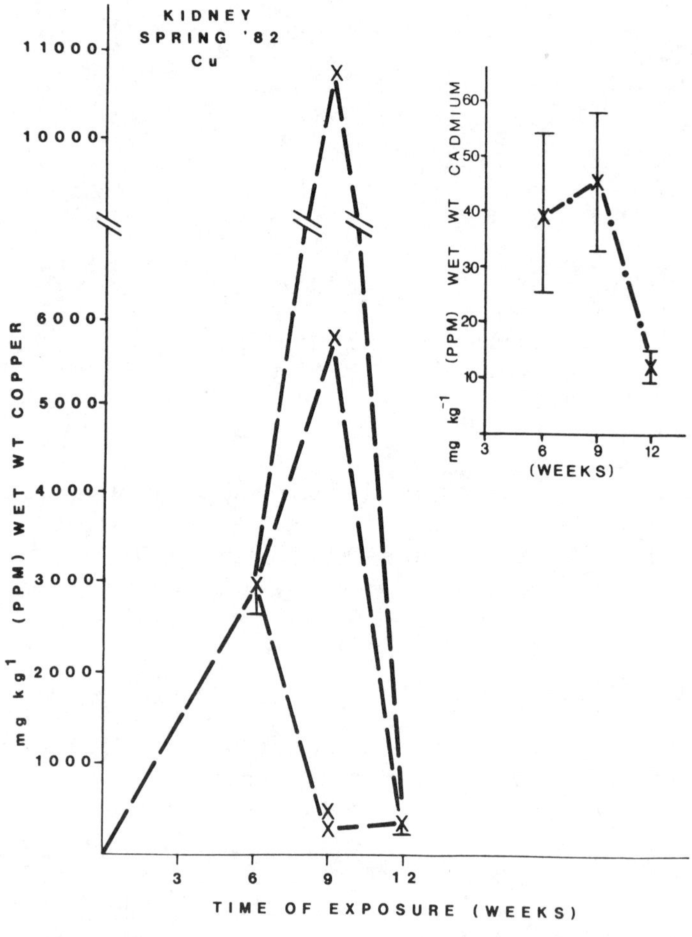

Fig. 2. Concentrations of copper and cadmium in kidney tissue of P. magellanicus exposed to 20 μg/L Cu, Spring 1982. Crosses at 6 and 12 wk indicate arithmetic mean and ticked vertical lines below crosses, standard error. At 9 wk, data are shown individually because of the large disparity in the Cu-exposed scallops at that time (see text).

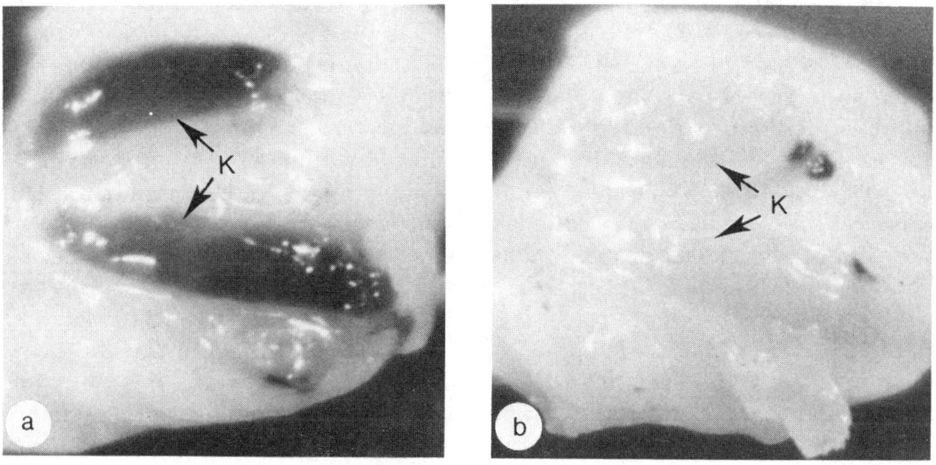

Fig. 3. Dissected view of kidneys of *P. magellanicus* exposed to cadmium (a) or to copper (b) for 7 wk + 2-wk depuration, Winter 1983. Exposure concentration of both metals was 20 μg/L. Viscera, gonad, and gills were removed for clarity, leaving the kidneys attached to the adductor muscle. Control scallop kidneys look the same as those of the Cd-exposed scallop in (a). A small portion of the digestive gland proximal to the kidney was left attached as a visual aid in locating the kidney in the Cu-exposed scallop (b).

After a 2-wk depuration, there was no reduction of cadmium levels in the kidney of Cd-exposed scallops (Table 1). In studying the effects of exposing *Placopecten* to drilling muds, Liss (1979) found that the kidneys continued to accumulate chromium even during a 2-wk period of clearing. He attributed this phenomenon to a clearing of metal from other tissues by the kidney. This explanation is very likely the case with the cadmium-exposed scallops in this study, where the kidneys continued to take up cadmium during the 2-wk period in clean water (Table 1). The cadmium excretion rate has been reported to be 18 times slower than uptake in another bivalve, *M. edulis* (George & Coombs, 1977).

After the 7-wk exposure of scallops to 20 μg/L Cd in Winter 1983 (water temperature 7.9°C), cadmium concentrations in the kidney averaged 653 mg/kg wet wt; yet after a 6-wk exposure to half that concentration in Spring 1982 (water temperature 13.9°C), kidney concentrations of that metal averaged 730 mg/kg wet wt, or twice the bioconcentration factor. This observation supports that of Coombs (1979), who noted in his review that an increase in temperature usually increases Cd uptake, except in *Mytilus* species; no tissue was specified. Bryan (1973), examining seasonal variations of metal concentrations in the kidney and digestive gland of two other scallop species (*Pecten maximus* and *Chlamys opercularis*), found that metal concentrations tended to fall in the spring; he associated the higher metal concentrations of late fall and winter with the low phytoplankton productivity at that time of year. Similarly, Frazier (1975) observed a 50% reduction in Cd and Cu body burdens in *Crassostrea virginica* in the summer months. Carmichael *et al*. (1979) have suggested that the kidney concretions normally may alternate between accumulation and resorption, a possibility that could explain apparently contradictory findings among bivalve species. A pattern for *Placopecten* may be discovered when a study of seasonal metal levels in the kidney and digestive gland of that animal has been completed for a single population (V. Zdanowicz, pers. comm.).

Of the enzyme activities examined in kidney tissue, only IDH provided a clear indication of metabolic dysfunction. This enzyme of the citric acid cycle is a mitochondrial marker that generates reducing equivalents necessary to drive reactions ultimately for ATP synthesis. Its significantly lower activity in the Cu-exposed scallops (Table 2) confirms an earlier observation in the Spring 1982 study. In that preliminary work, kidney IDH in Cu-exposed scallops was significantly lower than in controls for all three exposure periods: 6, 9, and 12 wk ($p<0.02$, <0.05, <0.001, respectively). In Winter 1983, kidney MDH activity in Cu-exposed scallops was also lower than controls, although only at $p<0.05$. It is probably that the MDH drop was due to lower activity of the mitochondrial isomer, another citric acid cycle enzyme; with the low IDH, it suggests an impaired mitochondrial functioning.

The increased activity of kidney IDH in the Cd-exposed scallops that had cleared for 2 wk (Table 3), as compared with activity before clearance and shortly after, might reasonably be interpreted as indicating a reversible, lesser stress than that seen in the Cu-exposed scallops. A clear demonstration of this kind of response may be seen in Stebbing's work (1981a). In observing the effects of metal challenge on a hydroid colony (*Campanularia flexuosa*), he found that when the growth-rate mechanism had equilibrated to an inhibitory load, removal of that load resulted in a transient increase in growth rate.

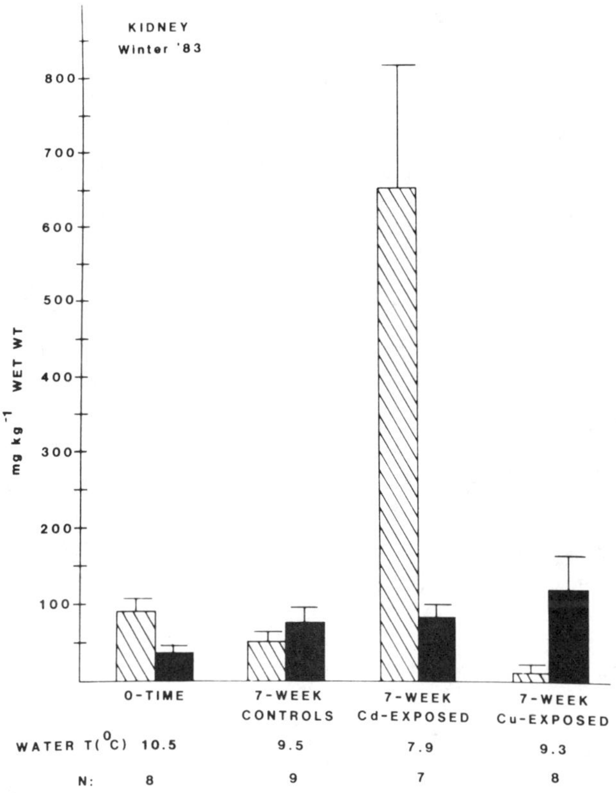

Fig. 4. Metal concentrations (arithmetic means) in the kidney of P. magellanicus exposed to 20 µg/L cadmium or copper for 7 wk, Winter 1983. Hatched bars represent cadmium and solid bars, copper. Vertical ticked lines indicate standard error. The same conventions are followed in Figures 4, 5, and 6.

The kidney concretions of Mytilus (George & Coombs, 1977) and of Argopecten irradians (Carmichael & Fowler, 1981) are a very efficient mechanism for sequestering cadmium; the same is apparently true for Placopecten. The analogous mechanism(s) for sequestering overloads of copper in this scallop are far less successful in maintaining homeostasis.

Table 1. Cadmium concentrations (mg/kg wet wt) in kidneys of cadmium-exposed (20 μg/L) *P. magellanicus*, Winter 1983.

	N	Controls	N	Cd-exposed	N	Cu-exposed
7-wk exposure:						
	8	55.2* (5.3)	5	653† (166)	6	11.0† (7.2)
7-wk exposure + 1-wk clearance:						
	8	56.7 (12.7)	6	641† (81)	6	8.2†† (2.9)
7-wk exposure + 2-wk clearance:						
	7	86.5 (8.1)	5	837† (42)	5	6.7† (2.8)

* Data are expressed as arithmetic mean ± (SE)
† Highly significant difference from controls ($p<0.001$)
†† Significant different from controls ($p<0.02$)

Table 2. IDH activity in kidneys of metal-exposed (20 μg/L) *P. magellancus*, Winter 1983

	N	Controls	N	Cd-exposed	N	Cu-exposed
7-wk exposure:						
	8	32.8* (2.8)	7	28.7 (4.1)	6	8.6†† (1.8)
7-wk exposure + 1-wk clearance:						
	6	36.7 (2.4)	6	25.9†† (1.3)	5	22.1 (7.5)
7-wk exposure + 2-wk clearance:						
	7	41.7 (3.8)	6	56.0 (3.8)	5	12.0† (1.3)

* μmoles substrate turnover/min/mg biuret protein: arithmetic mean ± (SE)
† Highly significant difference from controls ($p<0.001$).
†† Significant difference from controls ($p<0.005$).

Metal uptake in digestive gland and gill

Other organs also accumulated significant amounts of the metals to which the scallops were exposed. Neither digestive gland (Fig. 5) nor gill (Fig. 6), however, concentrated cadmium to the same degree as the kidney (190, 27, and 650 mg/kg wet wt, respectively). Assuming that the metal-saturated concretions had been evacuated from the kidneys of the Cu-exposed scallops (Figs. 3b & 4; Table 1), a comparison of metal concentrations in that organ with those in other tissues would not be valid. Concentrations of copper in the digestive gland of the Cu-exposed scallops, however, are seven times higher than those in the gill (140 and 19 mg/kg wet wt, respectively). Even in the cleared kidney, copper concentrations averaged higher (135 mg/kg) than in the gill. In this respect, *Placopecten* differs from at least one other bivalve species: Roesijadi (1980) found the gill to be the primary organ for copper concentration in the clam *Protothaca staminea*.

Table 3. Metal concentrations in gonads of metal-exposed (20 μg/L) *P. magellanicus*, before and after depuration periods, Winter 1983.

	N	Controls	N	Cd-exposed	N	Cu-exposed
Copper concentrations						
7-wk exposure:	15	7.3* (2.1)	16	2.7†^{95} (0.4)	9	23.0†^{98} (1.7)
7-wk exposure + 1-wk clearance:	10	3.5 (0.6)	8	3.9 (0.7)	16	21.9† (4.1)
7-wk exposure + 2-wk clearance:	8	4.6 (0.7)	9	2.8†^{95} (0.3)	7	40.5† (4.5)
Cadmium concentrations						
7-wk exposure:	15	0.9* (0.1)	16	5.6†† (0.9)	12	1.8 (0.4)
7-wk exposure + 1-wk clearance:	10	10 (0.1)	8	7.2† (1.0)	9	1.5 (0.9)
7-wk exposure + 2-wk clearance:	8	1.1 (0.5)	9	7.8† (1.1)	7	32.5 (1.3)

* Arithmetic mean ± (SE)
† Highly significant difference from controls ($p<0.001$)
†† Significant difference from controls ($p<0.005$)
†^{98} Significant difference from controls ($p<0.02$)
†^{95} Significant difference from controls ($p<0.05$)

Effects and uptake in gonad

The gonad of scallops exposed to copper or cadmium accumulated significantly higher concentrations of those metals than did the controls (Fig. 7). In the Cu-exposed group, gonad copper concentrations were six times higher than the controls (24 and 4 mg/kg wet wt, respectively). In the Cd-exposed group, gonad concentrations were seven times higher than controls (7 and 1 mg/kg wet wt). There was no clearing of either metal from the gonad after 2 wk in uncontaminated sea water (Table 3).

Striking differences in gonadal condition were observed in the metal-exposed scallops, especially after the 2-wk clearance period, on January 28 (Fig. 8). In a typical control animal (Fig. 8a), the gonad had begun to develop, becoming opaque as the gametes formed. In the Cu-exposed scallops (Fig. 8b), the gonad was flaccid and watery. (Note the transparent kidney, also typical of the Cu-exposed scallops.) By contrast, the gonad was full and ripe in a Cd-exposed scallop (Fig. 8c), far in advance of the usual spawning season for *Placopecten*, which ranges from mid-July through October, depending on population, site, temperature, and nutrient availability (*e.g.*, Dickie, 1955; Posgay & Norman, 1958; Naidu, 1970; Culliney, 1974; MacKenzie *et al*., 1978). A close-up (Fig. 8d) of the same Cd-exposed scallop, a female, shows the marked channeling in the mature gonad that both precedes spawning, as the oviducts form, and accompanies it (Posgay & Norman, 1958; Naidu, 1970). Not all Cd-exposed scallops had so ripe a gonad as this specimen, but the cadmium group as a whole appeared to be maturing (30%) more rapidly than controls (20%), as determined by the number of discernibly differentiated gonads (Table 4). Naidu (1970) has observed that spawning in *Placopecten* can be initiated by prolonged exposure to physical shocks. Chemical challenge may also provoke the same phenomenon; hormetic effects of exposure to sublethal levels of cadmium have been reported for a colonial hybrid, *C. flexuosa* (Stebbing, 1981b) and for brown algae, *Pelvetia caniculata* and *Ascophyllum nodosum* (Strömgren, 1980). Such a response should be considered not as an increase in growth rate (the parameter used), but as a diversion of metabolic energy (Stebbing, 1980) from somatic growth to reproductive effort.

The copper group, on the other hand, seemed unable to divert energy effectively to gamete production (5% differentiation, Table 4; Fig. 8b); gonad weights supported this observation (Table 5). In the freshwater fish species *Pimphelas promelas*, copper prevented spawning at only 33 μg/L (Mount, 1968), although the fish showed no other sign of stress. In marine animals, exposure to 50 μg/L copper has been reported to suppress both the growth of oöcytes and vitellogenesis in larger oöcytes in *M. edulis* (Myint & Tyler, 1982). Similarly, exposure to silver (10 μg/L) reduced the number of larval releases in the marine snail *Crepidula fornicata* (Nelson *et al*., 1983). A significant observation in the latter study was that the silver exposure induced an uptake of copper higher than that of control snails. The same phenomenon was observed in silver-exposed (10 μg/L) M. edulis, in which copper was accumulated to levels not only higher than controls, but also higher than in M. edulis concurrently exposed to 10 μg/L copper (Calabrese et al., 1984).

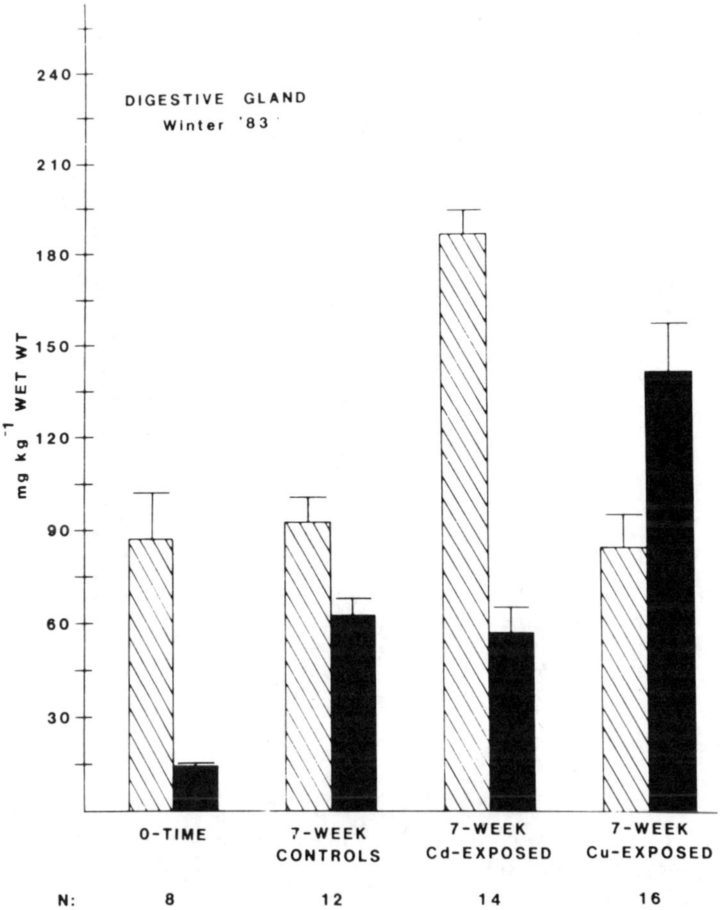

Fig. 5. Metal concentrations in the digestive gland of P. magellanicus exposed to 20 μg/L cadmium or copper for 7 wk, Winter 1983.

In the Winter 1983 study reported here, scallops having discernibly differentiated gonads were too few in number for statistical treatment. The clear suggestion, however, is that sublethal copper inhibited gamete maturation and that cadmium stimulated it.

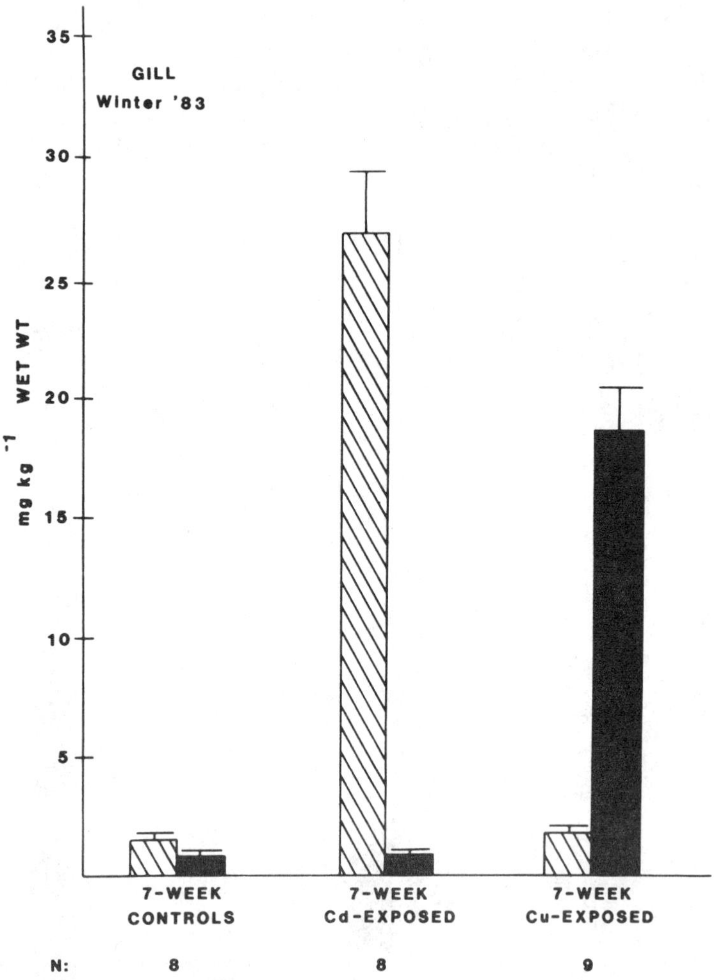

Fig. 6. Metal concentrations in the gill of P. magellanicus exposed to 20 µg/L cadmium or copper for 7 wk, Winter 1983.

Effects in adductor muscle

Of the two metal-exposed groups of scallops, only the copper group provided clear evidence of a weakened metabolism in the phasic adductor muscle. All four biochemical parameters measured (GDH, PK, MDH, glycogen) were significantly lower in that group than in control animals (Table 6), although activities of the three enzymes returned to normal during the two periods of clearing. The only abnormality in the measured enzyme activities in the cadmium group of scallops was PK activity, which was significantly lower than controls ($p<0.02$) after the 7-wk exposure. Although PK did not differ from that in controls after the first week of clearing, it fell significantly below controls again after the second week of clearing ($p<0.01$).

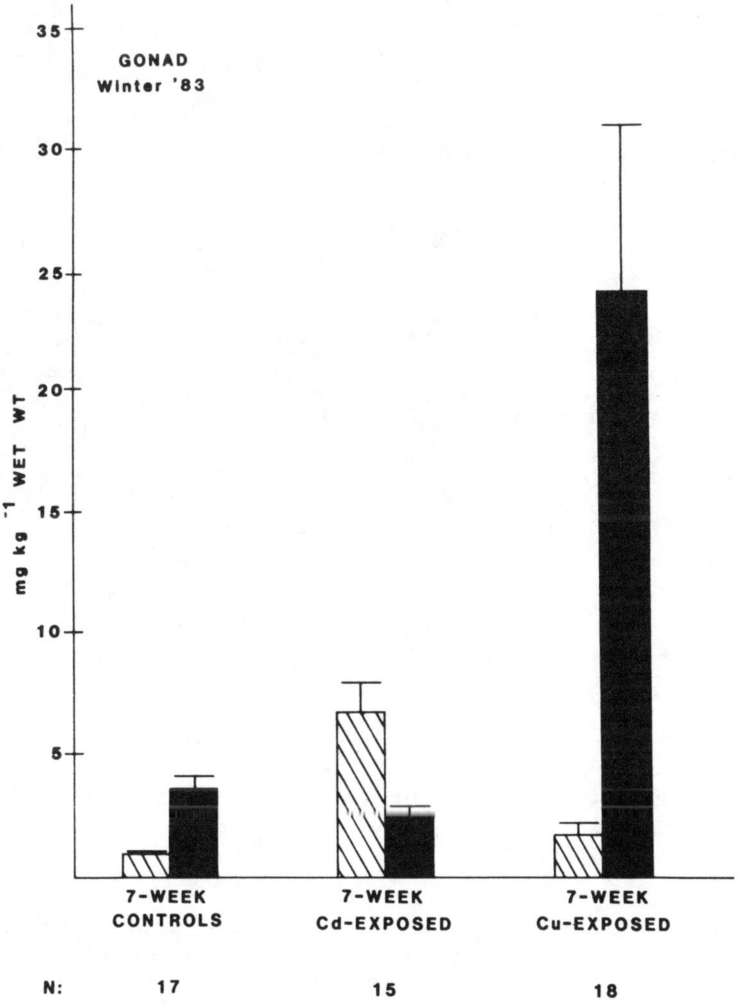

Fig. 7. Metal concentrations in the gonad of *P. magellanicus* exposed to 20 μg/L cadmium or copper for 7 wk, Winter 1983.

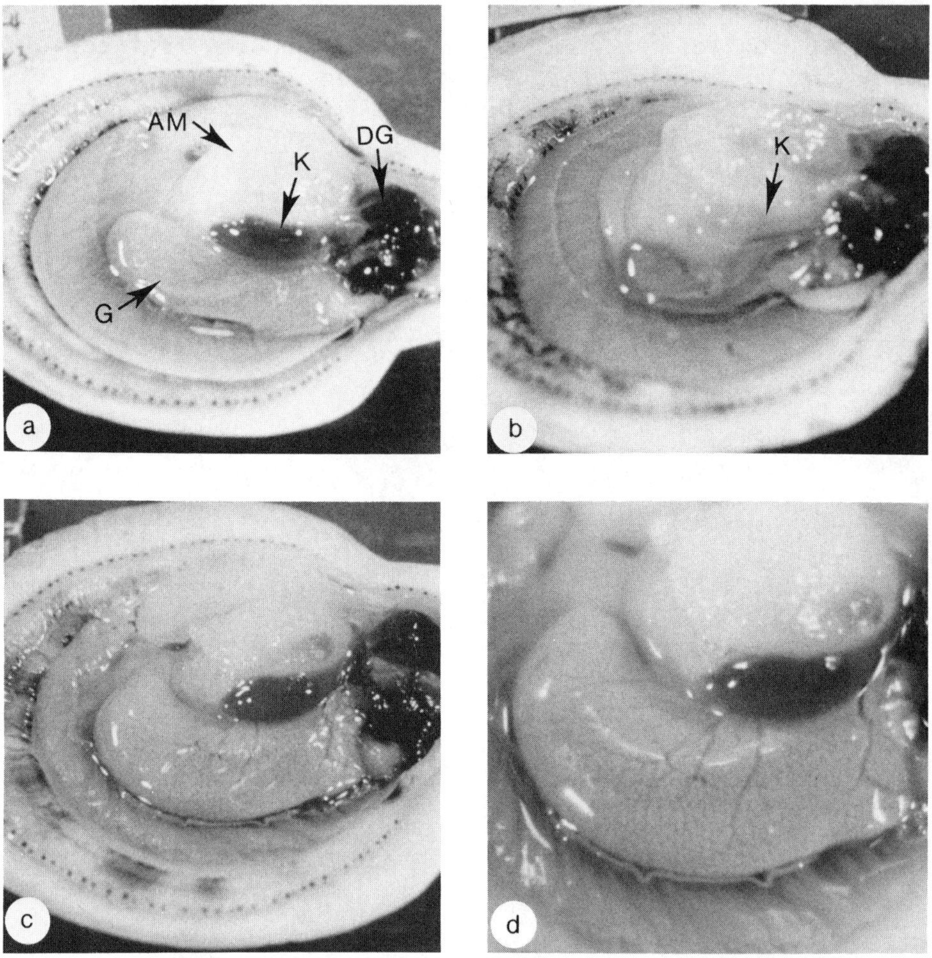

Fig. 8. Partially dissected view of P. magellanicus exposed to 20 μg/L cadmium or copper, Winter 1983; (a) = control, (b) = Cu-exposed, (c) = Cd-exposed, and (d) = close-up of (c). Arrows indicate gonad (G), adductor muscle (AM), kidney (K), and digestive gland (DG).

Muscle glycogen, however, was the single most useful parameter in that tissue for revealing the opposite nature of the effects of these two metals. Highly significant differences from controls were found in all three groups of Cu-exposed scallops, and in the two groups of clearing Cd-exposed scallops (Fig. 9). Muscle glycogen levels in the two clearing cadmium groups were also significantly lower than the earlier 7-wk Cd-exposure group ($p<0.005$). Free simple sugars remained at a low level (120-150 μg glucose/g wet wt) throughout the study, unchanged in both controls and metal-exposed scallops. Because the Cd-exposed scallops remained healthy and fed well throughout the periods of exposure and clearance, and because of the apparent hormetic effect of cadmium on gamete maturation, we construe the loss of muscle glycogen in the two clearing cadmium groups as an early transfer of energy reserves to the gonad.

Table 4. Comparative rates of gonad differentiation in metal-exposed (20 μg/L) *P. magellanicus*, Winter 1983.

	Controls				Metal-exposed		
	Total N scallops	Scallops with discernibly differentiated gonads			Total N scallops	Scallops with discernibly differentiated gonads	
		N	As % of total N			N	As % of total N
7-wk exposure (to January 1983):							
Cadmium	13	2	15.4	19.3	23	7	30.4
Copper	13	3	23.1		20	1	5.0
7-wk exposure + 1-wk clearance:							
Cadmium	10	2	20.0	20.0	20	6	30.0
Copper	10	2	20.0		16	1	6.3
7-wk exposure + 2-wk clearance:							
Cadmium	13	3	32.1	24.0	21	6	28.6
Copper	12	3	25.0		16	3	8.2

Table 5. Gonad weights (g) for sexually differentiated *P. magellanicus* in total period of metal exposures (7 wk + 1- and 2-wk depuration), Winter 1983.

	N	Controls	N	Cd-exposed	N	Cu-exposed
Male	4	1.99* (0.28)	8	1.93 (0.14)	2	1.34 (0.31)
Female	11	1.46 (0.10)	10	1.69 (0.08)	3	0.97 (0.25)

*Arithmetic mean ± (SE)

Table 6. Muscle GDH, MDH, PK, and glycogen in *P. magellanicus* after 7-wk exposure to cadmium or copper (20 μg/L), Winter 1983.

	Controls N = 16		Cd-exposed N = 14		Cu-exposed N = 12	
GDH*	58.4	(5.1)	50.0	(3.6)	37.1††	(1.9)
MDH	2817	(202)	2691	(184)	2102†98	(151)
PK	826	(55)	674†98	(40)	536††	(34)
Glycogen**	395	(27)	305	(34)	191†	(41)

* Enzyme units are μmoles of substrate turnover/min/mg biuret protein; all data are expressed as arithmetic mean ± (SE).
** Glycogen units are μg glucose liberated by hydrolysis under standard assay conditions (see Methods).
† Highly significant difference from controls ($p<0.001$).
†† Significant difference from controls ($p<0.005$).
†98 Significant difference from controls ($p<0.02$).

In the Cu-exposed groups, the reverse was seen (Fig. 9): muscle glycogen levels in both the 7-wk exposure and the 1-wk clearance groups were much lower than in controls ($p<0.001$). Either the animals were not feeding or assimilating food well, or what nutrients were assimilated went to metabolic needs other than building up reserves for gamete production. Muscle glycogen in the 2-wk clearance group, however, was at higher levels than either of the two earlier copper groups, although still lower than the controls. It is possible that the 2-wk clearance group of Cu-exposed scallops had begun to feed or assimilate food more efficiently than the two earlier copper groups, but it is more probable that the higher muscle glycogen levels were the result of gamete resorption. We favor the second explanation because of the flaccid, watery appearance of gonads from that group (Fig. 8b). Gamete resorption in response to stress has also been reported for *A. irradians* (Sastry, 1966) and for *M. edulis* (Gabbott & Bayne, 1973).

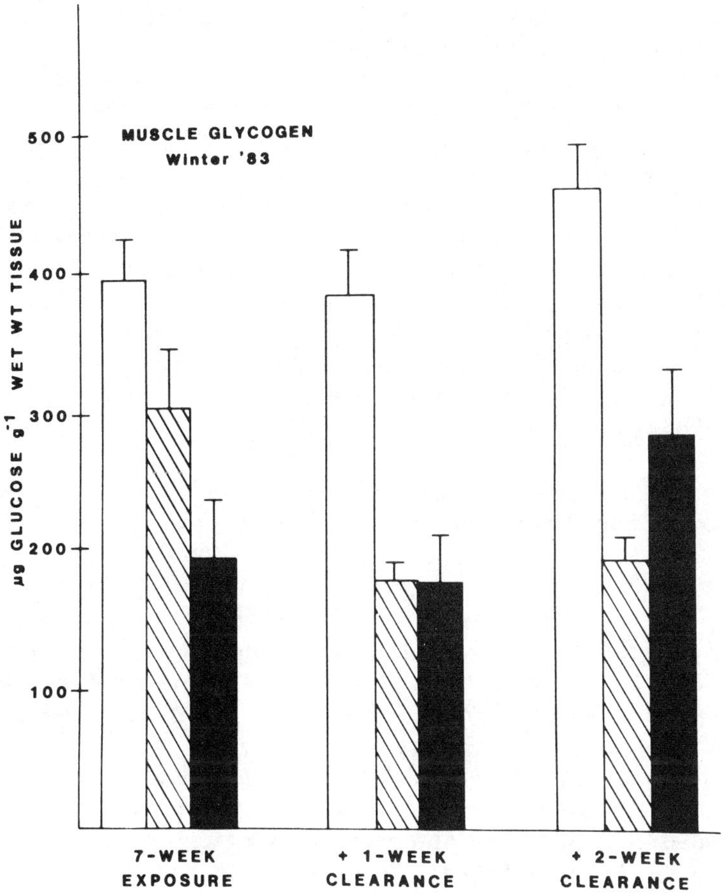

Fig. 9. Glycogen concentrations (arithmetic means) in adductor muscle of *P. magellanicus* exposed to 20 µg/L cadmium (hatched bars) or copper (solid bars), Winter 1983. Open bars are controls. Ticked vertical lines indicate standard error.

SPRING 1983

On the basis of the Spring 1982 and Winter 1983 studies, there was a marked difference in *Placopecten*'s susceptibility to the toxic effects of copper at different stages of the reproductive cycle. We therefore repeated the winter study's experimental design in the spring of 1983, using scallops with fully differentiated gonads. These scallops were collected from a site SE of Montauk Point, Long Island, as we were unable to obtain animals from the Block Island population used in the two prior studies.

During the exposure and clearance periods, however, it became obvious that these animals were under some pre-existing stress. First, there were mortalities, ten times as many in the Cu-exposure group (24.3%) as in either the Cd-exposed (2.7%) or the control group (2.5%); this was not the case with the Spring 1982 exposure with the Block Island scallops, in which no losses were sustained. Temperature and salinity were within the same range for both spring studies. Second, by the end of the 7-wk exposure period, three Cd-exposed scallops had already spawned, and during the subsequent clearance periods most of the Cd-exposed and several of the controls spawned. Finally, histological examination of the pericardial glands of control scallops showed abnormal cellular morphology (M.P. Morse, pers. comm.).

As a consequence, the chemical and biochemical results could not properly be compared to those of Winter 1983. The effects of the two metals on gonad development, however, provided some corroboration for the earlier study. Because the scallop gonads were sexually differentiated at the beginning of the study, sufficient numbers of both sexes were available to determine statistical differences between the metal-exposed scallops and controls. Gonad weights of the Cu-exposed scallops were significantly lower than controls (Table 7), confirming the impression formed in the Winter 1983 study. More importantly, histological analysis of fixed gonad sections by the point-counting technique of Lowe *et al*. (1982) showed that Cu-exposed scallops had significantly lower counts ($p<0.001$) of mature and ripening gametes than did control or Cd-exposed scallops (Table 8). The Cd-exposed group was not statistically different from controls. (Data for the three spawning scallops in the Cd-exposed group were not included.)

Nutritive stress in *M. edulis* with already ripe gonads has been shown to provoke gamete recession, along with a loss of protein from mantle tissue (Gabbott & Bayne, 1973). Considering the evidence for gamete resorption in Cu-exposed scallops in the Winter 1983 study, the observations of Spring 1983 indicate that *P. magellanicus*, when exposed to copper, also can resorb gametes from fully differentiated gonads.

In summary, overloads of the essential trace-element copper in *Placopecten magellanicus* produce strongly inhibitory effects in the reproductive system and metabolic dysfunction in the kidney. Overloads of the toxic metal

Table 7. Gonad weights (g) for *P. magellanicus* in metal-exposure study (7-wk + 1- and 2-wk depuration), Montauk population, Spring 1983.

	N	Controls		N	Cd-exposed		N	Cu-exposed	
7-wk exposure (to 3 May 1983):									
Male	11	4.34*	(0.43)	10	4.48	(0.49)	13	3.40	(0.40)
Female	6	6.05	(0.86)	10	6.12	(0.95)	6	3.33†95	(0.37)
7-wk exposure + 1-wk clearance:									
Male	10	3.91	(0.66)	5	2.95	(0.64)	10	1.86††	(0.17)
Female	6	6.22	(1.05)	9	3.77	(0.37)	5	1.83††	(0.60)
7-wk exposure + 2-wk clearance:									
Male	7	5.98	(1.47)	6	3.32	(0.38)	6	2.94†	(0.55)
Female	7	6.14	(1.38)	3	11.91	(1.73)	1	2.76	

* Arithmetic means ± (SE)
† Highly significant difference from controls ($p<0.001$)
†† Significant difference from controls ($p<0.01$)
†95 Significant difference from controls ($p<0.05$)

Table 8. Histological assessment* of gamete development in metal-exposed (20 μg/L) *P. magellanicus*, Montauk population, Spring 1983.

	Controls N = 10		Cd-exposed N = 9		Cu-exposed N = 12	
% ripe gametes:	80.2**	(2.6)	87.1	(3.8)	11.3†	(6.6)
% ripe + developing gametes:	88.4	(4.7)	89.2	(2.7)	35.3†	(5.1)

* Data from R.J. Thompson and B.A. MacDonald, Marine Science Research Laboratory, Memorial University, St. John's NFLD
** Arithmetic mean ± (SE)
† Highly significant difference from controls ($p<0.001$)

cadmium are apparently successfully sequestered in the kidney, and produce little observable effect other than to stimulate a precocious gamete maturation.

ACKNOWLEDGMENTS

We thank Dr. Ravenna Ukeles and Gary Wikfors (NMFS, Milford, CT) for providing algal cultures to the scallops in all three experimental exposures. We also thank Caroline A. and Paul P. Yevich (USEPA, Narragansett, RI) for participating in the Winter 1983 study. We are especially grateful to Dr. Ray J. Thompson and Bruce A. MacDonald (Marine Science Research Laboratory, Memorial University, St. John's NFLD) for their examination of scallop gonad sections in the Spring 1983 study, and to Dr. M. Patricia Morse (Marine Science Institute, Northeastern University, Nahant, MA) for her evaluation of scallop percardial glands from the same study.

Note: Use of trade names is to facilitate description, and does not constitute endorsement by the National Marine Fisheries Service.

LITERATURE CITED

Ansell, A.D. 1974. Seasonal changes in biochemical composition of the bivalve *Chlamys septemradiata* from the Clyde Sea area. Mar. Biol. 25: 85-99.

Bryan, G.W. 1973. The occurrence and seasonal variation of trace metals in the scallops *Pecten maximus* (L.) and *Chlamys opercularis* (L.). J. mar. biol. Ass. U.K. 53: 145-166.

Bücher, T. and G. Pfleiderer. 1955. Pyruvate kinase from muscle. pp. 435-440. *In*: Methods in Enzymology, Vol. I. S.P. Colowick and N.O. Kaplan (eds.). Academic Press, New York.

Calabrese, A., J.R. MacInnes, D.A. Nelson, R.A. Greig, and P.P. Yevich. 1984. Effects of long-term exposure to silver or copper on growth, bioaccumulation, and histopathology in the blue mussel *Mytilus edulis*. Mar. Environ. Res. 11: 253-274.

Carmichael, N.G. and B.A. Fowler. 1981. Cadmium accumulation and toxicity in the kidney of the bay scallop *Argopecten irradians*. Mar. Biol. 65: 35-43.

Carmichael, N.G., K.S. Squibb, D.W. Engel, and B.A. Fowler. 1980. Metals in the molluscan kidney: Uptake and subcellular distribution of ^{109}Cd, ^{54}Mn, and ^{65}Zn by the clam Mercenaria mercenaria. Comp. Biochem. Physiol. 65A: 203-206.

Carmichael, N.G., K.S. Squibb, and B.A. Fowler. 1979. Metals in the molluscan kidney: A comparison of two closely related bivalve species (Argopecten) using X-ray microanalysis and atomic absorption spectroscopy. J. Fish. Res. Board Can. 39: 1149-1155.

Coombs, T.L. 1979. Cadmium in aquatic organisms. pp. 93-139. In: The Chemistry, Biochemistry, and Biology of Cadmium. M. Webb (ed.). Elsevier/ North Holland Biomedical Press, Amsterdam.

Coombs, T.L., and S.G. George. 1978. Mechanisms of immobilization and detoxication of metals in marine organisms. pp. 179-187. In: Physiology and Behaviour of Marine Organisms. D.L. McLusky and A.J. Berry (eds.). Pergamon Press, Oxford.

Culliney, J.L. 1974. Larval development of the giant scallop Placopecten magellanicus (Gmelin). Biol. Bull. 147: 321-332.

Dickie, L.M. 1955. Fluctuations in the abundance of the giant scallop, Placopecten magellanicus (Gmelin), in the Digby area of the Bay of Fundy. J. Fish. Res. Board Can. 12(6): 797-857.

Dickie, L.M. 1958. Effects of high temperature on survival of the giant scallop. J. Fish. Res. Board Can. 15: 1189-1211.

Dickie, L.M. and J.C. Medcof. 1963. Causes of mass mortalities of scallops (Placopecten magellanicus) in the southwestern Gulf of St. Lawrence. J. Fish. Res. Board Can. 20: 451-482.

Doyle, L.J., N.J. Blake, C.C. Woo, and P. Yevich. 1978. Recent biogenic phosphorite: Concretions in mollusc kidneys. Science 199: 1431-1433.

Engel, D.W. and B.A. Fowler. 1979a. Factors influencing the accumulation and toxicity of cadmium to marine organisms. Environ. Health Perspectives 28: 81-88.

Engel, D.W. and B.A. Fowler. 1979b. Copper and cadmium-induced changes in the metabolism and structure of molluscan gill tissue. pp. 239-256. In: Marine Pollution: Functional Responses. W.B. Vernberg, A. Calabrese, F.P. Thurberg, and F.J. Vernberg (eds.). Academic Press, New York.

Fowler, B.A., N.G. Carmichael, and K.S. Squibb. 1981. Factors affecting trace-metal uptake and toxicity to estuarine organisms. II. Cellular mechanisms. pp.

145-163. *In*: Biological Monitoring of Marine Pollutants. F.J. Vernberg, A. Calabrese, F.P. Thurberg, and W.B. Vernberg (eds.). Academic Press, New York.

Frazier, J.M. 1975. The dynamics of metals in the American oyster, *Crassostrea virginica*. I. Seasonal effects. *Ches. Sci.* 16(3): 162-171.

Gabbott, P.A. 1975. Storage cycles in marine bivalve molluscs: A hypothesis concerning the relationship between glycogen metabolism and gametogenesis. pp. 191-211. *In*: The Biochemistry, Physiology, and Behaviour of Marine Organisms in Relation to Their Ecology. H. Barnes (ed.). Aberdeen University Press.

Gabbott, P.A. and B.L. Bayne. 1973. Biochemical effects of temperature and nutritive stress on *Mytilus edulis* L. J. mar. biol. Ass. U.K. 53: 269-286.

George, S.G. 1982. Subcellular accumulation and detoxication of metals in aquatic animals. pp. 3-52. *In*: Physiological Mechanisms of Marine Pollutant Toxicity. W.B. Vernberg, A. Calabrese, F.P. Thurberg, and F.J. Vernberg (eds.). Academic Press, New York.

George, S.G. and T.L. Coombs. 1977. The effects of chelating agents on the uptake and accumulation of cadmium by *Mytilus edulis*. Mar. Biol. 39: 261-268.

George, S.G., B.J.S. Pirie, and T.L. Coombs. 1980. Isolation and elemental analysis of metal-rich granules from the kidney of the scallop *Pecten maximus* (L.). J. Exp. Mar. Biol. Ecol. 42: 143-156.

Gould, E. 1983. Seasonal biochemical patterns for a single population of sea scallops, *Placopecten magellanicus*, and their use in interpreting field data. International Council for the Exploration of the Sea, C.M. 1983/E:57 (Marine Environmental Quality Committee).

Greig, R.A., T.K. Sawyer, E.J. Lewis, and M.E. Galasso. 1982. A study of metal concentrations in relation to gill color and pathology in the rock crab. Arch. Environ. Contam. Toxicol. 11: 539-549.

Grimm, F.C. and D.G. Doherty. 1961. Properties of the two forms of malic dehydrogenase from beef heart. J. Biol. Chem. 236: 1980-1985.

Liss, R.G. 1979. Uptake and depuration of metals by sea scallops (*Placopecten megallanicus*) exposed to drilling mud. M.S. Thesis. Massachusetts Institute of Technology.

Lowe, D.M., M.N. Moore, and B.L. Bayne. 1982. Aspects of gametogenesis in the marine mussel *Mytilus edulis* L. J. mar. biol. Ass. U.K. 62: 133-145.

MacKenzie, C.L., Jr., A.S. Merrill, and F.M. Serchuk. 1978. Sea scallop resources off the northeastern U.S. Coast, 1975. Mar. Fish. Rev. 40(2): 19-23.

Mason, A.Z. and J.A. Nott. 1981. The role of intracellular biomineralized granules in the regulation and detoxification of metals in gastropods with special reference to the marine prosobranch Littorina littorea. Aquat. Toxicol. 1: 239-256.

Mount, D.I. 1968. Chronic toxicity of copper to fathead minnow (Pimephales promelas Rafinesque). Water Res. 2: 215-223.

Mount, D.I. and W.A. Brungs. 1967. A simplified dosing apparatus for fish toxicology studies. Water Res. 1: 21-29.

Myint, U.M. and P.A. Tyler. 1982. Effects of temperature, nutritive and metal stressors on the reproductive biology of Mytilus edulis. Mar. Biol. 67: 209-223.

Naidu, K.S. 1970. Reproduction and breeding cycle of the giant scallop Placopecten magellanicus (Gmelin) in Port au Port Bay, Newfoundland. Can. J. Zool. 48: 1003-1012.

Nelson, D.A., A. Calabrese, R.A. Greig, P.P. Yevich, and S. Chang. 1983. Long-term silver effects on the marine gastropod Crepidula fornicata. Mar. Biol. 12: 155-165.

Pirie, B.J.S. and S.G. George. 1979. Ultrastructure of the excretory system and heart of Mytilus edulis. J. mar. biol. Ass. U.K. 59: 819-829.

Posgay, J.A. and K.D. Norman. 1958. An observation on the spawning of the sea scallop, Placopecten megallanicus (Gmelin), on Georges Bank. Limnol. Oceanogr. 3(4): 478.

Ridlington, J.W. and B.A. Fowler. 1979. Isolation and partial characterization of a cadmium-binding protein from the American oyster (Crassostrea virginica). Chem.-Biol. Interact. 25: 127-138.

Robinson, W.E., W.E. Wehling, M.P. Morse, and G.C. McLeod. 1981. Seasonal changes in soft-body component indices and energy reserves in the Atlantic sea scallop, Placopecten magellanicus. Fish. Bull. 79(3): 449-458.

Roesijadi, G. 1980. Influence of copper on the clam Protothaca staminea: Effects on gills and occurrence of copper-binding proteins. Biol. Bull. 158: 233-247.

Roesijadi, G. 1981. The significance of low molecular weight, metallothionein-like proteins in marine invertebrates: Current status. Mar. Environ. Res. 4: 167-179.

Sastry, A.N. 1966. Temperature effects in reproduction of the bay scallop, *Aequipecten irradians* Lamarck. Biol. Bull. 130: 118-134.

Schmidt, E. 1974. Glutamate dehydrogenase UV-assay. pp. 650-656. *In*: Methods of Enzymatic Analysis, Vol. 2, 2nd English Ed. H. Bergmeyer (ed.). Verlag Chemie, Weinheim.

Stebbing, A.R.D. 1980. Increase in gonozooid frequency as an adaptive response to stress in *Campanularia flexuosa*. pp. 27-32. *In*: Developmental and Cellular Biology of Coelenterates. P. Tardent and R. Tardent (eds.). Elsevier/North Holland Biomedical Press, Amsterdam.

Stebbing, A.R.D. 1981a. The kinetics of growth control in a colonial hydriod. J. Mar. Biol. Assoc. U.K. 61: 35-63.

Stebbing, A.R.D. 1981b. Hormesis - stimulation of colony growth in *Campanularia flexuosa* (hydrozoa) by copper, cadmium, and other toxicants. Aquat. Toxicol. 1: 227-238.

Strömgren, T. 1980. The effect of lead, cadmium, and mercury on the increase in length of five intertidal fucales. J. Exp. Mar. Biol. Ecol. 43: 107-119.

Taylor, A.C. and T.J. Venn. 1979. Seasonal variation in weight and biochemical composition of the tissues of the queen scallop, *Chlamys opercularis*, from the Clyde Sea area. J. Mar. Biol. Assoc. U.K. 59:605-621.

Thurberg, F.P., A. Calabrese, E. Gould, R.A. Greig, M.A. Dawson, and R.K. Tucker. 1977. Response of the lobster, *Homarus americanus*, to sublethal levels of cadmium and mercury. pp. 185-197. *In*: Physiological Responses of Marine Biota to Pollutants. F.J. Vernberg, A. Calabrese, F.P. Thurberg, and W.B. Vernberg (eds.). Academic Press, New York.

Walker, G. 1977. "Copper" granules in the barnacle *Balanus balanoides*. Mar. Biol. 39: 343-349.

HISTOPATHOLOGICAL EFFECTS OF CADMIUM AND COPPER ON THE SEA SCALLOP *PLACOPECTEN MAGELLANICUS*

Carolyn A. Yevich and Paul P. Yevich

SUMMARY

Sea scallops, *Placopecten magellanicus*, were exposed to 20 μg/L of cadmium and 20 μg/L copper for a period of 7 wk and depurated for 2 wk. Specimens were taken at the beginning of the experiment, at the end of 7-wk exposure and after 2 wk of depuration for histopathological examination.

The control animals were within normal seasonal limitation throughout the experiment. During the exposure period, both the copper- and cadmium-exposed animals had histopathological changes in the digestive diverticula, reproductive tract, tentacles, and kidneys. There was an improvement in the histological condition of these organs during the depuration period, with the exception of the kidneys of copper-exposed animals and the digestive diverticula of cadmium-exposed animals. The kidneys of copper-exposed scallops were very dilated with loss of folds. The epithelium showed extensive metaplastic changes consisting of basophilic cuboidal cells lacking the clear cytoplasm and concretions seen in the control scallops. Based on the above pathology and the mortality data, the kidney lesions could have caused the death of the animals by disruption of the excretory function.

INTRODUCTION

To determine the toxicity of copper and cadmium on sea scallops, *Placopecten magellanicus*, Gould *et al.* (this volume) conducted a multidisciplinary study (physiology, biochemistry, and chemistry) to examine the mechanisms and rates of uptake, detoxification, and clearance of these two heavy metals. In conjunction with the above studies, we examined these scallops histopathologically.

MATERIALS AND METHODS

Sea scallops 9.5-11 cm were collected from an area southeast of Block Island, RI. They were exposed in a flow-through proportional diluter system (Mount & Brungs, 1967) to 20 µg/L of copper and 20 µg/L cadmium in separate sea water tanks at the Milford Laboratory (Gould *et al.*, this volume). Representative samples from each treatment were sent to the Histopathology Unit of the Environmental Research Laboratory, Narragansett for examination (Table 1).

Table 1. Number of sea scallops submitted for histopathological examination.

Date	# of Scallops	Exposure Data
11/10/82	10	Pre-exposure Controls
1/12/83	10	Controls
(7-wk exposure)	9	20 µg/L cadmium
	7	20 µg/L copper
1/26/83	9	Controls
(2-wk depuration)	9	20 µg/L cadmium
	7	20 µg/L copper

In preparation for histopathological examination, the animals were shucked live into Helly's fixative, allowed to firm up for 20 min, and then cut saggitally to allow penetration of the tissues by the fixative. After approximately 20 h of fixation, the tissues were trimmed in such a way that each of the organ systems was present in at least two separate pieces, which resulted in an average of eight blocks per animal. These tissues were placed in cassettes, washed in running water for 24 h to remove excess fixative, and then stored in 70% ethanol until processed. The processing steps included dehydration in Technicon's S-29*, and

*Use of Technicon reagents and equipment and Paraplast does not imply endorsement by the Federal Government.

clearing in UC-670* on the Autotechnicon duo*, and infiltration by and embedding in Paraplast*. Sections were cut at 6 mu on an A.O. rotary microtome and stained with hemotoxylin and eosin. The above procedures are further described by Yevich and Barszcz (1981).

RESULTS

All the control animals were normal throughout the experiment. The lesions seen in the exposed animals are summarized in Table 2 and described below.

Kidneys

The kidney mass consists of folds that are lined with clear columnar epithelial cells (Fig. 1), which may contain a small number of yellow/brown or gray concretions. Occasionally there were accumulations of amebocytes and debris between the folds of the kidney epithelium.

Eighty-eight percent of the cadmium-exposed scallops had a scattered foci of hyperplasia of the clear columnar epithelial cells of the kidneys. There were light gray to brown concretions in the epithelial cells and accumulations of concretions and debris between the folds of the kidney epithelium.

All but one of the copper-exposed scallops at each sampling period had metaplastic lesions in the kidney epithelium, although the extent of the lesion varied among the animals. The least-affected kidneys had enlarged cells and some disrupted cell walls. The more severly affected kidneys were lined with a very basophilic low columnar to squamous epithelium (Fig. 2). There was also dilation of the kidney lumen resulting in a decrease in the number of folds. Very few concretions were observed.

Digestive diverticula

The digestive diverticulae consists of ducts that come from the stomach and end in blind tubules. Some tubules are lined with cuboidal to columnar digestive cells that are filled with yellow/green granules. Other tubules have cells containing vacuoles with large yellow/green accumulations of waste products (Fig. 3).

At 7 wk, about one half of the cadmium-exposed animals had some dilated tubules and ducts, and a decrease in the amount of granules in the digestive cells (Fig. 4). Eighty-

Table 2. Histopathological comparison of the cadmium- and copper-exposed sea scallops.

Organ and degree of involvement	Exposure Data and Percent Involved			
	7 wk		7 wk + 2 wk depuration	
	Cadmium 9 animals	Copper 7 animals	Cadmium 9 animals	Copper 7 animals
Kidney				
normal	11%	0	55%	14%
moderate lesions	88%	28%	44%	28%
severe lesions	0	71%	0	57%
Digestive diverticula				
normal	44%	28%	22%	42%
moderate lesions	55%	57%	77%	57%
severe lesions	0	28%	0	0
Reproductive tract				
normal	44%	42%	55%	42%
moderate lesions	55%	57%	44%	57%
severe lesions	0	0	0	0
Tentacles				
increase in red granules	66%	0	44%	0
decrease in red granules	0	85%	0	85%

five percent of the copper-exposed scallops had the same type of histopathological changes, plus degeneration and necrosis of some of the tubules in a few animals. There was also an increase in the number of waste-containing cells. At 9 wk, 77% of the cadmium-exposed scallops had lesions in the digestive diverticula while only 57% of the copper-exposed animals had lesions.

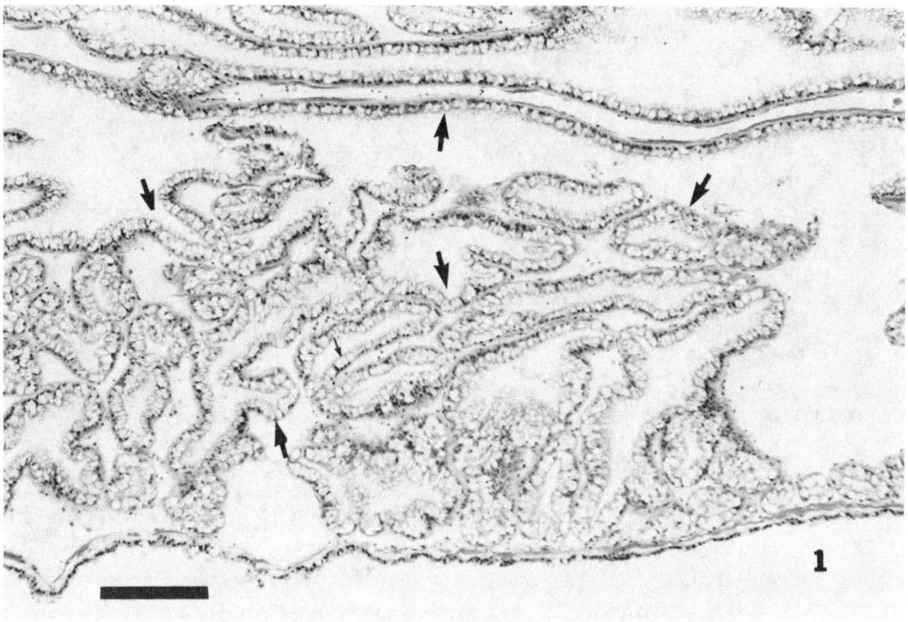

Fig. 1. Portion of kidney from a control sea scallop. Arrows indicate numerous folds lined with a clear columnar epithelium. H & E stain. Scale Bar = 200 μm.

Reproductive Tract

The reproductive tracts of the control sea scallops had active germinal epithelium and developing spermatogonia or oogonia (Fig. 5). There was a reduced amount of active germinal epithelium and resulting gonadal products, plus dilation of the follicles, in both the cadmium- and copper-exposed animals at the end of the exposure period. At the end of the 2-wk depuration period, the cadmium-exposed animals were showing gonadal differentiation, but the follicles were not as full as in the controls. Four of the 7 copper-exposed animals remained undifferentiated with little active germinal epithelium (Fig. 6). Table 3 shows the percent of differentiation at different stages of the experiment.

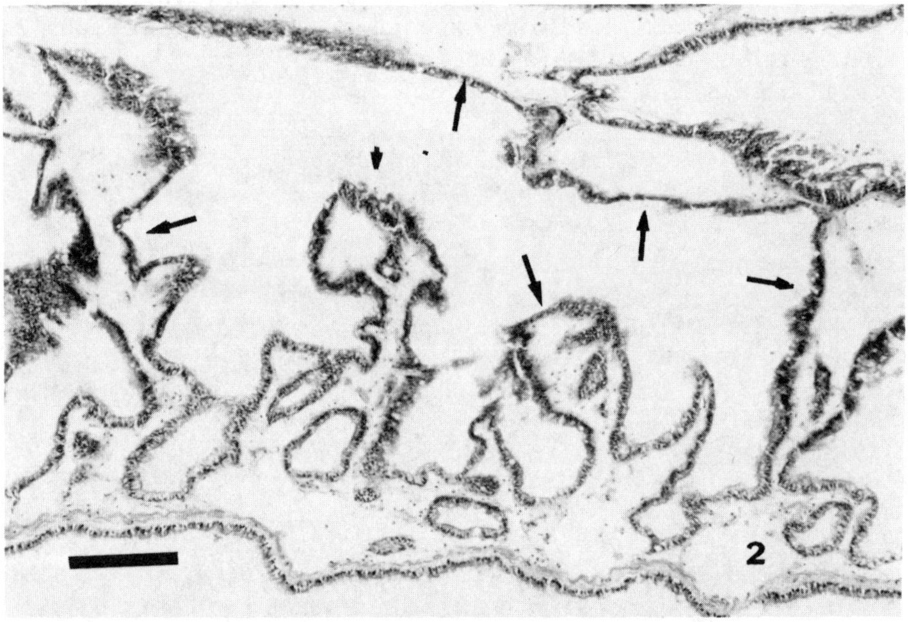

Fig. 2. Portion of kidney of a sea scallop exposed to 20 µm/L of copper for 7 wk. Note loss of fold due to dilation of the kidney mass. Arrows indicate area of basophilic metaplastic epithelium. H & E stain. Scale bar = 200 µm.

Tentacles

The tentacles of scallops have both tactile and chemosensory epithelial cells. The tactile cells have a tuft of cilia at the apical end and the chemosensory cells contain red-staining granules. The tentacles of the control scallops were normal throughout the experiment. However, the cadmium-exposed scallops had an increase in the amount of red granular material in the chemosensory portion of the tentacles and the copper-exposed animals had a decrease.

DISCUSSION

This experiment was terminated 2 wk earlier than planned because of the extensive deaths of the copper-exposed scallops. Twenty-two percent of the copper-exposed scallops died during the exposure period and 7% during the depuration period (Gould _et al_., this volume). The histopathological findings of this study indicated that the kidney lesions were the probable cause of death of the copper-exposed animals due to impairment of the excretory function. Kidney lesions were also noted in winter flounder (Baker, 1969) and in mummichogs (Lewis & Lewis, 1971) that were exposed to copper. Copper has been shown to cause extensive kidney damage in sheep, consisting of functional impairment with degeneration and necrosis of the proximal convoluted tubules (Gopinath _et al_., 1974).

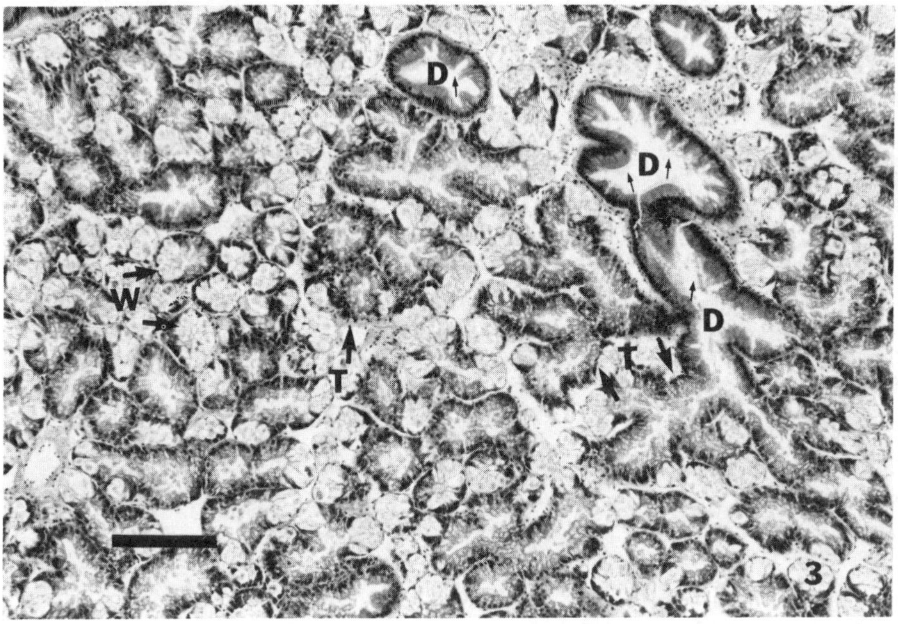

Fig. 3. Portion of the digestive diverticula of a control sea scallop. Arrow D indicates normal columar epithelium lining the ducts. Arrow T indicates tubules with columnar digestive cells filled with granules. Arrow W. indicates intact waste-containing cells. H & E stain. Scale bar = 200 μm.

In prior research, the digestive diverticula was shown to be the organ most affected. Copper was shown to affect the digestive diverticula of Asiatic freshwater clams (Martin, 1971) and oysters (Fujiya, 1960). At ERL,N, bay scallops, _Argopecten irradians_, were exposed to 10 ppb of copper for 20 wk. Beginning at the 11th week, histopathological changes were observed in the digestive diverticula, with the ducts being more involved than the tubules (Yevich, pers. comm.). The blue mussel, *Mytilus edulis*, exposed to copper also showed changes in the digestive diverticula (Calabrese *et al.*, 1984). In those experiments, the damage to the duct epithelium and the tubules was more severe than that noted in the present study, but this may be due to a longer exposure period.

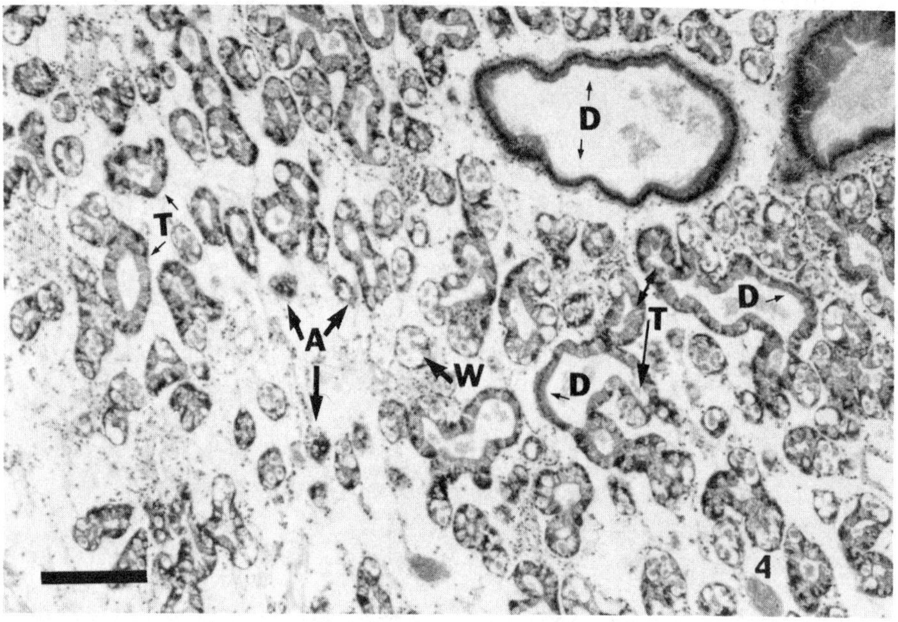

Fig. 4. Portion of the digestive diverticula from a sea scallop exposed to 20 µg/L of copper for 7 wk. Note the dilated duct (D) and tubules (T), atrophy of the tubules (A) with loss of granules, and loss of cellular structure of the waste-containing cells (W). The increased amount of space between the tubules is due to the atrophy and destruction of the tubules. H & E stain. Scale bar = 200 µm.

Although the kidney was usually the primary target organ when bivalves (such as bay scallops, blue mussels, quahogs, and oysters) were exposed to various concentrations of cadmium for varying lengths of time, the digestive diverticula were also affected. When oysters were exposed to 15 µg/L of cadmium for 10 months and then allowed to depurate for 6 months, the lesions noted in the digestive diverticula were more severe at the end of the depuration period (Yevich, pers. comm.). This seems to indicate a latent effect of the cadmium on the digestive diverticula, as was noted in this study.

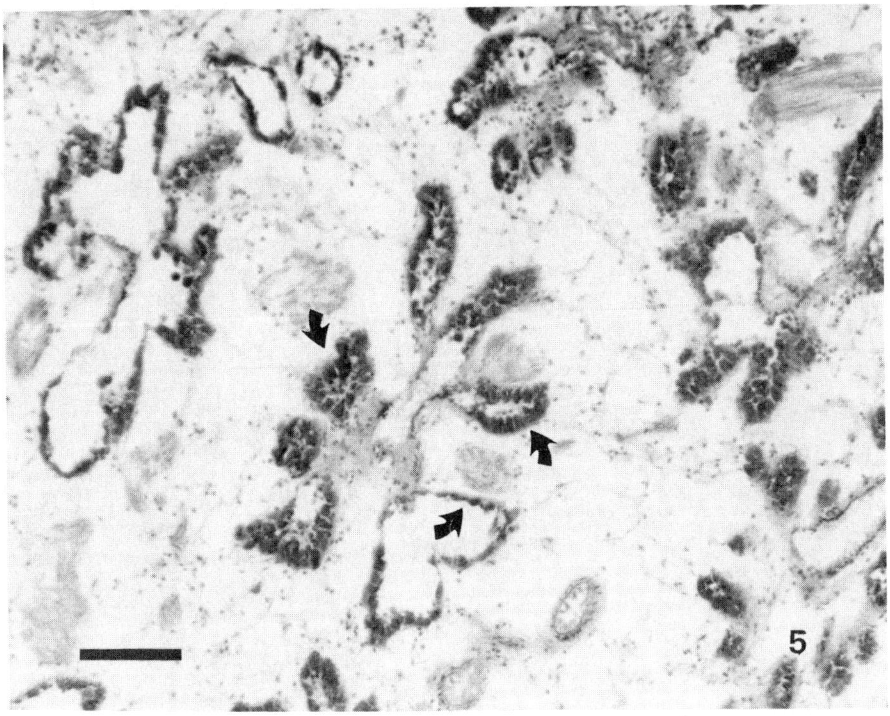

Fig. 5. Portion of the gonad of a control sea scallop at 9 wk. Notice the active germinal epithelium and developing ovagonia. H & E stain. Scale bar = 200 μm.

The long-term effect of copper and cadmium on the scallop reproductive tract is not known as there was only a partial recovery in the 2 wk post-exposure animals, particularly in the copper-exposed animals. Copper has been shown to have an adverse effect on reproduction in Daphnia magna, copepods, Artemia, annelids, oysters, fathead minnows, Atlantic salmon, rainbow trout, and other species of fish (Nriagu, 1979), and this may also hold true for the scallops.

While the chemical nature and exact purpose of the red-staining granules in the chemo-mechanical receptor cells of the tentacles is not known, they are part of the animal's early warning system against sediment and predators (Gutsell, 1930). How their increase or decrease would effect the scallop is not known.

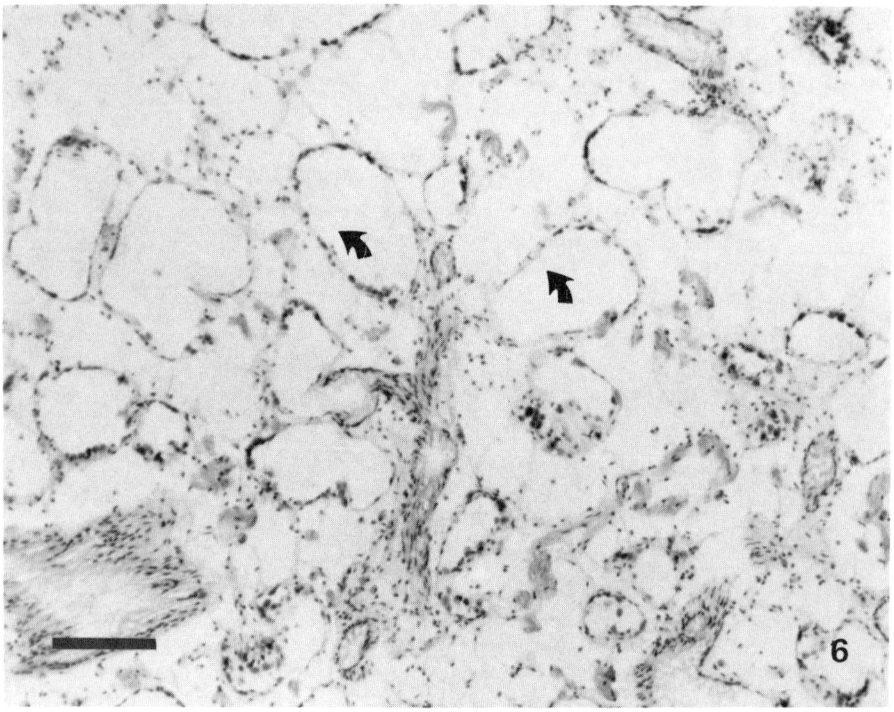

Fig. 6. Portion of the gonad of a sea scallop exposed to 20 μg/L of copper for 7 wk and depurated for 2 wk. Note lack of active germinal epithelium and dilated follicles. H & E stain. Scale bar = 200 μm.

Table 3. Percent differentiation of the reproductive tract of sea scallops.

Date	Exposure data	% differentiated Male	Female	% undifferentiated
11/10/82	pre-exposure controls	50%	30%	20%
1/12/83	controls	30%	50%	20%
	20 μg/L cadmium	22%	33%	45%
	20 μg/L copper		15%	85%
1/26/83	controls	33%	66%	0%
	20 μg/L cadmium	11%	88%	0%
	20 μg/L copper	14%	28%	57%

CONCLUSION

Histopathological changes were seen in the kidneys, reproductive tract, digestive diverticula, and tentacles of the exposed scallops. The lesions of the kidney in the copper-exposed scallops were considered irreversible and could have been the cause of death of the scallops from renal failure.

ACKNOWLEDGMENTS

Our thanks to Edith Gould of the Northeast Fisheries Center, Milford Laboratory, CT for allowing us to be a part of this experiment. Thanks to Esther Peters and Michael Casey for preparing the tissues for microscopic examination.

LITERATURE CITED

Baker, J.T.P. 1969. Histological and electron microscopical observations on copper poisoning in the winter flounder (*Pseudopleuronectes americanus*). J. Fish Res. Board Can. 26: 2785-2793.

Calabrese, A., J.R. MacInnes, D.A. Nelson, R.A. Greig, and P.P. Yevich. 1984. Effects of long-term exposure to silver or copper on growth, bioaccumulation, and histopathology of the blue mussel, *Mytilus edulis*. Mar. Environ. Res. 11: 253-274.

Fujiya, M. 1960. Studies on the effects of copper dissolved in sea water on the oyster. Bull. Jpn. Soc. Sci. Fish. 26: 462-468.

Gopinath, C., G.A. Hall, and J. McC. Howell. 1974. The effects of chronic copper poisoning on the kidneys of sheep. Res. Vet. Sci. 16: 57-69.

Gould, E., R.A. Greig, D. Rusanowsky, and B.C. Marks. Metal-exposed sea scallops, *Placopecten magellanicus* (Gmelin): A comparison of the effects and uptake of cadmium and copper. (This volume).

Gutsell, J.S. 1930. Natural history of the bay scallop. Bull. Bur. Fish. 46: 569-632.

Lewis, S.D. and W.M. Lewis. 1971. The effects of zinc and copper on the osmolarity of blood serum of the channel catfish, *Ictalurus punctatus* Rafinesque, and golden shiner, *Notemigonus crysoleuscas* Mitchill. Trans. Am. Fish. Soc. 100: 639-643.

Martin, S.G. 1971. An analysis of the histopathologic effects of copper sulfate on the Asiatic freshwater clam,

Corbicula sluminen (Muller). Ph.D. Thesis, University of Washington. 132 pp.

Mount, D.L. and W.A. Brungs. 1967. A simplified dosing apparatus for fish toxicological studies. Water Res. 1: 21-29.

Nriagu, J.O. (ed.). 1979. Copper in the Environment Part II: Health Effects. John Wiley and Sons, New York. 489 pp.

Yevich, P.P. and C.A. Barszcz. 1981. Preparation of aquatic animals for histopathological examination. Biological Methods Branch, U.S. Environmental Protection Agency, Cincinnati, OH. 38 pp.

PHYSIOLOGICAL AND HISTOPATHOLOGICAL EVALUATION OF THE TOXICITY OF HEXAVALENT CHROMIUM TO THE GRASS SHRIMP *PALAEMONETES PUGIO* HOLTHUIS

K. Ranga Rao, Philip J. Conklin, and Daniel G. Doughtie

SUMMARY

The toxicity of hexavalent chromium to newly-hatched grass shrimp larvae (zoea) varied in relation to the test salinity. Chromium (sodium chromate) was most toxic at 10°/$_{oo}$ S (96-h LC50: 3.00 ppm Cr), intermediate in toxicity at 20°/$_{oo}$ (96-h LC50: 5.88 ppm Cr), and least toxic at 30°/$_{oo}$ (96-h LC50: 13.55 ppm Cr). Intermolt (stage C) adult grass shrimp, none of which molted during 96-h tests, were considerably less sensitive to chromium at 10°/$_{oo}$ (96-h LC50: 14.63 ppm Cr) than were larvae; however, molting adult shrimp (96-h LC50: 4.86 ppm Cr) were nearly as sensitive as the larvae to chromium. Since the larval shrimp also molted during acute tests, the increased sensitivity to chromium appears to be largely molt-related.

Grass shrimp that were exposed to Cr^{6+} for 28 d accumulated considerable amounts of chromium. Of the tissues analyzed, hepatopancreas had the highest chromium levels (108 µg Cr/g dry wt) and abdominal muscle the lowest (8.4 µg Cr/g dry wt). Following a 1 wk depuration period, little if any accumulated chromium was lost from the shrimp.

Hexavalent chromium was found to be an effective inhibitor of regenerative limb growth in grass shrimp. A dose-related inhibition of limb regeneration (fifth periopod) by chromium was observed, with total blockage of premolt limb bud growth at the highest concentration tested (4.0 ppm Cr), and greater than 50% inhibition at a concentration of 0.5 ppm. Histopathological studies of severely retarded limb regenerates suggest that Cr^{6+} causes abnormal alterations in the mitochondria, rough endoplasmic reticulum, and nuclei of the blastemal epithelium. In addition, there was an apparent reduction in the incidence of blastema mitoses. Nerve axon regeneration, upon which basal limb growth is presumably dependent, was adversely affected as evidenced in the markedly swollen axonal processes. The significantly greater sensitivity of the limb regeneration assay is demonstrated by the fact that the 96-h LC50 (0.41 ppm Cr) was an order of magnitude lower than the 96-h LC50 (4.86 ppm) for sensitive molting grass shrimp.

In addition to these studies, histopathological/ultrastructural studies were performed on shrimp exposed to Cr^{6+} (0.5-4.0 ppm) for 28 d. The exposed shrimp frequently developed massive cuticular lesions, presumably initiated by pathological degeneration of the subcuticular epithelium, and these may have been a major cause of limb loss in many exposed animals. Of the four major internal organs studied (hepatopancreas, antennal glands, gills, and midgut), the severest histopathologial changes occured in the antennal glands and hepatopancreas. The changes in the former appeared to culminate in widespread necrosis and total organ dysfunction, while the hepatopancreas exhibited abnormal mitotic figures and hypertrophied nuclei. The gut epithelium frequently appeared hypertrophied, while the gill lamellae remained relatively unchanged during the exposure. Unusual intranuclear inclusions, most common in the antennal glands and hepatopancreas, occurred in all examined tissues and these may be important diagnostic features of chromium intoxication.

INTRODUCTION

Chromium is widely used in the electroplating industry and the manufacture of alloy steel, as well as in the textile and leather tanning industries (Towill *et al*., 1978). It is also used in cooling water systems and anti-corrosive coatings. Thus, chromium can enter the estuarine environment from discharge of industrial effluents, runoff, and sewage disposal, as well as natural enrichment by river transport. In addition, a significant proportion of the more than 65,000 tons of chromium compounds used annually in offshore oil and gas well drilling (API, 1978) enters the marine environment through discharge of waste drilling fluids.

Although chromium can exist in a number of different oxidation states, it occurs most commonly in the trivalent (Cr^{3+}) or hexavalent (Cr^{6+}) form. The former is generally less toxic, presumably because of its poor membrane permeability (Mertz, 1969) and lower solubility in water (Towill et al., 1978). Most of the toxicological studies have been focused on the relatively more toxic hexavalent chromium. Whereas the earlier work with Cr^{6+} dealt largely with freshwater organisms (Fromm & Schiffman, 1958; Trama & Benoit, 1960; Pickering & Henderson, 1966; Biesenger & Christensen, 1972; Benoit, 1976), more recent studies have examined its toxicity to marine and estuarine organisms (Olson & Harrel, 1973; Mearns et al., 1976; Moraïtou-Apostolopoulou & Verriopoulos, 1982 a,b; Pagano et al., 1983). In comparison to other heavy metals (e.g., cadmium and mercury), however, relatively little is known about the effect of long-term chromium exposure on crustaceans. Thus, we initiated a physiological and histopathological evaluation of Cr^{6+} toxicity in the grass shrimp, *Palaemonetes pugio* Holthuis.

MATERIALS AND METHODS

Grass Shrimp

Adult grass shrimp (*Palaemonetes pugio*) were collected from grass beds in Santa Rosa Sound, Gulf Breeze, Florida. The shrimp were held in 84.5-L glass aquaria containing aerated, recirculating sea water, and equipped with oyster shell biological filters. Stock animals were fed a daily diet of freshly hatched brine shrimp (*Artemia salina*) supplemented with Tetramin® staple fish food flakes. Grass shrimp were acclimated to the individual test conditions over at least 96 h prior to testing, and only non-ovigerous shrimp (22-28 mm overall length) were used in experiments. Determination of the stage in the molt cycle was done by microscopic examination of the uropods using previously described techniques (see Conklin & Rao, 1978).

Larval grass shrimp were obtained by selecting gravid shrimp with eggs in the later stages of development (determined under dissecting microscope), and placing individual animals in 2-L glass culture dishes (Carolina Biological Supply) containing 1 L of sea water and *Artemia* nauplii as food. Larvae usually hatched during the night, and were carefully removed from the bowls the following morning using a widebore pipette, and counted. All larvae used in tests were <24-h old (stage 1 zoea).

General Experimental Conditions

Adult grass shrimp were exposed individually in glass jars (8.5 cm dia. x 9.5 cm high) that contained 250 ml of test solution. Larval grass shrimp were tested in 2-L glass culture dishes containing 1 L of solution and 10 larvae per dish. Although the solutions were not aerated, they were renewed every 24 h to provide freshly oxygenated sea water and sea water/toxicant as well as to remove accumulated waste products. The dissolved oxygen levels and pH were measured just prior to renewing the solutions. All experiments were conducted in a controlled-temperature incubator at 20 ± 1°C under cool-white fluorescent illumination (12 h L:12h D) and $10°/_{oo}$ salinity (S), except for additional larval tests that were conducted at $20°/_{oo}$ and $30°/_{oo}$ to evaluate the interaction of salinity with chromium toxicity. Natural sea water (5 µm filtered) was used in all tests, and the required salinities were obtained by diluting sea water with distilled water. Feeding was discontinued during 96-h tests with adult shrimp, but during longer test exposures and all larval tests the animals were fed *ad libutum* with *Artemia* nauplii once daily.

The chromium test solutions were prepared from analytical grade sodium chromate ($Na_2CrO_4 \cdot 4H_2O$) purchased from Mallinckrodt Chemical Works, St. Louis, MO. A stock solution (1 g Cr^{6+}/L) was diluted with sea water to obtain the appropriate test concentrations. All concentrations reported are in ppm chromium (mg Cr^{6+}/L). Control shrimp for all tests were maintained in sea water alone.

Toxicity Tests

Intermolt (stage C) and very late premolt (stages D_2-D_4) grass shrimp were exposed to five concentrations of chromium as determined from range-finder tests. A minimum of 20 shrimp were exposed at each test concentration and also as controls. The test jars were examined every 24 h for mortalities, and evidence of molting (cast exuvia) for D_2-D_4 shrimp.

For larval toxicity tests, three replicate dishes containing 10 larvae (<24-h old initially) were tested at each of five chromium concentrations and for the controls. The larvae were examined at 24-h intervals, at which time dead animals were removed. The animals were considered to be dead when they became opaque (white) and exhibited no movement when taken up in a widebore pipette.

Limb Regeneration

Experiments were conducted to assess the effects of long-term exposure (28 d) of grass shimp to Cr^{6+} at $10°/_{oo}$ S and 20 ± 1°C, including the effect of chromium on the initiation and progression of regenerative limb growth. The experimental methods used to study limb regeneration were similar to those described earlier (Rao *et al.*, 1978). Initially, all grass shrimp used for regeneration experiments were in intermolt since adequate time prior to ecdysis is required for successful limb bud formation (Rao *et al.*, 1978). Immediately following removal of the left fifth pereiopod, the shrimp were placed in test solutions (0.5, 1, 2, and 4 ppm chromium) or sea water alone for the controls. Twenty-four animals were exposed at each chromium test concentration and as controls. On the 4th, 7th, 9th, and 11th day after limb removal, the shrimp were examined microscopically to assess premolt limb bud growth and progress in the molt cycle. The time until molting (ecdysis) and subsequent postmolt limb lengths were recorded. In order to compare regenerative growth for different shrimp, a regeneration index (R-value) was computed using the formula of Bliss (1956):

$$\text{R-value} = \frac{\text{length of limb bud}}{\text{length of carapace}} \times 100$$

Chromium Accumulation and Analysis

At the end of the 28-d exposure, surviving experimental shrimp and control shrimp were analyzed for chromium. Five control animals and five experimental animals from each chromium test concentration were sampled for determination of whole body chromium levels. The chromium accumulation in different tissues (abdominal muscle, cuticle, hepatopancreas, and cephalothorax) dissected from five control shrimp and five chromium-exposed (1.0 ppm) shrimp were also determined. In addition, another five shrimp (0.5 ppm chromium-exposed) were transferred to sea water and allowed to depurate for 1 wk. All shrimp were rinsed thoroughly with deionized water, and placed in acid (HNO_3)-cleaned polyethylene vials. The shrimp and tissue samples were freeze dried and weighed prior to subsequent determination of chromium.

The concentrations of chromium in whole shrimp, shrimp tissues, and test solutions were determined by atomic absorption spectrometry. Following digestion with perchloric, hydrofluoric, and nitric acids, the samples were analyzed for total chromium content on a Perkin-Elmer Model 603 spectrometer equipped with an HGA-2200 graphite furnace, and deuterium background corrector.

Analysis of Results

Data were analyzed by computer using the Statistical Analysis System (SAS) package (SAS Institute Inc., 1982). The LC50 and EC50 values and their corresponding fiducial limits were computed using the SAS Probit procedure; the SAS General Linear Model (GLM) procedure and Duncan's Multiple Range Test were used to test for significant ($p \leqq 0.05$ level) effects within treatment means.

Histopathological Studies

Four groups of 24 shrimp each were exposed to chromium concentrations of 0.5, 1.0, 2.0, and 4.0 ppm in $10^{o}/_{oo}$ sea water. One group of 24 shrimp maintained in $10^{o}/_{oo}$ sea water served as controls. At the beginning of the experiment the left fifth pereiopod was excised at or within 1 mm of the preformed breakage plane in the basi-ischium segment. Operated shrimp were immediately placed in the test solutions. During the 28-d exposure, test solutions were renewed every 24 h and the shrimp were fed live *Artemia* nauplii every 24 h. At the end of the experiment 10 shrimp from each of the five test solutions, including the controls, were prepared for electron microscopy. The basis for selection was that the control shrimp and experimental shrimp be in comparable stages in the molt cycle. In addition, all chromium-exposed shrimp with grossly inhibited limb buds were fixed. The gills, antennal glands, hepatopancreas, carapace, and limb buds were processed for ultrastructural examination using previously described methods (Bell *et al*., 1969; Doughtie & Rao, 1983a,b).

RESULTS

Toxicity

The toxicity of Cr^{6+} to *Palaemonetes* zoea varied with the salinity. The LC50 values were lower at $10^{o}/_{oo}$ than at 20 and $30^{o}/_{oo}$ (Fig. 1).

The tests with adult shrimp were conducted only at $10^{o}/_{oo}$. Molting shrimp (late premolt individuals undergoing ecdysis during the 96-h test) were more sensitive to Cr^{6+} than were intermolt (Stage C) shrimp. The 96-h LC50 for molting grass shrimp was about three times lower than that for intermolt shrimp (Table 1). In a long-term test, the LC50s decreased five-fold (from 14.6 ppm to 3.0 ppm) over a 28-d period (Table 2); much of this increased toxicity was molt-related.

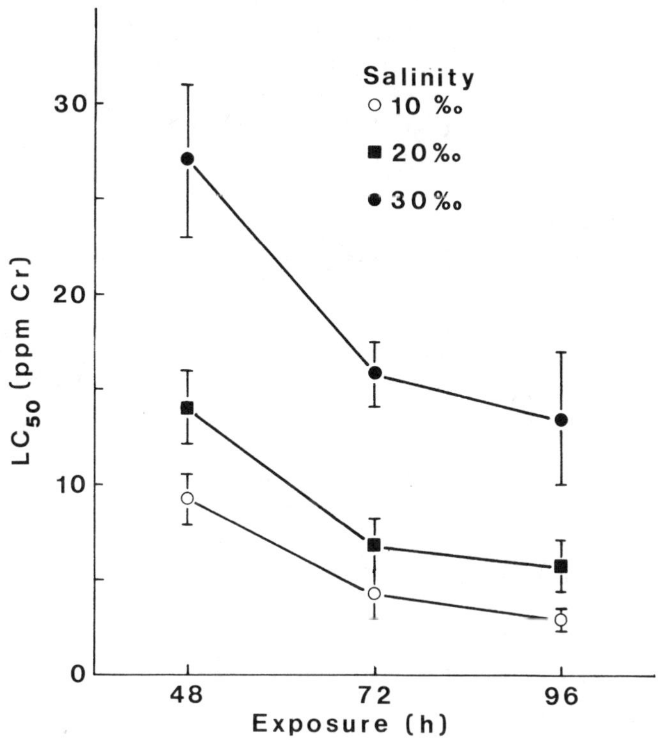

Fig. 1. Toxicity of hexavalent chromium to grass shrimp larvae (<24-h old initially) at different salinities.

The 28-d LC50 of Cr^{6+} for adult shrimp (3.0 ppm) was nearly identical to the 96-h LC50 for larval shrimp (2.6 ppm). The latter LC50 was only slightly lower than the 96-h LC50 for molting adult shrimp (4.86 ppm).

Table 1. Toxicity of hexavalent chromium (sodium chromate) to adult grass shrimp, *Palaemonetes pugio*, in relation to molt stage.

Toxicity Value [a]	Initial molt stage	
	C	D_2-D_4 [b]
96h LC10 (95% F.L.):	9.3 (5.51-11.28)	2.6 (0.32-3.61)
96h LC50 (95% F.L.):	14.6 (12.93-16.32)	4.9 (3.92-5.78)
96h LC90 (95% F.L.):	20.0 (17.99-23.72)	7.1 (6.14-9.33)

[a] All concentrations in ppm chromium (mg Cr^{6+}/L)
[b] All of these shrimp molted during 96-h tests

Table 2. Chronic toxicity of hexavalent chromium to adult grass shrimp *Palaemonetes pugio*[a], over a 28-day exposure period.

Days Exposed	LC50 (ppm)	95% fiducial limits
4	14.6	12.9-16.3
14	4.2	3.5-5.9
21	3.7	3.1-5.1
28	3.0	2.5-3.9

[a] All shrimp were in stage C (intermolt) at beginning of exposure.

Limb Regeneration

Following limb removal, control shrimp exhibited normal progression and growth of new limb buds; however, chromium-exposed shrimp exhibited dose-related inhibition of regenerative limb growth (Fig. 2). At the highest test concentration (4 ppm) there was essentially a total blockage of limb bud growth. The mean R-values for premolt limb bud growth measured on days 4, 7, 9, and 11 were all significantly ($p \leq 0.05$) lower in 0.5, 1.0, and 2.0 ppm chromium-exposed animals than in control animals (Fig. 3). Also, as expected from the reduction in premolt growth observed, the postmolt R-values for chromium-exposed shrimp were also significantly ($p \leq 0.05$) lower than control postmolt R-values (Fig. 4). The inhibition of limb regeneration was not accompanied by an alteration in the duration of intermolt cycle. There was no significant difference ($p<0.05$) between the time-to-molt in

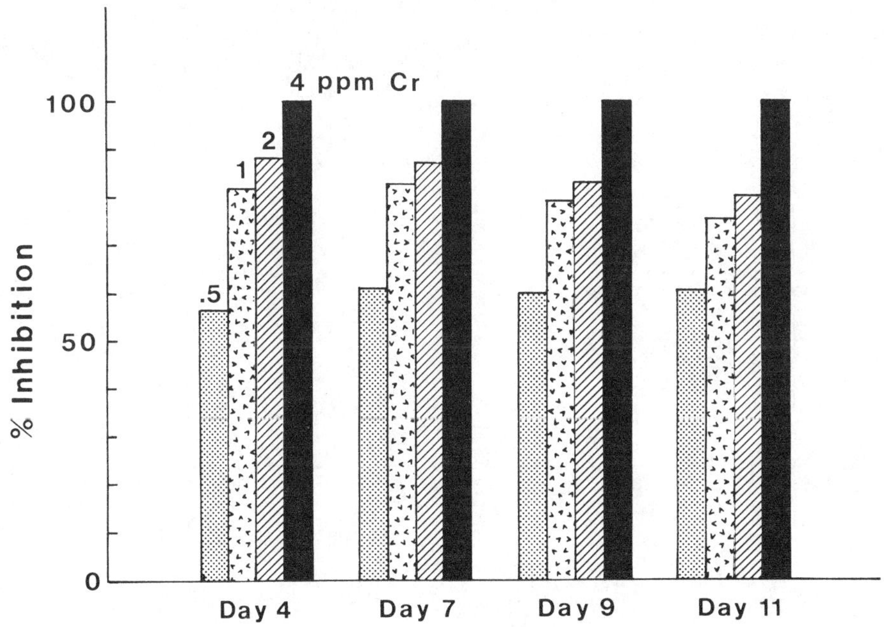

Fig. 2. Inhibition of regenerative limb growth in grass shrimp exposed to hexavalent chromium as compared to growth of limb buds in control shrimp.

Table 3. Accumulation of chromium by grass shrimp (*Palaemonetes pugio*) following a 28-day exposure to hexavalent chromium.

Exposure Concentration in ppm (mg Cr^{6+}/L)	Whole animal concentration in ppm (μg Cr/g dry wt)	Increased Cr content relative to controls
4.0	100.1 ± 18.5 [a]	208 x
2.0	81.4 ± 14.2	170 x
1.0	30.6 ± 3.2	64 x
0.5	23.0 ± 1.6	48 x
Control (<0.0002)	0.48 ± 0.15	-----

[a] Mean ± standard error

control shrimp (13.25 d; 95% f.l. = 11.99-14.51) and experimental shrimp exposed to chromium concentrations of 0.5 ppm (14.85 d; 13.24-16.46), 1 ppm (14.54 d; 13.04-16.06), 2 ppm 15.95 d; 13.97-17.94), and 4 ppm (15.55 d; 13.66-17.44).

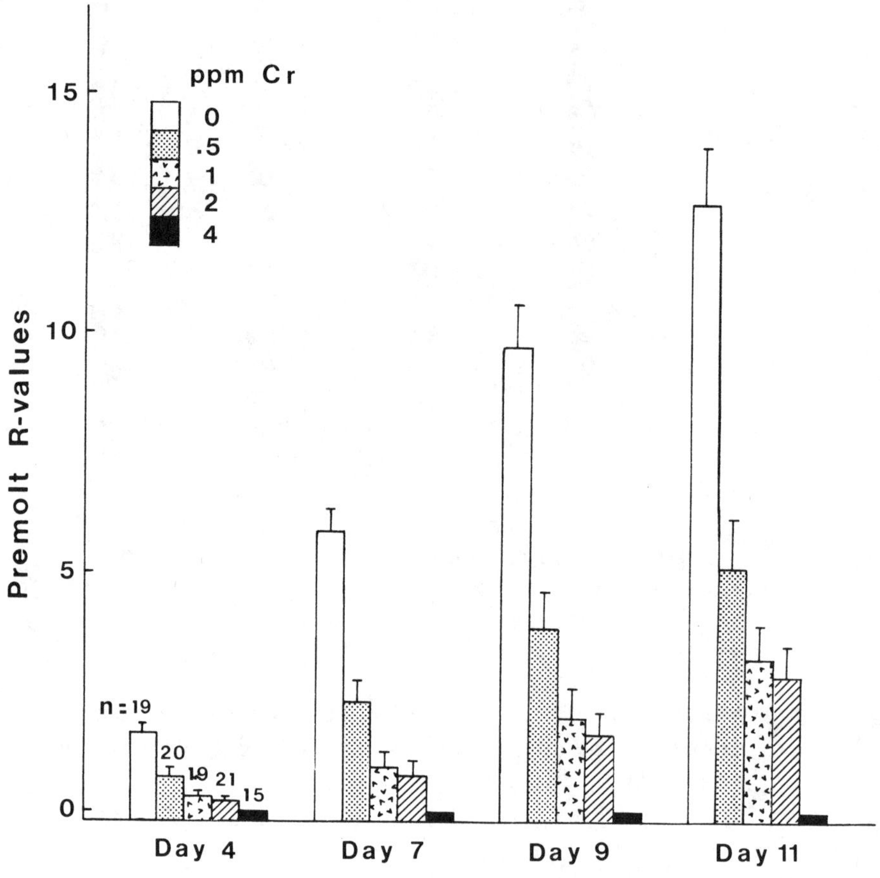

Fig. 3. Effect of hexavalent chromium on premolt regeneration of left fifth pereiopod of grass shrimp. The vertical bar represents the mean regenerative index on a given day following limb removal, and the vertical lines show the standard errors of their respective mean. R-values from animals that molted prior to Day 12 were not used in computation of mean R-values.

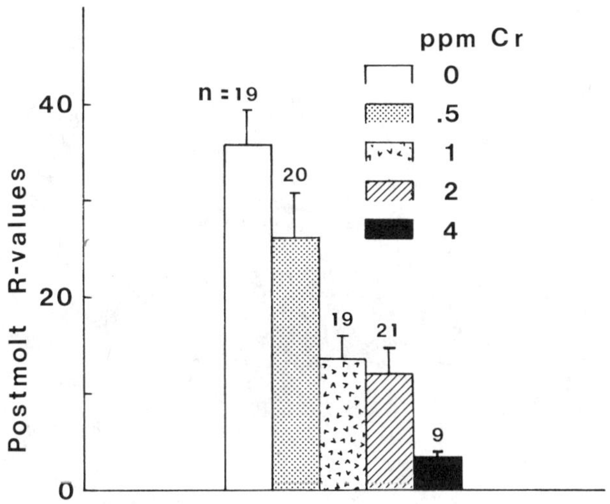

Fig. 4. Postmolt R-values for grass shrimp exposed to hexavalent chromium until completion of ecdysis. The vertical bar represents the mean postmolt R-values, while the vertical lines represent the standard error of the mean.

Accumulation of Chromium

There was a dose-related accumulation of chromium by grass shrimp (Table 3). After 28 d, hepatopancreas had the highest concentration of chromium and abdominal muscle the lowest. The chromium concentrations in hepatopancreas, cephalothorax, cuticle, and abdominal muscle from pooled tissue samples of five shrimp that were exposed to 1.0 ppm Cr^{6+} were 108.4, 67.8, 46.2, and 8.4 ppm (μg/g dry wt), respectively. Tissue concentrations of chromium for control shrimp were as follows: hepatopancreas, 4.9 ppm; cephalothorax, 0.8 ppm; cuticle, 3.6 ppm; and abdominal muscle, 0.8 ppm. There was no significant loss of accumulated chromium from grass shrimp that were exposed to 0.5 ppm Cr^{6+} for 28 d and then allowed to depurate in sea water for 1 wk. Following one week depuration, the whole-animal chromium concentrations were 21.78 ± 1.6 ppm ($\bar{x} \pm S.E._{\bar{x}}$, n = 5) compared to 23.0 ± 1.7 ppm for animals that were sampled at the end of 28-d exposure to chromium.

Histopathology

Cuticular Lesions

Long-term (28-d) exposure to Cr^{6+} (0.5-4.0 ppm) induced cuticular lesions (Doughtie *et al.*, 1983), primarily about

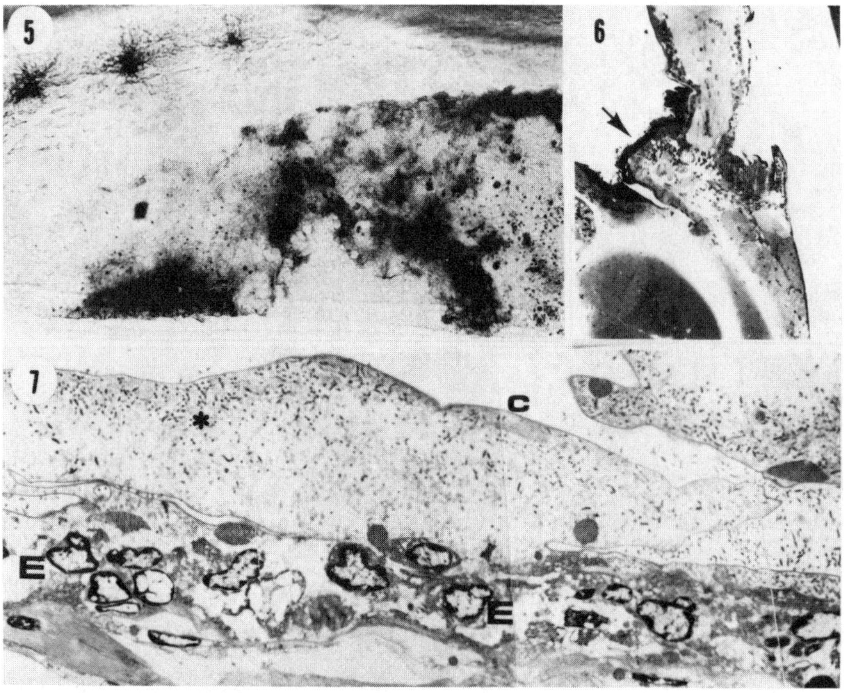

Fig. 5. Melanized lesion in cephalothoracic regions of the carapace from shrimp exposed to hexavalent chromium. x 55.

Fig. 6. Sagittal section through basal region of grass shrimp pleopod following exposure to chromium. Melanized lesion development (arrow) is apparent near the articulating area. x 25.

Fig. 7. Section through region of abdominal cuticular lesion from a postmolt chromium-exposed shrimp. The subcuticular epithelium (E) exhibits nuclear pyknosis, chromatin emargination, and extensive cytoplasmic disruption. Numerous bacteria (*) populate the area adjacent to the abnormally thin and elevated cuticle(c). x 500.

the articulations of the abdominal tergites, pereiopods, pleopods, and antennae in roughly 50% of the exposed animals (Figs. 5 & 6). These lesions macroscopically resemble "shell disease" syndrome and appear to originate as a result of degenerative changes in the subcuticular epithelium (Fig. 7). These changes included cytoplasmic vacuolization, mitochondrial swelling, chromatin emargination, and unusual nuclear inclusions similar to those in the major internal organs described below. Due to this focal epithelial degeneration, the process of premolt and early postmolt cuticle formation was markedly retarded, if not inhibited (Fig. 7). This inhibition was presumed to cause structural deficiencies and/or perforations to occur in the new exoskeleton, allowing a viaduct for pathogenic organisms and direct chromium influx.

Hepatopancreas and Midgut

A small percentage (<20%) of the chromium-exposed shrimp examined with the light microscope exhibited hepatopancreatic abnormalities (Doughtie & Rao, 1984), manifested primarily as hypertrophied nuclei, a marked loss of cytoplasmic density (Figs. 8 & 9), and an apparent decrease in mitoses in the tubule apices.

In most of the hepatopancreatic tissues examined ultrastructurally, the unusual spherical or ring-like intranuclear inclusions were common (Figs. 10 & 11) and the nucleolus was fragmented. In addition, mitotic figures were often hypertrophied and characterized by fragmented chromatin (Fig. 12). Generally, alterations in nuclear morphology appeared to precede the other cytoplasmic changes including swelling of mitochondria and rough endoplasmic reticulum, diminution of smooth basal endoplasmic reticulum, and loss of cytoplasmic density.

Moderate to severe hypertrophy of the midgut epithelium was present in 40% (8/20) of the shrimp exposed to 0.5 and 1.0 ppm chromium (Fig. 13). Intranuclear inclusions were relatively uncommon in the midgut, but diminution of the smooth basal endoplasmic reticulum was widespread.

Antennal Glands

A gradation of degenerative changes were noted in the labyrinths of 38 of the 40 examined chromium-exposed shrimp (Figs. 14-16). Low magnitude alterations included chromatin emargination and microvillar disruption. More severe

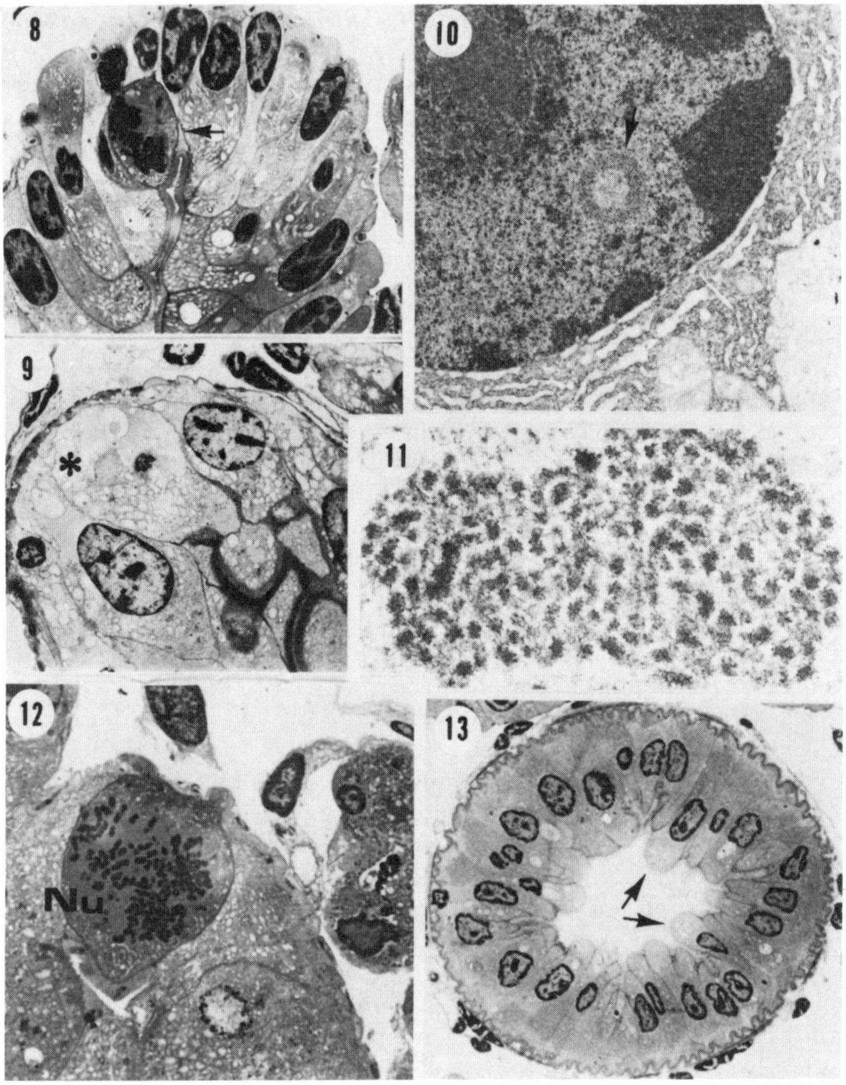

Fig. 8. Light micrograph of hepatopancreatic tubule apex from normal shrimp. Note the tight chromatin configuration in a mitotic nucleus (arrow). x 475.

Fig. 9. Hepatopancreatic tubule apex, from a chromium-exposed shrimp showing nuclear hypertrophy and marked loss of cytoplasmic density (*). x 475.

Fig. 10. Electron micrograph of hepatopancreatic "F-cell" from chromium-exposed shrimp. Note the intranuclear inclusion (arrow), characteristic of many hepatopancreatic nuclei. x 6000.

changes included nuclear hypertrophy (Fig. 15), presence of intranuclear inclusions, apparent sloughing of nuclei into lumenal channels, cytoplasmic vacuolization, a reduction in the extent of basal plasmalemmal infolding, and exfoliation of lumenal microvilli. In 17 of the 40 chromium-exposed shrimp, the labyrinth region of the antennal gland was macroscopically grey in color in contrast to the pale tan hue of normal shrimp; ultrastructurally the labyrinth region in the former was severely degenerate and/or necrotic, and characterized by the exfoliation of large regions of epithelium from the basement membrane (Fig. 16).

Gills

In terms of histopathological changes, the gills are the least affected of the four organs examined. Only 10 of the 40 chromium-exposed shrimp possessed moderately irregular profiles (Fig. 17). Intranuclear inclusions were rare (Fig. 18), but there was some loss of cytoplasmic density. In most cases there was marked diminution in the extent of the subcuticular plasmalemmal infoldings; an alteration of possible significance in the regulation of chromium.

Limb Regeneration: Normal Processes and Ultrastructure

Most limb regeneration processes in *Palaemonetes* are similar to those described for other Crustacea (*e.g.*, Adiyodi, 1972; Bazin, 1973, 1980; Hoarau, 1973). For comparative purposes a summary of these events is presented here. One to 2 d following limb excision (Fig. 19) in intermolt shrimp, the limb stump possesses a brownish scab composed of stratified, squamous, melanized hemocytes. This scab is usually sloughed by 3 d as the newly forming

---

Fig. 11. High magnification of hepatopancreatic intranuclear inclusion from chromium-exposed shrimp. Note "braided" appearance of constituent elements. x 45,000.

Fig. 12. Abnormally fragmented and hypertrophied metaphase nucleus (Nu) from hepatopancreas of chromium-exposed shrimp. x 475.

Fig. 13. Hypertrophied midgut epithelium from chromium-exposed shrimp. Note distension of cell apices into midgut lumen (arrows). x 300.

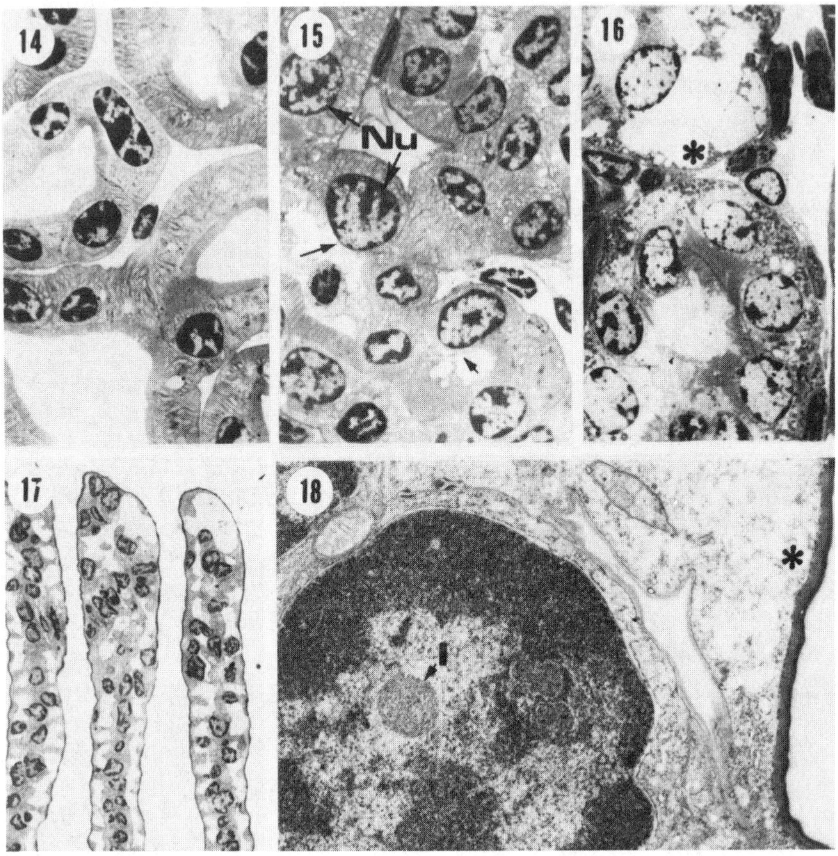

Fig. 14. Profile of antennal gland labyrinth from a normal grass shrimp. x 450.

Fig. 15. Moderate degenerative changes in labyrinth from grass shrimp exposed to hexavalent chromium. Note nuclear hypertrophy (N) and loss of lumenal microvilli (arrow). x 450.

Fig. 16. Severe degeneration and necrosis of labyrinth from chromium-exposed shrimp. Chromatin emargination is pronounced. Exfoliation of labyrinth epithelium (*) has already occurred. x 450.

Fig. 17. Sagittal section of branchial lamella from chromium-exposed shrimp. Moderate distension and irregularities are apparent in lamellar profiles. x 210.

Fig. 18. Nuclear inclusion (1) within the euchromatin of a lamellar epithelial cell. Subcuticular plasmalemmal infoldings (*) are absent. Cuticle (c). x 7000.

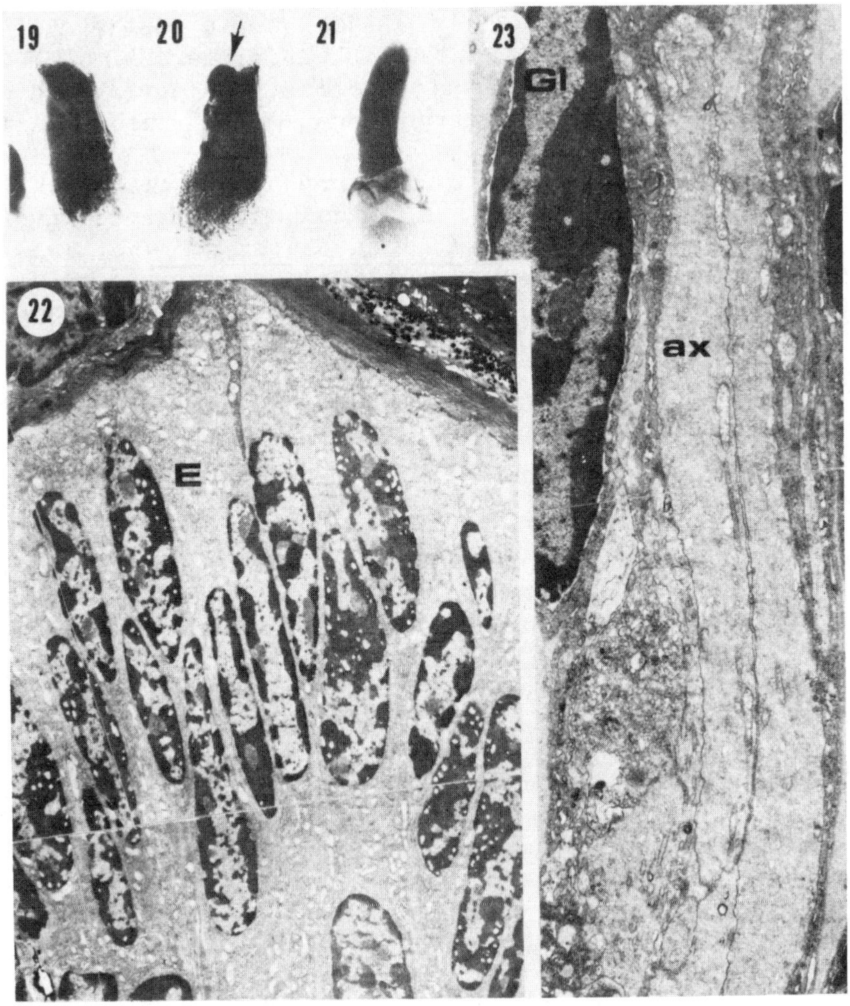

Fig. 19. Profile of control limb stump, 24 hours after limb removal. x 24.

Fig. 20. Profile of control limb stump, 72 hours after limb removal showing small protruding limb bud (arrow). x 24.

Fig. 21. Small (1.0 mm) limb regenerate in control animal approximately 7 days after limb removal. x 24.

Fig. 22. Electron micrograph of 96-hour limb bud blastema showing epithelial (E) cytoplasmic features that remain relatively unchanged through proecdysis. x 1,500.

Fig. 23. Regenerating nerve axons (ax) in a control, 7-day old limb regenerate. Axoplasm has considerable electron density and axons are uniformly distended. Glial cells (Gl). x 4500.

blastema expands outward (Fig. 20); by 4 d post-excision, a small limb bud papilla is usually visible. Limb growth proceeds rapidly during proecdysis (Fig. 21) and at ecdysis the limb regenerate is a virtual copy of the original limb, albeit smaller.

Histologically, the principal events of the first few days following limb removal are degenerative changes involving massive hemocytic invasion and phagocytosis of injured nerve, muscle, and epithelial cells. Subsequent to the formation of a protective scab, undifferentiated blastemal cells (derived primarily from the dedifferentiation of viable epithelium) form a multilayered sheet subjacent to the scab (Fig. 22). These cells soon begin to form a fibrous, presumably elastic cuticle characteristic of crustacean wound healing. Sagittal sections of 4-d old limb buds suggested major regenerative activity. Generally, the limb stump was occluded with frequently mitotic peripheral epithelial cells, numerous granular hemocytes, attenuated monolayers of fibroblast-like cells that ascended through the stump toward the limb bud apex, and regenerating axons.

During proecdysial limb growth, the ultrastructure of the peripheral epithelial cells and the regenerating axons changed little (Fig. 22); the epithelial cells were typically columnar and possessed extensive rough endoplasmic reticulum with the basal and medial cytoplasm. Additionally, mitochondria were numerous with narrow cristae projecting into a lucent matrix. The regenerating nerve axons (Fig. 23) originated as a dense bundle of large processes at a level approximately 0.5 mm subjacent to the original excision plane. As they ascended toward the limb bud apex, these processes divaricated into smaller (≅ 0.5 μm dia.), moderately dense bundles and finally discrete processes. The moderate electron density of the axons was derived in part from the numerous contained microtubules and dense axoplasm (Fig. 23).

Fine Structure of Chromium-Exposed Limb Buds

Only limb regenerates that exhibited marked inhibition of growth (Fig. 24) were examined for evidence of histopathology. Of these limb buds, eight measured between 1.5-3.0 mm in length (R-values: 15.8-29.4) and 10 were 1.0 mm or less (R-value = < 10.2). Fifty percent of the limb buds measuring between 1.5 and 3.0 mm appeared normal, whereas the remaining 50% showed significant abnormalities (*e.g.*, there were frequent incidences of localized epithelial morbidity and early nuclear pyknosis). Compared with the control animals, there was an apparent reduction of epithelial mitosis. The axons in these limb buds exhibited varying degrees of focal swelling and loss of axoplasm density (Fig. 25). The fine structure of severely inhibited limb buds (<1.0 mm) suggested a condition of stasis. The peripheral epithelium consisted of a monolayer of squamous cells (Fig. 24) and the limb stump lumen usually contained only a lattice of loosely arranged connective tissue. Further, hemocytic presence was negligible. Ultrastructurally, the epithelium lacked cytoplasmic and euchromatic density and possessed swollen mitochondria and rough endoplasmic reticulum (Fig. 26). Some nuclei possessed sperical inclusions (≅ 0.6 µm diameter) that resembled those found in the antennal glands of chromium-exposed shrimp (Doughtie & Rao, 1984).

DISCUSSION

Chromium Toxicity

Molting adult grass shrimp exhibited greater sensitivity to hexavalent chromium (sodium chromate) during acute toxicity tests than did non-molting shrimp. Likewise, grass shrimp larvae (<24-h old at start of test), all of which molted during 96-h exposures, were considerably more sensitive to chromium exposure than were intermolt adult grass shrimp. However, the 96-h LC50 for larvae was only slightly lower than the LC50 for molting adult grass shrimp, suggesting that larval sensitivity to chromium is also molt-related. Similar increases in molt-related toxicity have been noted in tests with chromium (postassium chromate) on *Daphnia pulex* (Lee & Buikema, 1979), heavy metals on *Crangon crangon* (Price & Uglow, 1979), and chlorophenols on *Palaeomonetes pugio* (Rao *et al*., 1981).

The toxicity of Cr^{6+} to larval grass shrimp was inversely related to salinity, being most toxic at the lowest salinity tested, as reported earlier for adult grass shrimp (Fales, 1978), juvenile *Callinectes sapidus* (Frank & Robertson, 1979), and *Rangia cuneata* (Olson & Harrel, 1973). The observed decreases in Cr^{6+} toxicity with increasing salinity may in part be due to increased reduction of Cr^{6+} to the less soluble trivalent forms (Towill *et al*., 1978).

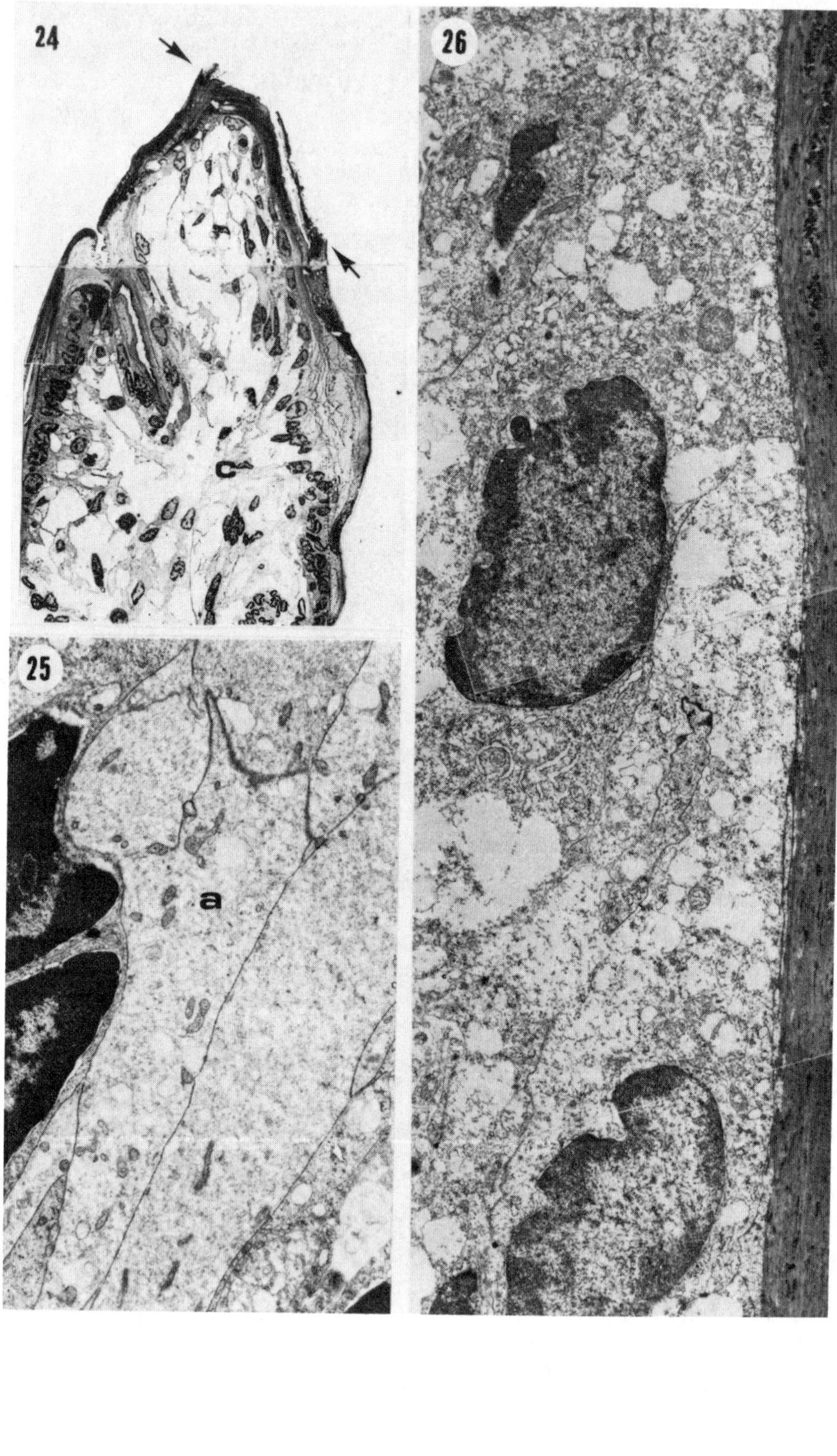
24
c
25
a
26

Fig. 24. Limb stump from shrimp exposed to 4.0 ppm Cr^{6+} for 28 days. No visible growth has occurred beyond the excision plane (arrows) and the stump lumen suggests a condition of stasis owing to the relative absence of cells other than occasional connective tissue (C). x 280.

Fig. 25. Distended and abnormally lucent axon regenerates (ax) in small limb bud of shrimp exposed to 2.0 ppm Cr^{6+} for 28 days. x 4500.

Fig. 26. Epithelial monolayer constituting portion of blastema from Cr^{6+}-inhibited limb regenerate (1.0 ppm/28 days). Note abnormal chromatin emargination, swollen mitochondria (m), and general disruption of cytoplasm. x 3000.

Accumulation of Chromium

The bioaccumulation of Cr^{6+} by grass shrimp was proportional to the chromium exposure levels. Such dose-related accumulation of chromium has been reported in the polychaete worms (*Hermione hystrix* (Chipman, 1967) and *Neanthes arenaceodentata* (Oshida & Word, 1982), as well as in the bivalve molluscs *Mytilus edulis* and *Crassostrea virginica* (Zaroogian & Johnson, 1983). It has also been reported that temperature and salinity can affect the uptake of chromium (Fales, 1979), with uptake of Cr^{6+} by grass shrimp being more rapid at higher temperatures and lower salinity. The increased uptake of chromium at lower salinities may help explain the higher toxicity of chromium to larval grass shrimp at lower salinities in this study.

Inhibition of Limb Regeneration

Hexavalent chromium caused a marked reduction in regenerative limb growth in grass shrimp. At the highest chromium test concentration (4.0 ppm) there was a complete inhibition of regenerative growth. The resultant EC50s for inhibition of regenerative growth were well below the LC50s derived from sensitive toxicity tests that utilized molting adult grass shrimp or larval grass shrimp. Since chromium-induced inhibition of regeneration was not accompanied by alterations in molt cycle duration, the inhibitory effect seems to be due to a direct action of chromium on the wound site or developing limb bud. This mode of action is similar to that suggested for chlorophenols (Rao *et al.*, 1981) and dithiocarbamates (Conklin & Rao, 1982) in other limb regeneration studies with grass shrimp. Certain other heavy metals, however, have been reported to cause a retardation of limb regeneration as well as a delay in the length of time-to-molting in fiddler crabs (Weis, 1976, 1977, 1978). Cadmium delays the onset of forelimb regeneration in the newt (*Triturus viridescens*) and causes an increased incidence of limb malformities (Mansen & O'Flaherty, 1978). Likewise, it has been reported that methylmercury causes an initial dose-dependent delay in newt limb regeneration, but that once sufficient bud growth was achieved the treated limbs appeared to regenerate at normal rates (Chang *et al.*, 1976). These findings, along with the results of the present study, suggest that the initial stages of limb regeneration, which include rapid cell division and subsequent proliferation of tissue, are the most critical.

Histopathology

The cuticular lesions found in grass shrimp exposed to Cr^{6+} are grossly indistinguishable from the "shell disease" syndrome, thought to be caused by bacterial and fungal infections (*e.g.*, Rosen, 1970; Cook & Lofton, 1973; Baross *et al.*, 1978; Lightner, 1979) as well as by absorbed heavy metals (Nimmo *et al.*, 1977). Crustacean shell disease is fairly widespread in some marine environments that have been heavily polluted with heavy metals, including chromium (*e.g.*, Pearce, 1972; Young & Pearce, 1975). In a recent study (Doughtie *et al.*, 1983) it was proposed that Cr^{6+} reaches the subcuticular epithelium by an endogenous and an exogenous route, whereupon it interferes with the normal processes of proecdysial cutical formation. Subsequent structural weaknesses and perforations develop in the cuticle of newly-molted shrimp, providing a route of entry for pathogenic organisms that perpetuate lesion development.

Of the major organs examined (gills, hepatopancreas, midgut, and antennal glands), the antennal glands were the most severely affected (see also Doughtie & Rao, 1984). The apparent state of irreversible necrosis that characterized the antennal glands of nearly 50% of the examined shrimp, including some of those exposed to the lowest tested concentration (0.5 ppm), were sufficient to dramatically affect the ionic/osmotic regulatory capabilities of the shrimp and hence account for the observed mortalities during the 28-d exposure. Previous studies have shown that other heavy metals such as copper and mercury cause extensive degenerative changes in the antennal glands of crayfish (Hubschman,

1965; Ellis & Fuller, 1979). Less dramatic, but no less significant, alterations were found to occur in the hepatopancreas and midgut of Cr^{6+}-exposed grass shrimp (see also Doughtie & Rao, 1984). In the hepatopancreas, the fragmentation and dispersion of nucleolar components, occurrence of abnormal mitotic figures, and widespread development of nuclear inclusions are of particular interest in view of cytogenetic (e.g., Stella et al., 1982) and mutagenic (e.g., DeFlora, 1981) action of Cr^{6+}. Further, the unusual intranuclear inclusions that were found in all examined tissues may be a useful diagnostic feature of chromium poisoning (Doughtie & Rao, 1984; Rao & Doughtie, 1984). The diagnostic utility of comparable inclusions is well known in the case of lead poisoning (Goyer et al., 1970).

The relative absence of pronounced degenerative changes in the gills of Cr^{6+}-exposed grass shrimp is in contrast to the findings of previous studies that the gills of Crustacea are a major target organ during exposure to other pollutants such as dithiocarbamates (Rao et al., 1982; Doughtie & Rao, 1983a) and different heavy metals (e.g., Bubel, 1976; Couch, 1977; Nimmo et al., 1977; Couch, 1978). In an earlier report (Doughtie & Rao, 1984) it was proposed that the absence of marked branchial histopathology might be due to the known capability of the gills to regulate chromium (Sather, 1967). This ability may be ultrastructurally indicated by the diminution in the degree of subcuticular plasmalemmal infolding, thereby decreasing the rate of chromium influx (Doughtie & Rao, 1984).

Hexavalent chromium seems to have a direct inhibitory effect on regenerative limb growth, as indicated by the swollen appearance of regenerating nerve axons and the abnormal changes within the organelles of the blastemal epithelial cells. The latter changes included the swelling of mitochondria and rough endoplasmic reticulum, as well as the emargination of nuclear chromatin and occasional development of intranuclear inclusions. The implications of abnormal changes in the regenerating axons within the stunted limb buds from chromium-exposed shrimp are clear in view of the findings that basal limb growth is dependent upon an intact peripheral nerve supply (see Bliss, 1960).

ACKNOWLEDGMENTS

This investigation was part of a project supported by Cooperative Grant Agreement CR-807417 from the US Environmental Protection Agency. We are thankful to Dr. Thomas

W. Duke, US EPA Environmental Research Laboratory, Gulf Breeze, for his agreement and support during this investigation. We are also grateful to Dr. Robert F. Shokes, Science Applications Inc., La Jolla, California, for performing the chromium analyses, and to Ms. Irene Wu for her assistance.

LITERATURE CITED

Adiyodi, R.G. 1972. Wound healing and regeneration in the crab, *Paratelphusa hycrodromus*. Int. Rev. Cytol. 32: 257-289.

API. 1978. Oil and gas well drilling chemicals. API. Bul. 13F. First Ed., Aug. 1978. American Petroleum Institute, Washington, DC, 9 pp.

Baross, J.A., P.A. Tester, and R.Y. Morita. 1978. Incidence, microscopy, and etiology of exoskeleton lesions in the tanner crab, *Chionocetes tanneri*. J. Fish. Res. Board Can. 35: 1141-1149.

Bazin, F. 1973. Formation de blastemès de régénérationdes péréiopodesautotomisés chez des Crabes *Carcinus maenas* privés de leurs glandes de mue. C.R. Acad. Sci. Paris 276: 2585-2588.

Bazin, F. 1980. Participation des hémocytes à la formation du blastème de régénération chez la Crabe *Carcinus maenas* (L.). Arch. Zool. Exper. Gen. 121: 49-62.

Bell, A.L., S.L. Barnes, and K.L. Anderson. 1969. A fixation technique for electron microscopy which provides uniform good preservation of the tissues of a variety of marine invertebrates. Biol. Bull. 173: 393.

Benoit, D.A. 1976. Toxic effects of hexavalent chromium on brook trout (*Salvelinus fontinalis*) and rainbow trout (*Salmo gairdneri*). Water Res. 10: 497-500.

Biesenger, K.E. and G.M. Christensen. 1972. Effects of various metals on survival, growth, reproduction, and metabolism of *Daphnia magna*. J. Fish. Res. Board Can. 29: 1691-1700.

Bliss, D.E. 1956. Neurosecretion and control of growth in a decapod crustacean. pp. 56-75. *In*: Bertil Hanström, Zoological papers in Honor of his Birthday, November 20, 1956. K.G Wingstrand (ed.). Zoological Institute, Lund, Sweden.

Bliss, D.E. 1960. Autotomy and regeneration. pp. 561-589. *In*: The Physiology of Crustacea. Vol. 1. T.H. Waterman (ed.). Academic Press, New York.

Bubel, A. 1976. Histological and electron microscopical observations on the effects of different salinities and heavy metal ions on the gills of Jeara nordmanni (Rathke) (Crustacea, Isopoda). Cell Tiss. Res. 167: 65-95.

Chang, L.W., L.L.M. MaK and A.H. Martin. 1976. Dose-dependent effects of methymercury on limb regeneration in newts (Triturus viridescens). Environ. Res. 11: 305-309.

Chipman, W.A. 1967. Some aspects of the accumulation of Cr-51 by marine organisms. pp. 931-941. In: Radioecological Concentration Processes. B. Aberg and F.P. Hungate (eds.). Pergamon Press, London.

Conklin, P.J. and K.R. Rao. 1978. Toxicity of sodium pentachlorophenate to the grass shrimp Palaemonetes pugio in relation to the molt cycle. pp. 181-192. In: Pentachlorophenol: Chemistry, Pharmacology, and Environmental Toxicology. K.R. Rao (ed.). Plenum Press, New York.

Conklin, P.J. and K.R. Rao. 1982. Effects of two dithicarbamates on the grass shrimp Palaemonetes pugio: Molt-related toxicity and inhibition of limb regeneration. Arch. Environ. Contam. Toxicol. 11: 431-435.

Cook, D.W. and S.R. Lofton. 1973. Chitinoclastic bacteria associated with shell disease in Penaeus shrimp and the blue crab (Callinectes sapidus). J. Wildl. Dis. 9: 154-159.

Couch, J.A. 1977. Ultrastructural study of lesions in gills of marine shrimp exposed to cadmium. J. Invertebr. Pathol. 29: 267-288.

Couch, J.A. 1978. Diseases, parasites and toxic responses of commercial penaeid shrimps of the Gulf of Mexico and South Atlantic Coasts of North America. Fish. Bull. 76: 1-44.

DeFlora, S. 1981. Study of 106 organic and inorganic compounds in the Salmonella/microsome test. Carcinogenesis 2: 283-289.

Doughtie, D.G., P.J. Conklin, and K.R. Rao. 1983. Cuticular lesions induced in grass shrimp exposed to hexavalent chromium. J. Invertebr. Pathol. 42: 249-258.

Doughtie, D.G. and K.R. Rao. 1983a. Ultrastructural and histological study of degenerative changes leading to black gills in grass shrimp exposed to dithiocarbamate biocide. J. Invertebr. Pathol. 41: 33-50.

Doughtie, D.G. and K.R. Rao. 1983b. Ultrastructural and histological study of degenerative changes in the antennal glands, hepatopancreas and midgut of grass shrimp exposed to two dithiocarbamate biocides. J. Invertebr. Pathol. 41: 281-300.

Doughtie, D.G. and K.R. Rao. 1984. Histopathological and ultrastructural changes in the antennal gland, midgut, hepatopancreas, and gill of grass shrimp following exposure to hexavalent chromium. J. Invertebr. Pathol. 43: 89-108.

Ellis, R.W. and E.G. Fuller. 1979. Mercury interaction in crayfish antennal gland and hepatopancreas. J. Invertebr. Pathol. 34: 308-309.

Fales, R.R. 1978. The influence of temperature and salinity on the toxicity of hexavalent chromium to the grass shrimp *Palaemonetes pugio* (Holthuis). Bull. Environ. Contamin. Toxicol. 20: 447-450.

Fales, R.R. 1979. The influence of temperature and salinity on the accumulation and elimination of chromium by the grass shrimp, *Palaemonetes pugio* (Holthuis). Bull. New Jersey Acad. Sci. 24: 18-21.

Frank, P.M. and P.B. Robertson. 1979. The influence of salinity on toxicity of cadmium and chromium to the blue crab, *Callinectes sapidus*. Bull. Environ. Contam. Toxicol. 21: 74-78.

Fromm, P.O. and R.H. Schiffman. 1958. Toxic action of hexavalent chromium on largemouth bass. J. Wildlife Managem. 22: 40-44.

Goyer, R.A., P. May, M.M. Cates, and M.R. Krigman. 1970. Lead and protein content of isolated intranuclear inclusion bodies from kidneys of lead poisoned rats. Lab. Invest. 22: 245-251.

Hoarau, F. 1973. Comportement de l'hypoderme et progression de la differenciation au cours de la regeneration d'un pereiopode chez l'isopode terrestre *Helleria brevicornis* ebner. Ann. d'Embryol. Morphogen. 6: 125-135.

Hubschman, J.H. 1965. Effects of copper on the crayfish *Orconectes rusticus* (Girard). II. Mode of toxic action. Crustaceana 12: 141-150.

Lee, D.R. and A.L. Buikema, Jr. 1979. Molt-related sensitivity of *Daphnia pulex* in toxicity testing. J. Fish. Res. Board Can. 36: 1129-1133.

Lightner, D.V. 1979. Fungal diseases of marine Crustacea. pp. 451-484. *In*: Pathogenesis of Invertebrate Microbial Diseases. E.W. Davidson (ed.). Allanheld, Osmun, Totowa, NJ.

Mansen, J.M. and E.J. O'Flaherty. 1978. Effects of cadmium on salamander survival and limb regeneration. Environ. Res. 16: 62-69.

Mearns, A.J., P.S. Oshida, M.J. Sherwood, D.R. Young, and D.J. Reish. 1976. Chromium effects on coastal organisms. J. Water Poll. Cont. Fed. 48: 1929-1939.

Mertz, W. 1969. Chromium occurrence and function in biological systems. Physiol. Rev. 49: 163-239.

Moraïtou-Apostolopoulou, M. and G. Verriopoulos. 1982a. Individual and combined toxicity of three heavy metals, Cu, Cd and Cr for the marine copepod _Tisbe holothuriae_. Hydrobiologia 87: 83-87.

Moraïtou-Apostolopoulou, M. and G. Verriopoulos. 1982b. Toxicity of chromium to the marine planktonic copepod _Acartia clausi_, Giesbrecht. Hydrobiologia 96: 121-127.

Nimmo, D.R., D.V. Lightner, and L.H. Bahner. 1977. Effects of cadmium on the shrimps _Penaeus duorarum_, _Palaemonetes pugio_, and _P. vulgaris_. pp. 131-183. _In_: Physiological Responses of Marine Biota to Pollutants. F.J. Vernberg, A. Calabrese, F.P. Thurberg, and W.B. Vernberg (eds.). Academic Press, New York.

Olson, K.R. and R.C. Harrel. 1973. Effect of salinity on acute toxicity of mercury, copper and chromium for _Rangia cuneata_ (Pelecypoda, Mactridae). Contrib. Mar. Sci. 17: 9-13.

Oshida, P.S. and L.S. Word. 1982. Bioaccumulation of chromium and its effect on reproduction in _Neanthes arenaceodentata_ (Polychaeta). Mar. Environ. Res. 7: 167-174.

Pagano, G., A. Esposito, P. Bove, M. DeAngelis, A. Rota, and G.G. Giordano. 1983. The effects of hexavalent and trivalent chromium on fertilization and development in sea urchins. Environ. Res. 30: 442-452.

Pearce, J.B. 1972. The effects of solid waste disposal on benthic communities in the New York Bight. pp. 404-411. _In_: Marine Pollution and Sea Life. M. Ruvio (ed.). Fishing Trading News, Ltd., London.

Pickering, Q.H. and C. Henderson. 1966. The acute toxicity of some heavy metals to different species of warmwater fishes. Air Water Pollut. (England) 10: 453-463.

Price, R.K.J. and R.F. Uglow. 1979. Some effects of certain metals on development and mortality within the molt cycle of _Crangon crangon_ (L.). Marine Environ. Res. 2: 287-299.

Rao, K.R., P.J. Conklin, and A.C. Brannon. 1978. Inhibition of limb regeneration in the grass shrimp, _Palaemonetes pugio_, by sodium pentachlorophenate. pp. 193-203. _In_: Pentachlorophenol: Chemistry, Pharmacology, and Environmental Toxicology. K.R. Rao (ed.). Plenum Press, New York.

Rao, K.R. and D.G. Doughtie. 1984. Histopathological changes in grass shrimp exposed to chromium, pentachlorophenol, and dithiocarbamates. Mar. Environ. Res. 14: 371-395.

Rao, K.R., D.G. Doughtie, and P.J. Conklin. 1982. Physiological and histopathological evaluation of dithiocarbamate toxicity to the grass shrimp, *Palaemonetes pugio*. pp. 413-445. *In*: Physiological Mechanisms of Marine Pollutant Toxicity. W.B. Vernberg, A. Calabrese, F.P. Thurberg, and F.J. Vernberg (eds.). Academic Press, New York.

Rao, K.R., F.R. Fox, P.J. Conklin, anc A.C. Cantelmo. 1981. Comparative toxicology and pharmacology of chlorophenols: Studies on the grass shrimp *Palaemonetes pugio*. pp. 37-72. *In*: Biological Monitoring of Marine Pollutants. F.J. Vernberg, A. Calabrese, F.P. Thurberg, and W.B. Vernberg (eds.). Academic Press, New York.

Rosen, B. 1970. Shell diseases of aquatic crustaceans. pp. 409-415. *In*: Symposium on Diseases of Fishes and Shellfishes. S.F. Sneiszko (ed.). Spec. Publ. No. 5, American Fisheries Society. Washington, DC.

SAS Institute Inc. 1982. SAS User's Guide: Statistics, 1982 Edition. SAS Institute, Cary, NC, 584 pp.

Sather, B.T. 1967. Chromium absorption and metabolism by the crab *Podophthalmus vigil*. pp. 389-510. *In*: Radioecological Concentration Processes. B. Aberg and F. Hungate (eds.). Pergamon Press, New York.

Stella, M., A. Montaldi, R. Rossi, G. Rossi, and A.G. Levis. 1982. Clastogenic effects of chromium on human lymphocytes *in vitro* and *in vivo*. Mutation Res. 101: 151-164.

Towill, L.W., C.R. Shriner, J.S. Drury, A.S. Hammons, and J.W. Holleman. 1978. Reviews on the environmental effects of pollutants: III. Chromium. EPA-600/1-78-023. U.S. Environmental Protection Agency. Cincinnati, Ohio. 288 pp.

Trama, F.B. and R.J. Benoit. 1960. Toxicity of hexavalent chromium to blue gills. J. Water Pollut. Cont. Fed. 32: 868-877.

Weis, J.S. 1976. Effects of mercury, cadmium, and lead salts on limb regeneration and ecdysis in the fiddler crab, *Uca pugilator*. U.S. Fish Wildlife Serv. Fish. Bull. 74: 464-467.

Weis, J.S. 1977. Limb regeneration in fiddler crabs: Species differences and effects of methyl mercury. Biol. Bull. 152: 263-274.

Weis, J.S. 1978. Interactions of methylmercury, cadmium, and salinity on regeneration in the fiddler crabs *Uca pugilator*, *U. pugnax* and *U. minax*. Mar. Biol. 49: 119-124.

Young, J.S. and J.B. Pearce. 1975. Shell disease in crab and lobster from New York Bight. Mar. Pollut. Bull. 6: 101-105.

Zaroogian, G.E. and M. Johson. 1983. Chromium uptake and loss in the bivalves *Crassostrea virginica* and *Mytilus edulis*. Mar. Ecol. Prog. Ser. 12: 167-173.

COMPARISON OF METAL METABOLISM AND METAL-BINDING PROTEINS IN THE BLUE CRAB AND THE AMERICAN LOBSTER

David W. Engel, Marius Brouwer, and Frederick P. Thurberg

SUMMARY

1. When exposed to cadmium via food, lobsters and blue crabs both actively accumulate the metal in the digestive gland, but when exposed to cadmium dissolved in sea water, only blue crabs accumulate large amounts in the gills.
2. Lobsters exposed to cadmium in the laboratory have only copper-metallothionein in the digestive gland and a small amount of cadmium-metallothionein in the gills, while blue crabs have a cadmium/zinc and a copper-metallothionein in the digestive gland and a cadmium-metallothionein in the gills. In the digestive glands of both exposed and control lobsters, cadmium was bound primarily to a 30,000 M_r protein and zinc to a 45,000 M_r protein. Neither of these proteins appear to be metallothioneins.
3. Both laboratory- and environmentally-exposed lobsters and blue crabs have been shown to have the same cytosolic distribution of metals (cadmium, copper, and zinc) in the digestive gland.
4. Preliminary experiments with lobsters have demonstrated that metallothioneins are not only important in metal detoxification, but are also involved in the molting process and hemocyanin synthesis.

INTRODUCTION

The American lobster, *Homarus americanus*, and the blue crab, *Callinectes sapidus*, have the capacity to concentrate trace metals, such as copper, zinc and cadmium, in their tissues. Because these animals are commercially important, some concern has been expressed about the cadmium concentrations measured in the digestive glands of animals from some locations. Lobsters collected from Belledune Harbor, New Brunswick, Canada, for example, had >400 μg/g wet weight of cadmium in the digestive gland (Uthe & Zitko, 1980; Ray & White, 1981). Blue crabs taken from Foundry Cove on the Hudson River in New York also had elevated cadmium levels (10-20 μg/g wet weight) in their digestive glands (Wiedow, 1981; Wiedow *et al*., 1982; Engel & Brouwer, 1984). Cadmium concentrations in the digestive glands of both species are of public health concern, since the organ is routinely consumed by area residents.

In addition to measurements of tissue concentrations and total body-burdens of cadmium, determinations also have been made of the low molecular weight cadmium-binding, metallothionein-like proteins in the gills and digestive glands of both lobster and blue crab (Ray & White, 1981; Wiedow, 1981; Wiedow *et al*., 1982; Engel & Brouwer, 1984, in press). Metallothioneins were first isolated from mammalian kidneys (Margoshes & Vallee, 1957) and were characterized as being low molecular weight, sulfur-containing, cadmium-binding proteins whose primary function was the detoxification of cadmium. More recently these proteins have been shown to bind copper and zinc, and are probably involved in normal metal metabolism (Brady, 1982). In other crustaceans the demonstrated characteristics of metallothionein-like proteins are low molecular weight (<10,000 M_r), heat stability, the ability to bind a variety of metals, and high concentrations of cysteine (25-30%) (Olafson *et al*., 1979; Roesijadi, 1981; Lerch *et al*., 1982; Otvos *et al*., 1982; Brouwer *et al*., 1984). These characteristics are similar to those described for mammalian metallothionein (Cherian & Goyer, 1978). These previous investigations primarily characterized metallothionein and metallothionein-like proteins as detoxifying agents. We hypothesize that metallothionein-like proteins in crustaceans are important both in normal physiological function and in metal detoxification.

In this paper we discuss the accumulation of cadmium by lobsters and compare the tissue distribution and intracellular partitioning of cadmium, copper, and zinc to our previous study of the blue crab (Engel, 1983; Engel & Brouwer, 1984). In addition, we discuss the possible physiological and detoxifying functions of metallothionein-like proteins in both species.

MATERIALS AND METHODS

Lobsters used in these experiments were collected in Long Island Sound, Connecticut. Two groups of lobsters were exposed for 60 d to 5 or 10 μg/L dissolved cadmium in a flowing water exposure system at the Milford Laboratory, and were fed chopped clam (*Spisula solidissima*) daily. A third group of unexposed lobsters was air-shipped to the Beaufort Laboratory. Upon arrival at Beaufort, the animals were kept in flowing sea water for 48 h so that animals injured during transport could be removed. The Beaufort lobsters were exposed to cadmium either via food (*i.e.*, cadmium-laden oysters) or via water over a period of 14 d in flowing water exposure systems at 15°C and 30°/$_{oo}$ salinity (these are the same techniques used with blue crabs, Engel & Brouwer, 1984). The concentration of cadmium in the water was 100 μg/L ± 20%. The concentrations of cadmium, copper, and zinc in the cadmium-enriched oysters were 88.5 ± 23, 19 ± 2, and 668 ± 67 μg/g wet weight, and for control oysters <0.3, 12 ± 2, and 430 ± 100 μg/g wet weight respectively (mean of five oysters ± SE).

The lobsters were killed by removing the carapace; the digestive gland, gills, and tail muscle were dissected out. Half of each tissue sample was used to determine total metal concentrations and the remainder was stored at -20°C until determining metal-binding proteins. Soluble metal-binding proteins were isolated from these tissues according to the modifed procedure of Ridlington and Fowler (1979). Tissues were homogenized in a Brinkman Polytron[2] homogenizer at high speed in 1.5 volumes of 10 mM phosphate buffer pH 8 and 5 x 10^{-4} M phenylmethyl sulfonylflouride (PMSF) at 4°C, and then centrifuged at 30,000 x g for 30 min at 4°C. The supernatant was heat-treated at 60°C for 10 min, cooled in ice to 0°C, and then centrifuged again at 30,000 x g for 30 min before being applied to a Sephadex G-75 gel filtration column (64 x 2.6 cm). The elutant was 10 mM phosphate buffer pH 8. The column was calibrated using molecular weight standards (blue dextran > 10^6 M_r, bovine serum albumin 6.7 x 10^4 M_r,

[2]Reference to trade names does not imply endorsement by the National Marine Fisheries Service, NOAA.

ovalbumin 4.3 x 10^4 M_r, chymotrypsinogen A 2.5 x 10^4 M_r, and ribonuclease 1.37 x 10^4 M_r). For further purification, fractions in the 10,000 M_r region of the G-75 elution profile were pooled and applied directly to either a DEAE Cellulose DE-52 or DEAE Sephacel ion-exchange column (2.6 x 15 cm). Proteins were eluted with a gradient generated from 800 ml 20 mM Tris and 800 ml 400 mM Tris, both at pH 8.6. Fractions were analyzed for absorbance at 280 and 254 nm, and metals were measured by flame aspiration atomic absorption spectrophotometry.

Tissue samples were analyzed for cadmium, copper, and zinc by using standard atomic absorption spectrophotometric techniques. Samples were dried to 90°C, wet-ashed in concentrated HNO_3, and then diluted and analyzed by flame aspiration atomic absorption spectrophotometry.

The National Bureau of Standards Oyster Reference Material # 1566 was used to calibrate the zinc, cadmium, and copper measurements. The certified concentrations in the Standard Reference Material were 852 ± 14 μg/g zinc, 3.5 ± 0.4 μg/g cadmium, and 63 ± 3.5 μg/g copper (mean ± 95% confidence interval). Mean concentrations for 10 replicate aliquots of the Standard measured in our laboratory were 843 ± 63 μg/g zinc, 3.4 ± 0.4 μg/g cadmium, and 60.2 ± 3.2 μg/g copper. Our mean values are in agreement with the specified elemental concentrations.

RESULTS AND DISCUSSION

In the tissues of lobsters exposed to 5 or 10 μg/L of dissolved cadmium, concentrations of cadmium and copper were highest in the digestive gland, intermediate in the gill tissue, and lowest in the muscle. Zinc was high in the digestive gland, but low in both gills and muscle. Cadmium concentrations in the digestive gland were similar at both 5 and 10 μg/L, but were significantly higher ($p<0.05$) than the controls by Student's T-test (Table 1). Cadmium in the gills also increased as the dosage increased, but the differences were not significant ($p>0.05$). Cadmium levels in the digestive glands of the control animals indicate that there is a significant background level, and exposure at 10 μg/L for 60 d increased this level by slightly less than a factor of two. Copper and zinc concentrations in the gill and digestive gland of cadmium-exposed lobsters did not change as a result of cadmium exposure (Table 1).

Sephadex G-75 elution profiles of the digestive glands from control and cadmium-exposed lobsters show the same number of components which bind cadmium, zinc, and copper (Fig. 1 & 2A). All three metals were associated with high molecular weight material eluting in the void volume (fractions 23-30), and with very low molecular weight fractions (<5,000 M_r) which contain amino acids and small peptides (fractions>65). A cadmium-containing peak also was present at an intermediate molecular weight of about 30,000 M_r (fractions 37-50). In addition, zinc was bound to a protein eluting at a molecular weight of about 45,000 M_r (fractions 30-40). Copper was the only metal associated with a protein peak in the 10,000 M_r region (fractions 45-57), which corresponds to the molecular weight of metallothionein-like proteins. This metallothionein-like protein differs from those observed in the digestive glands of blue crabs (Engel & Brouwer, 1984), which contained cadmium and zinc in addition to copper.

Table 1. Cadmium, copper, and zinc concentrations in tissues of lobsters, exposed to 5 or 10 μg/L cadmium in the laboratory for 60 days. (All values are μg/g wet weight of tissue ± SE).*

Tissue	Number of Individuals	Control	5 μg/L	10 μg/L
Cadmium				
Digestive Gland	(11)	9.4 ± 0.8	16.2 ± 1.8	18.2 ± 1.6
Gill	(11)	1.8 ± 1.0	3.0 ± 1.0	4.7 ± 1.1
Tail Muscle	(11)	<0.1	<0.1	<0.1
Copper				
Digestive Gland	(11)	618 ± 52	460 ± 62	645 ± 50
Gill	(11)	30 ± 2.1	27 ± 3.5	25 ± 2.5
Tail Muscle	(11)	5.1 ± 0.4	4.4 ± 0.7	4.3 ± 0.4
Zinc				
Digestive gland	(11)	23 ± 2.8	35 ± 3.0	32 ± 1.9
Gill	(11)	9.5 ± 0.4	15 ± 0.6	9.7 ± 1.9
Tail Muscle	(11)	15 ± 0.4	15 ± 0.6	15 ± 0.5

*The cadmium, copper, and zinc concentrations in lobster tissues presented in this table are the mean values for two separate groups of lobsters exposed to cadmium at the Milford Laboratory. Lobsters were fed surf clams which had tissue metal concentration of 0.1 ± 0.03 cadmium, 0.3 ± 0.04 copper, and 12 ± 0.4 zinc μg/g wet weight (from Hall *et al.*, 1978).

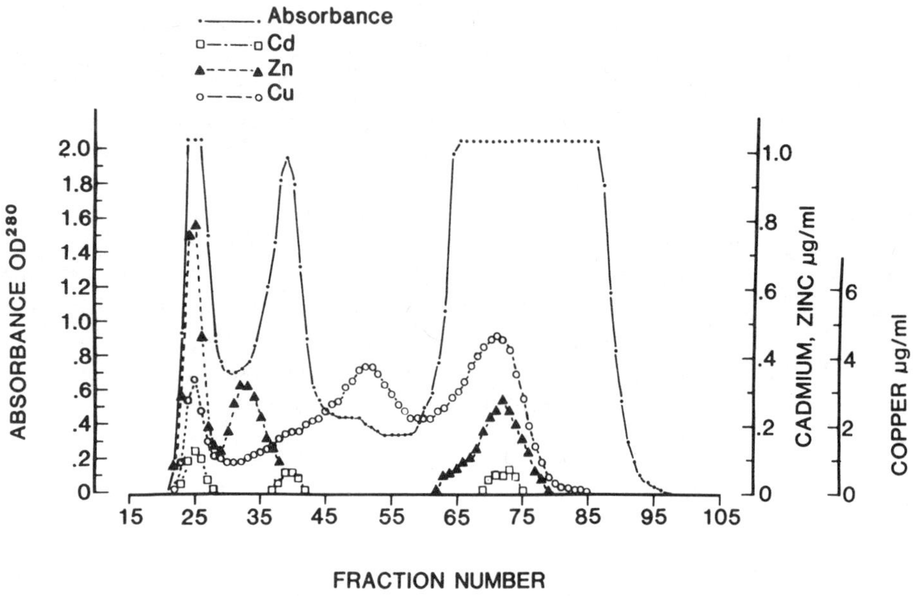

Fig. 1. Sephadex G-75 elution profile of the metal-binding proteins in the cytosol of the digestive gland from a control lobster collected from Long Island Sound near Milford, CT. Molecular weight decreases with increasing fraction number.

All samples of gill tissue from the group of lobsters exposed to 10 µg/L dissolved cadmium were pooled and the cytosolic fraction concentrated on an Amicon UM-2 filter and chromatographed on Sephadex G-75. The elution profile shows a single cadmium-containing protein peak with a small amount of copper (Fig. 2B). The protein eluted in the molecular weight range appropriate for metallothionein-like proteins, and was similar to the proteins demonstrated in blue crab gills exposed to dissolved cadmium (Engel & Brouwer, 1984).

In the second series of experiments, lobsters were exposed for 14 d either to cadmium dissolved in the water or to cadmium incorporated into oysters. Cadmium concentrations in the digestive glands of lobsters that were fed cadmium-rich oysters were about five times higher than the controls. Lobsters which were exposed to cadmium via the water had cadmium concentrations that were similar to the controls (Table 2). Concentrations of cadmium in the gills of lobsters exposed to cadmium via water were significantly higher ($p<0.05$) than the concentrations of cadmium in the gills of the controls. It appears, therefore, that lobsters readily accumulate cadmium from food into the digestive gland. In contrast, accumulation of metal into the digestive gland from water appears to be slow. Even though lobsters accumulate significant amounts of cadmium from the water into the gills, the levels remain low. Also, whether the source of cadmium was food or water, relatively low concentrations of cadmium were present in the edible muscle of the lobsters.

Table 2. The tissue concentrations of cadmium, copper, and zinc in the digestive glands, gills, and tail muscles of American lobsters exposed to 100 μg/L cadmium in the water or cadmium-laden oysters for 14 days. Metal concentrations are μg/g wet weight of tissues ± SE.

Treatment	Tissue	Number of Individuals	Cadmium	Copper	Zinc
Control	Digestive gland	(5)	7.5 ± 1.0	506 ± 56	99 ± 18
	Gill	(5)	0.7 ± 0.1	29 ± 5	16 ± 2
	Tail muscle	(5)	<0.1	8 ± 1	21 ± 2
Water	Digestive gland	(5)	10 ± 2	492 ± 174	420 ± 119
	Gill	(5)	2 ± 0.4	24 ± 7	18 ± 2
	Tail muscle	(5)	<0.1	5 ± 1	16 ± 1
Food	Digestive gland	(5)	41 ± 15	863 ± 326	200 ± 37
	Gill	(5)	1.4 ± 0.5	30 ± 7	18 ± 2
	Tail muscle	(5)	<0.1	6.8 ± 1	17 ± 0.5
Food and Water (40 day depuration)	Digestive gland	(4)	25 ± 4	334 ± 167	246 ± 3
	Gill	(4)	3 ± 1	32 ± 5	25 ± 3
	Tail muscle	(4)	<0.1	10 ± 2	23 ± 4

*All lobsters were fed oysters, but only lobsters in the food exposure groups received oysters containing elevated levels of cadmium. The mean metal concentrations for the normal and cadmium-laden oysters were cadmium 0.3 ± 0.02 and 89 ± 23; copper 12 ± 2 and 19 ± 2; and zinc 440 ± 114 and 668 ± 67 μg/g wet weight respectively.

In these experiments, rates of accumulation of cadmium from food or water could not be determined directly, because the modes of exposure were not equivalent. The data show that the tissue distribution of cadmium is influenced by the route of exposure (Table 2). Concentrations of copper and zinc in gill and muscle were in good agreement with those we had measured previously in exposed lobsters from Milford, Connecticut (Table 1 & 2). However, all 14-d exposed lobsters had significantly higher concentrations of zinc in the digestive glands than those exposed for 60 d to 5 or 10 μg/L cadmium and fed surf clams ($p<0.05$) (Table 1). These high zinc concentrations in the digestive gland were undoubtedly caused by the diet of oysters that are naturally high in zinc (Table 2). No such dramatic changes in zinc concentration in either gill or muscle tissue were demonstrated. In the digestive glands, copper concentrations that were high and extremely variable within treatment groups, showed no relation with cadmium exposure, and were in good agreement with those in the Milford group of lobsters.

Four lobsters (Beaufort group), two animals from each exposure group, were maintained in flowing sea water for 40 d and allowed to depurate. Since tissue concentrations of cadmium, copper, and zinc were similar for all animals, the data were pooled (Table 2). The cadmium concentrations in the digestive glands were significantly higher than in the controls ($p<0.05$), and intermediate between the two exposure groups that had been sampled immediately. The relatively high levels of cadmium in the digestive glands of lobsters exposed to cadmium in water may have been due to the redistribution of the cadmium body-burden during the depuration period. Retention of a substantial portion of the cadmium in the digestive gland is in agreement with McLeese *et al.* (1981), where lobsters removed from high cadmium environments retained a high percentage of the metal in their digestive glands for up to a year.

Sephadex G-75 gel chromatography also was done on the soluble protein or cytosolic fraction of digestive glands and gills of all three groups of exposed lobsters (Table 2), but only digestive glands had demonstratable metal-binding proteins (Fig. 3A). The partitioning of trace metals in the digestive gland cytosol from a lobster exposed to cadmium via food (Beaufort group) was similar to the profiles derived from lobsters exposed to cadmium via the water (Milford group) (Fig. 2A). In these experiments, even though the concentrations of zinc in the digestive glands were elevated, no zinc-containing, metallothionein-like proteins were demonstrated; only a zinc-containing protein at about 45,000 M_r was observed. Only copper-binding protein eluted in the region that was characteristic of metallothionein-like protein.

To further characterize these copper proteins, fractions (52-65) from the G-75 elution profile were pooled (Fig. 3A) and further purified on a DEAE cellulose ion-exchange column. The resulting profile showed a majority of the copper passing directly through the column (Fig. 3B). A second copper-containing protein eluted from the column at a conductivity of 1.1 mMho/cm and coincided with a strong OD absorbance peak at 250 nm. The ultraviolet spectra of the protein were consistent with those of a copper-thionein complex.

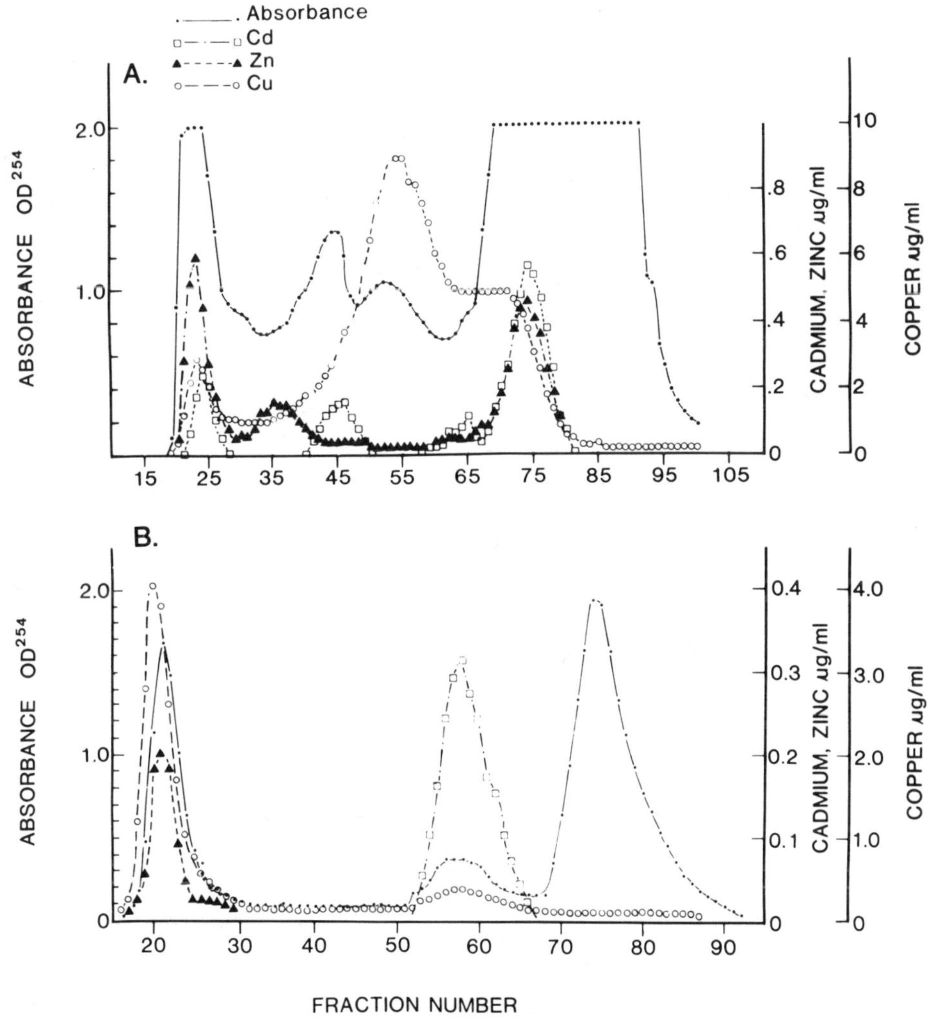

Fig. 2. Sephadex G-75 elution profile of the metal-binding proteins in the cytosol of the digestive gland (A) and gills (B) of lobster exposed for 60 days to 10 µg/L cadmium dissolved in sea water. The gills were a pooled sample from six exposed lobsters.

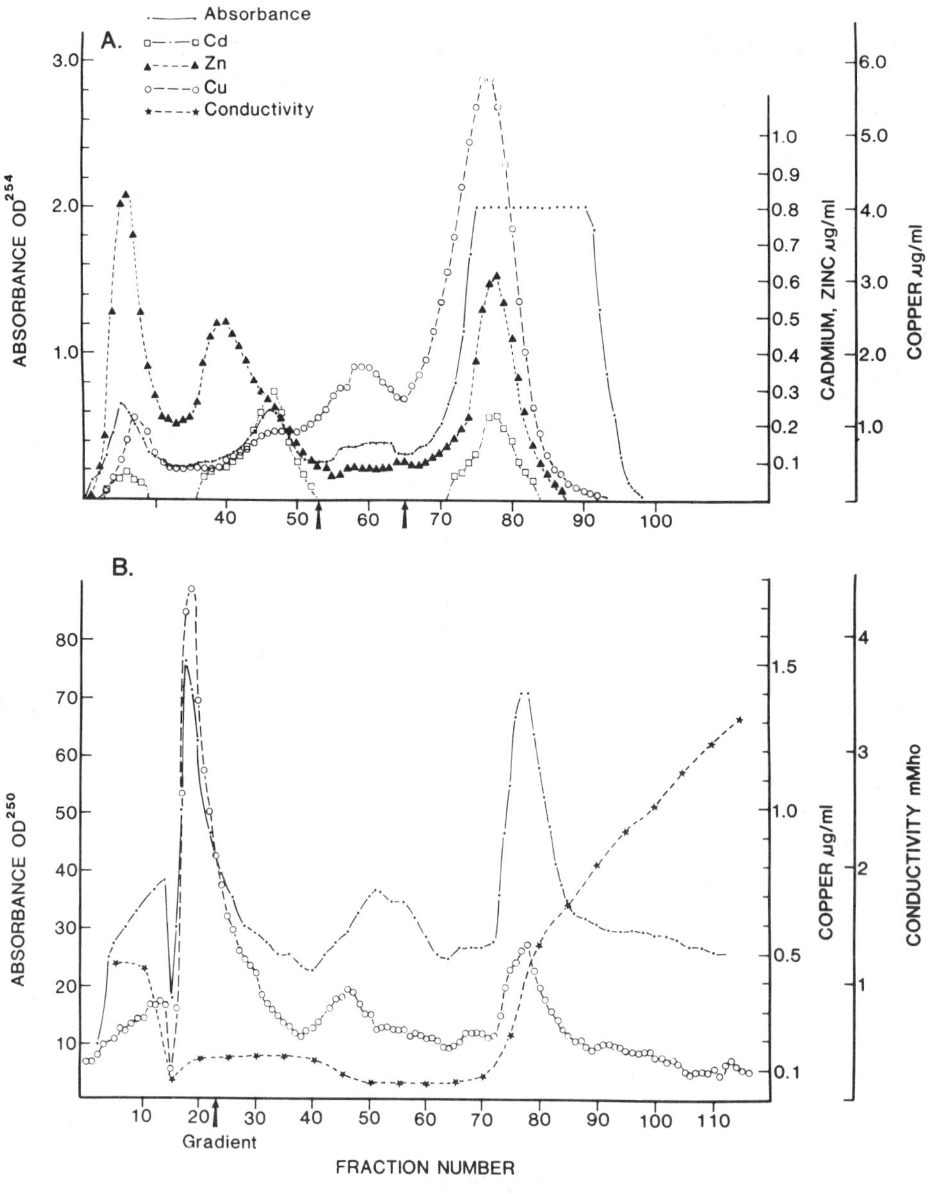

Fig. 3. (A) Sephadex G-75 elution profile of metal-binding proteins in the cytosol of the digestive gland from a lobster fed cadmium-laden oysters for 14 days. Arrows indicate fractions pooled for ion-exchange chromatography. (B) DEAE cellulose ion-exchange profile of the copper-binding proteins isolated by G-75 chromatography from the digestive gland of lobster (fractions 52-65).

By comparison, our laboratory data from blue crabs exposed to cadmium via food or water has shown that the tissue distribution of cadmium reflects the route of uptake (*i.e.*, digestive gland concentrations were highest when cadmium had been in food, gill concentrations were highest when cadmium had been dissolved in water [Engel, 1983]). It also was demonstrated that the Sephadex G-75 elution profiles from digestive glands of blue crabs exposed to cadmium via food contained metallothionein-like proteins in the 10,000 M_r range that bind copper, cadmium, and zinc (Fig. 4A), and a protein that binds only cadmium was present in the gills of crabs exposed to cadmium via water (Fig. 4B) (Engel & Brouwer, 1984). Additionally, further purification of these metal-binding proteins by DEAE-cellulose ion-exchange chromatography yielded different protein-metal combinations depending upon the tissue. In the digestive gland two distinct metal-binding proteins were present, one containing cadmium and zinc and another containing only copper (Fig. 4C); in the gills only a single cadmium-binding protein was present (Fig. 4D) (Engel & Brouwer, in press). The cadmium-binding proteins from both the digestive gland and gills also were characterized as metallothioneins (Brouwer *et al.*, 1984). Unlike blue crabs, the cytosolic metal distributions in the digestive glands of lobsters had a naturally-occurring copper-binding metallothionein-like protein occurring in the 10,000 M_r fractions on Sephadex G-75 (Figs. 1, 2A, & 3A). The observed cadmium-binding protein in lobster gills appears to be similar to that of the blue crab.

The molt cycle in crustaceans is a complex metabolic process that involves the breakdown and regeneration of tissues, as well as the mineralization of specific tissues. This process also may significantly influence the turnover and translocation of accumulated trace metals. For example, Zuckerkandl (1960) reported that the spider crab, *Maia squinado*, did not contain hemocyanin, the copper-containing respiratory protein, during the molting period. Correlations between blood copper concentration and molt stage were demonstrated in the shrimp, *Crangon vulgaris* (Djangmah & Grove, 1970). These data strongly suggest an intricate relationship between copper, hemocyanin, and molting. Since we have demonstrated that lobsters have large quantities of copper metallothionein-like proteins, there may be a connection between molting, copper, and metallothionein.

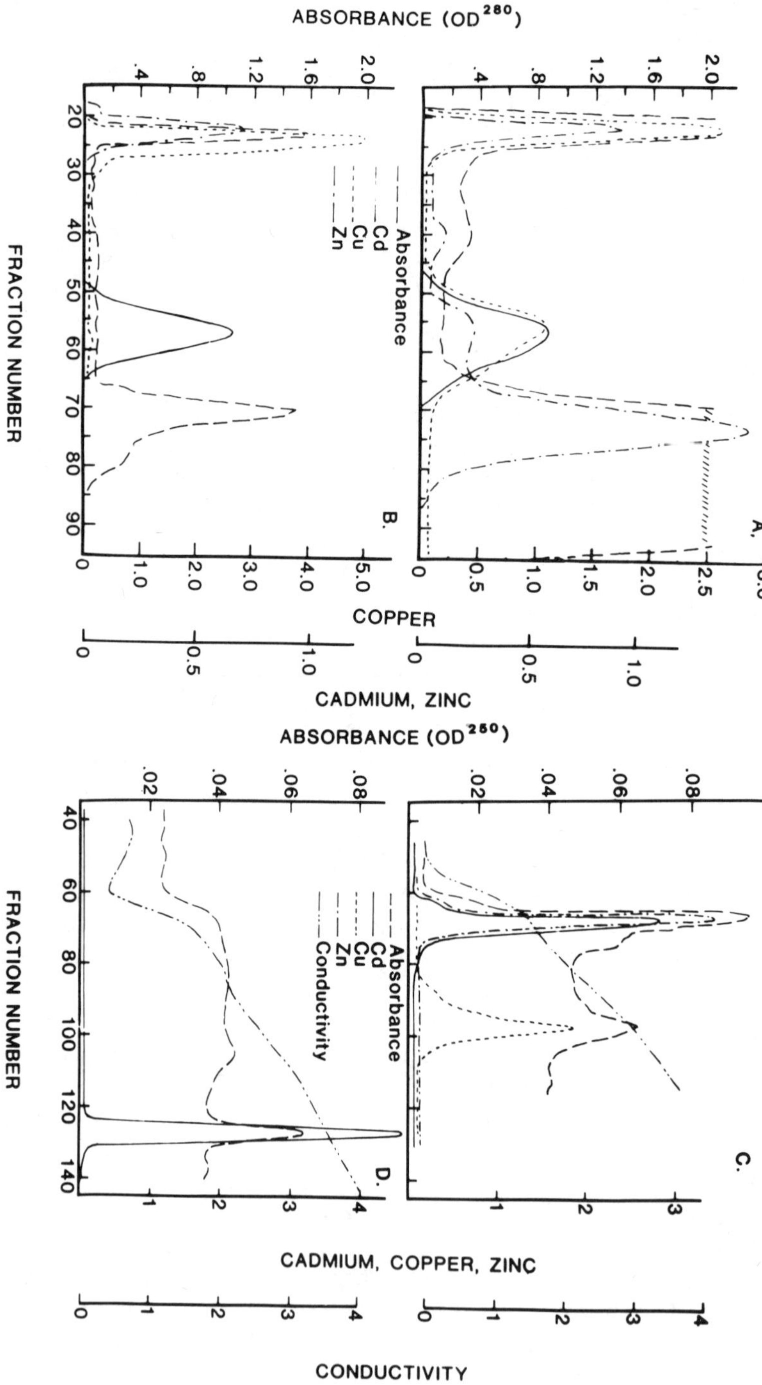
ABSORBANCE (OD280)
A.
B.
Absorbance
Cd
Cu
Zn
FRACTION NUMBER
COPPER
CADMIUM, ZINC
ABSORBANCE (OD250)
C.
D.
Absorbance
Cd
Cu
Zn
Conductivity
FRACTION NUMBER
CADMIUM, COPPER, ZINC
CONDUCTIVITY

Fig. 4. Sephadex G-75 elution profiles of the metal-binding proteins from the cytosol of digestive gland (A) and gills (B) of blue crabs exposed to cadmium from food (dosed oysters) (Engel & Brouwer, 1984) and water (100 µg/L) for 14 days and ion-exchange elution profiles of Sephadex G-75 pooled fractions of (fractions 47-65) digestive gland (C) and (fractions 50-65) gills (D). (Engel & Brouwer, in press).

Among the control lobsters in the food/water experiments was an individual that was very close to molting (i.e., stage D). While the tissue concentrations of trace metals were the same as in the other control lobsters (Table 2), the Sephadex G-75 elution profile was distinctly different (Fig. 5A). The zinc distribution in the profile showed only one major peak, in the very low molecular weight fractions, and copper followed the same distributional pattern observed previously with a distinct metallothionein-like protein peak at 10,000 M_r. Cadmium, however, was detected under the metallothionein-like protein peak and none in the 30,000 M_r range, which was quite different from all previously analyzed lobsters.

Purification of the copper/cadmium metallothionein-like protein peak on DEAE-cellulose produced an elution profile with a great deal of fine structure (Fig. 5B). Unlike the earlier profile (Fig. 3B), there are five distinct absorbance peaks at 250 nm. Because cadmium was present only in trace amounts in the G-75 fractions, it was not detected, but copper eluted as a single sharp peak coincident with the second major absorbance peak. The elution conductivity of this peak was about 2.5 mMho/cm, while the copper protein from the intermolt animals was about 1.3 mMho/cm. Such differences indicate that the two proteins are probably different, even though the absorbance spectra of the isolated proteins are both typical of metallothionein-like proteins. In addition to the copper protein, a calcium-binding protein also was demonstrated. The observed calcium peak coincided with a shoulder on the first absorbance peak and did not overlap with the copper protein peak. We have never been able to demonstrate a calcium protein in the 10,000 M_r range for intermolt lobsters. It seems reasonable, therefore, to assume that differences both in the copper-binding proteins and the presence of the calcium-binding protein are related to the molt cycle.

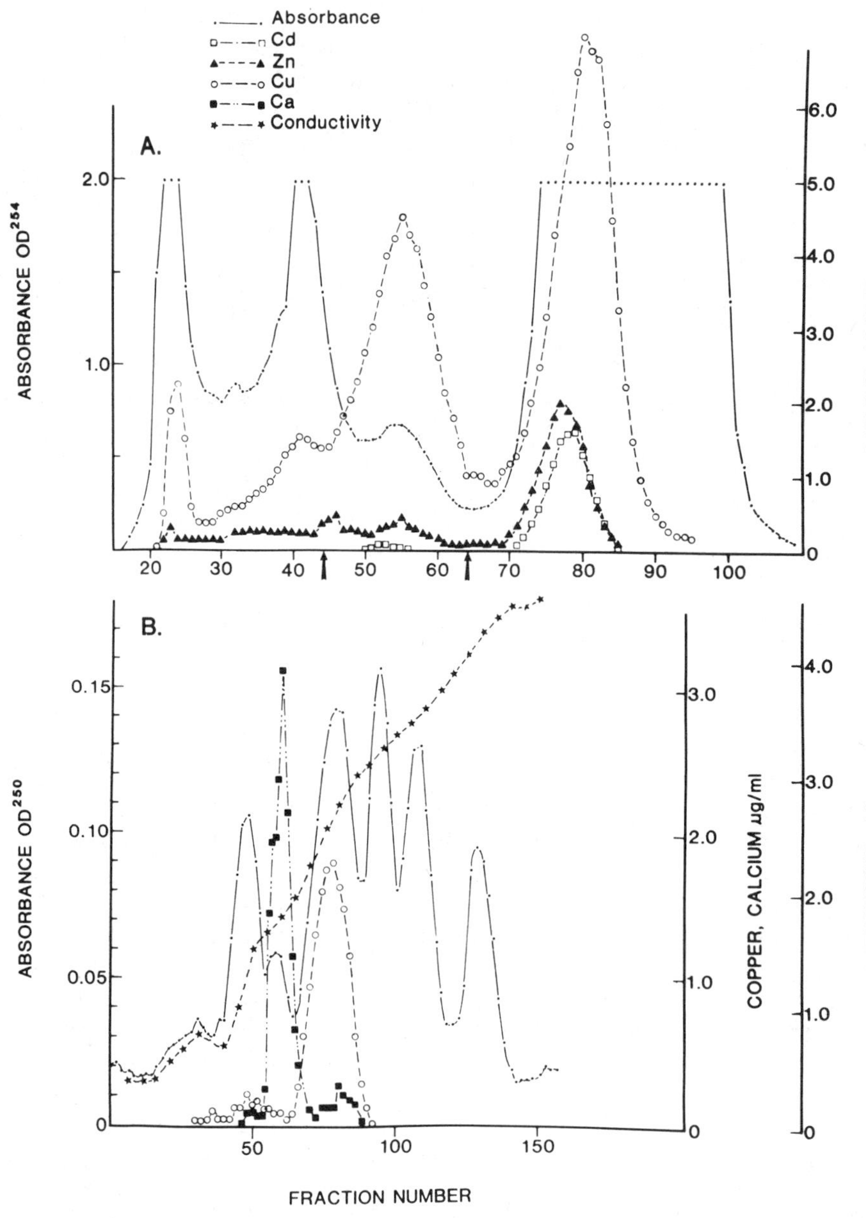

Fig. 5. (A) Sephadex G-75 elution profile of metal-binding proteins from the cytosol of digestive gland from a pre-molt lobster. Arrows indicate fractions pooled for ion-exchange chromatography. (B) DEAE cellulose ion-exchange elution profile of the low molecular weight proteins isolated by Sephadex G-75 chromatography of digestive gland cytosol (fractions 45-65).

Recent reconstitution experiments with lobster hemocyanin (Brouwer, unpubl. data) have demonstrated that copper can be transferred from metallothionein to apohemocyanin (*i.e.*, a hemocyanin molecule without copper, which cannot bind oxygen). Thus, the large quantities of copper-metallothionein in the digestive glands of lobsters may be required in hemocyanin synthesis and degradation pathways.

Such observations indicate that metallothionein-like proteins are not only actively involved in detoxifying accumulated trace metals, but also are involved in the normal metal metabolism and physiology of lobsters and other crustaceans. These data also have demonstrated that there is more than one type of metallothionein-like protein in an organism, and that depending upon the life-stage and physiological condition, the metal composition of these proteins can change. Therefore, it appears that metallothionein-like proteins are operative at many levels in the maintenance of normal organismal function.

ACKNOWLEDGMENTS

The authors would like to thank Drs. Celia and Joseph Bonaventura of the Duke University Marine Laboratory/Biomedical Center for their help in reviewing this manuscript. The research data presented in this paper was funded by NOAA's National Ocean Services, Ocean Assessment Division through a cooperative research program with the Southeast Fisheries Center's Beaufort Laboratory and the Duke University Marine Laboratory.

LITERATURE CITED

Brady, F.A. 1982. The physiological function of metallothionein. Trends Biochem. Sci. 7: 143-145.

Brouwer, M., T. Brouwer-Hoexum, and D.W. Engel. 1984. Cadmium accumulation by the blue crab, *Callinectes sapidus*: Involvement of hemocyanin and characterization of cadmium-binding proteins. Mar. Environ. Res. 14: 71-88.

Cherian, M.G. and R.A. Goyer. 1978. Metallothioneins and their role in the metabolism and toxicity of metals. Life Sci. 23: 1-10.

Djangmah, J.S. and D.J. Grove. 1970. Blood and hepatopancreas copper in *Crangon vulgaris*. Comp. Biochem. Physiol. 32: 733-745.

Engel, D.W. 1983. The intracellular partitioning of trace metals in marine shellfish. Sci. Total Environ. 28: 129-140.

Engel, D.W. and M. Brouwer. 1984. Cadmium-binding proteins in the blue crab, Callinectes sapidus: Laboratory-field comparison. Mar. Environ. Res. 14: 139-151.

Engel, D.W. and M. Brouwer. (in press). Trace metal-binding proteins in marine molluscs and crustaceans. Mar. Environ. Res.

Hall, R.A., E.G. Zook, and G.M. Meaburn. 1978. National Marine Fisheries Service Survey of Trace Elements in the Fishery Resource. NOAA Technical Report NMFS SSRF-721. 313 pp.

Lerch, K., D. Ammer, and R.W. Olafson. 1982. Crab metallothionein. Primary structures of metallothioneins 1 and 2. J. Biol. Chem. 257: 2420-2426.

Margoshes, M. and B.L. Vallee. 1957. A cadmium protein from equine kidney cortex. J. Am. Chem. Soc. 79: 4813-4814.

McLeese, D.W., S. Ray, and L.E. Burridge. 1981. Lack of excretion of cadmium from lobsters. Chemosphere 10: 775-778.

Olafson, R.W., A. Kearns, and R.G. Sim. 1979. Heavy metal induction of metallothionein synthesis in the hepatopancreas of the crab Scylla serrata. Comp. Biochem. Physiol. 62B: 417-424.

Otvos, J.D., R.W. Olafson, and I.M. Armitage. 1982. Structure of an invertebrate metallothionein from Scylla serrata. J. Biol. Chem. 257: 2427-2431.

Ray, S. and M. White. 1981. Metallothionein-like protein in lobsters (Homarus americanus). Chemosphere 10: 1205-1213.

Ridlington, S.W. and B.A. Fowler. 1979. Isolation and partial characterization of cadmium-binding protein from the American oyster (Crassostrea virginica). Chem. Biol. Interactions 25: 127-138.

Roesijadi, G. 1981. The significance of low molecular weight, metallothionein-like protein in marine invertebrates: Current status. Mar. Environ. Res. 4: 167-169.

Uthe, J.F. and V. Zitko. 1980. Cadmium Pollution of Belledune Harbor, New Brunswick, Canada. Can. Tech. Rep. Fish. Aquat. Sci. 962. 107 pp.

Wiedow, M.A. 1981. Distribution and binding of cadmium in the blue crab (Callinectes sapidus): Implications in human health. Ph.D. Thesis, New York University.

Wiedow, M.A., T.J. Kneip, and S.J. Garte. 1982. Cadmium binding proteins from blue crabs (*Callinectes sapidus*) environmentally exposed to cadmium. Environ. Res. 28: 164-170.

Zuckerkandl, E. 1960. Hemocyanine et cuivre chez un crustace decapode, dans leurs rapports avec le cycle d'intermue. Ann. Inst. Oceanogr., Monaco T-38, 1-122.

THE INFLUENCE OF METAL AND NONMETAL STRESSORS ON HEPATIC METAL-BINDING PROTEIN PRODUCTION IN BUFFALO SCULPIN, *ENOPHRYS BISON*

T.D. Sabourin, D.B. Gant, and L.J. Weber

SUMMARY

1. An inducible metallothionein-like protein occurs in the liver of the buffalo sculpin (*Enophrys bison*).

2. At 12°C, sublethal Cd administration by either ip injection or water exposure elicits synthesis of the metallothionein-like protein in liver tissue.

3. The combination of temperature and handling stress was found to induce synthesis of the metallothionein-like protein.

4. The amount and turnover of the induced hepatic protein is a function of acclimation temperature.

5. Preliminary data indicate a direct correlation between metallothionein-like protein concentration and total Cd burden in liver tissue.

INTRODUCTION

Coastal and estuarine fishes perform routine functions during extreme daily and seasonal changes in ambient environmental conditions. Physiological compensation allows for short- and long-term adaptation to environmental variables such as temperature. Sensitivity to potentially stressful anthropogenic materials may be influenced by the relative physiological disposition of a fish with regard to the state of this compensation.

These interactions are of consequence to researchers interested in indexing biochemical and physiological responses to certain pollutants in biomonitoring programs. One index which has received considerable attention is the relationship between the concentration of metals in sea water and in fish tissues, and the tissue content of metallothionein-like metal-binding proteins. This well documented relationship, between metal and protein contents, might also be responsible for the observed enhanced tolerance of fish to certain heavy metals (McCarter & Roch, 1983).

Virtually nothing is known of how various environmental conditions might affect the production and metabolism of metal-binding proteins. These proteins play an as-yet unidentified role in trace metal metabolism (Chen *et al*., 1974). Since effective water concentrations as well as tissue uptake of metals are affected by environmental conditions, much remains to be learned from the study of interactions among metals, environmental factors, and related biological responses. Our objective was to investigate the relationship between cadmium and metal-binding proteins in an estuarine fish with little to no prior exposure to heavy metals. For this study we chose buffalo sculpin, a commonly occurring fish in Yaquina Bay and the nearby coastal waters.

A preliminary characterization of the metallothionein-like metal-binding protein in the sculpin was conducted using traditional gel chromatographic procedures. This was followed by the development of a rapid assay technique for estimating the concentration of metal-binding protein content in buffalo sculpin tissues. The assay has been used extensively in mammalian metallothionein studies (Piotrowski *et al*., 1983). We used the assay in a determination of the rate and extent of hepatic metal-binding protein synthesis and observed the effect of temperature on these phenomena.

METHODS AND MATERIALS

Fish Collection and Maintenance

Buffalo sculpin (*Enophrys bison*) of mean weight 192 ± 118 (SD) grams (range 26-957 g) were collected by otter trawl from Yaquina Bay, Oregon, and placed in flowing water (12 ± 1°C, 30 ± 2 $^{\circ}/_{\circ\circ}$ S, pH 7.5) at the Oregon State University Marine Science Center, Newport. Sculpin were fed chopped squid three times per week throughout the experiments. Fish were held in the laboratory for a minimum of

2 wk prior to use in experiments. For temperature-related work, sculpin were acclimated to the new temperature (6 or 18°C) for a minimum 2-wk period prior to experiments to allow for thermal adaptation.

Cadmium Bioassay

Most experiments were carried out by intraperitoneal (ip) administration of $CdCl_2 \cdot 2H_2O$ prepared in 1.1% NaCl (saline). For these exercises, $CdCl_2$-injected, saline-injected, and uninjected fish were held in 200-L flowing sea water tanks. An initial Cd bioassay was performed to determine proper sublethal concentrations for administration during metal-binding protein induction experiments. Six groups of 10 sculpins each were injected with $CdCl_2$ or saline (0.25 µl of solution per gram wet weight) and placed in separate tanks of flowing sea water. Mortality was observed on a daily basis. Survivors were killed on day 28 by transfer to sea water containing 70 mg benzocaine/L. This bioassay was used to determine: a) the $CdCl_2$ concentration (ip administered) which was necessary to kill half of the buffalo sculpin in a group in 240 h (10-d LD50), b) tissue Cd distribution and levels, and c) tissue metallothionein-like proteins and/or tissue metal-binding capacity of the 28-d survivors. The 10-d median lethal dose of a single ip administration of $CdCl_2$ was 22 ± 2 mg $CdCl_2 \cdot 2H_2O$/kg wet weight (mean ± 95% CI, Spearman-Karber Method). No effect of Cd was observed at 6.25 mg/kg. Sublethal $CdCl_2$ doses were used in ensuing work.

Metal-Binding Protein Induction

Dose-response and time courses of metal-binding protein induction by $CdCl_2$ were determined at 6, 12, and 18°C. These temperatures represent the actual range of temperatures recorded annually in Yaquina Bay. Dose-response assays were performed by repeated ip administration of 0, 0.5, 1, 2, and 3 mg $CdCl_2 \cdot 2H_2O$/kg wet weight on days 1, 2, 3, 7, 9, and 11. Fish were killed on day 14 and livers were removed for analysis.

Time course of metal-binding protein induction was determined using a single injection of 6 mg $CdCl_2 \cdot 2H_2O$/kg wet weight. Livers were excised from fish, weighed, and assayed on days 1, 2, 3, 6, and 10 post-injection.

The effect of environmental salinity on metal-binding protein levels was observed by holding groups of sculpin for

28 d at 28-30, 15, and 5°/$_{oo}$ at 12°C. Livers were removed for analysis on day 28.

To assess whether ip injection of Cd into buffalo sculpin elicits a different biochemical response than water exposure to Cd, an experiment was performed with aquatic exposures to Cd. Eight fish each were placed in tanks containing 0, 0.1, and 1.0 mg $CdCl_2 \cdot 2H_2O$/L sea water at 12°C. Four fish were removed from each treatment on days 1 and 14 of the exposure period. Livers from these fish were removed for analysis.

Tissue Preparation

Whole tissues were excised, blotted dry, minced and divided for metal-binding protein and total Cd analysis, weighed, immediately frozen in liquid N_2, and stored at -20°C. All stored samples were assayed within 2 wk of the date of removal. Freezing whole tissue for this duration was found to have no statistically significant effect on metal-binding protein levels. For metal-binding protein assays, tissues were thawed and homogenized (Potter-Elvehjem tissue homogenizer) in 3 vols (V/W) of chilled 0.05 M Tris-HCl buffer (pH 8.6). All preparatory work was completed at 4°C.

For ^{203}Hg-saturation assays, the tissue homogenate was used directly. For gel filtration, the homogenate was further prepared by centrifugation at 10,000 x g for 10 min followed by a spin at 100,000 x g for 60 min to isolate the cytosol. This supernatant was used in column chromatographic procedures.

Gel Filtration

Samples were applied to a column (2.5 x 70 cm) of Sephadex G75-SF and eluted with Tris-HCl buffer (pH 8.6, 0.05 M) at a flow rate of 10 ml/h. Fractions were assayed for the presence of metal-binding proteins and Cd via their UV absorbance at 250 and 280 nm, direct protein analysis (Bradford, 1976), and by flameless atomic absorption spectrophotometry (AAS).

Metal Analyses

Weighed and homogenized buffalo sculpin tissues were calcinated for 20 h at 450°C. The ash was dissolved by refluxing in a solution of concentrated HNO_3 and 30% H_2O_2 (Van

Hoof & Van San, 1981). Samples were appropriately diluted with Milli-Q water and analyzed for Cd (and in certain cases, Zn) by an HGA graphite furnace by AAS. Gel fractions and water samples were analyzed directly by AAS.

Hg-Saturation

Tissue homogenates were assayed for mercury-binding capacity by the method of Piotrowski *et al*. (1973). Briefly, the homogenate was incubated with a pre-determined optimal amount of $^{203}HgCl_2$ for 20 min at 12°C. Following the addition of 10% trichloroacetic acid (0.30 ml/ml homogenate) to precipitate the high molecular weight proteins, samples were centrifuged at 3500 x g and the supernatant was counted for radioactivity by liquid scintillation of Biofluor with a counting efficiency of 49 ± 5%. Standard curves compared linearly with those derived from counting on an auto-gamma counter (efficiency = 38%) with an accuracy of ± 6%.

The Hg-saturation method was assessed for its usefulness with blood, intestine, muscle, gill, kidney, and liver tissues. These tissues were taken on day 15 from buffalo sculpin which had been previously ip injected with 0.4 mg $CdCl_2 \cdot 2H_2O$/kg wet weight on day 1 and 2.0 mg/kg wet weight on day 8.

RESULTS

Gel Filtration

The normal elution profile of cytosol fractions of liver tissue from *E. bison* controls is shown in Figure 1. Cadmium and zinc occur in three peaks associated with (1) the elution volume of the column (ca. 95 ml), (2) the position corresponding to approximately 9000 daltons mol wt (the metallothionein-like protein region [ca. 190 - 210 ml]), and (3) <3000 daltons mol wt. This elution profile corresponds to tissue levels of 12.1 ± 5.5 μg Cd/g liver wet wt and 151.7 ± 36.8 μg Zn/g liver wet wt.

Hg-Saturation

When metal-binding protein induction was stimulated by Cd administration, only liver tissue showed strong potential in use of the Hg-saturation method (Fig. 2). This occurred despite the fact that a significant amoung of Cd was also found in kidney and intestine; Cd was distributed primarily among liver, kidney, and intestinal tissues (Table 1). Liver tissue showed the Hg-saturation plateau region necessary to demonstrate that binding sites of the metallothionein-like protein were being saturated with Hg prior to TCA-precipitable moeities (Fig. 2). Figure 3 illustrates that metal-binding protein induction in liver tissue resulted from Cd administration. The liver Hg-binding capacity for Cd-exposed sculpins was elevated over that of saline-injected controls.

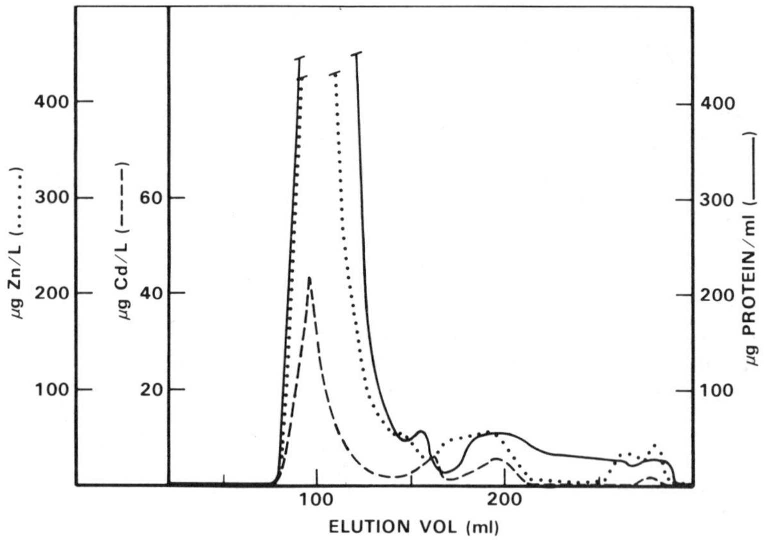

Fig. 1. Sephadex G75-SF elution profile of the cytosolic fractions of liver tissue from buffalo sculpin controls. Three-milliliter fractions were collected. Zinc, cadmium, and protein contents of the fractions are shown. Trypsinogen (ca. 24 K daltons) and lysozyme (ca. 13 K daltons) eluted at 150 and 180 ml, respectively.

The question of specificity of the Hg-saturation method in focusing on metallothionein-like protein was addressed by chromatography of TCA-supernatants. Liver TCA-supernatants from Cd-loaded fish were passed through a Sephadex G75-SF column at pH 8.6 (0.05 M Tris-HCl) and at pH 2.1 (0.05 M Na-citrate) (Fig. 4). The lower pH is the pH at which the assay takes place. The 9000 dalton mol wt regions contained approximately 84% of the bound ^{203}Hg at pH 8.6 and 93% at pH 2.1. Thus, the majority of the ^{203}Hg binds to the region associated with the metal-binding proteins.

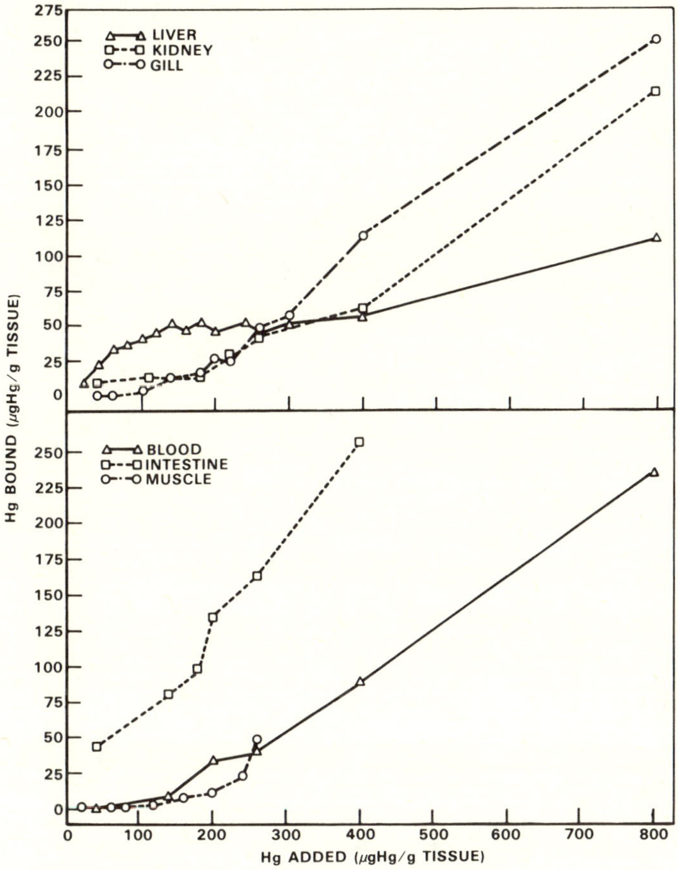

Fig. 2. Mercury levels in TCA-supernatants of various tissue homogenates as a function of mercury added. Buffalo sculpins were injected with 0.4 and 2.0 mg $CdCl_2$/kg weight on days 1 and 8, respectively, and sampled on day 15. N = 3 for kidney, blood, and muscle and N = 5 for liver, gill, and intestine.

Only a small amount of the metallothionein-like protein appeared to precipitate with the TCA. When resuspended TCA-pellet material (in 0.05 M Tris-HCl, pH 8.6) was eluted on a Sephadex column, only 7% of the total pellet ^{203}Hg was found in the protein fractions corresponding to 9000 daltons. This represents less than 0.5% of the ^{203}Hg used in the incubation.

Table 1. Cadmium concentration (µg Cd/g wet wt) in various tissues. Cadmium was administered on day 1 (0.4 mg/kg wet wt) and day 8 (2.0 mg/kg wet wt). Fish were killed on day 15 for analysis. N = 5.

Tissue	Cd	SE
Liver	46.8	12.9
Kidney	43.1	20.6
Intestine	27.5	3.7
Gill	2.8	1.8
Blood	0.4	0.2
Muscle	0.1	0.03

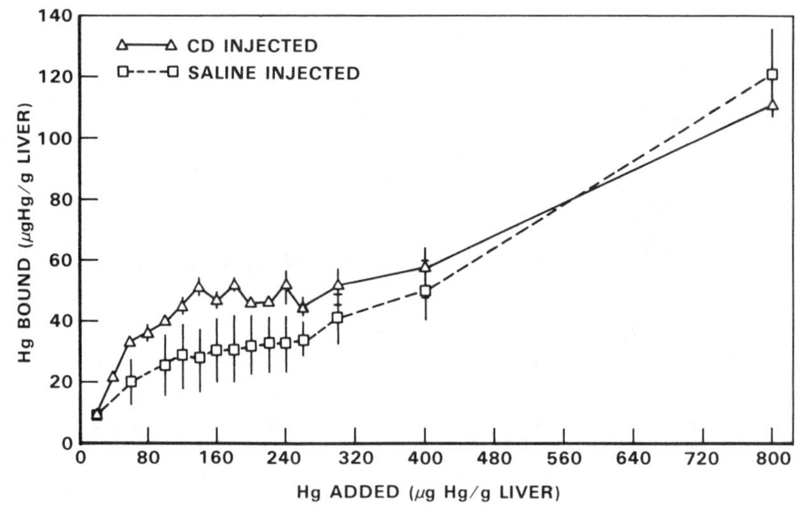

Fig. 3. Mercury levels in TCA-supernatants of liver homogenates as a function of mercury added. Curves represent Cd-injected (N = 5) and saline injected (N = 3) buffalo sculpins. Mean ± standard errors are given. P≦0.05 for the difference between means in the plateau regions of 140-240 g Hg added/g liver (Least Significant Difference Procedure).

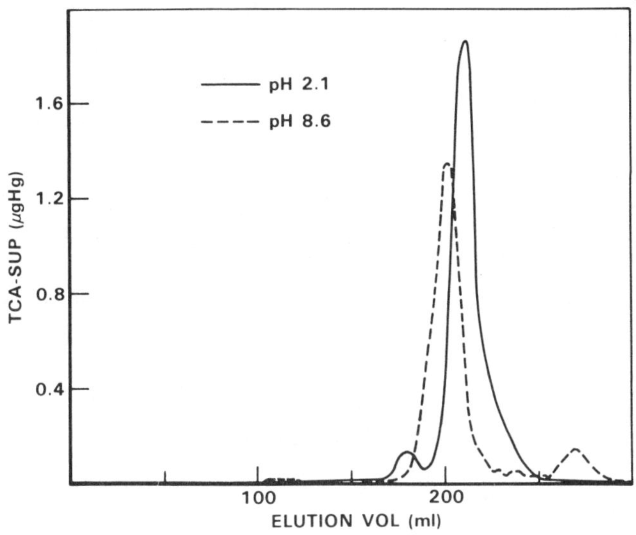

Fig. 4. Sephadex G75-SF elution profile of TCA-supernatants of samples representing the plateau region of the Hg-saturation assay. Mercury eluted primarily at the region corresponding to approximately 9000 daltons mol wt. Curves represent differing buffer pH: 8.6 (0.05 M Tris-HCl), 2.1 (0.05 M Na-citrate).

Metal-Binding Protein Induction and Cd/Zn Relationships in Liver and Kidney

Evidence of Cd retention via metal-binding proteins existed in liver and kidney tissues 28 d following Cd-administration (Table 2). A strong correlation occurred among administered Cd dose, liver and kidney Cd concentration, and Hg-binding capacity (HgBC). Liver HgBC was considerably higher than that of kidney. When HgBC data are expressed on a total tissue protein basis, the results are similar to those obtained on a weight basis. These results imply that the increase in metal-binding protein levels is not due simply to general protein synthesis. Interestingly, the tissue Cd content increase following Cd administration did not affect the total tissue concentration of Zn in liver and kidney. These data do not rule out the possibility of a shift in subcellular distribution of Zn.

Table 2. Cadmium and zinc concentrations, and mercury-binding capacity in liver and kidney of buffalo sculpin 28 days after a single administration of cadmium. N = 2-9. *p<0.05.

$CdCl_2$ Administered (mg/kg)	Cadmium (μg/g tissue)	Zinc (μg/g tissue)	HgBC (μg Hg/g tissue)	HgBC (μg Hg/mg protein)
		-- Liver --		
Control (Saline)	12.1 ± 5.5	151.7 ± 36.8	31.88 ± 11.03	2.63 ± 0.91
3.125	39.8 ± 27.3	226.2 ± 24.1	47.43 ± 17.08	4.30 ± 2.56
6.250	80.3 ± 31.4*	197.9 ± 20.1	74.85 ± 14.29*	6.24 ± 1.18*
		-- Kidney --		
Control (Saline)	12.2 ± 4.9	516.2 ± 158.5	8.98 ± 6.28	0.74 ± 0.52
3.125	35.1 ± 0.1	792.1 ± 252.8	12.30	1.40
6.250	40.8 ± 14.4*	561.6 ± 133.2	21.44 ± 5.77	1.96 ± 0.41

Induction by Salinity

Acclimation of *E. bison* to low salinity had a surprising effect. The mercury-binding capacity of liver homogenates from fish held at 5°/₀₀ was increased over HgBC from fish at higher salinities (Table 3). Further characterization of the low mol wt protein responsible for this increased Hg binding remains to be performed. Sculpin held at 30 and 15 °/₀₀ had identical blood sodium and chloride levels, and did not demonstrate a change in body weight over the 28-d period. The 5°/₀₀ group showed a 41 ± 3% drop in blood chloride levels and 8.2 ± 1.8% decline in body weight over the same time frame, evincing signs of stress. The decreased body weight was due to rejection of food despite tissue hydration. Total hepatic protein was similar among the three salinity groups.

Table 3. Hepatic mercury-binding capacity (μg Hg/mg protein) as a function of salinity. Fish held for 28 days at 12°C and the respective salinities. N = 7-10. *p<0.05.

Salinity (°/₀₀)	HgBC
28-30	0.96 ± 0.32
15	1.13 ± 0.30
5	4.72 - 1.03*

Induction by Cadmium and Temperature

Liver mercury-binding capacity was increased in saline-injected controls compared to uninjected controls, regardless of temperature (Fig. 5, $p \leqq 0.05$ in all cases, t-tests). Additionally, HgBC at 6 and 18°C was nearly double that of saline-injected controls at 12°C. The mercury-binding capacity of 12°C fish liver increased linearly with increased Cd-dosage (up to 12 mg $CdCl_2$/kg wet wt). Again, a temperature-related effect occurred, as 6 and 18°C acclimated fish demonstrated no further change in HgBC as increased Cd was administered. Liver Cd content increased proportionally under the same conditions of Cd dosage and no temperature-related differences occurred (Fig. 6). The data in Figures 5 and 6 indicate that the subcellular distribution of Cd was different at 6 and 18°C than at 12°C. Presumably, the rise in HgBC at 12°C (Fig. 5) was metallothionein-like protein which was available to bind the excess Cd.

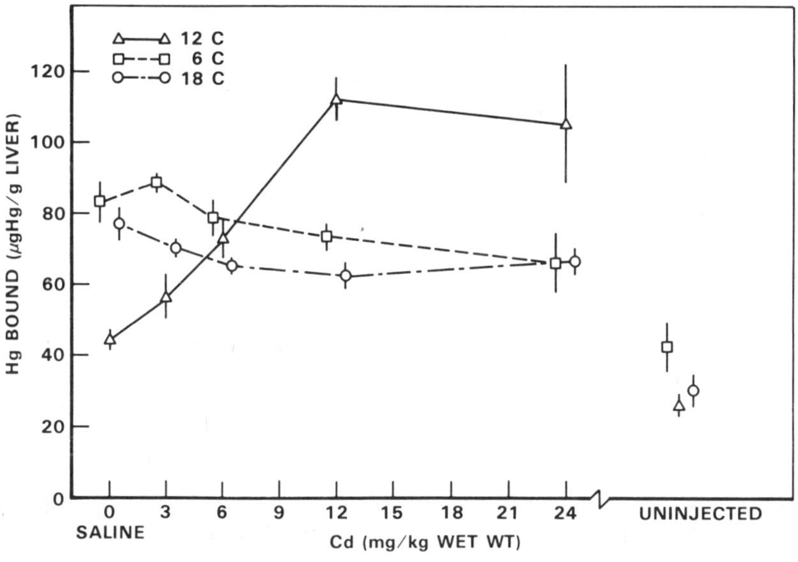

Fig. 5. Mercury-binding capacity (mean ± SE) of liver homogenates of buffalo sculpins repeatedly injected with Cd over a two-week period. N = 6. Total Cd administered is given by the abscissa.

The time course of (a) liver Cd content and (b) hepatic HgBC was followed after a single administration of 6 mg $CdCl_2$/kg wet wt. Accumulation of Cd by liver tissue was similar for all three temperatures during the first 24 h

(Fig. 7), but showed significant variance after this period. Significant mortality occurred between days 3 and 4 at 18°C, and all of the test fish at that temperature had died by day 6. The mortality can be attributed to a combination of high temperature stress and Cd. Liver Cd peaked after 2 d and then declined in the 12°C group. At 6°C, hepatic Cd levels gradually rose until day 6. No further increase in liver Cd occurred after 24 h in the 18°C group, and Cd concentration remained constant through day 4. A similar pattern was observed for HgBC at 18°C (Fig. 8).

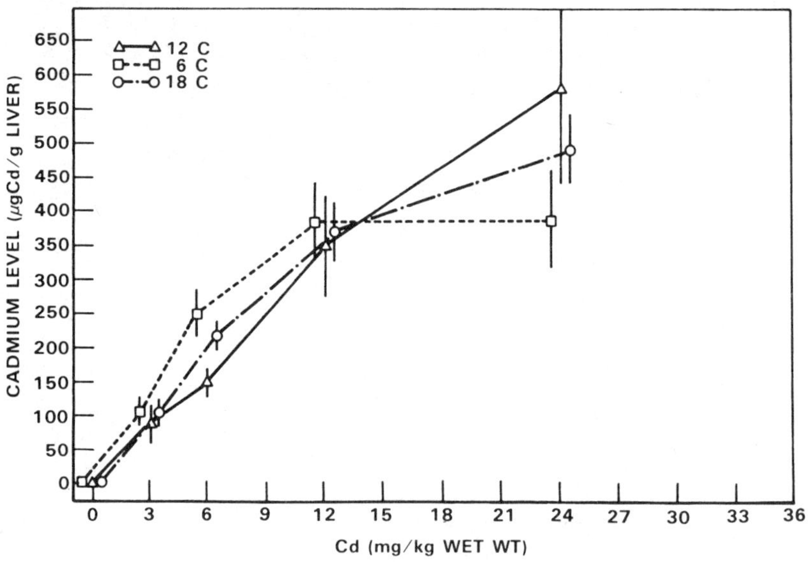

Fig. 6. Liver Cd levels (mean ± SE) of buffalo sculpins repeatedly injected with Cd over a two-week period. N = 6. Total Cd administered is given by the abscissa.

Mercury-binding capacity rose sharply during the first day after Cd administration at 6°C and remained at this level throughout the 10-d period. This response is not unlike that of the dose-response determinations, where saline-injection (and/or handling) only, caused an increase, which was not amplified by Cd administration. It is not known whether this response is attributable to Cd or the saline and/or handling.

The response at 12°C was an immediate increase in HgBC, peaking on day 2. A second rise occurred before day 10. This second rise took place despite a decrease in total liver Cd, indicating that a larger percentage of the Cd might be bound to the metal-binding protein on day 10.

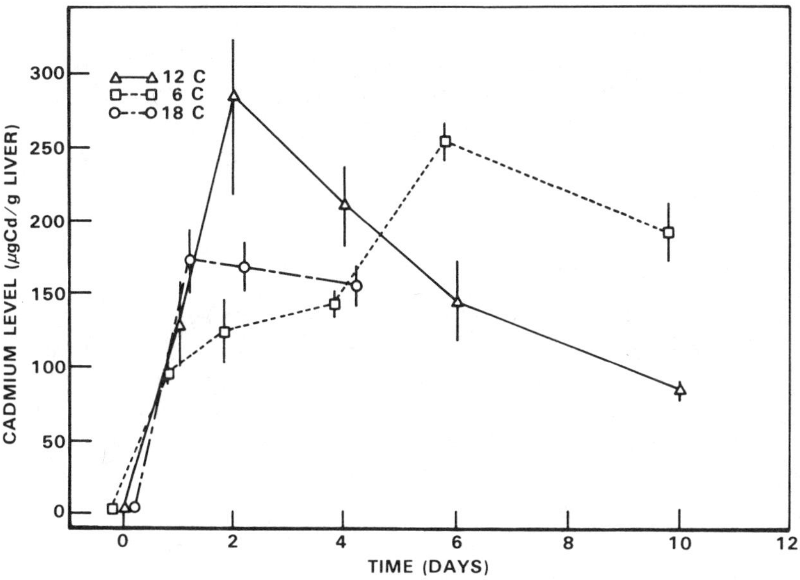

Fig. 7. Liver Cd levels (mean ± SE) of buffalo sculpins as a function of time after Cd administration. N = 6. Six mg Cd/kg body weight was given at time 0.

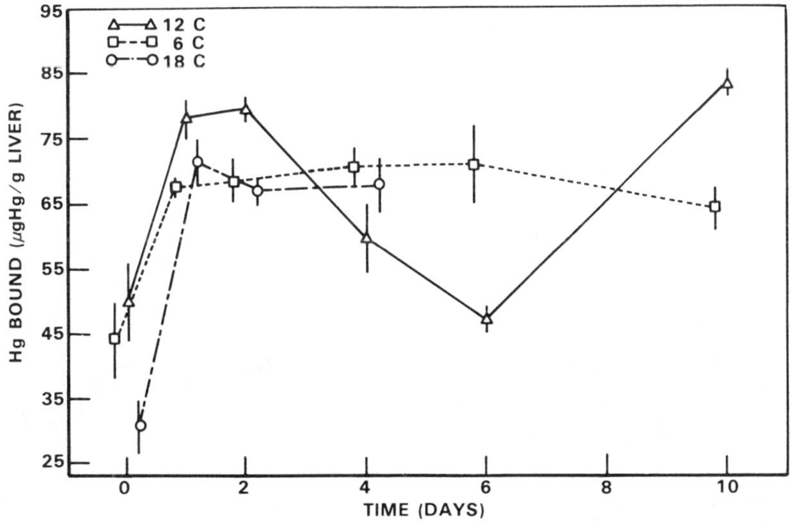

Fig. 8. Mercury-binding capacity (mean ± SE) of liver homogenates of buffalo sculpins as a function of time after Cd administration. N = 6.

These curves (Figs. 7 & 8) demonstrate that the time course of tissue distribution of Cd could be a function of temperature. It is interesting to speculate that at 18°C the kidney was overburdened with Cd, leading to the observed mortality. At low temperature (6°C), liver Cd levels rose gradually for a long period (6 d), possibly at the expense of kidney Cd levels.

Aquatic Exposure to Cadmium

When buffalo sculpin were exposed to water-borne $CdCl_2$, liver Cd concentration was significantly higher within 24 h (Table 4). Metal-binding protein levels did not differ from controls at hour 24, but were approximately double after 14 d of Cd-exposure. Although liver Cd levels increased directly with increasing water Cd concentrations, HgBC showed no further increase between 0.124 and 1.15 mg $CdCl_2$/L on day 14. Simultaneous gel filtration of liver cytosol from day 14 samples resulted in a peak of protein associated with the majority of the Cd at a Ve/Vo of approximately 2.0, identical to the peak associated with Cd-injected metal-binding protein. These data indicate that a metal-binding protein is synthesized in liver tissue of *E. bison* exposed to water-borne Cd, which has similar characteristics to that synthesized following ip administration.

Table 4. Liver cadmium concentrations and mercury-binding capacity of buffalo sculpin after 1 day (24 h) and 14 days of exposure to water-borne cadmium. N = 4. $^*p \leq 0.05$.

	Day 1		Day 14	
Exposure Concentration (mg/L)	Cd Concentration (μg Cd/g liver)	HgBC (μg Hg/g liver)	Cd Concentration (μg Cd/g liver)	HgBC (μg Hg/g liver)
0.016 + 0.001 (control)	1.5 ± 0.5	27.0 ± 2.9	2.2 ± 1.2	27.9 ± 1.3
0.124 + 0.025	16.5 ± 1.5*	29.7 ± 3.1	16.9 ± 1.5*	59.2 ± 6.3*
1.150 + 0.300	19.7 ± 1.4*	34.1 ± 4.5	34.3 ± 7.7*	62.3 ± 3.9*

DISCUSSION

Induction of hepatic metal-binding protein synthesis by cadmium was demonstrated in *E. bison* which had no previous exposure to heavy metals. The metal-binding protein of the sculpin liver has characteristics similar to those of metallothionein, including induced synthesis by cadmium, an apparent molecular weight between 6000 and 10,000 with a Ve/Vo on Sephadex G75-SF of approximately 2.0, an absorption maximum at 250 and not 280 nm when bound to Cd, and subcellular distribution in the cytosol. The magnitude of response to intraperitoneal administration of Cd, as quantified by the Hg-binding capacity of liver homogenates, was lower than that reported for mammals (Kotsonis & Klaassen, 1977). Assuming that the metal-binding protein of sculpin is metallothionein (mt) (binding 7 g-atoms of Hg/g molecular weight), control and Cd-exposed maxima at 12°C contained approximately 20 n moles mt/g liver and 79 n moles mt/g liver, respectively. Thus, the peak response observed represented a four-fold increase in metal-binding protein levels.

Induction of metal-binding protein was exhibited within 24 h of Cd injection. This rapid response is common to the higher vertebrates studied. Shaikh and Smith (1977) reported maximal cystine incorporation into liver metallothionein occurred within 12 h after a subcutaneous injection of Cd in rats. Similarly, Yoshikawa and Ohta (1981) demonstrated that metallothionein was induced for a 2-3 d period following Cd administration. However, the response to a single injection is temporary, as hepatic metallothionein levels decline approximately 3 d post-Cd administration in mammals. The 12°C acclimated fish in our investigation exhibited a similar pattern after a single administration of Cd, with an additional rise by day 10. Mercury-binding capacity at 6 and 18°C, however, did not decline after the second day of the monitoring period. These data indicate that the liver metal-binding proteins of the sculpin are rapidly produced and remain available for a considerable period. Further supporting evidence was observed in a group of sculpins whose liver HgBC remained elevated 28 d after administration of 6.25 mg $CdCl_2$/kg wet wt.

The time course of liver Cd content loosely followed that reported for mammals but demonstrated a lag of several days. Cadmium levels of liver tissue do not rapidly decline to background as in mammals (Yoshikawa & Ohta, 1981). These differences may signify a greater importance of liver tissue as a storage organ for metals in the buffalo sculpin. Twenty-eight days after initial Cd injection, mean liver Cd levels were equal to or higher than those found in kidney.

Introduced Cd was distributed primarily among liver, kidney, and intestinal tissues as expected. The intraperitoneal route of administration is likely to be responsible for this result. Gill epithelium has been shown to be an additional storage site for Cd, as Cd-bound metal binding protein (Reichert *et al*., 1979; Noël-Lambot, 1981). It appears that this circumstance occurs only under the condition of water-borne Cd exposure, as gill Cd and metal-binding protein contents after ip administration of Cd in the present study were minimal.

In our experiments, two apparent differences between aquatic exposure to Cd and intraperitoneal injection of Cd were found. First, the amount of Cd found in liver tissue was relatively low, even after 14 d of continuous aquatic exposure. Second, Cd did not induce metal-binding protein production until sometime after the first 24 h of aquatic exposure. The time lag can be attributed to the route of entry and lower liver Cd concentrations. Induction of metal-binding protein was apparent on day 14. The gel chromatogram of liver cytosol was similar between water-exposed and intraperitoneal-injected sculpins. Thomas *et al*. (1983) recently reported that absorbed water-borne Cd associates with two low molecular weight proteins, which are not metallothionein, in rainbow trout liver. Simultaneously, they determined the presence of Zn and Cu thionein in the same liver tissue. Extensive work is necessary to ascertain whether the route of exposure is a determinate of the type of hepatic low molecular weight proteins induced in buffalo sculpin.

Preliminary data indicate that the presence of Cd and the resulting increase in metal-binding proteins of liver tissue might effect a decline in tissue Zn levels. A considerable body of information currently exists to demonstrate changes in subcellular distribution of specific metals in fish (Jenkins *et al*., 1982). The next step would be to relate the observed changes in the subcellular distribution of a certain metal (*e.g.*, Cd) with those of another (*e.g.*, Zn). Such changes in distribution have previously been reported for liver and kidney of rats and mice (Suzuki, 1972). Additional evidence for fish has been found in a study with plaice (Coombs, 1975). The relationships between metals may have far reaching ramifications and are worthy of future research. They may play a casual role in the impairment of fish ion regulation and carbohydrate metabolism attributed to Cd (Larsson *et al*., 1976; Larsson *et al*., 1981).

The most intriguing results of the present investigation centered around the hepatic response of buffalo sculpins to various stresses. Stress related induction of metallothionein production was first reported in the liver of rats subjected to temperature extremes, heavy exercise, and CCl_4 intoxication (Oh *et al*., 1978). The vertebrate liver appears to play a primary role in response to various stresses. This response of increased hepatic metabolism may be associated with Zn mobilization, and further production of thioneins. Stresses of low salinity, handling, and handling and temperature combined elicited increased HgBC in *E. bison*. Diverse temperature effects were observed when the sculpins were exposed to $CdCl_2$. Phase lags in responses of liver Cd uptake and HgBC at 6°C can be attributed directly to the kinetics at low temperature. *In vivo* protein synthesis in fish liver cells is inversely related to temperature (Haschemeyer & Matthews, 1982). The no-effect dose of a single injection of $CdCl_2$ at 12°C (6 mg/kg wet wt) was lethal at 18°C. Lower amounts of Cd in liver tissue at this temperature despite a rapid and stable response of metal binding protein production, may indicate that Cd exerted mortality by distributing to other key tissues (*e.g.*, kidney). Alternatively, lower Cd levels may be lethal at 18°C. Maximal toxic effects of Cd have previously been associated with high temperature in aquatic species (Taylor, 1983).

Use of the simple and rapid ^{203}Hg-saturation method for estimating tissue metal-binding protein content in buffalo sculpin appears to be limited to the liver. The sensitivity, specificity, and reproducibility of the assay with hepatic tissue is adequate for purposes of large-scale sampling programs, where the time and sample intensive process of gel filtration is impractical. Extensive use of this procedure on kidney and gill tissues will require some modification of the assay methodology (See Patierno *et al*., 1983). Several methods for determination of metallothionein-like proteins have been employed with varying degrees of success (Onosaka & Cherian, 1982). Our results demonstrate that the choice of assay should initially be validated for the tissues of concern.

In order to gain insight into the biological fate and effects of trace metals in fish systems, future studies must thoroughly considered acclimation to metals and the relation to non-metal anthropogenic stressors, as well as environmental variables such as temperature and salinity. Brown *et al*. (1982) found DDT metabolites associated with the metallothionein-like metal-binding protein pool of white croakers. Simultaneously, a reduction of protein-bound metals was observed. It remains to be determined whether the presence of organics will stimulate production of metal-binding proteins or if organics simply bind to amino acid residues of metal-binding proteins (*e.g.*, cysteine) in a random process. We have demonstrated in this report that environmental variables play a noteworthy role in metal-binding protein and metal metabolism.

ACKNOWLEDGMENTS

This research was supported by grants from the National Institute of Health, NIH 5T32ES07060 and NIEHS-MFBC-ESO 1926.

REFERENCES

Bradford, M. 1976. A rapid and sensitive method for quantitation of microgram quantities of protein utilizing the principle of protein-dye binding. Anal. Biochem. 72: 248-254.

Brown, D.A., R.W. Gossett, and K.D. Jenkins. 1982. Contaminants in white croakers *Genyonemus lineatus* (Ayres, 1855) from the Southern California Bight: II. Chlorinated hydrocarbon detoxification/toxification. pp. 197-213. *In*: Physiological Mechanisms of Marine Pollutant Toxicity. W.B. Vernberg, A. Calabrese, F.P. Thurberg, and F.J. Vernberg (eds.). Academic Press, New York.

Chen, R.W., K.H. Eakin, and P.D. Whanger. 1974. Biological function of metallothionein. II. Its role in zinc metabolism in the rat. Nutr. Rept. Intern. 10: 195-200.

Coombs, T.L. 1975. The significance of multielement analyses in metal pollution studies. pp. 187-195. *In*: Ecological Toxicology Research. A.D. McIntyre and C.F. Mills (eds.). Plenum Press, New York.

Haschemeyer, A.E.V. and R.W. Matthews. 1982. Effects of temperature extremes on protein synthesis in liver of toadfish, *Opsanus tau*, *in vivo*. Biol. Bull. 162: 18-27.

Jenkins, K.D., D.A. Brown, G.P. Hershelman, and W.C. Meyer. 1982. Contaminants in white croakers *Genyonemus lineatus* (Ayres, 1855) from the Southern California Bight: I. Trace metal detoxification/toxification. pp. 177-196. *In*: Physiological Mechanisms of Marine Pollutant Toxicity. W.B. Vernberg, A. Calabrese, F.P. Thurberg, and F.J. Vernberg (eds.). Academic Press, New York.

Kotsonis, F.N. and C.D. Klaassen. 1977. Comparison of methods for estimating hepatic metallothionein in rats. Toxicol. Appl. Pharmacol. 42: 583-588.

Larsson, A., B.-E. Bengtsson, and C. Haux. 1981. Disturbed ion balance in flounder, Platichthys flesus L. exposed to sublethal levels of cadmium. Aquat. Toxicol. 1: 19-35.

Larsson, A., B.-E. Bengtsson, and O. Svanberg. 1976. Some haematological and biochemical effects of cadmium on fish. pp. 35-45. In: Effects of Pollutants on Aquatic Organisms. A.P.M. Lockwood (ed.). Cambridge University Press, New York.

McCarter, J.A. and M. Roch. 1983. Hepatic metallothionein and resistance to copper in juvenile coho salmon. Comp. Biochem. Physiol. 74C: 133-137.

Noël-Lambot, F. 1981. Presence in the intestinal lumen of marine fish of corpuscles with a high cadmium-, zinc- and copper-binding capacity: A possible mechanism of heavy metal tolerance. Mar. Ecol. Prog. Ser. 4: 175-181.

Oh, S.H., J.T. Deagen, P.D. Whanger, and P.H. Weswig. 1978. Biological function of metallothionein. V. Its induction in rats by various stresses. Am. J. Physiol: Endocrinol. Metab. Gastrointest. Physiol. 3: E282-285.

Onosaka, S. and M.G. Cherian. 1982. Comparison of metallothionein determination by polarographic and cadmium-saturation methods. Toxicol. Appl. Pharmacol. 63: 270-274.

Patierno, S.R., N.R. Pellis, R.M. Evans, and M. Costa. 1983. Application of a modified ^{203}Hg binding assay for metallothionein. Life Sci. 32: 1629-1636.

Piotrowski, J.K., W. Bolanowska, and A. Sapota. 1973. Evaluation of metallothionein content in animal tissues. Acta. Biochem. Polonica. 20: 207-215.

Reichert, W.L., D.A. Federighi, and D.C. Malins. 1979. Uptake and metabolism of lead and cadmium in coho salmon (Oncorhynchus kisutch). Comp. Biochem. Physiol. 63: 229-234.

Shaikh, Z.A. and J.S. Smith. 1977. The mechanism of hepatic and renal metallothionein biosynthesis in cadmium-exposed rats. Chem.-Biol. Interactions. 19: 161-171.

Suzuki, Y. 1972. Effects of cadmium-binding protein on metal distribution in supernatant of the liver homogenate of rat. Ind. Health. 10: 52-55.

Taylor, D. 1983. The significance of the accumulation of cadmium by aquatic organisms. Ecotox. Environ. Saf. 7: 33-42.

Thomas, D.G., J.F. de L.G. Solbe, J. Kay, and A. Cryer. 1983. Environmental cadmium is not sequestered by metallothionein in rainbow trout. Biochem. Biophys. Res. Commun. 110: 584-592.

Van Hoof, F. and M. Van San. 1981. Analysis of copper, zinc, cadmium and chromium in fish tissues. A tool for detecting metal caused fish kills. Chemosphere. 10: 1127-1135.

Yoshikawa, H. and H. Ohta. 1981. Interaction of metals and metallothionein. pp. 11-23. *In*: Biological Roles of Metallothionein. E.C. Foulkes (ed.). Elsevier North Holland, Inc., Amsterdam.

EFFECT OF HEAVY METALS ON THE INTESTINAL ABSORPTION OF L-LEUCINE AND L-METHIONINE IN THE TOADFISH *OPSANUS TAU*

R. Socci and A. Farmanfarmaian

SUMMARY

Heavy metals are known to inhibit enzymes and transport systems in various animal tissues. We examined the effect of several heavy metal chlorides ($HgCl_2$, CH_3HgCl, $CdCl_2$, and $SrCl_2$) on the intestinal absorption of the essential amino acids, L-leucine and L-methionine, by the toadfish, *Opsanus tau*, using *in vitro* and *in vivo* techniques.

The *in vitro* technique was validated against the *in vivo* method and it was found to give comparable amino acid uptake rates.

At 0.25 mM L-leucine, $HgCl_2$ produced significant inhibition of leucine uptake at all levels tested (2.5-20 ppm). CH_3HgCl significantly inhibited leucine uptake at 5, 10, and 20 ppm. No significant inhibition was produced by either $CdCl_2$ or $SrCl_2$ for the incubation time and concentration range tested (2.5-20 ppm at 10 min). The inhibitions produced by $HgCl_2$ were found to be qualitatively and quantitatively different from the effects of the other heavy metal compounds at each toxicant concentration.

At 25 mM L-leucine, only 20 ppm $HgCl_2$ produced significant inhibition of leucine uptake.

At 0.25 mM L-methionine, $HgCl_2$ produced significant inhibition of uptake at 5, 10, and 20 ppm; while CH_3HgCl significantly inhibited uptake at 2.5, 5, 10, and 20 ppm. The inhibition of uptake by $HgCl_2$ at all levels tested (2.5-20 ppm) was found to be significantly greater than that produced by CH_3HgCl at each metal concentration.

Of the heavy metals tested, $HgCl_2$ appears to be the most potent inhibitor of the intestinal amino acid transport system. Reduction in the uptake of essential amino acids has far reaching consequences for immune defenses, growth, development, and reproduction in fish and other aquatic organisms. From this point of view, $HgCl_2$ should be regarded as a most hazardous ecotoxicant if the level of its occurrence in the environment or the food causes a luminal accumulation of 2 ppm in the intestine of fish or macroinvertebrates.

By comparing results from the *in vitro* and *in vivo* techniques, separate mechanisms of action were suggested for $HgCl_2$ and CH_3HgCl.

INTRODUCTION

Heavy metals, such Hg, Cd, Zn, and Cu, are introduced into coastal waters from various sources and accumulate to high levels in the sediments (Waldhauer *et al.*, 1978; O'Connor & Stanford, 1979; Meyerson *et al.*, 1981). The main route of heavy metal accumulation in higher invertebrates and vertebrates is the gut. Beyond the primary and secondary producers in the food chain the important vector is food, although drink can contribute appreciably in the case of marine teleosts (Pentreath, 1973; Bryan, 1976; Pentreath, 1976a,b,c,d; Noël-Lambot, 1981). The intestinal epithelium acts as the first barrier to the random entry of substances into the organism from the gut, and thus the mucosal epithelium is exposed to heavy metals from the lumen of the gut. The effects of heavy metals on digestive and metabolic enzyme activities have been examined (Jackim *et al.*, 1970; Jackim, 1974; Renfro *et al.*, 1974; Henderson *et al.*, 1979; Sastry & Gupta, 1979), but less is known of their effects on the absorptive processes of the intestine. For this reason, we have investigated the effect of heavy metals in the form of $HgCl_2$, CH_3HgCl, $CdCl_2$, and $SrCl_2$ on the intestinal absorption of the essential amino acids, L-leucine and L-methionine, by the toadfish, *Opsanus tau*.

Materials and Methods

Toadfish were obtained from the Marine Resources Department of the Marine Biological Laboratory (MBL) at Woods Hole. At MBL, they were maintained in aerated running sea water tanks at 17-20°C. They were fed live or cut up killifish (*Fundulus* sp.). At Rutgers University, fish were maintained in aerated recirulating sea water tanks fitted with dacron and charcoal filters in a room regulated at 17 ± 1°C. Tanks were checked on a daily basis and cleaned as needed. A natural light cycle was maintained in all cases. The fish were starved for 24-48 h prior to experimentation to clear the midgut.

The *in vivo* experimental procedure has been described previously (Farmanfarmaian *et al.*, 1972). Briefly, toadfish were anesthetized with MS-222 (0.5 g/L ethyl m-aminobenzoate methanesulfonate, Sigma), transferred to a surgical rack, and maintained at 20 ± 1°C with oxygenated sea water containing a low dose (0.1 g/L) of MS-222 flowing over the gills. It has previously been demonstrated that MS-222 at 0.1 g/L does not have a significant effect on fish intestinal absorption (DiBenedetto & Farmanfarmaian, 1975). A defined region of the midgut (just below the entrance of the bile duct to approximately 7 cm below it) was exposed, cannulated, washed with fish Ringer, and ligated to produce an isolated gut segment with intact circulation. An isosmotic solution of fish Ringer containing appropriate concentrations of ^{14}C-labelled amino acid and ^{3}H-inulin (a water marker) in the presence or absence of the heavy metal was inserted through a cannula with a three-way valve. An initial sample of approximately 0.5 ml of this incubation solution was removed after about 0.5 min of manual mixing with a syringe. The solution was then mechanically mixed with an oscillating pump stirrer (described by Barkalow, 1978) to mimic the peristaltic motion of the gut. At the end of the incubation period a terminal sample was taken and the isolated segment of the gut was then excised, drained, and weighed in a defined, standardized manner. Duplicate samples from the initial and terminal incubation medium were counted in a three-channel liquid scintillation spectrometer. After correction for background, quench, and ^{3}H-inulin space changes, the absorption of amino acid was computed from the difference in ^{14}C counts of initial and terminal samples. Uptake of amino acid was computed in μmoles/g fresh tissue/h in the absence (control) and the presence (experimental) of heavy metal, and the difference was expressed as percent (%) inhibition by the heavy metal at a given concentration.

The *in vitro* method was developed and validated against the *in vivo* method (see results section and Farmanfarmaian & Socci, 1984). Briefly, 0.5 cm segments of intestinal tissue were cut from the region of the midgut just caudal to the bile duct entrance and randomized in a flask containing oxygenated fish Ringer at 20°C. Aliquots of tissue were weighed and incubated in 5 ml or more of the desired oxygenated solution on a shaker bath at 20°C for a defined time (see the results section). After incubation, the tissues were rapidly blotted, weighed, and extracted in 0.1 N nitric acid for 6 h or overnight. Duplicate extract samples were counted for ^{14}C and ^{3}H and the uptake was computed as described above. All radioisotopes were purchased from New England Nuclear Co., Boston. The fish Ringer was buffered at pH 7.3 with Tris chloride. Heavy metal concentrations given in ppm were computed from molar ratios.

The heavy metal concentrations chosen were based on levels reported in nature or expected in the lumen of the gut (O'Connor & Stanford, 1979; Meyerson *et al*., 1981; Noël-Lambot, 1981; Farmanfarmaian & Socci, 1984).

The concentrations of leucine, 0.25 and 25 mM, were selected on the basis of our experience with leucine transport kinetics; 0.25 mM is close to the K_t (concentration needed to maintain half maximal velocity), while 25 mM is sufficient to sustain the transport at nearly maximum velocity (Socci *et al*., 1983).

Data were subjected to an unpaired t-test analysis according to Sokal and Rohlf (1969).

RESULTS

The experiments presented here were performed using the *in vitro* technique, which was validated against the *in vivo* procedure (see the controls in Table 1A,B). It can be seen that there is no significant ($p>0.05$) difference between the uptake rates as determined by the two techniques for 0.25 mM L-leucine. In the presence of 10 ppm $HgCl_2$, both procedures demonstrated comparable and significant ($p<0.05$) inhibition (64%) of uptake from their respective controls (Table 1A). It is important to note that the *in vitro* procedure allows mucosal and serosal exposure of the tissue, while with the *in vivo* technique only the mucosa is exposed in short incubations.

The effect of tissue aging, due to *in vitro* processing time, on uptake rates was examined in paired experiments (Table 2). A significant ($p<0.02$) reduction of 30% in the uptake rate was observed with a processing time of 80 minutes as compared to one of 30 minutes. This indicates the importance of rapid processing of tissue with the *in vitro* technique to obtain uptake rates comparable to those of the *in vivo* procedure.

Table 1. Comparison of 0.25 mM L-leucine uptake under defined *in vivo* and *in vitro* conditions by the intestine of the toadfish during a 10 min incubation in the presence and absence of $HgCl_2$ and CH_3HgCl.

1A.

	Leucine Uptake (µmoles/g/h)[1]	
	In Vivo	*In Vitro*
Control	2.75 ± 0.31(8)	2.41 ± 0.15(7)
Experimental (10 ppm $HgCl_2$)	1.01 ± 0.07(8)	0.86 ± 0.01(5)
% inhibition	-64	-64

[1]Uptake values given as means ± SE (N). Differences between *in vivo* and *in vitro* were not significant ($p>0.05$).

1B.

	Leucine Uptake (µmoles/g/h)[2]	
	In Vivo	*In Vitro*
Control	2.72 ± 0.26(7)[a]	2.75 ± 0.21(6)[a]
Experimental (20 ppm CH_3HgCl)	1.83 ± 0.1(7)[b]	1.50 ± 0.07(6)[c]
% inhibition	-33	-45

[2]Uptake values given as means ± SE (N). Uptake values with different superscripts were significantly different ($p<0.05$).

The effect of several heavy metal compounds on the *in vitro* uptake of 0.25 mM L-leucine is presented in Table 3. $HgCl_2$ is a potent inhibitor of L-leucine absorption and produced significant ($p<0.05$) inhibition (20-67%) of uptake at all levels tested (2.5-20 ppm). CH_3HgCl significantly inhibited uptake at 5 ppm (22%), 10 ppm (30%), and 20 ppm (46%), but not at 2.5 ppm. No significant ($p>0.05$) effect was produced by either $CdCl_2$ or $SrCl_2$ at any level tested (2.5-20 ppm). The inhibitions produced by $HgCl_2$ were found to be significantly ($p<0.05$) different from the effects of the other heavy metal compounds at each toxicant concentration. The control flasks were not significantly different ($p>0.05$).

Table 2. Tissue aging due to *in vitro* processing time in paired experiments using toadfish.

Tissue Processing Time (min)	Leucine Uptake (μmoles/g/h)[1]	% Change[2]	p
30	2.75 ± 0.21(6)	-27	0.02>p>0.01
80	2.00 ± 0.16(6)		
35	2.76 ± 0.15(6)	-32	p<0.01
80	1.89 ± 0.07(6)		

[1]Fish were dissected and tissues randomized in oxygenated Tris-FTR. Tissues for the first set of the paired experiments were processed immediately and the second set was processed later as shown in the table. Incubation time 10 min; 0.25 mM L-leucine in Tris-FTR; Uptake given as mean ± SE (N).

[2]% difference between the first set and the second set in the pair.

At 25 mM L-leucine, the effect of several heavy metals on *in vitro* uptake was examined (Table 4). Only 20 ppm $HgCl_2$ produced significant ($p<0.05$) inhibition of uptake (18%). Neither CH_3HgCl nor $CdCl_2$ produced significant inhibition at any level tested (2.5-20 ppm). $SrCl_2$ produced significant stimulation of uptake at 20 ppm.

The generality of mercury inhibition of essential amino acid absorption was investigated by using L-methionine. The effect of $HgCl_2$ and CH_3HgCl on the uptake of 0.25 mM L-methionine is presented in Table 5. $HgCl_2$ significantly ($p<0.05$) inhibited uptake at 5 (41%), 10 (60%) and 20 (74%) ppm. Even though inhibition was repeatedly observed at 1.25 (3%) and 2.5 (13%) ppm, these values were not statistically significant at the level of $p<0.05$ due to tissue variability. CH_3HgCl produced significant inhibition at 2.5 (15%), 5 (26%), 10 (41%) ppm, and 20 (52%) ppm, but none at 1.25 (0.7%) ppm. It should be noted that inhibition by CH_3HgCl at 2.5 ppm is not statistically significant if compared to the pooled controls from both series. A statistical comparison between the $HgCl_2$ and CH_3HgCl series showed that the differences were significant ($p<0.05$) at each heavy metal concentration tested, except among the control values.

Table 3. The effect of several heavy metal compounds on the *in vitro* uptake of 0.25 mM L-leucine by toadfish intestine during a 10 min incubation.

Compound	Heavy Metal Concentration in ppm: 0 (control)	2.5	5	10	20
	Leucine uptake in µmoles/g/h				
$HgCl_2$	2.93 ±0.15 (0)	2.34 ±0.10 (20) *	1.75 ±0.05 (40) *	1.34 ±0.05 (54) *	0.98 ±0.04 (67) *
CH_3HgCl	2.75 ±0.21 (0)	2.74 ±0.08 (0.4)	2.15 ±0.11 (22) *	1.94 ±0.05 (30) *	1.50 ±0.07 (46) *
$CdCl_2$	2.51 ±0.09 (0)	2.32 ±0.14 (8)	2.46 ±0.13 (2)	2.40 ±0.08 (4)	2.27 ±0.09 (10)
$SrCl_2$	2.70 ±0.08 (0)	2.76 ±0.13 (+2)	2.96 ±0.10 (+10)	2.97 ±0.14 (+10)	2.74 ±0.07 (+2)

Uptake values are given as mean ± SE; n = 6 for each category. *Indicates values are significantly different from control ($p<0.05$). Values in () are % inhibition; (+) indicates % stimulation.

DISCUSSION

This study examined the effects of several heavy metals on the absorption of L-leucine and L-methionine *in vitro*. $HgCl_2$ was found to be the most potent inhibitor of the intestinal absorption of L-leucine and L-methionine when compared to the other heavy metals tested. CH_3HgCl_2 showed inhibition of uptake which was significantly ($p<0.05$) lower than that of $HgCl_2$ when compared at the same heavy metal concentration (Tables 3, 4, & 5). Cadmium had no appreciable effect and strontium had a slight stimulatory effect at the concentrations tested. These observations are not related to the high concentration of leucine in this series since exactly the same lack of significant effect was observed for $CdCl_2$ and $SrCl_2$ at 0.25 mM leucine (see Table 3 and also Farmanfarmaian *et al.*, 1982).

Table 4. The effect of several heavy metal compounds on the *in vitro* uptake of 25 mM L-leucine by toadfish intestine during a 10 min incubation.

Compound	Heavy Metal Concentration in ppm				
	0 (control)	2.5	5	10	20
	Leucine Uptake in µmoles/g/h				
$HgCl_2$	62.18 ±2.83 (0)	58.29 ±2.63 (6)	60.74 ±0.86 (2)	59.02 ±1.93 (5)	51.26 ±0.63 (18) *
CH_3HgCl	60.05 ±1.39 (0)	60.76 ±2.34 (+1)	64.97 ±2.06 (+8)	63.54 ±1.53 (+6)	57.61 ±0.70 (4)
$CdCl_2$	62.96 ±1.13 (0)	64.24 ±1.46 (+2)	55.27 ±1.51 (12)	59.18 ±2.22 (6)	60.44 ±1.50 (4)
$SrCl^2$	57.61 ±0.95 (0)	58.97 ±0.64 (+2)	59.97 ±2.88 (+4)	61.67 ±2.81 (+7)	66.82 ±2.18 (+16) *

Uptake values are given as mean ± SE; n = 6 for each category. *Indicates values are significantly different from control ($p<0.05$). Values in () are % inhibition; (+) indicates % stimulation.

Heavy metal inhibition of amino acid absorption by the intestinal and renal epithelium of aquatic organisms has been previously reported (Miller & Kinter, 1979; Pritchard, 1979; Socci *et al.*, 1979; Farmanfarmaian *et al.*, 1980; Pritchard & Miller, 1980; Miller, 1981; Socci *et al.*, 1981; Farmanfarmanian *et al.*, 1982). We have examined the inhibition of L-leucine absorption by $HgCl_2$ and CH_3HgCl using both *in vitro* and *in vivo* techniques (Table 1A,B). Under *in vivo* conditions, only the mucosal side is initially exposed to the toxicants. By contrast, the *in vitro* method described above exposes both the mucosal and the serosal side of the tissue during the incubation period. Table 1A shows that under *in vitro* conditions the absorption of leucine or its inhibition by $HgCl_2$ does not differ from the *in vivo* method, implying that the exposure to the serosal side has no further effect on the absorptive process over and above that observed for the mucosal side alone under *in vivo* conditions. Since we have previously established that $HgCl_2$ penetrates the intestinal wall very slowly (Farmanfarmaian & Socci, 1984), we interpret these results to mean that the observed inhibition of leucine is primarily due to $HgCl_2$ action on the leucine transporter at the brush border membrane facing the lumen of the intestine. By contrast Table 1B shows that CH_3HgCl inhibition of leucine transport is significantly greater *in vitro* as compared to *in vivo*. In the control set of these experiments, and those provided in Table 1A, leucine absorption did not differ significantly. Since we have previously demonstrated that CH_3HgCl rapidly penetrates the intestinal tissue (Farmanfarmaian & Socci, 1984), we propose that the *in vitro* method provides for both mucosal and serosal penetration of this toxicant and the amino acid transport function is impaired due to the intracellular effect of CH_3HgCl. This proposal, however, does not rule out a mucosal as well as an intracellular effect of CH_3HgCl.

Table 5. The effect of $HgCl_2$ and CH_3HgCl on the *in vitro* uptake of 0.25 mM L-methionine by toadfish intestine during a 10 min incubation.

Compound	Heavy Metal Concentration in ppm					
	0 (control)	1.25	2.5	5	10	20
	Methionine Uptake in μmoles/g/h					
$HgCl_2$	3.98 ±0.29 (0)	3.86 ±0.06 (3)	3.45 ±0.08 (13)	2.34 ±0.10 (41) *	1.59 ±0.05 (60) *	1.03 ±0.03 (74) *
CH_3HgCl	4.63 ±0.17 (0)	4.60 ±0.25 (0.7)	3.95 ±0.18 (15) *	3.42 ±0.13 (26) *	2.73 ±0.10 (41) *	2.21 ±0.14 (52) *

Uptake values are given as mean ± SE; n = 5 or 6 for each category. *Indicates values are significantly different from control ($p<0.05$). Values in () are % inhibition.

In conclusion, $HgCl_2$ and CH_3HgCl both inhibit the absorption of L-leucine and L-methionine by the intestine of the toadfish and other fish tested (Farmanfarmaian & Socci, 1984). $HgCl_2$ is the more potent inhibitor. The two forms of mercury appear to act by different mechanisms. $HgCl_2$ acts directly on the neutral amino acid transporter from the luminal side, while CH_3HgCl acts indirectly on the transporter or its bioenergetic coupling from within the cell. Although Cd^{2+} and Sr^{2+} possess the same divalent charges as Hg^{2+}, they do not attack sulfhydryl groups on transporter or enzymatic proteins with the same avidity as Hg^{2+} (Rothstein, 1970). This may be due to the difference in ionic radii of these divalent species and the general charge distribution.

Inhibition of leucine uptake at 25 mM was observed only at 20 ppm $HgCl_2$. This is due to high levels of diffusion and facilitated diffusion of leucine at such high concentrations. We have previously shown that these fluxes tend to mask the inhibitory effects of mercury which are mainly on the active transport component of the uptake (Farmanfarmaian & Socci, 1984). Under natural feeding conditions, however, only low levels of amino acids are expected in the gut since the processes of peptide hydrolysis and amino acid absorption are tandem reactions and high levels of nutrilites are not accumulated.

We have concentrated our efforts mainly on leucine absorption in the toadfish in this paper. However, there is no question that the observed inhibition of amino acid uptake by mercury is a general phenomenon extending to all the essential amino acids and sugars and to many species of aquatic organisms, including those of immediate commercial and recreational value (Miller & Kinter, 1979; Pritchard, 1979; Socci *et al.*, 1979; Farmanfarmaian *et al.*, 1980; Miller, 1981; Farmanfarmaian *et al.*, 1982; Socci *et al.*, 1983; Farmanfarmaian & Socci, 1984). The experiments presented or cited in this paper are for the most part acute in design and do not directly demonstrate that in contaminated coastal waters aquatic species chronically suffer a reduced level of amino acid absorption and the consequent expected reduction in growth, fecundity, and immunological defense. We are currently engaged in examining the chronic effects of mercury contamination of food, both in laboratory experiments and field observations.

ACKNOWLEDGMENTS

Supported in part by NOAA Sea Grant No. NA82AA-D-00065 and the Center for Coastal and Environmental Studies and Busch Fellowship from the Bureau of Biological Research of Rutgers University to R. Socci.

We are grateful to Dr. Norbert Psuty, Director of the Center for Coastal and Environmental Studies (Rutgers University) and his staff for continued administrative support of this project. We acknowledge the help of V. Iannaccone, V. Klimek, and K. Pugliese in conducting some of the experiments presented in this paper and the Kimya-Lara-Parvin Foundation for aid.

LITERATURE CITED

Barkalow, D.T. 1978. The effect of mucosal sodium and exogenous adenosine triphosphate on net intestinal absorption of glycine or alanine: In vivo studies on toadfish midgut and hamster jejunum. Doctoral Dissertation. Rutgers University, New Brunswick, NJ.

Bryan, G.W. 1976. Some aspects of metal tolerance in aquatic organisms. pp.7-34. In: Effects of Pollutants on Aquatic Organisms. A.P.M. Lockwood (ed.). Cambridge University Press, Cambridge.

DiBenedetto, F.E. and A. Farmanfarmaian. 1975. Intestinal absorption of naturally occurring sugars in rainbow trout (*Salmo gairdneri*) under free swimming conditions. Comp. Biochem. Physiol. 50A: 555-559.

Farmanfarmaian, A., A. Ross, and D. Mazal. 1972. In vivo intestinal absorption of sugar in the toadfish (marine teleost, *Opsanus tau*). Biol. Bull. 142: 427-445.

Farmanfarmaian, A. and R. Socci. 1984. Inhibition of essential amino acid absorption in marine fishes by heavy metals. Mar. Environ. Res. 14: 185199.

Farmanfarmaian, A., R. Socci, and V. Iannaccone. 1982. Interaction of heavy metals with intestinal transport mechanisms in echinoderms. pp. 339-344. In: Proceedings of the International Echinoderm Conference, Tampa Bay. J.M. Lawrence (ed.). Balkema Press, Rotterdam.

Farmanfarmaian, A., R. Socci, and T. Polidore. 1980. Mechanism of heavy metal inhibition of amino acid transport in intestine of marine fish. Biol. Bull. 159: 458.

Henderson, G.R., W.H. Huang, and A. Askari. 1979. Transport ATPase--Different modes of inhibition of the enzyme by various mercury compounds. Biochem. Pharmacol. 28: 429-433.

Jackim, E. 1974. Enzyme responses to metals in fish. pp. 59-65. In: Pollution and Physiology of Marine Organisms. F.J. Vernberg and W.B. Vernberg (eds.). Academic Press, New York.

Jackim, E., M. Hamlin, and S. Sonis. 1970. Effects of metal poisoning on five liver enzymes in killifish (Fundulus heteroclitus). J. Fish. Res. Board. Can. 27: 383-390.

Meyerson, A.L., G.W. Luther, III, J. Krajewski, and R.L. Hires. 1981. Heavy metal distribution in Newark Bay sediments. Mar. Poll. Bull. 12: 244-250.

Miller, D.S. 1981. $HgCl_2$ inhibition of nutrient transport in teleost fish small intestine. J. Pharmacol. Exp. Ther. 216: 70-76.

Miller, D.S. and W.B. Kinter. 1979. Pathways of cycloleucine transport in killifish small intestine. Am. J. Physiol. 237: E567-E572.

Noël-Lambot, F. 1981. Presence in the intestinal lumen of marine fish of corpuscles with a high cadmium-, zinc-, and copper-binding capacity: A possible mechanism of heavy metal tolerance. Mar. Ecol. Prog. Ser. 4: 175-181.

O'Connor, J.S. and H.M. Stanford (eds.). 1979. Chemical Pollutants of the New York Bight; Priorities for Research. National Oceanic and Atmospheric Administration, Boulder.

Pentreath, R.J. 1973. The accumulation and retention of ^{65}Zn and ^{54}Mn by the plaice, Pleuronectes platessa L. J. Exp. Mar. Biol. Ecol. 2: 1-18.

Pentreath, R.J. 1976a. Some further studies on the accumulation and retention of ^{65}Zn and ^{54}Mn by the plaice, Pleuronectes platessa L. J. Exp. Mar. Biol. Ecol. 21: 179-189.

Pentreath, R.J. 1976b. The accumulation of organic mercury from sea water by the plaice, Pleuronectes platessa L. J. Exp. Mar. Biol. Ecol. 24: 121-132.

Pentreath, R.J. 1976c. The accumulation of inorganic mercury from sea water by the plaice, Pleuronectes platessa L. J. Exp. Mar. Biol. Ecol. 24: 103-119.

Pentreath, R.J. 1976d. The accumulation of mercury from food by the plaice, Pleuronectes platessa L. J. Exp. Mar. Biol. Ecol. 25: 51-65.

Pritchard, J.B. 1979. Toxic substances and cell membrane function. Federation Proc. 38: 2220-2225.

Pritchard, J.B. and D.S. Miller. 1980. The teleost kidney in evaluation of xenobiotic toxicity and elmination. Federation Proc. 39: 3207-3212.

Renfro, J.L., B. Schmidt-Nielsen, D. Miller, D. Benos, and J. Allen. 1974. Methylmercury and inorganic mercury: Uptake, distribution, and effect on osmoregulatory mechanisms in fishes. pp. 101-102. In: Pollution and Physiology of Marine Organisms. F.J. Vernberg and W.B. Vernberg (eds.). Academic Press, New York.

Rothstein, A. 1970. Sulfhydryl groups in membrane structure and function. pp. 135-176. In: Current Topics in Membranes and Transport. F. Bronner and A. Kleinzeller (eds.). Academic Press, New York.

Sastry, K.V. and P.K. Gupta. 1979. The *in vivo* effect of mercuric chloride on some digestive enzymes of a freshwater teleost fish, *Channa punctatus*. Bull. Environ. Contamin. Toxicol. 22: 9-16.

Socci, R., V. Iannaccone, and A. Farmanfarmaian. 1981. Membrane transport of the essential amino acid leucine in the gut of *Arbacia punctulata*--Effect of Hg^{++} and other heavy metals. Biol. Bull. 161: 317.

Socci, R., V. Iannaccone, and A. Farmanfarmaian. 1983. Kinetics of L-leucine transport inhibition by mercury compounds. Fed. Proc. 42: 1255.

Socci, R., J. Mercuro, and A. Farmanfarmaian. 1979. Heavy metal effects on intestinal absorption of nutrients in the toadfish, *Opsanus tau*. Biol. Bull. 157: 396.

Sokal, R.R. and F.J. Rohlf. 1969. Biometry. W.H.F. Freeman and Co., San Francisco.

Waldhauer, R., A. Matte, and R.K. Tucker. 1978. Lead and copper in the waters of Raritan and lower New York bays. Mar. Poll. Bull. 9: 3842.

AGE-DEPENDENT RESPONSE DIFFERENCES OF *FUNDULUS HETEROCLITUS* EMBRYOS FOLLOWING CHRONIC EXPOSURE TO MERCURY

J.R. Sharp and J.M. Neff

SUMMARY

Zero-, 1-, 2-, and 5-day old *Fundulus heteroclitus* embryos were exposed continuously to mercuric chloride (MC) and methylmercuric chloride (MMC) until all embryos had hatched or died. When compared to other teleosts, *F. heteroclitus* embryos are moderately tolerant to mercury contamination. MMC is more toxic than MC to developing embryos. Short-term embryonic responses as measured by four-day mortality and percent abnormal development among 12-day old survivors were lower when continuous exposure was initiated in progressively older embryos. This represents age-differential embryonic sensitivity to mercury among the four ages examined. When longer-term effects of exposure such as total hatching success and viability of hatch were assessed, all embryos exposed prior to the completion of gastrulation (less than 5-day old) were approximately equally sensitive to mercury exposure.

INTRODUCTION

In the mid to late 1970's, several review papers summarized the current knowledge about effects of environmental stressors to the early development of fishes (Rosenthal & Alderdice, 1976; McKim, 1977; Macek & Sleight, 1977; and Birge *et al*., 1979). The theme of these publications was that, in general, the early life history stages of most freshwater and marine fishes are the most sensitive (or among the most sensitive) stages to natural and man-made stressors. As a consequence, many investigators began describing the embryotoxic nature of a variety of pollutants and comparing these results to the existing information concerning responses in adult fish.

Information on the effects of mercury compounds on the reproductive capacities and embryological development of teleosts is slight when compared to the literature describing acute toxicity, bioaccumulation, and physiological responses of adult fishes. Akiyama (1970) reported that tolerance of various developmental stages of the medaka *Oryzias latipes* to phenylmercuric acetate and methoxyethylmercuric acetate increased with advancing development up to the time of hatching.

The teratogenic nature of mercury in fishes has been reported by several investigators. Heisinger and Green (1975) reported that exposure of *Oryzias latipes* embryos to 30 μg Hg^{++}/L (= parts per billion), as MC, produced several circulatory anomalies, including blood vessel deterioration, hemorrhaging, and loss of circulatory pigmentation. Dial (1978) reported that exposure to MMC produced reduced and stunted growth, various cardiac anomalies, and spinal curvature in embryos of *O. latipes* exposed to 60 μg Hg^{++}/L and greater. Weis and Weis (1977a) demonstrated differences in the teratogenic nature of 30 μg Hg^{++}/L as MC to two different blastula stages of *Fundulus heteroclitus*. Early blastula (Stage 12, Armstrong & Child, 1965) embryos had a lower percent axis formation and greater percent of forebrain defects than Stage 14 blastulae. In addition, spinal deformities were observed among eleutheroembryos (= fry, Balon, 1975) following exposure to MC. These results support the view that certain earlier life history stages are more sensitive than adult stages to environmental stressors. Jackim *et al*. (1970) and Klaunig *et al*. (1975) reported 96-h LC50s for *F. heteroclitus* adults following acute exposure to MC, as 230 and 2010 μg Hg^{++}/L, respectively; 7.7-67 times greater than the teratogenic dose for this species. Weis and Weis (1977b) also described the teratogenic nature of MMC to *F. heteroclitus* embryos and concluded that MMC is more teratogenic than MC. Furthermore, they proposed that 1-day old embryos are more sensitive than 0- and 2-day old embryos, with regard to abnormal development.

Kihlstrom and Hulth (1972) and Kihlstrom *et al*. (1973) showed that at concentrations of phenylmercuric acetate as low as 15 μg Hg^{++}/L, both spawning success and hatchability of embryos of *Brachydanio rerio* decreased with increasing concentration. In the more tolerant carp (*Cyprinus carpio*), total hatching success was dramatically reduced in 2-3 mg Hg^{++}/L as mercuric chloride (Huckabee & Griffith, 1974).

Fundulus heteroclitus, a dominant estuarine cyprinodont found along the Atlantic seaboard, has been used extensively in embryological investigations. Indeed, much research has focused on the teratogenic nature of mercury to this species (Weis & Weis, 1977a,b) as well as embryonic survival following exposure to mercury for varying lengths of time (Sharp & Neff, 1980, 1982). However, relatively little information on the effects of mercury on short-term mortality, total incidence of anomalous development, hatching success, and viability of hatch following static renewal exposure of specific-aged embryos has been published.

This paper describes the toxic effect of MC and MMC following static renewal exposure of four different-aged *Fundulus heteroclitus* embryos. The embryonic responses assessed include 96-h mortality, total anomalous development, hatching success, and the percent of the hatch exhibiting spinal curvature.

MATERIALS AND METHODS

Adult *Fundulus heteroclitus* were obtained from the National Marine Fisheries Service Laboratory at Beaufort, North Carolina, and maintained in the laboratory under environmental conditions conducive to year-round breeding (Linden *et al*., 1979b; Sharp & Neff, 1980).

Gametes were obtained from several adults following the procedure described by Trinkaus (1967). More precise information regarding the procedure is provided by Sharp and Neff (1980). Pooled sperm and ova from several adults were used to more precisely represent the response of the population as a whole and to counter the effect of a single adult's genetic variability or the state of maturation the gametes might be undergoing.

Experiments were designed to determine the embryotoxic effects of mercury, as mercuric chloride (MC) and methylmercuric chloride (MMC), to *Fundulus heteroclitus* in relation to the age of the embryos initially and continuously exposed. Zero-day old (4-8 cell, Stage 4-5 [Armstrong & Child, 1965]), 1-day old (mid-glastula, Stage 12-13), 2-day old (mid-neurula, Stage 19-20), and 5-day old (well pigmented/beating heart, Stage 25-27) were subjected to the chronic, static renewal bioassay. Embryos were exposed to mercury-free water until the appropriate age was reached. At the time of exposure, embryos were transferred to aqueous solutions of mercuric chloride or methylmercuric chloride. Experimental concentrations included 0 (control), 4, 10, 20, 30, 40, 60, 80, 100 μg Hg^{++}/L (= parts per billion). For all treatment conditions, four replicates of 25 embryos each were maintained in 5-cm diameter all glass culture dishes, containing 25 ml of well aerated 20 $^o/_{oo}$S artificial sea water (Instant Ocean) adjusted to the appropriate mercury concentration. Embryos were incubated at 25°C under a light:dark photoperiod of 16h:8h. Linden *et al*. (1979b) demonstrated this salinity-temperature regime to be within the optimal range of these factors for maximum hatching success of embryos resulting from our laboratory-maintained adult population.

All water was replaced daily. Freshly prepared media were analyzed for initial mercury concentration and for loss of mercury after 24 h, using cold vapor mercury spectrophotometry.

Embryos were examined daily. All mortalities were recorded and removed. Although a qualitative and quantitative assessment of specific abnormalities was made, results are presented here as percent of living embryos expressing some abnormal appearance in morphology on day 13 of development. This day of development was selected because previous experimentation had demonstrated that, under the incubation conditions described above, hatching begins within the next 48 h. All surviving embryos were allowed to continue development until all had either hatched or died. Percent total hatching success was determined, as well as the percent of the hatch exhibiting spinal curvature.

All percent data were initially transformed using the arcsine of the square root of the percent response. The data were analyzed using the General Linear Model of SAS 76 (Barr *et al*., 1976) which yields a multiple regression analysis of variance. A two-factor second-order polynomial equation was used, in which the entire data set was regressed on the quadratic function of mercury concentration and embryonic age initially exposed:

$$\hat{Y} = b_0x_0 + b_1x_1 + b_{11}x_1^2 + b_2x_2 + b_{22}x_2^2 + b_{12}x_1x_2$$

where:

Y = the estimated response

x_0 = the intercept

x_1 and x_2 = the linear effect of mercury concentration and embryonic age, respectively

x_1^2 and x_2^2 = the quadratic effect of mercury concentration and embryonic age, respectively
x_1x_2 = the linear x linear interaction between mercury concentration and embryonic age
b_0, b_1, etc. = the regression coefficients

Procedure Scatter of SAS 76 was used to plot contours from the regression equations. To produce the predicted percent contours, the data were retransformed from the arcsine $\sqrt{x}$ to the percentage values which were plotted.

Following the analysis of variance, Duncan's New Multiple Range Test ($\alpha = 0.05$; Ott, 1977) was used to determine where significant differences in embryonic response occurred among the mercury concentrations tested (of each mercury compound) for a given age group, and for the same mercury concentration tested among the different age groups. Likewise, this test was employed to determine compound-dependent response differences between mercuric chloride and methylmercuric chloride, following comparable embryonic exposure.

Four EC50 (median effective dose) values were determined when possible by probit analysis following correction for the natural (control) adverse response, using Abbott's formula (Finney, 1971). The EC50s determined were 96-h LC50, 12-d old AD50 (concentration resulting in 50% abnormal development of surviving 12-d old embryos), SH50 (concentration restricting successful hatch to 50%), and VH50 (concentration restricting viable hatch to 50%). Embryos expressing spinal curvature were not considered viable. Thus, percent viability was determined by subtracting the number of afflicted eleutheroembryos from the total number hatched and dividing by the total number of embryos initially exposed (= 25).

RESULTS

Mercury Analysis of Water

The actual mercury concentrations of exposure media at the time of dosing were within 10% of the nominal values. Thus, nominal values will be used throughout the remainder of this paper. Approximately 25 and 35% loss of mercuric chloride (MC) and methylmercuric chloride (MMC), respectively, was observed following 24 h of incubation (Sharp & Neff, 1982). Adsorption, volatilization, and embryonic accumulation are generally regarded as the processes associated with mercury loss from exposure media.

96-h Mortality

For each mercury compound, the quadratic effect of mercury, the linear and quadratic effect of age, and the interaction of mercury and age were highly significant in affecting mortality (Table 1).

Early cleavage stage embryos (0-day old) were more sensitive to mercury exposure than more advanced stages, indicating a progressive increase in mercury tolerance with advancing embryonic development (Fig. 1). Exposure of 0-day embryos to MC and MMC resulted in marked increases in mortality at concentrations of 60 µg Hg^{++}/L and greater and 40 µg Hg^{++}/L and greater, respectively. Comparable increases in mortality were shifted to higher concentrations as progressively older embryos were exposed. In most cases, embryonic exposure to MMC resulted in greater mortality than comparable exposure to MC.

Duncan's New Multiple Range Test ($\alpha = 0.05$) demonstrated that mortality of 0-day old embryos was significantly increased at concentrations of 60 µg Hg^{++}/L and greater and 40 µg Hg^{++}/L and greater following exposure to mercuric chloride and methylmercuric chloride, respectively. It is statistically evident that tolerance to mercury increases with advancing development, although the percent decrease in mortality is lower following exposure to MMC than MC.

Compound-dependent response differences (between comparably treated MC and MMC embryos) for the most sensitive 0-day old embryos, occurred at 40, 60, and 80 µg Hg^{++}/L, indicating the greater toxic nature of MMC. As exposure was delayed, there was a progressive increase in survival following exposure to each mercury compound, with the resulting compound-dependent response differences, between MC and MMC, occurring at progressively higher concentrations (Sharp & Neff, 1982).

Anomalous Development

For each mercury compound, all linear, quadratic, and interactive effects were highly significant, except for the quadratic effect of mercury (Table 2).

It is apparent that the early cleavage stage embryos (0-day old) are more sensitive to mercury than more advanced stages, with respect to abnormal development patterns (Fig. 2). Increases in the percent abnormal development among 0-day old embryos following exposure to each compound were

Table 1. Multivariate regression analysis of 96-hour survival of _F. heteroclitus_ following continuous exposure of 0-, 1-, 2-, and 5-day old embryos to mercuric chloride and methylmercuric chloride. DF - degrees of freedom; Hg - mercury concentration; Age - embryonic age when continuous exposure was initiated. Adapted, in part, from Sharp and Neff (1980, 1982). Reprinted courtesy of Marine Environmental Research and Environmental Biology of Fishes.

	DF	F	P>F	R	Coefficient for Regression Equations
Mercuric Chloride (MC)					
Treatment	5	67.49	0.0001	0.89	
Error	90				
Corrected Error	95				
Interactive Analysis					
Hg	1	0.14	0.7086		-0.043
Age	1	27.06	0.0001		12.483
Hg^2	1	23.18	0.0001		-0.005
Age^2	1	39.47	0.0001		-2.595
Hg x Age	1	83.86	0.0001		0.151
Intercept					73.947
Methylmercuric Chloride (MMC)					
Treatment	5	116.14	0.0001	0.94	
Error	106				
Corrected Error	111				
Interactive Analysis					
Hg	1	0.00	0.9521		-0.006
Age	1	15.79	0.0001		7.844
Hg^2	1	56.34	0.0001		-0.007
Age^2	1	14.29	0.0003		-1.390
Hg x Age	1	75.16	0.0001		0.138
Intercept					74.418

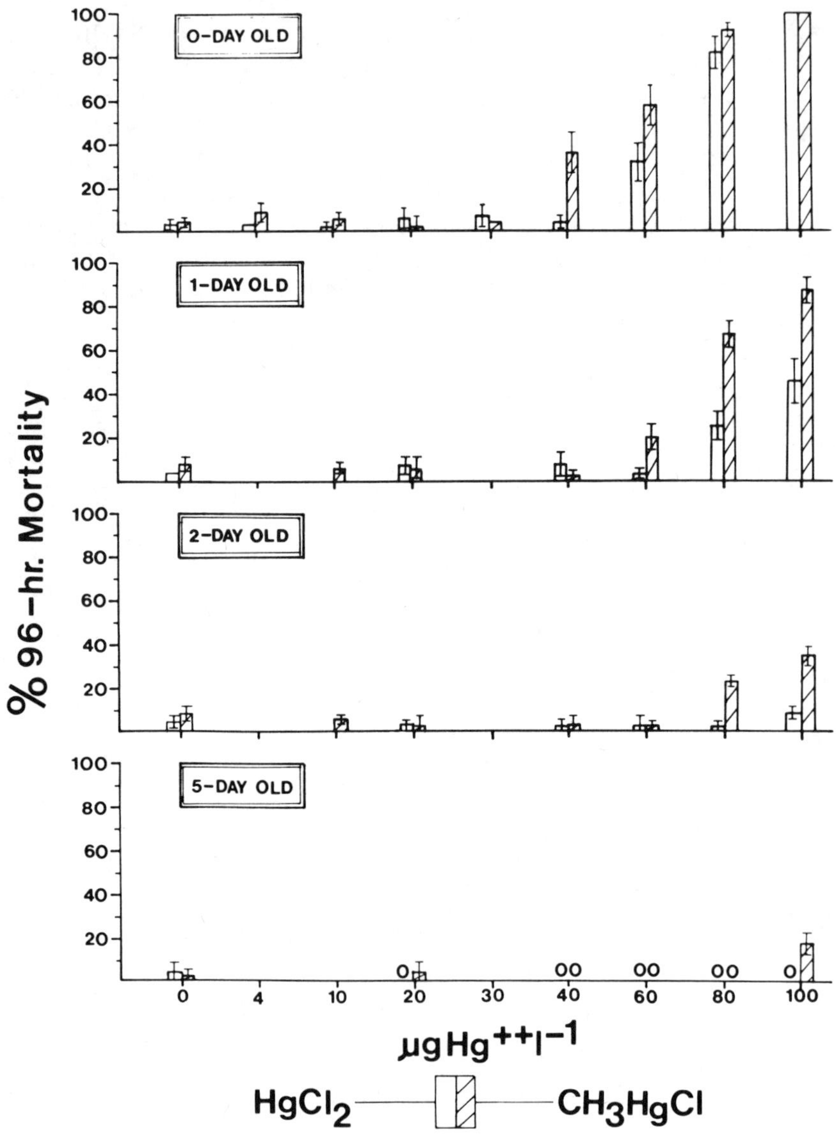

Fig. 1. The percent 96-h mortality of 0-, 1-, 2- and 5-day old embryos of _F. heteroclitus_ continuously exposed to mercury, as $HgCl_2$ (MC) and CH_3HgCl (MMC). Each bar represents the mean response of four replicates of 25 embryos each. Vertical lines represent the standard deviation of the replicates. Adapted, in part, from Sharp and Neff (1980, 1982). Reprinted courtesy of Marine Environmental Research and Environmental Biology of Fishes.

Table 2. Multivariate regression analysis of percent abnormal development of *F. heteroclitus* among 12-day old survivors following continuous exposure of 0-, 1-, 2-, and 5-day old embryos to mercuric chloride and methylmercuric chloride. DF - degrees of freedom; Hg - mercury concentration; Age - embryonic age when continuous exposure was initiated.

	DF	F	P>F	R	Coefficient for Regression Equations
Mercuric Chloride (MC)					
Treatment	5	126.72	0.0001	0.99	
Error	97				
Corrected Error	102				
Interactive Analysis					
Hg		100.83	0.0001		0.832
Age		13.52	0.0004		-6.166
Hg^2		0.32	0.5757		-0.0005
Age^2		17.38	0.0001		1.340
Hg x Age		162.02	0.0001		-0.178
Intercept					2.647
Methylmercuric Chloride (MMC)					
Treatment	5	149.51	0.0001	0.98	
Error	106				
Corrected Error	111				
Interactive Analysis					
Hg		98.55	0.0001		1.089
Age		60.01	0.0001		-16.678
Hg^2		1.91	0.1695		-0.001
Age^2		67.56	0.0001		3.237
Hg x Age		109.17	0.0001		-0.166
Intercept					7.264

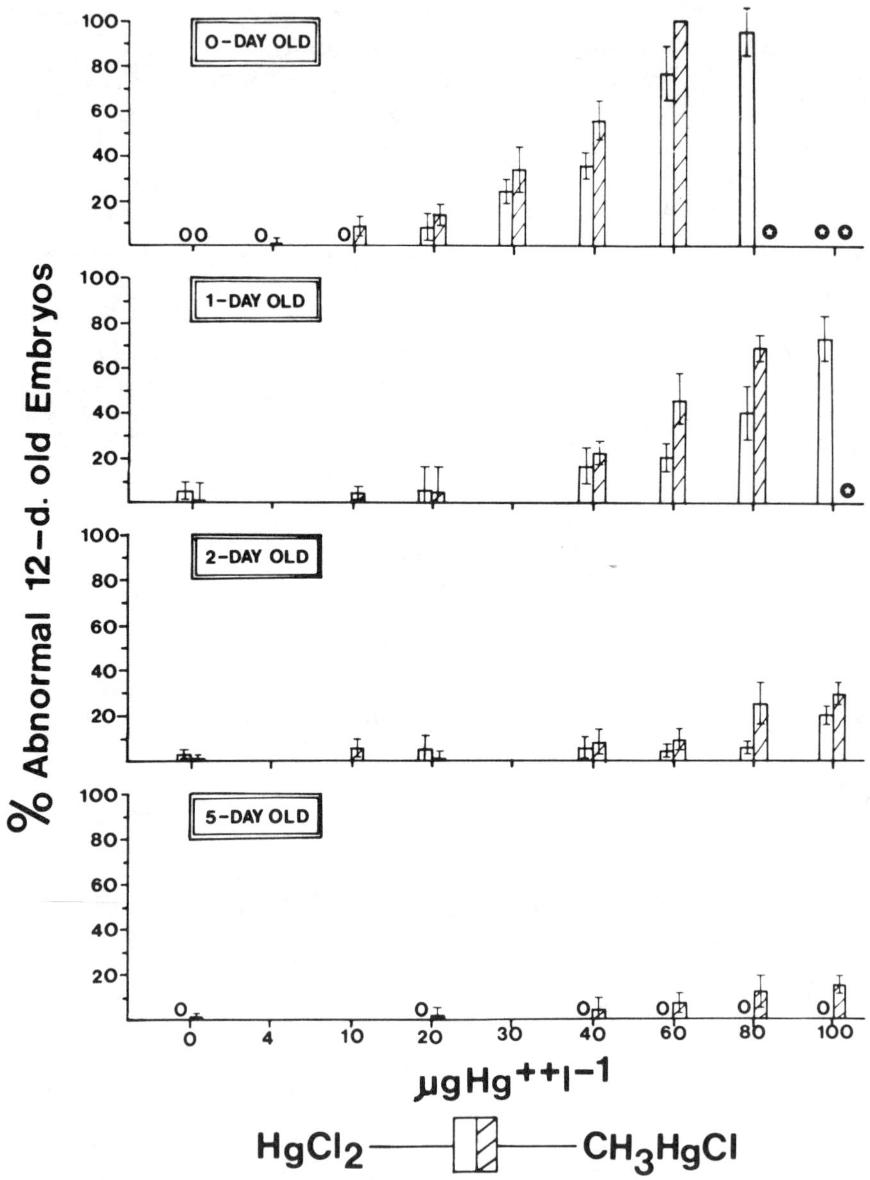

Fig. 2. The percent anomalous development of 12-d old embryos of F. heteroclitus following continuous exposure of 0-, 1-, 2-, and 5-d old embryos to ($HgCl_2$) MC and (CH_3HgCl) MMC. Each bar represents the mean response of four replicates. Vertical lines represent the standard deviation of the replicates. Enclosed stars indicate no survival occurred.

most conspicuous at concentrations greater than 20 µg Hg^{++}/L. As exposure was progressively delayed for 1, 2, and 5 d, the percent abnormality among surviving 12-day old embryos progressively decreased from that observed among 0-day old embryos, following comparable exposure to each mercury compound.

The Duncan's analysis (α = 0.05) of 0-day old initially-exposed embryos revealed significant increases in percent abnormality of 12-day old survivors at 30 µg Hg^{++}/L and greater and 20 µg Hg^{++}/L and greater concentrations for MC and MMC, respectively. Significant decreases in percent abnormality among 1-day old embryos, when compared to the 0-day old embryonic response, were noted at 40, 60, and 80 µg Hg^{++}/L MC and 40 and 60 µg Hg^{++}/L MMC. Additional reductions in percent abnormality were noted among the 2-day and 5-day old embryos for both mercury compounds.

Compound-dependent response differences between MC- and MMC-treated embryos were noted at 40 and 60 µg Hg^{++}/L for 0-day old embryos. With advancing embryonic age, the first concentration effecting compound-dependence, with regard to this embryonic response, was progressively increased.

Figure 3 depicts response-surface diagrams for predicted 96-h percent survival and percent abnormal development among 12-day old embryos, with respect to embryonic age initially exposed (Y-axis) and mercury concentration (X-axis) for each mercury compound. For both embryonic responses and each mercury compound, percent contours are gradually shifted up, indicating increased mercury tolerance with advancing embryonic development. The upward shift of contours for both responses is greater for the projected response to MMC than to MC, indicating the greater toxic nature of the organic mercury.

Total Hatching Success

Table 3 summarizes the multivariate regression analysis of total hatching success of 0-, 1-, 2-, and 5-day old embryos exposed continuously to MC and MMC. This statistical analysis revealed that the linear effect of mercury and the interactive effect of mercury and age were highly significant for exposure to inorganic mercury. The linear and quadratic effects of mercury, as MMC, were highly significant.

Figure 4 illustrates that in total hatching success, unlike the previous embryonic responses discussed, delaying the initiation of exposure for 1 and 2 d does not seem to

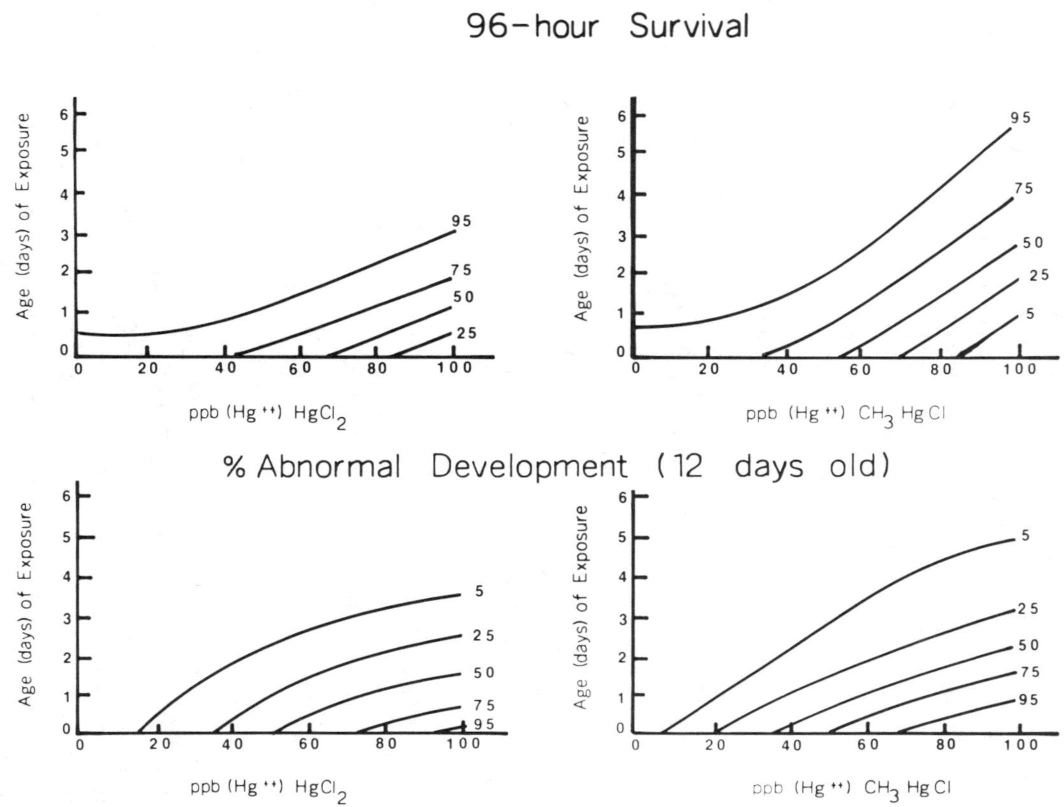

Fig. 3. Response-surface diagrams estimating percent 96-h survival and percent abnormal development of 12-d old embryos of F. heteroclitus with respect to mercury concentration and embryonic age initially exposed to $HgCl_2$ (MC) and CH_3HgCl (MMC).

Table 3. Multivariate regression analysis of total hatching success of embryos of *F. heteroclitus* following continuous exposure of 0-, 1-, 2-, and 5-day old embryos to mercuric chloride (A) and methylmercuric chloride (B). DF - degrees of freedom; Hg - mercury concentration; Age - embryonic age when continuous exposure was initiated. Adapted in part from Sharp and Neff (1980). Reprinted courtesy of Marine Environmental Research.

	DF	F	P>F	R	Coefficient for Regression Equations
Mercuric Chloride (MC)					
Treatment	5	145.47	0.0001	0.96	
Error	86				
Corrected Error	91				
Interactive Analysis					
Hg	1	51.24	0.0001		-0.878
Age	1	0.55	0.4603		-1.812
Hg^2	1	0.03	0.8663		0.0001
Age^2	1	3.98	0.0492		0.817
Hg x Age	1	9.66	0.0026		0.061
Intercept					74.606
Methylmercuric Chloride (MMC)					
Treatment	5	145.17	0.0001	0.93	
Error	110				
Corrected Error	115				
Interactive Analysis					
Hg	1	92.98	0.0001		-1.168
Age	1	0.03	0.8531		0.432
Hg^2	1	8.09	0.0053		0.003
Age^2	1	2.00	0.1596		0.602
Hg x Age	1	0.09	0.7604		-0.005
Intercept					73.972

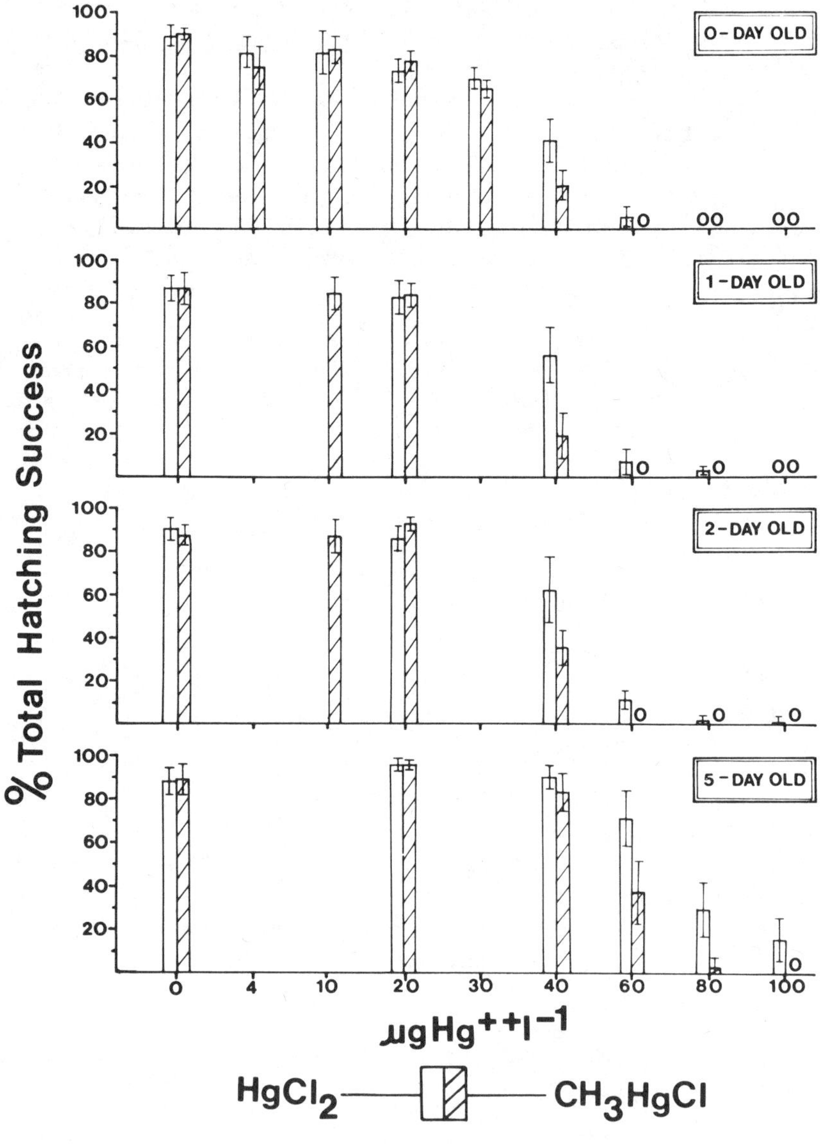

Fig. 4. The percent total hatching success of 0-, 1-, 2-, and 5-d old embryos of F. heteroclitus continuously exposed to mercury, as $HgCl_2$ (MC) and CH_3HgCl (MMC). Each bar represents the mean response of four replicates of 25 embryos each. Vertical lines represent the standard deviation of the replicates. Adapted, in part, from Sharp and Neff (1980). Reprinted courtesy of Marine Environmental Research.

elicit a less severe embryonic response than that of the 0-d embryos. This supports the view that longer-term effects must be assessed. Five-day old embryos did, however, respond more favorably in hatching success following exposure to each compound. Increases in hatching success among the 5-day old embryos were most apparent at concentrations of 40 μg Hg^{++}/L and greater.

The Duncan's analysis ($\alpha = 0.05$) of the hatching success data revealed significant decreases in hatchability at 20 μg Hg^{++}/L and greater for 0-day old embryos exposed to each compound. Although 96-h survival was relatively high at intermediate concentrations, total hatching success was low. Significant increases in total hatching success among the 1-day old embryos exposed to MC, when compared to the response of the 0-day old embryos, occurred only at 20 and 40 μg Hg^{++}/L. No further significant increases in total hatching success were determined for 2-day old embryos following exposure to MC. Five-day old embryos exposed to 20 μg Hg^{++}/L and greater as MC hatched at a significantly greater percent than the 0-, 1-, and 2-day old embryos. One-day old embryos exposed to MMC did not demonstrate any significant increases in hatching success from that of the 0-day old embryos. Two-day old embryos hatched at a significantly greater percent following exposure to 20 and 40 μg Hg^{++}/L than younger-staged embryos. This was comparable to the response of 1-day old embryos following exposure to MC, further indicating the higher toxicity of MMC to *F. heteroclitus* embryos. No hatching was observed among the 0-, 1-, and 2-day old embryos exposed to 60, 80, and 100 μg Hg^{++}/L as MMC. Hatching success of 5-day old embryos exposed to MMC was significantly higher than that of earlier staged embryos at 40 and 60 μg Hg^{++}/L. Significant compound-dependent response differences were found for only the 40 μg Hg^{++}/L-exposed 0- and 1-day old embryos, the 40 and 60 μg Hg^{++}/L-exposed 2-day old embryos, and the 60, 80, and 100 μg Hg^{++}/L-exposed 5-day old embryos. In all cases demonstrating differences in hatching success between MC- and MMC-exposed embryos, the MMC-treated embryos had a lower percent hatch.

Hatching began on day 14 of development and continued for the next 18 d. There were no apparent effects of mercury compounds or concentrations used on the rate of hatching.

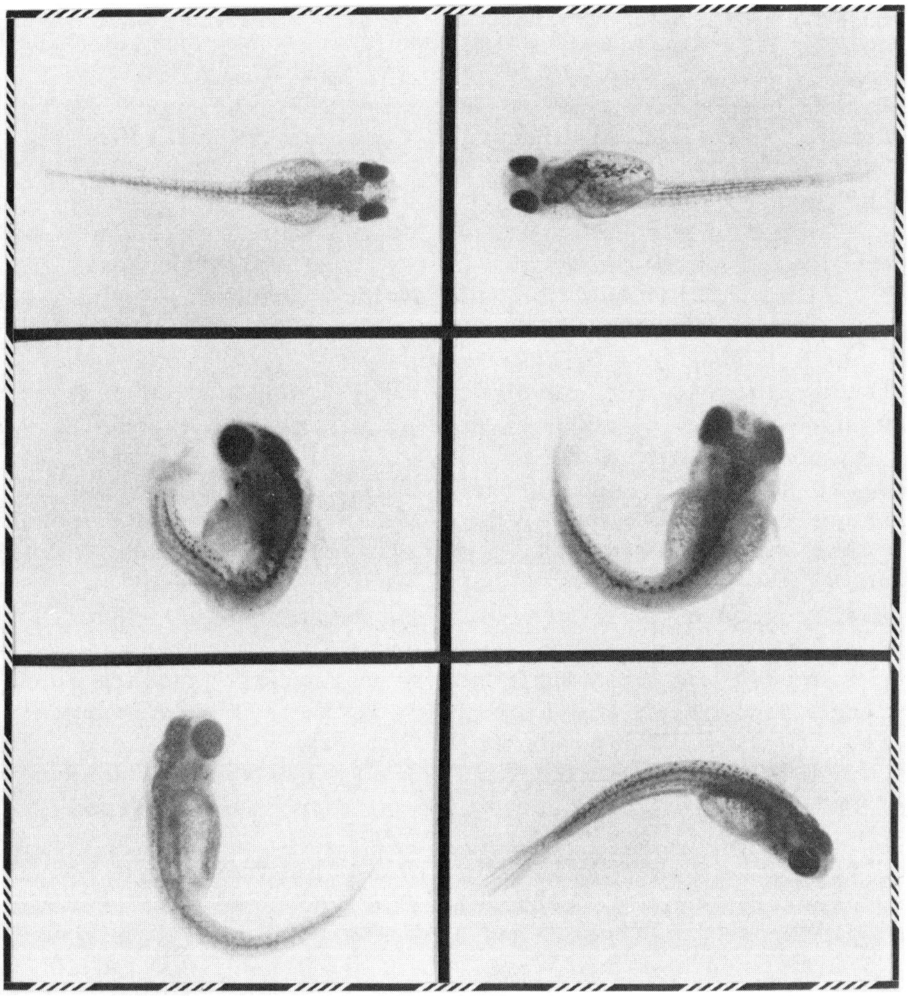

Fig. 5. Lateral curvature of the spine of F. heteroclitus following embryonic exposure to mercury. Upper two eleutheroembryos represent normal (control) embryos.

Table 4. Multivariate regression analysis of percent incidence of lateral spinal curvature among eleutheroembryos of *F. heteroclitus* following continuous exposure of 0-, 1-, 2- and 5-day old embryos to mercuric chloride and methylmercuric chloride. DF - degrees of freedom; Hg - mercury concentration; Age - embryonic age when continuous exposure was initiated. Adapted in part from Sharp and Neff (1980). Reprinted courtesy of Marine Environmental Research.

	DF	F	P>F	R	Coefficient for Regression Equations
Mercuric Chloride (MC)					
Treatment	5	93.50	0.0001	0.94	
Error	73				
Corrected Error	78				
Interactive Analysis					
Hg	1	25.13	0.0001		0.790
Age	1	2.76	0.1009		5.644
Hg^2	1	15.66	0.0002		0.007
Age^2	1	2.36	0.1288		-0.926
Hg x Age	1	31.44	0.0001		-0.149
Intercept					-9.114
Methylmercuric Chloride (MMC)					
Treatment	5	126.57	0.0001	0.97	
Error	68				
Corrected Error	73				
Interactive Analysis					
Hg	1	79.52	0.0001		0.979
Age	1	2.17	0.1452		2.532
Hg^2	1	31.06	0.0001		0.013
Age^2	1	0.97	0.3289		-0.316
Hg x Age	1	69.10	0.0001		-0.224
Intercept					-5.195

Spinal Curvature and Viability

Figure 5 illustrates the degree of variability in the expression of scoliosis exhibited by emerging embryos. There was no apparent influence of exposure treatment on the degree of lateral curvature. For both the MC- and MMC-exposed embryos, the linear and quadratic effect of mercury and the interactive effect of age and mercury were highly significant (Table 4).

It is apparent that with increasing mercury concentration, the percent scoliosis among eleutheroembryos increased (Fig. 6). Likewise, exposure to MMC resulted in a greater mean percent response than comparable embryonic exposure to MC. As observed with hatching success, favorable response (significant decreases in percent spinal curvature) occurred only when post-neural stage _F. heteroclitus_ embryos were exposed to each mercury compound under the conditions described above. The Duncan's analysis ($\alpha = 0.05$) for the percent of scoliotic embryos within the 0-day old exposure group revealed significant increases in this anomaly at concentrations of 30 to 60 µg Hg^{++}/L and 20 to 40 µg Hg^{++}/L, as MC and MMC, respectively. Significant decreases in the response of the 1- and 2-day old embryos exposed to MC, when compared to the 0-day old embryos, occurred only following exposure to 40 µg Hg^{++}/L. Likewise, 2-day old embryos exposed to 20 µg Hg^{++}/L as MMC had a significant decrease in percent response when compared to 0-day old exposed embryos. Five-day old embryos demonstrated further significant decreases in this adverse response following exposure to 40 and 60 µg Hg^{++}/L as MC and 40 µg Hg^{++}/L as MMC. Compound-dependent response differences among 0-day old embryos were demonstrated at 20 and 40 µg Hg^{++}/L, while both the 1- and 2-day old embryos demonstrated compound dependence only at 40 µg Hg^{++}/L. No compound-dependent response differences were observed for the 5-day old embryos.

Figure 7 shows response-surface diagrams for predicted total hatching success and percent spinal curvature, with respect to embryonic age initially exposed (Y-axis) and mercury concentration (X-axis) used for each compound. For both embryonic responses and each mercury compound, percent contours at the lower and intermediate mercury concentrations are nearly vertical up to age day 2. This indicates that age at time of exposure, with regard to these long-term sublethal effects, is not as important as with short-term effects.

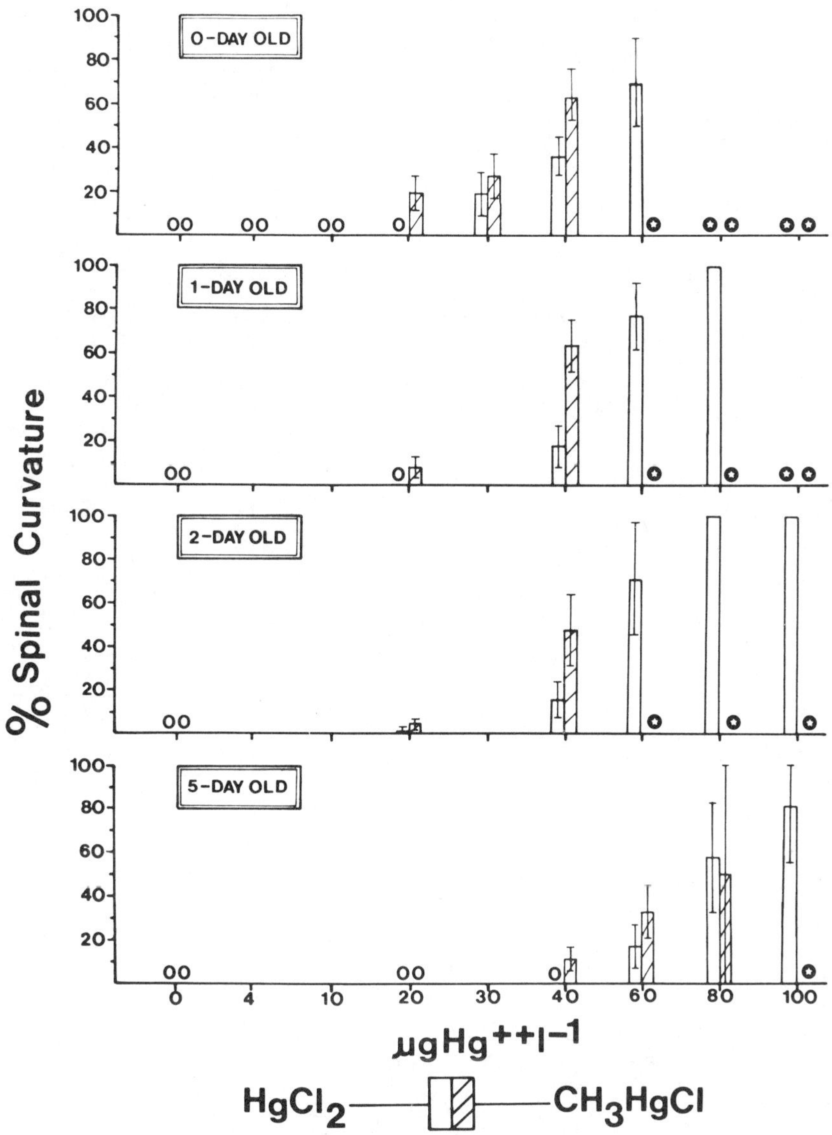

Fig. 6. The percent of eleutheroembryos of F. heteroclitus exhibiting spinal curvature following exposure of 0-, 1-, 2-, and 5-d old embryos to mercury, as $HgCl_2$ (MC) and CH_3HgCl (MMC). Each bar represents the mean response of four replicates. Vertical lines represent the standard deviation of the replicates. Enclosed stars indicate no hatching occurred. Adapted in part, from Sharp and Neff (1980). Reprinted courtesy of Marine Environmental Research.

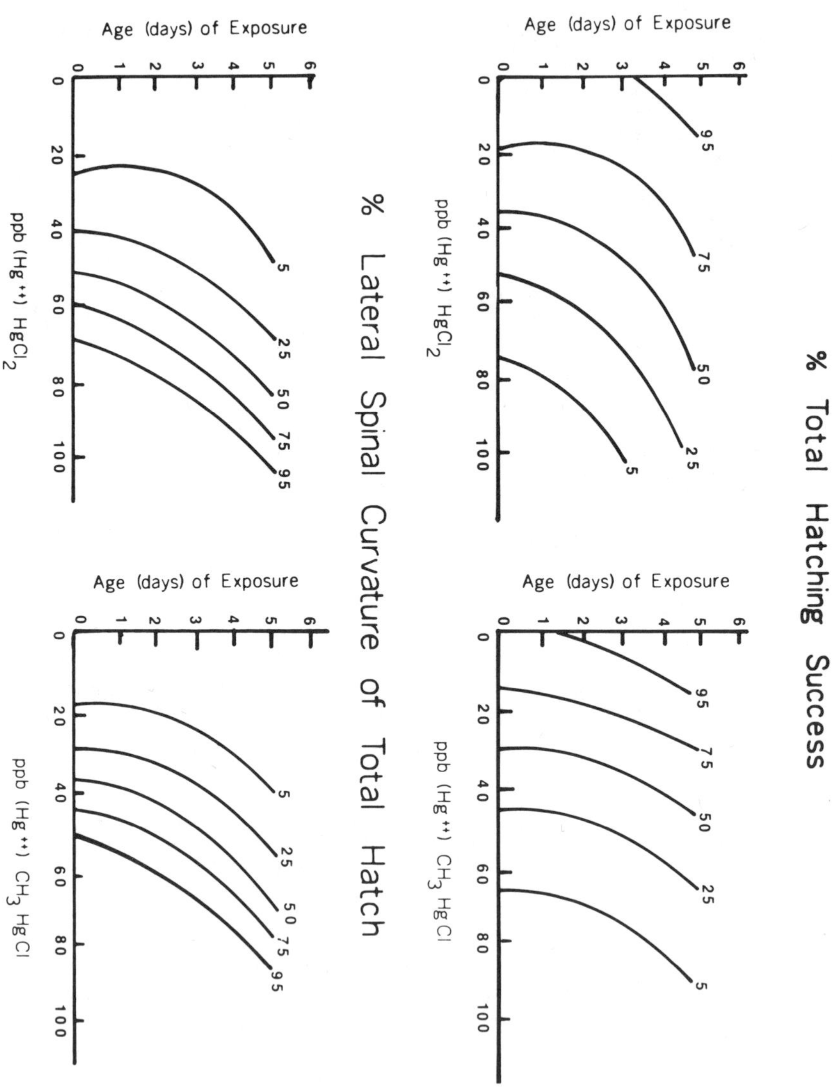

Fig. 7. Response-surface diagrams estimating percent total hatching success and percent of hatch exhibiting spinal curvature of *F. heteroclitus*, with respect to mercury concentration and embryonic age initially exposed to $(HgCl_2)$ MC and (CH_3HgCl) MMC.

Table 5 summarizes the embryonic responses following MC and MMC intoxication in terms of generated EC50 values. It is obvious that in all cases MMC is more toxic than MC to developing F. heteroclitus embryos, as indicated by a lower dosage requirement to effect a 50% adverse response. When short-term survival and total anomalous development were assessed, there were apparent increases in embryonic tolerance to mercury as progressively older embryos were initially exposed. Mortality and abnormal development were so reduced in the older-aged embryos that accurate EC50 values could not be determined. When the longer-term effects of mercury exposure to embryos were evaluated (i.e., total hatching success and viability of hatch), only the oldest embryos (5-d old) demonstrated increased tolerance to mercury contamination.

Table 5. EC50 values estimated from experimental data: 96-h LC50 (50% lethal concentration), AD50 (concentration producing 50% abnormality among surviving 12-d old embryos), SH50 (concentration limiting hatching success to 50%), VH50 (concentration limiting viable hatch to 50%). Adapted, in part, from Sharp and Neff (1980, 1982). Reprinted courtesy of Marine Environmental Research and Environmental Biology of Fishes.

Embryonic Age (days) initially exposed	EC50 ($\mu g\ Hg^{++}/L$)							
	96-h LC50		AD50		SH50		VH50	
	MC	MMC	MC	MMC	MC	MMC	MC	MMC
0	67.2	51.2	46.3	38.2	37.2	32.2	32.1	28.0
1	120.9	72.7	86.2	72.3	39.24	36.8	31.2	*
2	*	*	*	*	41.2	32.9	32.5	*
5	*	*	*	*	70.9	53.0	61.4	48.9

*insufficient data for EC50 determination.

DISCUSSION

For 96-h mortality, the 0-day old embryos of Fundulus heteroclitus are more sensitive than older embryos, and the embryos are more sensitive to MMC than MC. The LC50 values for 1-day old embryos were much greater than those values determined for the early cleavage groups, particularly following MC exposure. Furthermore, Sharp and Neff (1980, 1982) reported that 4-8 cell stage embryos are 3-30 times more sensitive than adult F. heteroclitus following an acute exposure to MC, with regard to reported 96-h LC50 values for adults.

Nearly all initial mortality occurred during the first four days of development. There was a trend of increasing yolk sac rupture with increasing mercury concentration. All other mortalities were the result of a less dramatic cessation of developmental processes. In many, axis formation did not occur and embryos were nothing more than amorphous masses of undifferentiated cells. Rosenthal and Alderdice (1976) attribute this to the stress-stressor syndrome, stating that if energy is sufficient only for cell maintenance, cellular differentiation could be inhibited and even dedifferentiation could occur. The first four days of development of this species are characterized by the vital processes of cellular migration, differentiation, and the initiation of organogenesis.

Only minor additional mortality was recorded during the intermediate stages of development (day 5-12). This decrease in mortality during the intermediate phases of development may be the result of the completion of most of the vital processes mentioned above. With the onset of hatching, mortality again increased. Thus, nearly all mortalities occurred prior to day 5 and after day 12 of development. Nearly all mortality of the more resistant 1-, 2-, and 5-day old embryos and embryos exposed to less toxic Hg^{++} concentrations occurred after the onset of hatching. This biomodal distribution in mortality has been described several times in fish exposed to natural stressors and pollutants (Kinne & Kinne, 1962; Siefert *et al*., 1973; Dethlefsen, 1977; Sharp *et al*., 1984).

Percent anomalous development among 12-day old embryos was used as a measure of the teratogenic effects of mercury in developing *F. heteroclitus* embryos. Increasing Hg^{++} concentration, as MC or MMC, or initiating continuous exposure at progressively earlier embryonic stages resulted in progressive increases in the percent incidence of anomalous development. As seen with 96-h mortality, there appears to be an increase in embryonic tolerance to mercury among older embryos as determined by decreasing percent anomalous development among 12-day old survivors. Also, MMC is more teratogenic than MC. For details regarding specific abnormalities (*e.g.*, stunted growth, cardiac malformations, etc.) associated with mercury exposure on fish development refer to Weis and Weis (1977a,b), Dial (1978), and Birge *et al*. (1979).

A second period of increased mortality occurred near the end of embryogeny (past day 12 of development), coinciding with and continuing through the hatching process. Consequently, although embryonic tolerance may have been great up to the beginning of the hatching period, this second period of increased sensitivity would result in decreased hatching success. This in turn would affect the recruitment and ecological success of the entire population.

Hatching began on day 14 of development and continued for 18 d. Sharp *et al.* (1979, 1984) reported similar time frames in *F. heteroclitus* hatching following exposure to the water-soluble fraction of a #2 fuel oil and the mud-aqueous fraction of a used drilling mud. There was no consistent change in the rate of hatching as a result of the embryonic age at exposure, the concentration used, or the mercury compound tested.

DiMichele and Taylor (1980, 1981) and DiMichele *et al.* (1981) reported that hatching of *F. heteroclitus* was inhibited when dissolved oxygen levels exceeded 6 mg. We used well aerated sea water when we changed solutions daily, which could account for the extended period for the completion of the hatching process.

Several investigators have reported an increase in mortality with the onset of hatching. Because there is wide variety of environmental stressors known to elicit such a response, a brief review of the main developmental processes occurring near the initiation of hatching seems in order. During these latter stages of development, the intra-embryonic circulatory system is completed, and the oral and opercular cavities open. The opening of cavities is directly related to the release of the hatching enzyme, chorionase (Kaighn, 1964). At this time, embryonic activity increases, resulting in a "rolling" of the embryo within the chorion. It is believed that successful emergence of embryos is both a biochemical and motor phenomenon. The release of the enzyme through the mouth and opercular cavities, the thorough distribution of the enzyme as a result of embryonic movement, and the mechanical rupturing of the chorion are all essential to the hatching process. DiMichele and Taylor (1980) suggested that external conditions of hypoxia may be an additional ingredient necessary for the hatching process of *F. heteroclitus*. Inhibition of the hatching enzyme and/or reduced embryonic motor processes could account for the decrease in total hatching success and increased mortality of mercury-exposed embryos during the normal hatching period. In the present investigation, normal embryonic movement within the chorion decreased with increasing mercury concentration. Incomplete hatching of embryos as a result of natural or pollutant stress has been reported (Rosenthal & Alderdice, 1976). Presumably this is due to the decrease in embryonic movement, which is necessary for the even distribution of the hatching enzyme, resulting in a higher concentration of chorionase at the cephalic region. This hydrolyzes the chorion in one spot only, allowing the embryos to emerge only partially. In most reported cases, the embryos never complete hatching and eventually die.

Dethlefsen (1977) attributed the second period of increased mortality of *Gadus morhua* embryos exposed to DDE and DDT to the increased efficiency in the transport of these pesticides to more vulnerable embryonic tissues resulting from the completion of the embryonic circulatory system. The increased mortality and decreased hatching success of *F. heteroclitus* embryos following exposure to mercury also could be attributed to an increase in mercury distribution. In addition, the opening of the oral and opercular cavities would increase the total exposed embryonic surface area to mercury-rich perivitelline fluids, leading to a possible increase in the influx of mercury to the embryo.

Teleost embryos are extremely variable in their hatching response following exposure to mercury. *F. heteroclitus* appears to be intermediate in its tolerance to mercury in respect to total hatchability following embryonic exposure to mercury. Inhibition of hatching in zebrafish, *Brachydanio rerio*, was reported at exposure concentrations as low as 50 µg Hg^{++}/L as phenylmercuric acetate (Kihlstrom *et al.*, 1973), whereas in carp, *Cyprinus carpio*, hatching success was decreased only at concentrations of 2-3 mg Hg^{++}/L as mercuric chloride (Huckabee & Griffith, 1974).

We have shown in this investigation that the point during embryonic development at which exposure is initiated has little effect on total hatching success and resulting SH50 values, despite differences in shorter-term embryonic sensitivity.

The incidence of spinal curvature and the resulting VH50 values probably offer the best estimate of population or ecosystem effects of mercury toxicity to developing *F. heteroclitus* embryos. Other investigators have reported an increase in spinal abnormalities following embryonic exposure to various pollutants. These include oil and oil dispersants (Linden *et al.*, 1979a: *Fundulus heteroclitus* - #2 fuel oil; Linden, 1974: *Clupea harengus* - BP 1100X and Finasol SC), heavy metals (McCulloch, 1977: *Fundulus similis* - cadmium; Weis & Weis, 1977a: *F. heteroclitus* - lead; Dial, 1978: *Oryzias latipes* - mercury), and sulfuric water pollutants (Kinne & Rosenthal, 1967: *Clupea harengus* - sulfuric acid and ferric sulfate). As with mortality, anomalous development, and total hatching success, the incidence of this anomaly was greater following exposure to methylmercuric chloride than mercuric chloride under comparable conditions of exposure.

Bengtsson (1975) states that pollutant-mediated spinal damage may be the result of neuromuscular damage or demineralization of vertebral components. Merle and Mayer (1975a,b) showed that exposure of the fathead minnow, *Pimephales promelas*, and the brook trout, *Salvelinus fontinalis*, to toxaphene resulted in a decrease in collagen synthesis by skeletal components. They suggested that the effect of toxaphene on collagen synthesis was due to a reduction in Vitamin C levels. Bengtsson (1979) offers an excellent review of skeletal deformities in fishes and the application of these for monitoring pollution. Regardless of the cause, an increase in non-viable offspring could severely affect the recruitment of new individuals into the population. These embryos would be at a distinct disadvantage in the procurement of food and in avoiding predation.

ACKNOWLEDGMENTS

We would like to express our appreciation to Jack Price, NMFS-NOAA, Beaufort, NC, for providing the adult breeding stock, to M.S. Sharp and Dr. C.L. McKenney, Jr. for assistance with data analysis and Ms. Margaret Taylor and Mrs. Vicki Howell for clerical assistance. Ms. Betty Black assisted in figure preparation. This research was supported in part by Grant No. OCE75-04890 from the National Science Foundation - International Decade of Ocean Exploration to Dr. J.M. Neff. This research was performed at Texas A & M University and represents part of the Ph.D. dissertation of J.R. Sharp.

LITERATURE CITED

Akiyama, A. 1970. Acute toxicity of two organic mercury compounds to the teleost, *Oryzias latipes*, in different stages of development. Bull. Japanese Soc. Sci. Fisheries 36: 563-570.

Armstrong, P.B. and J.S. Child. 1965. Stages in the normal development of *Fundulus heteroclitus*. Biol. Bull. 128: 143-166.

Balon, E.K. 1975. Terminology of intervals in fish development. J. Fish. Res. Board Can. 32: 1663-1670.

Barr, A.J., J.H. Goodnight, J.P. Sall, and J.T. Helwig. 1976. A User's Guide to SAS 76. SAS Institute, Inc., Raleigh, NC. 329 pp.

Bengtsson, B.E. 1975. Vertebral damage in fish induced by pollutants. pp. 23-30. In: Sublethal Effects of Toxic Chemicals on Aquatic Organisms. J.H. Koeman and J.J.T.W.A. Strik (eds.). Elsevier Scientific Publ., Amsterdam.

Bengtsson, B.E. 1979. Biological variables, especially skeletal deformities in fish, for monitoring marine pollution. Phil. Trans. R. Soc. Lond. B. 286: 457-464.

Birge, W.J., J.A. Black, A.G. Westerman, and J.E. Hudson. 1979. The effects of mercury on reproduction of fish and amphibians. pp. 629-655. In: The Biogeochemistry of Mercury in the Environment. Nriagu (ed.). Elsevier /North Holland Biomedical Press, Amsterdam.

Dethlefsen, V. 1977. The influence of DDE and DDT on the embryogenesis and the mortality of larvae of cod (*Gadus morhua* L.). Sonderdruck aus Bd. 25: 115-148.

Dial, N.A. 1978. Methylmercury: Some effects on embryogenesis in the Japanese medaka, *Oryzias latipes*. Teratology 17: 83-92.

DiMichele, L. and M.H. Taylor. 1980. The environmental control of hatching in *Fundulus heteroclitus*. J. Exp. Zool. 214: 181-187.

DiMichele, L. and M.H. Taylor. 1981. The mechanism of hatching in *Fundulus heteroclitus*: Development and physiology. J. Exp. Zool. 217: 73-79.

DiMichele, L., M.H. Taylor, and R. Singleton, Jr. 1981. The hatching enzyme of *Fundulus heteroclitus*. J. Exp. Zool. 216: 133-140.

Finney, D.J. 1971. Probit Analysis. Cambridge University Press, Cambridge. 333 pp.

Heisinger, F.J. and W. Green. 1975. Mercuric chloride uptake by eggs of the rice fish and resulting teratogenic effects. Bull. Environ. Contam. Toxicol. 14: 665-673.

Huckabee, J.W. and N.A. Griffith. 1974. Toxicity of mercury and selenium to the eggs of carp (*Cyprinus carpio*). Trans. Am. Fish. Soc. 104: 822-824.

Jackim, E., J.M. Hamlin, and S. Sonis. 1970. Effects of metal poisoning of five liver enzymes in the killifish (*Fundulus heteroclitus*). J. Fish. Res. Board Can. 27: 383-390.

Kaighn, M.E. 1964. A biochemical study of the hatching process in *Fundulus heteroclitus*. Develop. Biol. 9: 56-80.

Kihlstrom, J.E. and L. Hulth. 1972. The effect of phenylmercuric acetate upon the frequency of hatching eggs from the zebrafish. Bull. Environ. Contam. Toxicol. 7: 111-114.

Kihlstrom, J.E., C. Lundberg, and L. Hulth. 1973. Numbers of eggs and young produced by zebrafish (*Brachydanio rerio*, Ham.-Buch.) spawning in water containing small amounts of phenylmercuric acetate. pp. 101-105. *In*: Mercury Poisoning: II. MSS Information Corporation, New York.

Kinne, O. and E.M. Kinne. 1962. Rates of development in embryos of a cyprinodont fish exposed to different temperature-salinity-oxygen combinations. Can. J. Zool. 40: 231-253.

Kinne, O. and H. Rosenthal. 1967. Effects of sulfuric water pollutants on fertilization, embryonic development and larvae of the herring, *Clupea harengus*. Mar. Biol. 1: 65-83.

Klaung, J., S. Koepp, and M. McCormick. 1975. Acute toxicity of a native mummichog population (*Fundulus heteroclitus*) to mercury. Bull. Environ. Contam. Toxicol. 14: 534-537.

Linden, O. 1974. Effects of oil spill dispersants on the early development of the Baltic herring. Ann. Zool. Finnici 11: 141-148.

Linden, O., R. Laughlin, J.R. Sharp, and J.M. Neff. 1979a. The combined effect of salinity, temperature and oil on the growth patterns of embryos of the killifish, *Fundulus heteroclitus*. Mar. Environ. Res. 3: 129-144.

Linden, O., J.R. Sharp, R. Laughlin, and J.M. Neff. 1979b. Interactive effect of salinity, temperature and chronic exposure to oil on the survival and developmental rate of embryos of the estuarine killifish, *Fundulus heteroclitus*. Mar. Biol. 51: 101-109.

Macek, K.J. and B.H. Sleight. 1977. Utility of toxicity tests with embryos and fry of fish in evaluating hazards associated with chronic toxicity of chemicals to fishes. *In*: Aquatic Toxicology and Hazard Evaluation, ASTM STP 634. F.L. Mayer and J.L. Hamelink (eds.). American Society for Testing and Materials, Philadelphia.

McCulloch, W.L. 1977. Some physiological effects of cadmium on the early life stages of the estuarine teleost, *Fundulus similis*. M.S. Thesis, Texas A & M University.

McKim, J.M. 1977. Evaluation of tests with early life stages of fish for predicting long term toxicity. J. Fish. Res. Board Can. 34: 1148-1154.

Merle, P.M. and F.L. Mayer. 1975a. Toxaphene effects on growth and bone composition of fathead minnows, *Pimphales promelas*. J. Fish. Res. Board Can. 32: 593-598.

Merle, P.M. and F.L. Mayer. 1975b. Toxaphene effects on growth and bone structure of brook trout (Salvelinus fontinalis). J. Fish. Res. Board Can. 32: 609-613.

Ott, L. 1977. An Introduction to Statistical Methods and Data Analysis, Duxbury Press, North Scituate, MA. 730 pp.

Rosenthal, H. and D.F. Alderdice. 1976. Sublethal effects of environmental stressors, natural and pollutional, on marine fish eggs and larvae. J. Fish. Res. Board Can. 33: 2047-2065.

Sharp, J.R., R.S. Carr, and J.M. Neff. (in press). The influence of a chrome lignosulfonate drilling mud on the early life history of the mummichog, Fundulus heteroclitus. In: Wastes in the Oceans, Vol. 5, Chapter XX. I. Duedall (ed.).

Sharp, J.R., K.W. Fucik, and J.M. Neff. 1979. Physiological basis of differential sensitivity of fish embryonic stages to oil pollution. pp. 85-107. In: Marine Pollution: Functional Responses. W.B. Vernberg, F.P. Thurberg, A. Calabrese, and F.J. Vernberg (eds.). Academic Press, New York.

Sharp, J.R. and J.M. Neff. 1980. The effects of the duration of exposure to mercuric chloride on the embryogenesis of the estuarine teleost, Fundulus heteroclitus. Mar. Environ. Res. 3: 195-213.

Sharp, J.R. and J.M. Neff. 1982. The toxicity of mercuric chloride and methylmercuric chloride to Fundulus heteroclitus, in relation to exposure conditions. Environ. Biol. Fish. 7(3): 277-284.

Siefert, R.E., W.A. Spoor, and R.J. Syrett. 1973. Effects of reduced oxygen concentrations on norther pike (Esox lucius) embryos and larvae. J. Fish. Res. Board Can. 30: 849-852.

Trinkaus, J.P. 1967. Fundulus. pp. 113-137. In: Methods in Developmental Biology. F.H. Wilt and N.K. Wessells (eds.). Thomas Y. Crowell, New York.

Weis, J.S. and P. Weis. 1977a. Effects of heavy metals on development of the killifish, Fundulus heteroclitus. J. Fish. Biol. 11: 49-54.

Weis, P. and J.S. Weis. 1977b. Methylmercury teratogenesis in the killifish, Fundulus heteroclitus. Teratology 16: 317-325.

SEARCH FOR A PHYSIOLOGICAL COMPONENT OF METHYLMERCURY TOLERANCE IN THE MUMMICHOG, *FUNDULUS HETEROCLITUS*

J.S. Weis, P. Weis, M. Renna, and S. Vaidya

SUMMARY

Heavy metal tolerance can be acquired by both short-term physiological mechanisms and by long-term genetic selection. In view of our previous findings that Fundulus heteroclitus embryos from a polluted environment (Piles Creek - PC) are more tolerant to methylmercury (meHg) than embryos from a cleaner habitat (Southampton - SH), we investigated the short-term component of tolerance in four experiments:

1. PC females were maintained in clean water to see if meHg tolerance of subsequent clutches would be reduced relative to initial clutches. Data showed that tolerance was actually increased in later clutches. SH females were maintained in PC water and in water with low levels of meHg to see if subsequent clutches would show an increase in tolerance, but no subsequent clutches were produced.
2. Tolerance of larvae which were pre-exposed to 0.02 ppm as embryos was compared with that of littermate control groups which were not pre-exposed; the pre-exposed groups were found to be less resistant.
3. A fin regeneration assay compared the tolerance of adults pre-exposed to 0.01 and 0.005 ppm meHg with adults not pre-exposed. While those pretreated with 0.005 ppm and those not pretreated were equally affected by 0.05 ppm during regeneration,

the fish which had been pretreated with 0.01 ppm showed a greater reduction in fin regeneration rate, indicating that they were less tolerant.

4. Tolerance of PC adults (pre-exposed for their whole lives) was compared to that of SH adults in the fin regeneration assay. In both groups 0.05 ppm meHg slowed regeneration; the SH controls grew at a much greater rate than PC controls, and PC experimentals had a very high mortality rate which indicates lower tolerance. In 0.03 ppm meHg, SH fish showed an enhancement of growth (hormesis) not seen in the PC fish.

Therefore, no evidence for a short-term component to meHg tolerance in this species has been found, and it is likely that the observed tolerance is genetic.

INTRODUCTION

A number of studies have shown that there is an increased tolerance to metal pollution after pre-exposure. Possible short-term mechanisms for developing increased resistance after pre-exposure to metals include the synthesis of metal-binding proteins, or metallothioneins (MT) (Cherian & Coyer, 1978), small proteins rich in cysteine. Metals such as Hg, Cd, Zn, and Cu readily displace H from the free SH groups of these proteins. Since metals can induce synthesis of metallothionein, it would be expected that fish previously exposed to metals would have higher levels of these proteins (Noel-Lambot *et al*., 1978). The protective role of these proteins may function by preventing the metal from binding to various enzymes which take part in biological reactions. Pretreatment of experimental animals (*Salmo gairdneri*) with cadmium reduced the toxicity of higher doses, presumably by inducing the synthesis of MT (Pascoe & Beattie, 1979). A similar protective effect of pre-treatment with a sublethal dose of $HgCl_2$ was seen in eels which were later tested with higher doses (Bouquegneau, 1979). Dixon and Sprague (1981) and Duncan and Klavercamp (1983) have further documented this phenomenon.

Another short-term mechanism proposed for sequestering heavy metals is via uptake into lysosomes or other storage granules (Carpene & George, 1981).

In addition to these short-term mechanisms, chronically-exposed populations can develop resistance through selection for the more tolerant individuals. Russell and Morris (1972) found genetically greater resistance to copper in the brown alga *Ectocarpus* from environments with higher copper concentrations. Similarly, Stokes *et al*. (1973) isolated a nickel-and-copper resistant strain of the green alga *Scenedesmus* from a contaminated lake. Foster (1977) has reported that the mechanism of copper tolerance in *Chlorella vulgaris* from a Cu-polluted stream was by accumulating far less metal than a non-tolerant strain.

Bryan and colleagues (reviewed in Bryan, 1976) have studied the polychaete *Nereis diversicolor* and have found that populations from sediments rich in a particular metal were resistant to the particular metal. Animals that were tolerant to zinc were also less permeable to the metal than were non-tolerant ones. The differences were believed to be genetic since the resistance was not changed by exposing tolerant and non-tolerant worms to the sediment of the opposite type (Bryan & Hummerstone, 1971). Fraser *et al*. (1978) have demonstrated resistance to lead in the isopod *Asellus aquaticus* from sites with high lead levels. Brown (1976), studying *Asellus meridianus*, found that strains tolerant to lead and copper could accumulate high levels which proved lethal to non-tolerant animals. Tolerance persisted into subsequent laboratory-bred generations, indicating a genetic basis for the resistance.

Nevo *et al*. (1981) studied allozymic variation of phosphoglucomutase (PGM) genotypes in the shrimp *Palaemon elegans*. They found differential tolerance of some genotypes to mercury, suggesting that they are adaptive. Luoma (1977) discussed the potential use of toxicant resistance as a tool in assessing contaminant effects in ecosystems. He hypothesized that greater resistance to a toxicant in a population from one area, as compared to another area, is direct evidence that the toxicant is exerting selective pressure at the first site, and that the level of resistance of the organisms reflects the degree of contamination of the environment.

Rahel (1981) selected for zinc tolerance in a laboratory population of the flagfish, *Jordanella floridae*, and found increased resistance after one generation. However, continued selection for three more generations failed to increase tolerance further, implying that the zinc exposure eliminated the weakest individuals but did not cause genetic changes specifically related to zinc tolerance. Furthermore, shiners (*Notropus cornutus*) from a zinc-polluted stream were not more zinc tolerant (in terms of survival and swimming stamina) than those from unpolluted areas. These results and a literature search suggested to Rahel that although fish have the genetic potential to evolve metal tolerance, they are unable to do so rapidly enough to survive in contaminated environments.

Research in our laboratories on the mummichog, Fundulus heteroclitis, indicates otherwise. In studying effects of methylmercury (meHg) on embryos from a population from relatively unpolluted Montauk and Southampton, N.Y. (Long Island), we found considerable variation in susceptibility of eggs from different females (Weis et al., 1982). Some females produced very tolerant eggs, while others produced more susceptible eggs. Tolerance was associated with the female's fin ray count, suggesting a genetic basis. Susceptible clutches tended to incorporate higher levels of Hg. We then studied a population from Piles Creek, a tributary of the Arthur Kill, in Linden, N.J., an area heavily impacted by metal and oil pollution. The marsh surrounding the creek is ringed by industrial sites, including a generating station, an oil refinery, a sewage plant, several chemical plants, and a major highway. The creek heads at the marsh edge and therefore does not receive diluting freshwater from outside the industrial area (Luther et al., 1982). In this population very few females produced susceptible eggs, and most clutches were tolerant with respect to the production of embryonic malformations (Weis et al., 1981).

In order to investigate the short-term physiological component of this enhanced meHg tolerance, experiments were performed to answer the following questions:

1. Would maintenance of Southampton (SH) fish in low levels of meHg or in Piles Creek (PC) water cause them to produce more resistant clutches of eggs and, conversely, would maintenance of PC fish in clean water cause them to produce less resistant eggs?
2. Would embryonic exposure to a non-teratogenic dose of meHg cause greater tolerance in the larvae after hatching?
3. Would short-term exposure of SH adults to low levels of meHg enhance their subsequent tolerance as measured by fin regeneration rate, a process which has previously been shown to be retarded by meHg (Weis & Weis, 1978)?
4. Would PC adults be more tolerant to meHg than SH adults? If physiological mechanisms play a major role in the tolerance, one would expect that having lived in polluted PC would have stimulated

protective mechanisms and would have increased the degree of tolerance. Again, fin regeneration was the parameter examined.

MATERIALS AND METHODS

Initial vs Subsequent Clutches of Eggs

Gravid female mummichogs collected from Bullhead Bay, SH, and from PC were stripped and the eggs fertilized with milt stripped from males. After removing all non-cleaving eggs, part of each clutch was exposed to 0.05 ppm meHg (methods described in Weis *et al.*, 1981) at 24°C, 25°/oo S. After a week, embryos were examined under a stereomicroscope for malformations of the head, cardiovascular system, and skeletal system. The results were quantified by indices which grade the severity of the malformation (Weis *et al.*, 1981). The craniofacial index (CFI) grades convergence of the eyes on a scale of 0-6 (0 = normal, 1 = slight convergence, 2 = greater convergence to touching, 3 = synophthalmia, 4 = fused into one median eye with two lenses, 5 = cyclopia, 6 = anencephaly). The cardiovascular index (CVI) grades heart defects (0 = normal, 1 = slight deficiency in structure or function, 2 = tube heart, contracting, 3 = tube heart, not contracting, 4 = undifferentiated mass of beating cells, 5 = no discernable cardiac tissue). The skeletal index (SKI) grades skeletal defects (0 = normal, 1 = slight bend or kink, 2 = major bend - greater than 90° angle or more than one bend, 3 = stunted, 4 = very stunted but axis discernable, 5 = no axis, amorphous). The mean CFI, CVI, and SKI for each clutch were calculated. In addition, the percent of normal embryos was recorded for each clutch.

Identified females were maintained in the laboratory, SH fish in PC water (20°/oo) or clean water with 0.005 meHg, and PC fish in clean artifical sea water (Rila® Mix -20°/oo). Subsequent clutches of eggs stripped from the same individual could then be exposed to 0.05 ppm meHg and their tolerance compared with that of the first clutch. We have previously shown that the male used to fertilize the eggs does not affect their tolerance (Weis *et al.*, 1982).

Effects of Embryonic Pre-exposure on Larval Tolerance

Adult *F. heteroclitus* were collected from Bullhead Bay, SH, and eggs and sperm stripped into fingerbowls of sea water. Clutches of eggs were maintained separately, as before. For the pre-exposure, part of the clutch was treated with 0.02 ppm meHg (ICN Pharmaceuticals, Plainview, NY) from a stock solution frequently assayed by atomic absorption spectrophotometry. This is a level that produces negligible amounts of embryonic abnormalities. Exposure started after eggs began to cleave and all non-cleaving eggs had been removed. Embryos were kept in 50 ml filtered sea water (25°/$_{oo}$) at 24°C and water was changed daily for the first 4 d of development. Controls were also subjected to water changes. After hatching, which was stimulated by changing the water daily starting on day 12, larvae from each treatment and control group were subdivided and transferred to either 0.05 ppm meHg or clean water. Each group contained 10 larvae from the same clutch and the same embryonic treatment. Larvae were unfed and maintained at 24°C in 100 ml sea water which was changed and redosed daily, at which time any dead larvae were removed and recorded. They were maintained up to 10 d. For each group an LT50 (time to 50% death) was calculated, and a mean LT50 for pretreated and for non-pretreated groups was derived. These were compared by t-test.

Adult Pre-exposure Experiments

Fish 40-60 mm in standard length were collected from Bullhead Bay, SH. Three groups were pre-exposed for 1 wk to 0.00, 0.01 or 0.005 ppm meHg in aerated, unfiltered sea water (25°/$_{oo}$) at ambient temperature (25°C) in all-glass aquaria. Water was changed and redosed after 4 d. After 7 d, fish from both the pre-exposure groups and the control group had the bottom half of their caudal fin amputated just distal to the caudal peduncle. Each group was then subdivided into untreated controls and experimentals which were exposed to 0.05 ppm meHg in 25°/$_{oo}$, ambient temperature. Water was changed and redosed twice weekly during the 2-wk experiment. The fish were fed commercial flake food. Regenerates were measured by ocular micrometer in a stereomicroscope over the 14-d period. The mean regenerated length of the pre-exposed fish (experimentals and controls) was compared to that of the fish which had not been pre-exposed. Each group had 15-21 fish.

The experiment was repeated with pre-exposure concentrations of 0.01 and 0.001 ppm and a pre-exposure time of 4 d. After the last measurement, the fish were frozen. Later, samples of caudal peduncle (n = 3 per group) were analyzed for total Hg by cold-vapor atomic absorption spectrophotometry (Hatch & Ott, 1968).

Population Differences

Fish were collected from SH and PC and acclimated to lab conditions. Amputations, maintenance, and measurements were described previously except that fish were in artificial sea water (Rila Mix) at 20°/$_{oo}$, 20°C. In the first experiment fish were exposed to 0.05 ppm meHg, and in the second to 0.03 ppm meHg. In addition to t-test, analysis of covariance (ANCOVA) was used to compare regenerative growth curves. ANCOVA can test for differences in elevation (Y-intercept) and for differences in slope. The Y-intercept represents the time needed for would healing and blastema formation, while the slope represents the rate of fin elongation (Zimmerer, 1980).

RESULTS

Initial vs. Subsequent Clutches of Eggs

The SH fish maintained in 0.005 ppm meHg failed to give any additional clutches of eggs to compare with the original. SH fish maintained in PC water did likewise.

PC fish maintained in clean water sometimes did give additional clutches. These varied from a few days to one month after the first clutch had been stripped. For data analysis, all subsequent clutches were pooled together (n = 16) and compared with all initial clutches (n = 11). While some showed increased and some showed decreased tolerance, the general trend was to become more tolerant. Average CFIs went from 0.40 to 0.11, CVIs from 1.24 to 0.69, and SKIs from 1.33 to 0.79. Mann-Whitney U test (Zar, 1974) showed that differences for CVI and SKI were significant ($U = 141.5$, $p<0.01$, and $U = 129.5$, $p<0.05$, respectively). The overall percent of normal fish went from 18.2 in initial clutches to 45.7 in later clutches, but the variance was quite large and this difference was not significant.

Effects of Embryonic Pre-exposure on Larval Tolerance

All control larvae, whether pretreated or not, had very little (<10%) mortality during the 10 d, and had not reached an LT50 when the experiments were terminated. The LT50 values for the meHg-treated larvae (Table 1) show that pre-exposure to meHg reduced the time to 50% death, indicating that these larvae were less tolerant than those which had not been pre-exposed.

Table 1. LT50 of *F. heteroclitus* larvae in methylmercury (meHg).

Embryonic Treatment	Larval Treatment	(n)	LT50
Control	0.05 ppm meHg	22	5.32 ± 0.31 (SE)
0.02 ppm meHg	0.05 ppm meHg	15	4.33 ± 0.24*

*Significantly different by t-test, $p<0.001$.

Adult Pre-exposure Experiments

The regeneration rates of control and meHg-exposed fish are seen in Figure 1. The meHg groups regenerated significantly slower than controls, in agreement with previous work (Weis & Weis, 1978). The group that had been pre-exposed to 0.005 ppm meHg regenerated in mercury (0.005 meHg → 0.05 meHg) at a rate comparable to those which had not been pre-exposed, while the group that had been pre-exposed to 0.01 ppm (0.01 meHg → 0.05 meHg) regenerated more slowly than all the others. At day 9, the t-values for controls (C → C) vs. C → meHg was 2.50 ($p<0.05$); controls vs. 0.005 meHg → meHg was 2.10 ($p<0.05$); and controls vs 0.01 meHg → meHg was 3.04 ($p<0.005$). For statistical purposes the C → C group was combined with the 0.01 → C and 0.005 → C groups, since all three grew at a similar rate. The data indicate that while pretreatment with 0.005 ppm had no effect on the subsequent tolerance, pretreatment with 0.01 ppm made the fish more susceptible to subsequent exposure. When the experiment was repeated, those pre-exposed to 0.001 ppm were retarded to the same degree as those not pre-exposed while those pre-exposed to 0.01 ppm were again retarded to a greater degree. For example, on day 14 the controls measured 4.43 ± 0.08 mm while C → meHg was 3.60 ± 0.21 mm, 0.001 → meHg was 3.63 ± 0.18 mm, and 0.01 meHg was 3.07 ± 0.21 mm.

Data for Hg uptake (Table 2) indicate a dose-dependent increase in Hg burden in relation to pre-exposure level, but the differences are not significant.

Population Differences

In experiment 1, 0.05 ppm meHg reduced the regenerative growth in both populations, but not always significantly (Table 3 and Fig. 2). The PC fish exhibited 85% mortality during the 2-wk period, while the SH fish had only 15% mortality. This is evidence that the PC fish, rather than being more tolerant, are less tolerant to meHg at the adult stage. Furthermore, the SH controls regenerated significantly faster than the PC controls. When this experiment was repeated, the PC fish in 0.05 ppm meHg had 67% mortality, while the SH fish had 22% mortality.

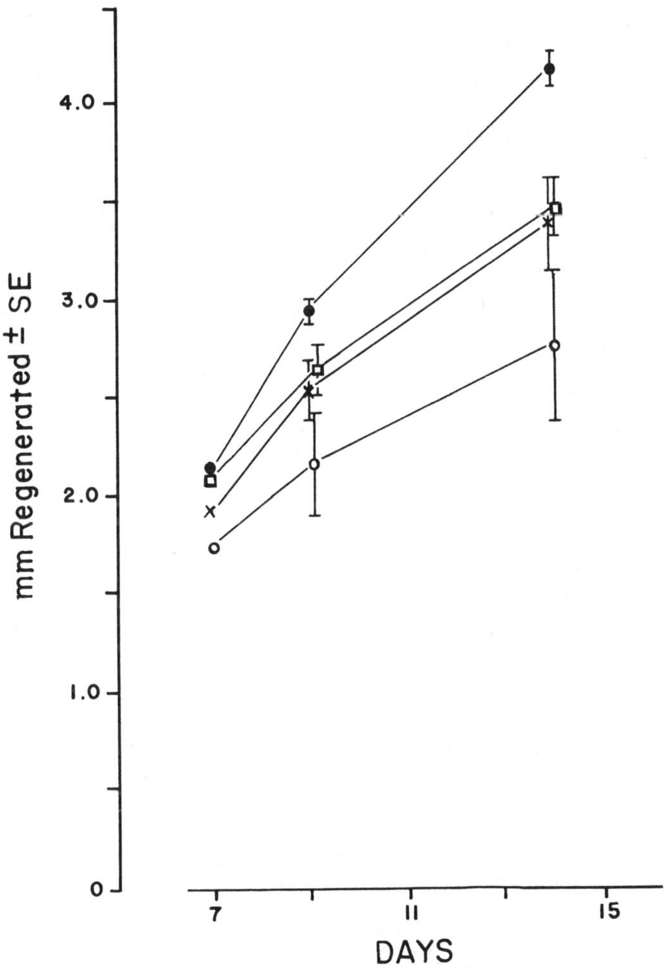

Fig. 1. Pre-exposure effects on regeneration of F. heteroclitus caudal fins (mm ± SE) □ = 0.005 -- 0.05 ppm meHg; ○ = 0.01 -- 0.05 ppm meHg; x = control -- 0.05 ppm meHg; ● = pooled controls (C--C, 0.01 ppm meHg -- C, and 0.005 ppm meHg -- C). See text for explanation.

Table 2. Uptake of Hg by *F. heteroclitus* during fin regeneration.

Pretreatment (ppm meHg)	Treatment (ppm meHg)	(n)	Hg (x ± SD) (ppm total Hg)
Control	Control	3	0.09 ± 0.03
0.001	Control	3	0.10 ± 0.01
0.01	Control	3	0.52 ± 0.11
Control	0.05	3	7.75 ± 1.98
0.001	0.05	3	8.31 ± 1.19
0.01	0.05	3	9.80 ± 3.67

Table 3. Caudal fin regeneration in two populations of *F. heteroclitus*. Mean growth in mm ± SE.

Exp.	Day	n	Growth	n	Growth	t-test	F*
		0.05 ppm meHg		Controls			
1 Piles Creek Population							
	7	10	0.20 ± 0.19	11	0.73 ± 0.13	2.30 $p<0.05$	F_1 = 0.103 $p>0.05$
	11	6	0.27 ± 0.24	11	0.91 ± 0.21	2.01 $p>0.05$	F_2 = 3.96 $p>0.05$
	14	3	0.53 ± 0.44	10	0.94 ± 0.22	0.83 $p>0.05$	
1 Southampton Population							
	7	17	0.91 ± 0.14	14	1.16 ± 0.09	1.50 $p>0.05$	F_1 = 4.40 $p<0.05$
	11	16	1.44 ± 0.19	14	2.15 ± 0.17	2.78 $p<0.01$	F_2 = 17.71 $p<0.001$
	14	16	1.50 ± 0.23	11	2.52 ± 0.20	3.36 $p<0.005$	
		0.03 ppm meHg		Controls			
2 Piles Creek Population							
	7	20	0.59 ± 0.09	20	0.86 ± 0.08	2.24 $p<0.05$	F_1 = 0.059 $p>0.05$
	11	19	1.28 ± 0.17	20	1.48 ± 0.17	0.83 $p>0.05$	F_2 = 3.37 $p>0.05$
	14	19	1.80 ± 0.26	19	2.17 ± 0.24	1.05 $p>0.05$	
2 Southampton Population							
	7	21	1.65 ± 0.12	18	1.27 ± 0.10	2.43 $p<0.05$	F_1 = 0.399 $p>0.05$
	11	20	3.10 ± 0.15	18	2.51 ± 0.15	2.78 $p<0.01$	F_2 = 16.21 $p<0.0005$
	14	20	3.96 ± 0.20	18	3.48 ± 0.14	1.97 $p>0.05$	

ANCOVA comparisons between the Piles Creek and Southampton populations.

Exp.	Condition	Covariance			
1	0.05 ppm meHg	F_1 = 0.106	$p>0.05$	F_2 = 3.62	$p>0.05$
	Control	F_1 = 10.60	$p<0.0025$	F_2 = 45.73	$p<0.0005$
2	0.03 ppm meHg	F_1 = 11.64	$p<0.0005$	F_2 = 126.07	$p<0.0005$
	Control	F_1 = 8.02	$p<0.01$	F_2 = 45.63	$p<0.0005$

*F^1 = comparison between slopes F^2 = comparison between Y intercepts

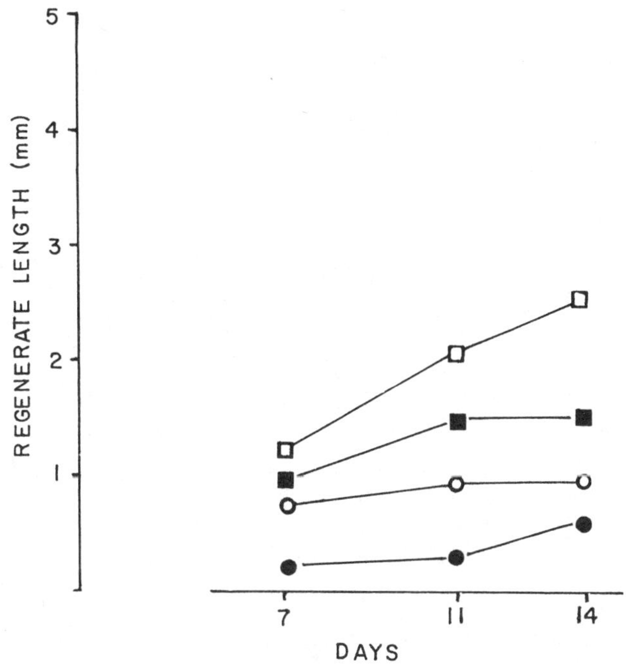

Fig. 2. Population differences in caudal fin regeneration of 0.05 mg/L meHg-treated and untreated fish. ○,● = Piles Creek population; □,■ = Southampton population; open figures = control; filled figure = treated.

In experiment 2, the exposure to 0.03 ppm had no significant effect on the regeneration of PC fish, but did significantly increase the rate of regeneration in the SH fish. This phenomenon was confirmed in a repeat experiment. Again, SH controls regenerated faster than their PC counterparts (Table 3 and Fig. 3).

DISCUSSION

Overall, eggs of PC fish maintained in clean water became slightly more tolerant than original clutches. This is the opposite of what would be expected if embryo tolerance is a short-term response to the environmental conditions of the female. If a short-term physiological response were involved, one would expect PC fish kept in clean water to produce less tolerant eggs. One would expect SH fish kept in polluted water to produce eggs with greater tolerance. Unfortunately, they produced no eggs at all, so there is no answer to this half of the question.

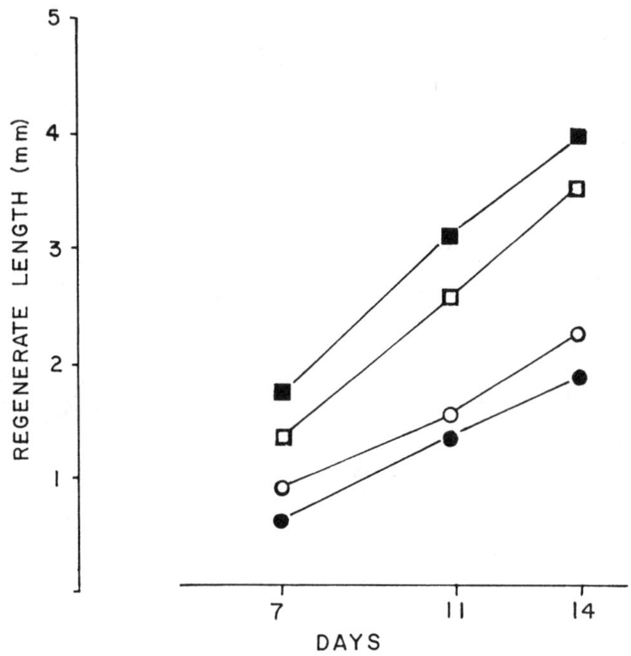

Fig. 3. Population differences in caudal fin regeneration of 0.03 mg/L meHg-treated and untreated fish, keyed as in Figure 2.

In the experiments involving larval toxicity, the data indicate that embryonic pre-exposure to meHg decreased rather than increased larval tolerance. It would appear that this pre-exposure failed to activate a protective mechanism. This is supported by the work of Weis (1984) which showed that embryos exposed to meHg did not have higher levels of metallothionein than control embryos.

In several previous studies with larval fish, metal tolerance was enhanced by pre-exposure. Beattie and Pascoe (1978) showed that treatment of rainbow trout (*S. gairdneri*) eggs with cadmium made the larvae more resistant to cadmium. Similarly, Spehar *et al*. (1978) found that larvae of the flagfish (*Jordanella floridae*) initially exposed as embryos to Zn and Cd were much more tolerant than those not pre-exposed.

There have also been reports of pre-exposure failing to produce an increase in tolerance. For example, Green *et al*. (1976) found that pre-exposing white shrimp, *Penaeus setiferus*, to 0.5-1.0 ppb Hg had no effect on the LC50 or on growth and molting in response to higher concentrations.

The concentration chosen for pre-exposure can have important effects on the outcome. Roesijadi _et al_. (1982) found that while pre-exposure of _Mytilus edulis_ to 0.5 ppb Hg enhanced the subsequent tolerance, pre-exposure to 5.0 ppb did not. It may be that the one pre-exposure concentration which we selected was not appropriate for inducing mechanisms for increased tolerance, but rather weakened the larvae by increasing the body burden and reducing their tolerance. Vaidya (unpubl.) found that other pre-exposure doses did not increase larval tolerance.

In both experiments involving pre-exposure effects on fin regeneration, neither pre-exposure dose enhanced the tolerance of killifish to meHg as measured by fin regeneration rate. The lower pre-exposure doses (0.001 and 0.005 ppm) made no difference on their subsequent tolerance, while the higher dose (0.01 ppm) made them more susceptible to the growth retarding effects of meHg. This is similar to what was observed in the embryo-larval pre-exposure experiments, in that no evidence for enhanced tolerance after pre-exposure was observed. Rather, the increased amount of total exposure had an additive effect and weakened the fish, making them more susceptible to subsequent exposure. In this series of experiments the uptake of Hg was measured; the differences in pretreatment did not result in significant differences in uptake among the treated fish. Perhaps the body burden by itself is not the critical component in such an experiment. The additional time - the pre-exposure period - may be what is required for enhancing the deleterious effects of meHg.

In the population experiments, the rate of fin regeneration was depressed in both populations by treatment with 0.05 ppm meHg. However, the PC population exhibited a much greater mortality, indicating a lower tolerance. In the controls, the rate of fin regeneration was significantly faster in the SH fish, which may be considered evidence that the PC fish are stressed and weakened. The slower growth of PC controls may be due to a shift of energy into detoxification mechanisms and a resulting decrease in energy available for growth.

Exposure to 0.03 ppm, while having no significant effect on regeneration of the PC fish, had a stimulatory effect on regenerative growth of the SH fish. Such a stimulatory effect of low levels of toxicants has been observed before and has been termed "hormesis" or "sufficient challenge." Pickering and Gast (1972) observed an increase in egg production in fathead minnows in 13 μg Cd/L. Laughlin *et al*. (1981) found enhancement of growth in crabs exposed to water-soluble fractions of oil. Low doses of toxicants are hypothesized to cause an overcompensation by homeostatic regulatory control mechanisms (Stebbing, 1981a) which are responsible for the growth enhancement. This response was seen only in the SH population. This population is from a relatively clean environment so they are not severely stressed by pollutants. A concentration of 0.03 ppm meHg could function as a sufficient challenge for them. On the other hand, the PC fish are from an environment in which pollutants are continuously present to stress their homeostatic mechanisms, and therefore, they may not be capable of hormesis. The ability to counteract toxicants can be considered a measure of health (Stebbing, 1981b); these data therefore support the concept of poor health in the PC population.

These experiments show that the PC adults are less tolerant of meHg and that they appear to be weakened and stressed. This is in contrast to the greater tolerance exhibited by PC embryos, but is similar to the stressed and weakened condition observed by Rahel (1981) in fish from a metal-contaminated area. Since PC adults were less meHg-tolerant than the unpolluted population (despite ample opportunity to develop protective mechanisms), it is likely that their protective mechanisms are saturated by their natural environment and are not capable of further expansion.

We have found no evidence that short-term physiological events are involved in meHg tolerance in this species. Pre-exposure did not increase larval or adult tolerance, although it is possible that an appropriate dose was not found. However, the lowered tolerance of pre-exposed larvae and adults, as well as that of the PC fish which have been exposed to a broad spectrum of pollution for their whole lives, is evidence that exposure causes weakness rather than increased tolerance. Therefore, it seems likely that the enhanced tolerance seen in the PC embryos has a genetic basis and it not caused directly by the female's immediate environment.

F. heteroclitus in a clean environment has a broad plasticity and therefore a high probability of adapting. It has one of the highest genetic variations of all teleosts (Smith & Fujio, 1982). Mitton and Koehn (1975) felt that the polygynous mating system of this species allowed for rapid evolutionary responses to a variable environment. Differential isozyme patterns were found in Long Island vs.

Piles Creek fish (Heber, 1981), but these differences were correlated with geographic differences in genotypes of this species (Place & Powers, 1978) and were probably not pollution-related. Tolerance has been correlated with the female's fin ray count, a characteristic which has a major genetic component (Weis *et al.*, 1982).

In experiments on fiddler crabs, it was found that pre-exposure to low levels of meHg did not enhance tolerance to higher levels in *Uca pugilator*. However, a population of *U. pugnax* from Piles Creek was more tolerant than crabs from a cleaner environment (Callahan & Weis, 1983). Therefore, this tolerance is probably genetic rather than physiological in nature. Also, it may be that the methylated form of mercury is not able to activate physiological protective mechanisms in these organisms. This may be an additional reason why this particular pollutant is especially damaging.

ACKNOWLEDGMENTS

Supported by NOAA Office of Sea Grant NA81AA-D-0065. This is publication no NJ83-123 of the NJ Marine Sciences Consortium.

LITERATURE CITED

Beattie, J. and D. Pascoe. 1978. Cadmium uptake by rainbow trout *Salmo gairdneri* eggs and alevins. J. Fish Biol. 13: 631-637.

Bouquegneau, J.M. 1979. Evidence for the protective effect of metallothioneins against inorganic mercury injuries to fish. Bull. Environ. Contam. Toxicol. 23: 218-219.

Brown, B.E. 1976. Observations on the tolerance of the isopod (*Asellus meridianus* Lac.) to copper and lead. Water Res. 10: 555-559.

Bryan, G.W. 1976. Some aspects of heavy metal tolerance in aquatic organisms. pp. 7-34. *In*: Effects of Pollutants on Aquatic Organisms. A. Lockwood (ed.). Cambridge Univ. Press, Cambridge.

Bryan, G.W. and L.G. Hummerstone. 1971. Adaptation of the polychaete *Nereis diversicolor* to estuarine sediments containing high concentrations of heavy metals. I. General observations and adaptation to copper. J. Mar. Biol. Assoc. U.K. 51: 845-863.

Callahan, P. and J.S. Weis. 1983. Methylmercury effects on regeneration and ecdysis in fiddler crabs (*Uca pugilator*, *U. pugnax*) after short-term and chronic pre-exposure. Arch. Environ. Contam. Toxicol. 12: 707-714.

Carpene, E. and S. George. 1981. Absorption of cadmium by gills of *Mytilus edulis* (L.). Molec. Physiol. 1: 23-34.

Cherian, M.G. and R.A. Goyer. 1978. Metallothioneins and their role in the metabolism and toxicity of metals. Life Sci. 23: 1-10.

Dixon, D.G. and J.B. Sprague. 1981. Copper bioaccumulation and hepatoprotein synthesis during acclimation to copper by juvenile rainbow trout. Aquat. Toxicol. 1: 69-81.

Duncan, D.A. and J.F. Klavercamp. 1983. Tolerance and resistance to cadmium in white suckers (*Catostomus commersoni*) previously exposed to cadmium, mercury, zinc, or selenium. Can. J. Fish. Aquat. Sci. 40: 128-138.

Foster, P.L. 1977. Copper exclusion as a mechanism of heavy metal tolerance in a green alga. Nature 269: 322-323.

Fraser, J.D., D. Parker, and E. Verspoor. 1978. Tolerance to lead to the freshwater isopod *Asellus aquaticus*. Water Res. 12: 637-641.

Green, F., J. Anderson, S. Petrocelli, D. Presley, and R. Sims. 1976. Effect of mercury on the survival, respiration and growth of postlarval white shrimp, *Penaeus setiferus*. Mar. Biol. 37: 75-81.

Hatch, W.R. and W.L. Ott. 1968. Determination of submicrogram amounts of mercury by atomic absorption spectrophotometry. Anal. Chem. 40: 2085-2087.

Heber, M.A. 1981. Comparative embryonic methylmercury susceptibility and reproductive biology of two populations of the mummichog (*Fundulus heteroclitus*). Master's Thesis, Rutgers Univ., Newark. 96 pp.

Laughlin, R.B., Jr., J. Ng, and H. Guard. 1981. Hormesis: A response to low environmental concentrations of petroleum hydrocarbons. Science. 211: 705-707.

Luoma, S.N. 1977. Detection of trace contaminant effects in aquatic ecosystems. J. Fish. Res. Board Can. 34: 436-439.

Luther, G., A.L. Meyerson, K. Rogers, and F. Hall. 1982. Tidal and seasonal variations of sulfate ion in a New Jersey marsh system. Estuaries 5: 189-196.

Mitton, J.B. and R. Koehn. 1975. Genetic organization and adaptive response of allozymes to ecological variables in *Fundulus heteroclitus*. Genetics 79: 97-111.

Nevo, E., T. Perl, A. Beiles, and D. Wool. 1981. Mercury selection of allozyme genotypes in shrimp. Experientia 37: 1152-1154.

Noël-Lambot, F., Ch. Gerday, and A. Disteche. 1978. Distribution of cadmium, zinc and copper in liver and gills of the eel *Anguilla anguilla* with special reference to metallothioneins. Comp. Biochem. Physiol. 61C: 177-187.

Pascoe, D. and J.H. Beattie. 1979. Resistance to cadmium by pretreated rainbow trout alevins. J. Fish. Biol. 14: 303-308.

Pickering, Q.H. and M.H. Gast. 1972. Acute and chronic toxicity of cadmium to the fathead minnow (*Pimephales promelas*). J. Fish. Res. Board Can. 29: 1099-1106.

Place, A.R. and D. Powers. 1978. Genetic bases for protein polymorphisms in *Fundulus heteroclitus* (L.). I. Lactate dehydrogenase (Ldh - B), malate dehydrogenenase (Mdh - A), glucosephosphate isomerase (Gpi - B), and phosphoglucomutase (Pgm - A). Biochem. Genet. 16: 577-591.

Rahel, F.J. 1981. Selection for zinc tolerance in fish: Results from laboratory and wild populations. Trans. Amer. Fish. Soc. 110: 19-28.

Roesijadi, G., A.S. Drum, J.M. Thomas, and G.W. Fellingham. 1982. Enhanced mercury tolerance in marine mussels and relationship to low molecular weight, mercury-binding protein. Mar. Poll. Bull. 13: 250-253.

Russell, G. and O. Morris. 1972. Ship fouling as an evolutionary process. pp. 719-730. *In*: Proc. 3rd Internat. Cong. on Mar. Corrosion and Fouling. Washington, D.C.

Smith, P.J. and Y. Fujio. 1982. Genetic variation in marine teleosts: High variability in habitat specialists and low variability in habitat generalists. Mar. Biol. 69: 7-20.

Spehar, R., E. Leonard, and D. Defoe. 1978. Chronic effect of cadmium and zinc mixtures on flagfish (*Jordanella floridae*). Trans. Amer. Fish. Soc. 107: 354-360.

Stebbing, A.R.D. 1981a. Hormesis - stimulation of colony growth in *Campanularia flexuosa* (Hydrozoa) by copper, cadmium, and other toxicants. Aquat. Toxicol. 1: 227-238.

Stebbing, A.R.D. 1981b. Stress, health and homeostasis. Mar. Poll. Bull. 12: 326-329.

Stokes, P., T. Hutchinson, and K. Krauter. 1973. Heavy metal tolerance in algae isolated from contaminated lakes near Sudbury, Ontario. Can. J. Bot. 51: 2155-2168.

Weis, J.S., P. Weis, and M. Heber. 1982. Variation in response to methylmercury by killifish (*Fundulus heteroclitus*) embryos. pp. 109-119. *In*: Aquatic Toxicology and Hazard Assessment: Fifth Conf. ASTM STP 766. J. Pearson, R. Foster, and W. Bishop (eds.). Amer. Soc. for Testing and Materials.

Weis, J.S., P. Weis, M. Heber, and S. Vaidya. 1981. Methylmercury tolerance of killifish (*Fundulus heteroclitus*) embryos from a polluted vs. non-polluted environment. Mar. Biol. 65: 283-287.

Weis, P. 1984. Metallothionein and mercury tolerance in the killifish *Fundulus heteroclitus*. Mar. Environ. Res. 14: 153-166.

Weis, P. and J.S. Weis. 1978. Methylmercury inhibition of fin regeneration in fishes and its interaction with salinity and cadmium. Estuar. Coast. Mar. Sci. 6: 327-334.

Zar, J.H. 1974. Biostatistical Analysis. Prentice-Hall, Englewood Cliffs. 620 pp.

Zimmerer, E.J. 1980. Environmental effects on growth and fin regeneration in Poeciliid fishes. Ph.D. Dissertation, Rutgers Univ. Newark, NJ. 132 pp.

THE IMPACT OF DREDGED MATERIAL ON THE BIOENERGETICS OF THE POLYCHAETE *NEPHTYS INCISA*

D. Michael Johns and Ruth Gutjahr-Gobell

SUMMARY

The effects of dredged material on the physiology of the polychaete *Nephtys incisa* were studied. Juvenile *N. incisa* were exposed to various combinations of control and Black Rock Harbor (BRH) sediment for a 10-d period.

Nephtys incisa juveniles were found to curtail burrowing activity in order to avoid contact with BRH sediment. Since the primary food of this species is obtained from ingested sediment, the decrease in burrowing activity probably reduced energy intake and caused a reduction in physiolgical rates.

The overall effect of exposure to BRH sediment was a significant reduction in the net growth efficiency of *N. incisa* juveniles.

INTRODUCTION

Effective, short-term measurements, which can adequately detect biological responses to the effects of environmental contaminants, has recently been requested by several international commissions (ICES,1978, IMCO *et. al.*, 1980; McIntyre & Pearce, 1980). To be of value the measurements must have relevance to ecological fitness. In addition, these relatively short-term laboratory tests (usually <28 d) must have a predictive capability for estimating the degree of ecological change which will occur.

A technique which may satisfy the preceding criteria is the determination of biological energy balance (Edwards,

1978; Johns & Pechenik, 1980; Capuzzo & Lancaster, 1981; Johns & Miller, 1982; McKenney, 1982), along with its various modifications including scope for growth (Warren & Davis, 1967; Bayne, 1975). Previous studies using this principle have found reasonable correlations between changes in energy balance or scope for growth and changes in population fitness (Bayne *et al.*, 1979; Gilfillan, 1980). In a series of detailed field studies (*e.g.* Gilfillan & Hanson, 1975; Gilfillan *et al.*, 1976; Gilfillan & Vandermeulen, 1978), a reduced scope for growth was found in *Mya arenaria* from oil-impacted sites when compared to individuals from nearby, relatively-clean reference populations. These data were correlated with and predictive of eventual changes in population structure at the impacted sites, including reductions in yearly individual growth rate and population density.

This paper describes efforts to develop a biological energy budget for the polychaete *Nephtys incisa* and to determine if the dredged material has any effect on this energy budget. *Nephtys incisa* is a dominant infaunal macroinvertebrate in Long Island Sound benthic communities (Sanders, 1956, 1958; Carey, 1962) and any impact from dredged disposal on this species could change the present community structure.

The research presented in this paper was undertaken as part of a five-year research program jointly conducted by the U.S. Army Corps of Engineers and the U.S. Environmental Protection Agency. The program's general research plan is to study the effects of dredged material from Black Rock Harbor, Connecticut (near Bridgeport) on a Long Island Sound benthic community. One purpose of this program is to develop, and then field verify, reliable biological effects measurements which subsequently can be used in other ocean disposal sites.

MATERIALS AND METHODS

Experimental Organisms

Nephtys incisa were collected from Long Island Sound in the vicinity of the proposed dump site. Bottom samples were collected using a Smith-MacIntyre grab and were initially sieved on board the research vessel. Worms which passed through a 2-mm sieve but were retained on a 0.37-mm sieve were all juveniles (Carey, 1962). These juvenile worms were placed in a single, unsieved Smith-MacIntyre grab sample for transport back to the laboratory, where they were resieved with a 0.37-mm mesh using ambient sea water.

In the laboratory, individual worms were visually separated into relative size classes for experiments. Although somewhat subjective, careful visual separation leads to a coefficient of variation in dry weight at the start of an experiment of only 24%. A more accurate sizing technique, such as wet weight determinations employing a wet weight to dry weight regression curve, would not only be time consuming, but the handling may cause physical damage to the worms.

All experiments were conducted at 20 ± 1°C. Laboratory acclimation for worms collected at 20 ± 3°C was a minimum of three days to allow for adjustments to laboratory holding conditions. For worms collected at temperatures below 17°C, acclimation included a 2°C increase/day until 20°C was attained.

Sediment Sources and Exposure Systems

The sediment used as the control was collected from the Long Island Sound site where the worms were collected. This sediment and Black Rock Harbor (BRH) sediment were press-sieved through a 0.50-mm sieve and stored at 4°C until used.

In the experiments in which worms were exposed to only the solid phase of the sediments, approximately 400-ml of the appropriate substrate mixture was placed in a 150 x 75 mm dish and then put in a 20°C water bath (Fig. 1). Sea water at 20°C flowed through the treatment bowls at approximately 50 ml/min. Where both control and Black Rock Harbor sediments were used in a treatment, the two sediments were combined in a volume to volume ratio.

In the particulate phase assays, worms were exposed to either control or BRH suspended particles with a nominal concentration of 200 mg/L (Control: 204.8 ± 6.7; BRH: 198 ± 31.3) after being placed in 100% control and 100% BRH sediments, respectively. The particles were delivered using the composite dosing system described by Rogerson _et. al._ (1984). Sea water (20°C), with the appropriate concentration of suspended particles, was allowed to flow through the treatment bowls at approximately 35 ml/min (Fig. 1). In order to maintain the particles in suspension as long as possible, a small crystalization dish with a stir bar was placed in the middle of the exposure dish.

In both the solid phase and particulate phase assays, 15 worms of approximately the same size were placed in each treatment. All treatment exposures were for 10-d. After exposure, the worms were sieved out of the mud, counted, and saved for physiological measurements.

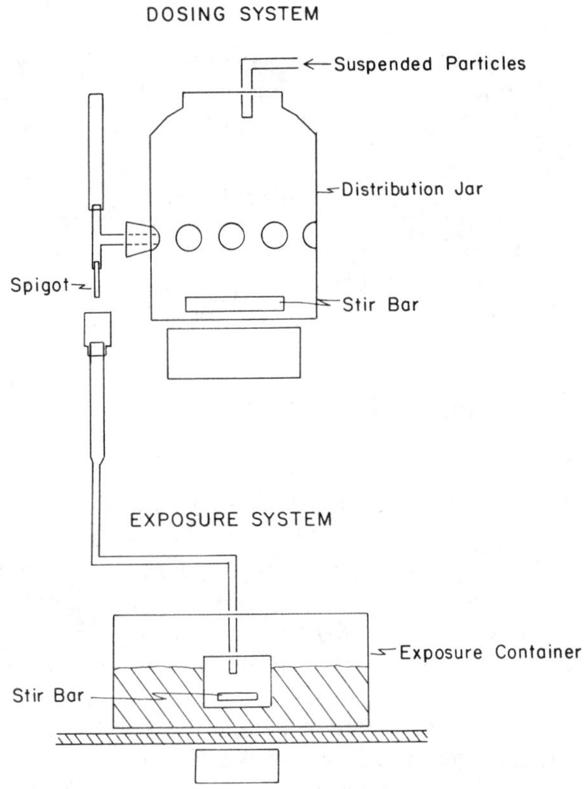

Fig. 1. Schematic of the suspended distribution and dosing system used to expose juvenile Nephtys incisa to control and Black Rock Harbor Sediment. In the solid-phase assays, only the exposure system was used. In the particulate-phase assays, both the dosing system (for particle distribution) and exposure system were employed.

During the experiment, worms were offered prawn flakes (Aquatic Diet Technology, Inc. Brooklyn, NY) as a food source. Previous research (Schauer & Pesch, in prep.) has indicated that laboratory-held N. incisa grew better in the presence of prawn flakes than without. It was unclear, however, whether the prawn flakes were used directly as food or if they acted as a substrate for bacteria which was utilized as food.

Physiological Measurements

Although the fundamental integration of biochemical and physiological mechanisms is complex and difficult to measure, the net response can be measured at the whole organism level. Bioenergetic analysis compares the major anabolic and catabolic processes that occur and allows an evaluation of the relative partitioning of available energy between growth and maintenance requirements.

The energy budget of the juvenile stage of an organism can be described by the following formula:

$$C = P + R + E + F \qquad (1)$$

where C is the total energy consumed, P is the amount of energy converted into tissue, R is the amount of energy used for maintenance as measured via aerobic respiration, E is the amount of energy lost through ammonia excretion, and F is the amount of energy lost through feces.

In this study, production, respiration and excretion were measured directly; while no measurements could be made directly of consumption and fecal production rates. Thus, a balanced energy budget for *Nephtys incisa* could not be derived, although an estimate of the relative partitioning of energy between production and maintenance could be made. The measured parameter allowed for the calculation of net growth efficiency using the following formula:

$$\text{Net Growth Efficiency} = P/P + R + E \times 100\% \qquad (2)$$

where P + R + E is an estimation of energy assimilated.

PRODUCTION

In all experiments juvenile worms were used, so changes in dry weight reflect only changes in somatic growth (production). At the beginning of an experiment, a subsample of individuals (10-15) was taken to estimate initial worm weight. The worms were quickly rinsed in deionized water, dried at 60°C for 24 h, and then weighed to the nearest μg on a Perkin-Elmer AD-2Z Autobalance (Perkin-Elmer Corp., Norwalk, CT). At the appropriate sampling times, 10-15 worms were taken from the experimental conditions to assess their physiological state. Individual respiration and excretion rates were determined prior to rinsing, drying, and weighing. Changes in growth were computed as being the difference in dry weight from the beginning to the end of an experiment. Energy values for the tissue were determined using a wet oxidation technique in the presence of an acid-dichromate mixture (Maciolek, 1962).

RESPIRATION

Routine rates of oxygen consumption were measured using a 5cc syringe, containing a small amount of sediment (0.5 - 0.75cc), as a respirometer (Fig. 2). An individual worm was placed in each syringe with 1.5cc of air-equilibrated sea water. After approximately 1-1.5 h, the oxygen concentration of 1cc of the sea water was determined by injecting the sample onto a Radiometer oxygen probe (Radiometer, Copenhagen, Denmark) fitted with a Radiometer thermostated sampling cell. The syringe was then refilled to 1.5cc, and the new oxygen concentration recalculated. This procedure was repeated up to three times per worm. A control syringe respirometer containing an appropriate amount of sediment was treated similarly in each experiment to account for dissolved oxygen losses due to biological and chemical oxygen demand of the sediment.

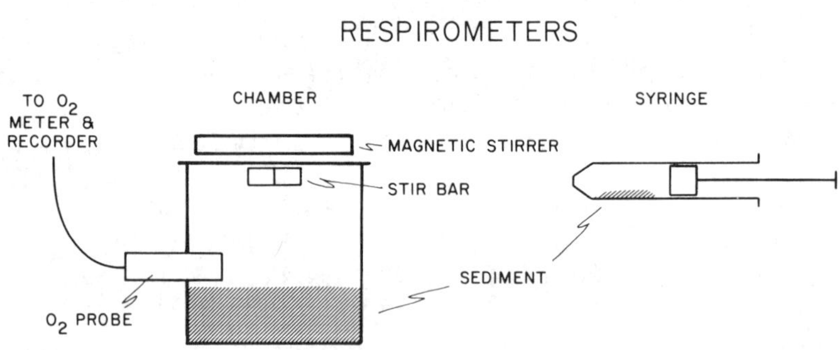

Fig. 2. Schematic of respirometers used to determine the oxygen consumption rates of *Nephtys incisa* juveniles.

To test the validity of oxygen consumption rates obtained from the syringe respirometers, oxygen consumption rates of *N. incisa* juveniles in stender dishes containing approximately 25 ml of sediment were measured. Each 35 x 50 mm stender dish had a hole drilled in the side which was fitted with a Radiometer oxygen probe (Fig. 2). Declining oxygen tension in the dish was monitored using a strip chart recorder. The water within the dish was stirred using a magnetic stir bar held at the top of the respirometer by a water-driven magnetic stirrer. Control runs were made with sediment alone to determine the amount of oxygen depletion due solely to the biological oxygen demand of the sediment. Comparison of the two methods indicated that there was no significant difference in measuring respiration rates in syringes and in stender dishes (syringe: 1.59 ± 0.61 μl O_2/mg/h; stender dish: 1.33 ± 0.54 μl O_2/mg/h).

In all experiments, oxygen consumption rates were reported as weight-specific rates. Previous research with this species has indicated little change in the weight-specific rates for the sizes of worms used for each assay (Johns, unpubl. data). When reported, the energy expended during routine metabolism was considered to be the average oxygen consumption rate (μl/h) during the experimental period multiplied by the appropriate energy equivalent for oxygen (3.40 cal/mg oxygen consumed; Elliott & Davison, 1975).

EXCRETION

Following their use for respiration measurements, individual worms were placed in 6 ml of filtered sea water to determine ammonia excretion rates. After approximately 3 h, ammonia concentrations were determined according to the technique of Bower and Holm-Hansen (1980). Calories lost per unit time as excreted ammonia were calculated by multiplying the excretion rate (μg NH_4N/h) during the experimental period by the heat of formation of ammonia (0.62 cal/mg; Elliott & Davison, 1975).

Burrowing Activity

Burrowing activity of *N. incisa* juveniles for the solid-phase assays was estimated using narrow, glass-walled aquaria. The aquaria were 14 x 1 x 16 cm (Fig. 3; Davis, 1979). Typically, an aquarium was filled with the appropriate variation of sediment and allowed to equilibrate in a flowing sea water bath. Following this equilibration period (12 h), two juvenile worms were randomly placed on the sediment. At the end of the experimental period, depths of the three deepest burrows were measured. Plastic sheets were then laid against the glass walls of the aquarium and the visible burrows traced. Total burrow lengths were calculated from these tracings using a Numonics 1224 electronic digitizer (Numonics Corp., Lansdale, NJ). In the particulate-phase assays, a tracing was made of those burrows visible on the side of the exposure dish.

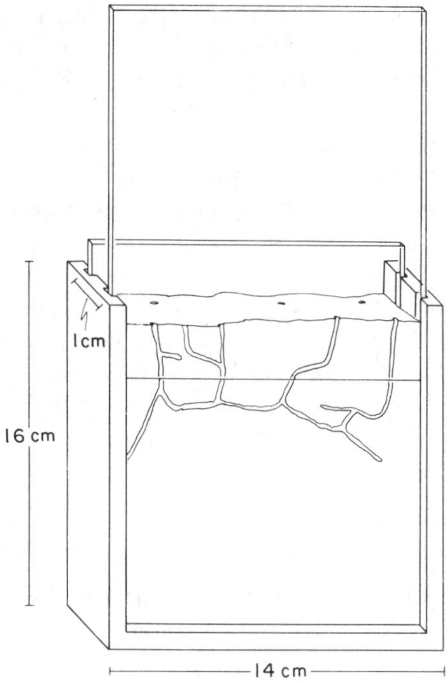

Fig. 3. Schematic of the narrow, glass-walled aquarium used to determine burrowing activity and maximum burrow depth.

As the narrow aquaria restrict worm movements to essentially two dimensions, data collected from these aquaria do not necessarily reflect the normal patterns and depths that might be expected in the field from the sizes of worms used. They do, however, provide a relative index of the worm activity in the various treatment conditions.

Statistical Analysis

A one-way analysis of variance was computed to determine the effects of the various treatments on the physiological function of *N. incisa* juveniles (Snedecor & Cochran, 1967). If significant differences (at $p = 0.05$) were found, a Tukey's posterior comparison was used to determine where the differences existed (Snedecor & Cochran, 1967). Data on burrow length and maximum burrow depth were rank transformed prior to one-way analysis of variance (Conover & Iman, 1981).

RESULTS

Production

Exposure to 100% BRH sediment for 10 d had a significant effect on the growth of juvenile *Nephtys incisa* (Table 1). Worms maintained in 100% control and 50% control/50% BRH sediment conditions showed an average gain, respectively, of 24 and 9% of their dry weight during the 10-d exposure, while those worms in the 100% BRH treatment lost an average of 16% of their dry weight during the same period. Results of the three experiments indicate that while there is some variation in the magnitude of weight change from experiment to experiment, the ranking between the three treatments remains the same (Table 1).

Juvenile *N. incisa* in the particulate phase assays exhibited similar changes in dry weight (Table 2). In both experiments, worms from the control/ control treatments (suspended particles/original sediment) were significantly heavier at the end of the 10-d exposure than were those exposed to BRH/BRH. Worms from the BRH/control treatments were not significantly different from the control/control worms. Those worms from the control/BRH exposure were not significantly different from the control/control treatment in the first experiment, but were significantly smaller in the repeat experiment (Table 2).

Maintenance Costs

Respiratory Expenditures

Weight-specific respiration rates of *N. incisa* juveniles exposed to 100% control sediment were significantly higher than those of juvenile worms exposed to 100% BRH sediment (Table 3). Worms from the 50% control/50% BRH treatment exhibited oxygen consumption rates intermediate to those from the 100% control and 100% BRH conditions.

A similar pattern was found for those worms from the particulate-phase experiments. In this assay, *N. incisa* juveniles from the control/control treatment had significantly higher respiration rates than those worms from the

Table 1. Changes in dry weight (mean ± SD) of *Nephtys incisa* juveniles exposed to various mixtures of control and Black Rock Harbor (BRH) sediments for 10 days. Means having the same grouping (Gp) letter are not significantly different ($p = 0.05$).

EXPERIMENT NO.	TREATMENT (% Control/BRH)	INITIAL DRY WT (mg)	FINAL DRY WT (mg)	Gp	CHANGE IN WT (mg; %)
1	100% Control	1.389 ± 0.205	1.468 ± 0.132	A	+0.08; + 6
	50% Con/50% BRH	1.389 ± 0.205	1.342 ± 0.280	A	-0.05; - 4
	100% BRH	1.389 ± 0.205	0.899 ± 0.094	B	-0.49; -35
2	100% Control	1.397 ± 0.328	1.938 ± 0.156	A	+0.54; +39
	50% Con/50% BRH	1.397 ± 0.328	1.780 ± 0.511	A	+0.38; +27
	100% BRH	1.397 ± 0.328	1.393 ± 0.110	B	.00; 0
3	100% Control	1.448 ± 0.204	1.822 ± 0.324	A	+0.37; +26
	50% Con/50% BRH	1.448 ± 0.204	1.506 ± 0.375	A	+0.06; + 4
	100% BRH	1.448 ± 0.204	1.274 ± 0.110	B	-0.17; -12

Table 2. Changes in dry weight (mean ± SD) of *Nephtys incisa* juveniles exposed to various combinations of 100% sediment and suspended particles for 10 days. Means having the same grouping (Gp) letter are not significantly different ($p = 0.05$).

EXPERIMENT NO.	TREATMENT (Particle/ Sediment)	INITIAL DRY WT (mg)	FINAL DRY WT (mg)	Gp	CHANGE IN WT (mg; %)
1	Control/Control	3.125 ± 0.601	3.686 ± 0.729	A	+0.56; +18
	Control/BRH	3.125 ± 0.601	3.581 ± 0.802	A	+0.46; +15
	BRH/Control	3.125 ± 0.601	3.494 ± 0.767	A	+0.40; +12
	BRH/BRH	3.125 ± 0.601	2.896 ± 0.884	B	-0.23; - 7
2	Control/Control	3.429 ± 0.922	4.660 ± 0.340	A	+1.23; +40
	Control/BRH	3.429 ± 0.922	3.902 ± 0.708	B	+0.47; +14
	BRH/Control	3.429 ± 0.922	4.205 ± 0.678	A	+0.78; +23
	BRH/BRH	3.429 ± 0.922	3.560 ± 0.671	B	+0.13; + 4

BRH/BRH exposure (Table 4). Respiration rates of worms from the BRH/control treatment were also significantly higher than those of the BRH/BRH-treated worms, but were not statistically different from the control/control individuals. Worms from the control/BRH treatment exhibited oxygen consumption rates similar to those in the BRH/BRH treated worms (Table 4).

Table 3. Weight-specific respiration rate (mean ± SD) of *Nephtys incisa* juveniles exposed to various mixtures of control and Black Rock Harbor (BRH) sediments for 10 days. Means having the same grouping (Gp) letter are not significantly different (at $p = 0.05$). (n): Number of determinations, all three experiments combined.

TREATMENT (% Control/BRH)	RESPIRATION RATE ($\mu l\ O_2/mg/h$)	(n)	Gp
100% Control	1.65 ± 0.77	(62)	A
50% Control/50% BRH	1.36 ± 0.63	(56)	A,B
100% BRH	1.10 ± 0.73	(44)	B

Table 4. Weight-specific respiration rates (mean ± SD) of *Nephtys incisa* juveniles exposed to various combinations of sediment and suspended particles for 10 days. Means having the same grouping (Gp) letter are not significantly different (at $p = 0.05$). (n): Number of determinations, both experiments combined.

TREATMENT (Particle/Sediment)	RESPIRATION RATE ($\mu l\ O_2/mg/h$)	(n)	Gp
Control/Control	0.75 ± 0.20	(42)	A
Control/BRH	0.62 ± 0.25	(42)	B
BRH/Control	0.86 ± 0.25	(42)	A
BRH/BRH	0.66 ± 0.28	(44)	B

In both the solid-phase and particulate-phase experiments, maintenance costs were higher in the control treatments (100% control; control/control) than they were in the Black Rock Harbor treatments (100% BRH; BRH/BRH) (Table 7). Estimated maintenance costs for the worms from the other treatments were generally intermediate between these two extremes.

Excretory Energy Losses

As with weight-specific respiration rates, weight-specific ammonia excretion rates of *N. incisa* juveniles exposed to 100% BRH and BRH/BRH conditions were significantly lower than those rates of worms exposed to 100% control and control/control (Tables 5 & 6). The other treatments were intermediate between these two values.

Table 5. Weight-specific ammonia excretion rate (mean ± SD) of *Nephtys incisa* juveniles exposed to various mixtures of control and Black Rock Harbor (BRH) sediments for 10 days. Means having the same grouping (Gp) letter are not significantly different (at p = 0.05). (n): Number of determinations, all three experiments combined.

TREATMENT (% Control/BRH)	EXCRETION RATE (µg NH_4N/mg/h)	(n)	Gp
100% Control	0.011 ± 0.003	(62)	A
50% Control/50% BRH	0.010 ± 0.002	(56)	A,B
100% BRH	0.008 ± 0.004	(44)	B

Table 6. Weight-specific ammonia excretion rate (mean ± SD) of *Nephtys incisa* juveniles exposed to various combinations of 100% sediment and suspended particles for 10 days. Means having the same grouping (Gp) letter are not significantly different (at p = 0.05). (n): Number of determinations, both experiments combined.

TREATMENT (Particle/Sediment)	EXCRETION RATE (µg NH_4N/mg/h)	(n)	Gp
Control/Control	0.012 ± 0.003	(42)	A
Control/BRH	0.010 ± 0.002	(42)	B
BRH/Control	0.013 ± 0.004	(42)	A
BRH/BRH	0.010 ± 0.004	(44)	B

Energy losses through ammonia excretion were not important when compared to other energy costs involved in maintenance (*i.e.*, respiratory expenditures; Table 7). However small, there were significant differences in energy lost

Table 7. Energy values (mean ± SD) for production (P), maintaince costs (R), and excretion (E) of *Nephtys incisa* juveniles exposed to various combinations of control and Black Rock Harbor (BRH) sediments. J = Joules.

TREATMENT (%Control/BRH)	P Growth (J)	Gp	R Respiratory Energy Expenditure (J)	Gp	E Excretory Energy Expenditure (J)	Gp
100% Control	+4.85 ± 0.97	A	12.6 ± 3.8	A	0.035 ± 0.009	A
50% Control/50% BRH	+2.01 ± 0.48	B	8.9 ± 3.6	A,B	0.029 ± 0.008	A,B
100% BRH	-2.30 ± 0.36	C	5.6 ± 2.1	B	0.020 ± 0.006	B
(Particle/Sediment)						
Control/Control	+9.77 ± 1.86	A	14.1 ± 4.7	A	0.077 ± 0.018	A
Control/BRH	+4.62 ± 0.92	B	11.2 ± 3.9	B	0.064 ± 0.018	B
BRH/Control	+6.44 ± 1.42	A	15.8 ± 5.0	A	0.084 ± 0.027	A
BRH/BRH	-1.73 ± 0.52	C	9.8 ± 3.1	B	0.054 ± 0.012	B

when the excretion rates of worms maintained in 100% control and control/control were compared to 100% BRH and BRH/BRH. Since the ammonia excretion rates were determined with worms out of sediment, the values presented here should only represent an approximate rate. Because of the uncertainty of these values no O:N ratios were calculated.

Partitioning of Energy Resource

The above data was used to calculate the efficiency with which available energy was partitioned between production of new tissue (P) and maintenance (R+E). Comparison of net growth efficiencies of *N. incisa* juveniles from the different treatments showed that worms maintained in 100% control sediment were most efficient (+27%) at converting available energy to new tissue (Table 8; Fig.4). Worms maintained in 100% BRH were the least efficient (-24%) and, in fact, lost weight during the exposure period. In the 50% control/50% BRH treatment, worms converted energy into tissue with an efficiency of +17%, which was not statistically different from the efficiency of worms maintained in the 100% control condition.

Table 8. Net growth efficiencies (mean ± SD) of *Nephtys incisa* juveniles exposed to various combinations of control and Black Rock Harbor (BRH) sediments.

TREATMENT (% CONTROL/BRH)	NET GROWTH EFFICIENCY (%)	Gp
100% Control	+27 ± 9	A
50% Control/50% BRH	+17 ± 9	A
100% BRH	-24 ± 30	B
(Particle/Sediment)		
Control/Control	+39 ± 10	A
Control/BRH	+32 ± 16	A
BRH/Control	+27 ± 7	A
BRH/BRH	-28 ± 40	B

A similar pattern of net growth efficiency was found in the particulate-phase assays. *N. incisa* juveniles maintained under control/control conditions exhibited the highest net growth efficiency (+39%), while those in the BRH/BRH treatment had a negative net growth efficiency (-28%)

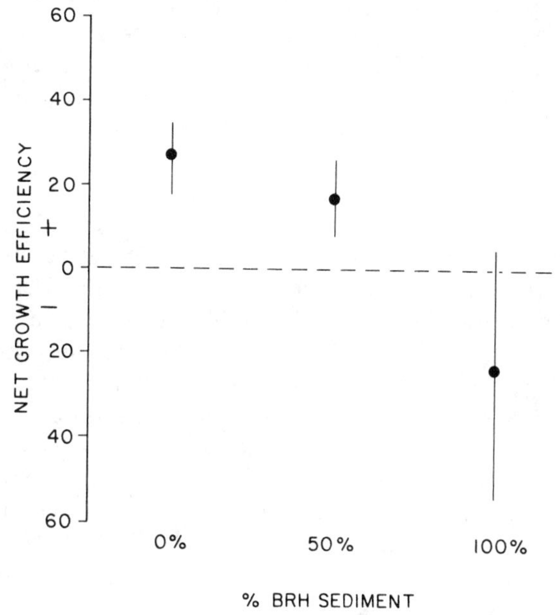

Fig. 4. Net growth efficiencies (mean ± SD) of Nephtys incisa juveniles exposed to various mixtures of control and Black Rock Harbor (BRH) sediment for 10 days.

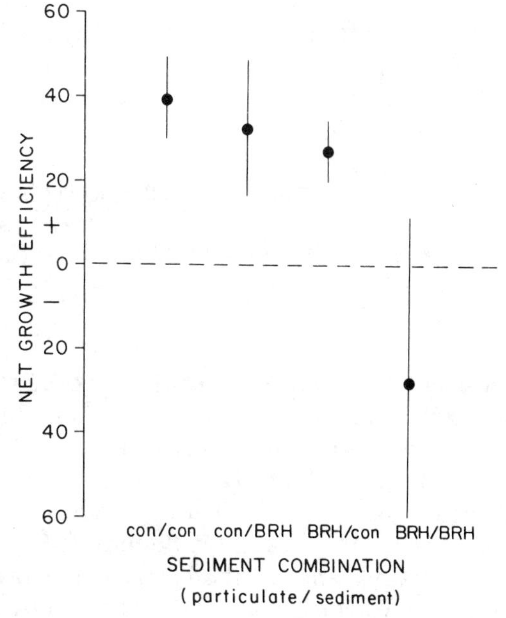

Fig. 5. Net growth efficiencies (mean ± SD) of Nephtys incisa juveniles exposed to various combinations of sediment and suspended particles for 10 days.

(Table 8; Fig. 5). Net growth efficiencies of worms kept in control/BRH and BRH/control exposures were +32 and +37% respectively. These values were not significantly different from those of the worms in the control/control conditions.

Burrowing Activity

Both burrow length and maximum depth of burrowing in the 100% control treatment were significantly different from similar measurements taken from the 100% BRH exposure (Table 9). The total length of visible burrows was approximately 3 times greater in the 100% control condition than it was in 100% BRH (Fig. 6). The depth to which the worms burrowed was 3 times greater in the 100% control sediment (30.4 mm) than in 100% BRH (9.5 mm).

Results of measurements of burrowing activity in the particulate-phase experiments were similar to the preceding results, although somewhat complicated by the addition of the suspended particles. *N. incisa* from the control/control condition produced significantly more burrows and burrowed to a significantly greater depth than those worms maintained in BRH/BRH (Table 9; Fig. 7).

Burrowing activity in the other particulate-phase treatments depended on whether the control sediment was in the upper particulate layer or present as the original sediment. In the BRH/control treatment, most of the burrowing activity (77%) was concentrated in the original sediment and the depth of burrowing was not significantly different from that found in the control/control treatment (Table 9; Fig. 7). In the control/BRH exposure most of the burrowing activity (59%) was found in the particulate layer which formed during the 10-d exposure. Maximum burrow depth in this treatment was only 17 mm which is very similar to the maximum burrow depth found in the BRH/BRH treatment.

DISCUSSION

Nephtys incisa is a non-selective deposit feeder typically found in soft sediments. It does not build permanent tubes, but rather burrows indiscriminately, injesting sediment and associated microorganisms as a food source (Davis, 1979). This species is considered a member of the equilibrium assemblage in Long Island Sound (Rhoads & Germano, 1982), and is usually associated with sediments that are oxygenated to depths of up to 10 cm. The ecological effects of errant burrowing behavior of *N. incisa* on local sediments are the provision of vertical particle mixing and the

Table 9. Burrow length and maximum burrow depth (mean ± SD) of *Nepthys incisa* juveniles exposed to various combinations of sediment and suspended particles for 10 days. % in sediment: Percentage of burrow length that is found below the boundary of the original sediment.

TREATMENT (% Control/BRH)	BURROW LENGTH (cm)	Gp	% IN SEDIMENT	MAXIMUM BURROW DEPTH (mm)	Gp
100% Control	42.2 ± 9.9	A	-	30.4 ± 7.3	A
50% Control/50% BRH	N.D.		-	N.D.	
100% BRH	14.2 ± 4.7	B	-	9.5 ± 2.3	B
(Particle/Sediment)					
Control/Control	43.1 ± 8.9	A	58.7	29 ± 2.0	A
Control/BRH	21.8 ± 1.1	B	40.8	17.0 ± 2.4	B
BRH/Control	36.9 ± 6.2	A	77.0	31.2 ± 6.1	A
BRH/BRH	9.3 ± 0.8	C	31.3	16.8 ± 2.5	B

N.D. - Not Determined.

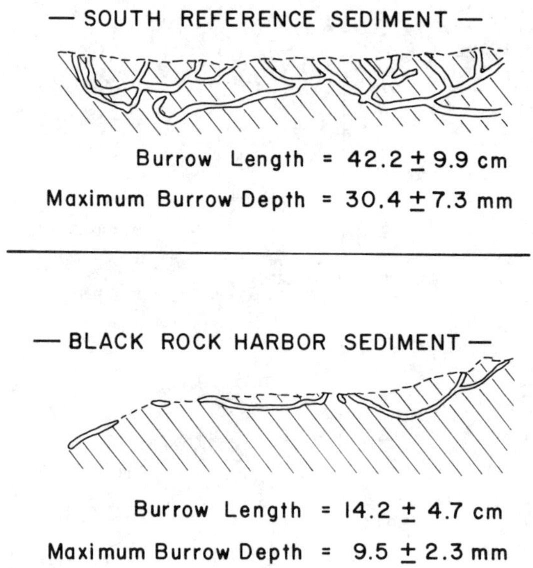

Fig. 6. Drawing of burrows created by _Nephtys incisa_ juveniles exposed to control or Black Rock Harbor sediments for 10 days.

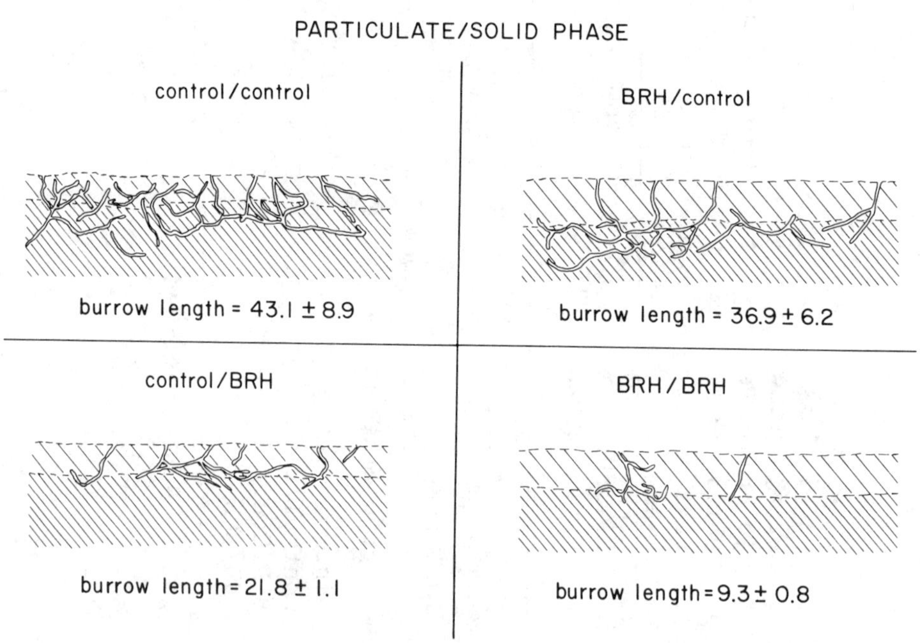

Fig. 7. Drawing of burrows created by _Nephtys incisa_ juveniles exposed to various combinations of control and Black Rock Harbor sediment for 10 days.

enhancement of pore water exchange (Rhoads & Germano, 1982). Much of the burrowing activity occurs at the redox boundary, a zone typically high in microorganism productivity (Yingst & Rhoads, 1980).

The data presented here from the control treatments indicate that *N. incisa* juveniles perform much the same organism-sediment roles as do the adult stages. Burrowing activity during the 10-d experimental period was extensive, with burrows penetrating to 30 mm in depth. Respiratory pumping also appeared to be effective in pore water exchange, as the burrow tubes had an observable "halo" of light sediment which is an indicator of oxidized sediment (Aller & Yinst, 1978).

Net growth efficiencies for *Nephtys incisa* juveniles maintained in control sediments are within the the range of efficiencies reported for polychaetes under a variety of conditions. For instance, Carey (1962) estimated population production and respiration values for *N. incisa* from a Long Island Sound study site. Using data from his Table 9, a mean net growth efficiency of 36% is calculated (population production/population production + respiration). Net growth efficiency for omnivorous polychaetes (both indiscriminate sediment ingesters, as well as surface detrital feeders) have been found to be generally between 14 and 40% (Kay & Brafield, 1973; Tenore & Gopalan, 1974; Neuhoff, 1979).

The effects of exposure to Black Rock Harbor sediment are two-fold. First, *Nephtys incisa* juveniles greatly reduce their burrowing activity when exposed to 100% BRH sediment. This is graphically demonstrated in Figures 6 and 7, where worms maintained in BRH sediment for 10 d demonstrated almost no burrowing activity and no depth penetration. Similar results were found in the lugworm, *Arenicola cristata* exposed to Kepone-contaminated sediments (Rubinstein, 1979).

In the suspended-particulate assays, where a choice between sediments was offered as settling occurred, *N. incisa* appeared to avoid BRH sediment. Comparison of the control/BRH treatment to the BRH/control treatment illustrates this point. Almost all of the burrowing activity in the control/BRH treatment was in the upper, settled layer. This activity occurred only after the control sediment layer was formed by settling particles. Worms from the BRH/control exposure exhibited an opposite activity level. In this treatment, there was extensive reworking of the lower, control sediment, while there was very little activity in the upper, sedimented BRH material. The only burrows present in this upper layer were a few tubes fairly vertical to the surface (when compared to control/control and control/BRH). These tubes probably provided the worms with a supply of oxygenated water to meet their respiratory requirements.

The dramatic decrease in burrowing activity of *N. incisa* juveniles that occurred during exposure to BRH sediment indicated that the worms were probably curtailing physiological functions. If the primary energy source of this species is ingested sediment, then curtailing burrowing activity has the effect of starving the individual. This is, in essence, the second effect BRH sediment has on the species. This effect is brought about by the failure of worms to consume a sufficient amount of energy for normal physiological and biochemical processes. *Nephtys incisa* exposed to BRH sediments generally exhibited lowered maintenance costs, most probably in response to reduced energy intake and reduced burrowing activity. As evidenced by reduced growth rates, individuals in the BRH sediment had to catabolize some tissue in order to provide the energy requirements for routine metabolism, despite the lowered maintenance costs. Catabolism of tissue to meet energy requirements is a common phenomenon in organisms that are below their required maintenance ration (Pandian, 1975).

In those treatments where there was a choice between BRH and control sediments there was an increase in burrowing activity above that seen in 100% BRH sediment; with a majority of the activity occurring in the control sediment (Fig. 5). The activity increase noted in the BRH/control and control/BRH sediments, if reflective of increased ingestion rates, is apparently sufficient to provide enough energy for both maintenance costs and growth. Worms under these sediment conditions had lower, but not significantly different, net growth efficiencies than did worms maintained in the control sediment (Table 8).

The responses of *N. incisa* to BRH sediment were probably due to at least one of two factors. Either the worms were responding to a difference in particle size or they were reacting to contaminants within the BRH sediment. While the first possibility can not be completely discounted, Carey (1962) found no correlation between the occurrence of *N. incisa* populations in Long Island Sound and particle size distribution, despite significant differences in granulometry among the various sites. The second possibility appears to be the most likely factor affecting *N. incisa*. Black Rock Harbor sediment is a multi-contaminated sediment that contains a wide variety of anthropogenic chemicals. Qualitative analysis of BRH sediments indicated the presence of high concentrations of PCBs, polynuclear aromatic hydrocarbons, and heavy metals (including Cu, Pb, Cd, and Cr) (Rogerson *et al*. 1984). Despite this large spectra of contaminants, BRH sediment is not acutely toxic to *N. incisa* juveniles.

The exposure conditions presented in this paper do not represent realistic exposure regimes for *Nephtys incisa* in Long Island Sound. Rather, they were employed to produce a sublethal response. Current research within the joint Corps of Engineers-Environmental Protection Agency's program is aimed at establishing the exposure regimes that *N. incisa* would be exposed to near the Black Rock Harbor dump site. Once these limits are established, long-term laboratory studies with environmental conditions that more closely approximate field conditions will be conducted.

ACKNOWLEDGMENTS

Although not totally inclusive, we would like to thank the following researchers: Drs. J. Gentile, G. Pesch, J. Scott, D. Peddicord, and T. Dillon; Mr. W. Nelson, and Ms. C. Pesch. Dr. P. Schauer and Mr. M. Balboni for the use of and help with the benthic exposure systems. Dr. W. Davis for use of the "ant farms" and Dr. J. Helshe for advice on statistical procedures. We would especially like to thank Dr. A. Calabrese and Mr. B. Alix of the NMFS in Milford, CT for boat time on the R/V Shang Wheeler. This research was supported by Cooperative Agreement CR809956 between the USEPA and New England Aquarium to DMJ.

LITERATURE CITED

Aller, R.C. and J.Y. Yingst. 1978. Biogeochemistry of tube-dwellings; A study of the sedentary polychaete *Amphitrite ornata*. J. Mar. Res.: 201254.

Bayne, B.L. 1975. Aspects of physiological condition in *Mytilus edulis* (L.), with special reference to the effects of oxygen tension and salinity. pp. 213-238. *In*: Proceedings of 9th European Marine Biology Symposium. H. Barnes (ed.). Aberdeen University Press, Scotland.

Bayne, B.L., M.N. Moore, J. Widdows, D.R. Livingstone, and P. Salkeld. 1979. Measurement of the responses of individuals to environmental stress and pollution:

Studies with bivalve molluscs. Phil. Trans. Roy. Soc. Ser. B. 286: 563-581.

Bower, C.E. and T. Holm-Hansen. 1980. A salicylate-hypochlorite method for determining ammonia in seawater. Can. J. Fish. Aquat. Sci. 37: 794-798.

Capuzzo, J.M. and B.A. Lancaster. 1981. Physiological effects of south Louisana crude oil on larvae of the American lobster (*Homarus americanus*). pp. 405-424. *In*: Biological Monitoring of Marine Pollutants. F.J. Vernberg, A. Calabrese, F.P. Thurberg, and W.B. Vernberg (eds.). Academic Press, New York.

Carey, A. Jr. 1962. An ecologic study of two benthic population in Long Island sound. Ph. D. Thesis. Yale University, New Haven, CT. 61 pp.

Conover, W.J. and R.L. Iman. 1981. Rank transformations as a bridge between parametric and nonparametric statistics. The American Statistician 35: 124-129.

Davis, W.R. 1979. The burrowing, feeding and respiratory activities of *Nephtys incisa* Malmgren, 1865 (Polychaeta: Annelida). Ph. D. Thesis. University of South Carolina. 103 pp.

Edwards, R.R.C. 1978. Effects of water-soluble oil fractions on metabolism, growth, and carbon budget of the shrimp *Crangon crangon*. Mar. Biol. 46: 259-265.

Elliott, J.M. and W. Davison. 1975. Energy equivalents of oxygen consumption in animal energetics. Oecologia. (Berl.) 19: 195-201.

Gilfillan, E.S. 1980. The use of scope-for-growth measurements in monitoring petroleum pollution. *In*: Biological Effects of Marine Pollution and the Problems of Monitoring. A.D. McIntyre, and J.B. Pearce (eds.). Rapp. P.-V. Reun. Cons. Int. Explor. Mer. 179: 71-75.

Gilfillan, E.S. and S.A. Hanson. 1975. Effects of paralytic shellfish poisoning toxin on the behavior and physiology of marine invertebrates. pp. 367-375. *In*: Proc. of First Int. Conf. Tos. Dinoflag. V.R. LoCicero, (ed.). The Mass. Sci. and Tech. Foundation.

Gilfillan, E.S., D. Mayo, S. Hanson, D. Donovan, and L.C. Jiany. 1976. Reduction in carbon flux in *Mya arenaria* caused by a spill of No. 6 fuel oil. Mar. Biol. 37: 115-123.

Gilfillan, E.S. and J.H. Vandermeulen. 1978. Alterations in growth and physiology of soft shell clams. *Mya arenaria* chronically oiled with Bunker C from Chedabucto Bay, Nova Scotia, 19701976. J. Fish. Res. Board Can. 35: 630-636.

ICES. 1978. On the feasibility of effects monitoring. Cooperative Research Report No. 75. ICES. Charlottenlund. Denmark. 42 pp.

IMCO/FAO/UNESCO/WMO/WHO/IAEA/UN/UNEP Joint Group of Experts on the Scientific Aspects of Marine Pollution (GESAMP). 1980. Monitoring Biological Variables Related to Marine Pollution. Reports and Studies No. 11. United Nations Educational, Scientific and Culture Organization. 21 pp.

Johns, D.M. and D.C. Miller. 1982. The use of biogenergetics to investigate the mechanisms of pollutant toxicity in crustacean larvae. pp. 261-288. In: Physiological Mechanisms of Marine Pollutant Toxicity. W.B. Vernberg, A. Calabrese, F.P. Thurberg, and F.J. Vernberg (eds.). Academic Press, New York.

Johns, D.M. and J.A. Pechenik. 1980. Influence of the water-accommodated fraction of No. 2 fuel oil on energetics of *Cancer irroratus* larvae. Mar. Biol. 55: 247-254.

Kay, D.G. and A.E. Brafield. 1973. The energy relations of the polychaete *Neanthes (=Nereis) virens* (Sars). J. Anim. Ecol. 42: 673-692.

Maciolek, J.L. 1962. Limnological organic analysis by quantitative dichromate oxidation. Fisheries Res. Rep. No. 60. pp. 161.

McIntyre, A.D. and J.B. Pearce (eds). 1980. Biological effects of Marine Pollution and the Problems of Monitoring. Rapp. P.-V. Reun. Cons. Int. Mer 179. 346 pp.

McKenney, C.L. Jr. 1982. Interrelationships between energy metabolism, growth dynamics, and reproduction during the life cycle of *Mysidopsis bahia* as influenced by sublethal endrin exposure. pp. 447-476. In: Physiological Mechanisms of Marine Pollutant Toxicity. W.B. Vernberg, A. Calabrese, F.P. Thurberg and F.J. Vernberg. (eds.). Academic Press, New York.

Neuhoff, H.G. 1979. Influence of temperature and salinity on food conversion and growth of different *Nereis* species (Polychaeta, Annelida). Mar. Ecol. Prog. Ser. 1: 255-262.

Pandian, T.J. 1975. Mechanisms of heterotrophy. pp. 61-250. In: Marine Ecology, Vol. II. Physiological Mechanisms, Part 1. O. Kinne (ed.). Wiley, Chichester.

Rhoads, D.C. and J.D. Germano. 1982. Characterization of organism-sediment relations using sediment profile imaging: An efficient method of remote ecological monitoring of the seafloor (Remots (TM) System). Mar. Ecol. Prog. Ser. 8: 115-128.

Rogerson, P., S. Schimmel, and G. Hoffman. 1984. Chemical and Biological Characterization of Black Rock Harbor Dredged Material. Technical Report D-83-XX, prepared by U.S. Environmental Protection Agency, Narragansett, R.I., for the U.S. Army Engineer Waterways Experiment Station, CE. Vicksburg, MS.

Rubinstein, N. 1979. A benthic bioassay using time-lapsed photography to measure the effect of toxicants of the feeding behavior of lug-worms (Polychaeta: Arenicolida). pp. 341-351. In: Marine Pollution: Functional Responses. W.B. Vernberg, F.P. Thurberg, A. Calabrese, and F.J. Vernberg (eds.). Academic Press, New York.

Sanders, H.L. 1956. Oceanography of Long Island Sound. X. The biology of bottom marine communities. Bull. Bingham. Oceanog. Collection 15: 345-414.

Sanders, H.L. 1958. Benthic studies in Buzzards Bay. I. Animal-sediment relationships. Limnol. Oceanog. 3: 245-253.

Snedecor, G.W. and W.G. Cochran. 1967. Statistical Methods. Iowa State University Press, Ames. 593 pp.

Tenore, K.R. and U.K. Gopalan. 1974. Feeding efficiencies of the polychaete *Nereis virens* cultured on hard-clam tissue and oyster detritus. J. Fish. Res. Board Can. 31: 1675-1678.

Warren, C.E. and G.E. Davis. 1967. Laboratory studies on the feeding, bioenergetics and growth of fish. pp. 175-214. In: The Biological Basis of Freshwater Fish Production. S.D. Gerking (ed.). Blackwell, Oxford.

Yingst, J.Y. and D.C. Rhoades. 1980. The role of bioturbation in the enhancement of microbial turnover rates in marine sediments. pp. 407-421. In: Marine Benthic Dynamics. K.R. Tenore, and B.C. Coull (eds.). University of South Carolina Press, Columbia.

INTERACTIONS OF CHLORINE-PRODUCED OXIDANTS (CPO) AND SALINTY IN AFFECTING LETHAL AND SUBLETHAL EFFECTS IN THE EASTERN OR AMERICAN OYSTER, *CRASSOSTREA VIRGINICA* (GMELIN), INFECTED WITH THE PROTISTAN PARASITE, *PERKINSUS MARINUS*

Geoffery I. Scott, Douglas P. Middaugh, and Tommy I. Sammons

SUMMARY

Exposure of oysters, infected with the protistan parasite, *Perkinsus marinus*, to high salinity and high salinity, chlorine-produced oxidants (CPO) conditions was quite toxic. The combination of high salinity and CPO exposure generally increased toxicity above levels measured in high salinity controls and was generally correlated with parasitism by the protistan parasite, *P. marinus*.

Low salinity and the combination of low salinity and CPO exposure resulted in a significant reduction in *P. marinus* related oyster mortalities.

Low salinity, CPO exposure reduced oyster mortalities from *P. marinus* infections by as much as 56%. The observed lower mortality rates in oysters exposed to low salinity, CPO regimes appeared to be related to reductions in tissue glycogen and free amino acid levels, which may have limited the intensification of the disease to lethal hypnospore infection levels. Earlier studies by several authors (Lange, 1970; Block *et al.*, 1977; Soniat & Koenig, 1982) have indicated that exposure to low salinity or CPO reduces free amino acid levels in oysters and that free amino acid levels decrease with increased parasitism by *P. marinus*. Results from this study indicate that the combined effects of low salinity and CPO exposure lower mortality resulting from *P. marinus* infections perhaps due to the reduction of free amino acid levels in adult oysters. Reductions in free amino acid levels would prevent and limit the spread and intensification of the disease, thus increasing oyster survival.

Low salinity alone reduced oyster mortalities from *P. marinus* infection, but only by 25%. Reduced free amino acid levels were probably responsible for limiting mortality from *P. marinus* due to a delay and limitation of infection to lethal levels.

INTRODUCTION

Chlorine has been widely used to disinfect sewage plant effluents and potatable drinking water supplies and to control antifouling organisms at power plants (White, 1972; Hall *et al*., 1981). Typically, sewage waste waters are chlorinated at a specified level (*i.e.*, 2 mg/L chlorine residual) for specified contact time (*i.e.*, 30 min) to lower coliform bacterial densities to levels safe for discharge in aquatic environments. Discharges into estuarine habitats of sewage effluent containing excess chlorine concentrations may be potentially toxic to fish and invertebrates, particularly the eastern or American oyster, *Crassostrea virginica* (Gmelin) (Scott, 1983).

Chlorinated discharges may be acutely toxic to oyster larvae (Rosenberg *et al*., 1980; Roberts & Casey, in press). Roberts and Casey (in press) reported a 48-h median lethal concentration (LC50) of 23 μg/L chlorine-produced oxidant (CPO) concentration for oyster embryos exposed from fertilization to straight-hinge veliger stage. Rosenberg *et al*. (1980) reported a 96-h LC50 for post straight-hinge oyster larvae of 300 μg/L CPO and suggested that the 96-h LC50 of recently set spat was greater than 500 μg/L CPO.

Chlorinated effluents may be chronically toxic to adult oysters, causing significant mortality at CPO concentrations of 660 μg/L and significant, irreversible metabolic effects (decreased physiological performance) at CPO concentrations as low as 110 μg/L (Scott, 1979). Additionally, the sensitivity of oysters to chlorination exposure varies significantly from season to season, with adults being three to four times more sensitive to exposure during the summer (postspawning) than spring (prespawning). When oysters were compared under identical temperature, salinity, and CPO-exposure regimes, Scott *et al*. (1980) reported greater sublethal sensitivity in summer oysters as respiration rates were significantly higher in the summer than the spring. Both spring and summer groups were infected with the protistan parasite, *Perkinsus marinus*.

Studies on the epizootiology (the abrupt increase of a disease in animals other than humans) of *P. marinus* have indicated that this parasite is basically a summer disease, causing peak mortalities (>50%) in August and September when water temperatures and salinity values are maximal (Mackin 1962; Quick and Mackin, 1971). Temperatures greater than 25°C are required for high infection rates to develop, while temperatures less than 10°C have been shown to inhibit hypnospore growth (Andrews and Hewatt, 1957; Perkins, 1966). Salinity has also been shown to be important, as infection intensities and associated mortalities are greatest in high-salinity waters (Mackin and Wray, 1952; Ray, 1954). Low salinity exposure tends to delay development of heavy infection intensities, to depress infection intensities and the incidence of infection, and to significantly enhance survival (Mackin & Wray, 1952; Ray *et al*., 1953; Ray, 1954; Andrews & Hewatt, 1957).

As earlier discussions indicated, oysters are much more susceptible to chlorination exposure during summer months. This increased summer susceptibility to chlorine exposure may result from lessened resistance in the oyster as a result of *P. marinus* infection. Several authors (Couch, 1978; Sinderman, 1978) have described the interactions between increased levels of pollution and increased incidences of parasitism and disease in marine organisms.

The purpose of this study was to investigate the effects of chlorination exposure on the American or eastern oyster, *Crassostrea virginica* (Gmelin), infected with the protistan parasite, *Perkinsus marinus*, so as to ascertain any potential interactions between toxicant and parasite in defining the lethal and sublethal response of the oyster to combined toxicant/parasite effect. All tests were conducted at high (25°/$_{oo}$) and low (<10°/$_{oo}$) salinities, due to known interactions of high salinity in increasing oyster mortalities from enhanced *P. marinus* infection intensities, and low salinity in decreasing oyster mortalities due to reduced *P. marinus* infection intensities.

MATERIALS AND METHODS

Collection and Acclimation

Adult oysters (6-12 cm in height) were collected from Leadenwah Creek (latitude N 32°36'; longitude W 80°15'), a large tidal tributary of the North Edisto River Estuary in South Carolina, on 21 June 1982. Following collection, oysters were transported immediately to the laboratory where they were scrubbed and all fouling organisms removed. After numbering and measuring the height of individual oysters, 120 animals were placed in each test aquarium (circular tanks: 152.4-cm radius x 30-cm depth with a working volume of 310 L) and were acclimated to ambient light, salinity ($\bar{x}$ = 22.8°/₀₀), and temperature ($\bar{x}$ = 30.1°C) for 39 d. Unfiltered sea water was pumped in daily from the Stono River at high tide. Tanks were cleaned daily during acclimation. An additional 24 d of acclimation beyond the scheduled 15-d period were required because high rainfall in the area depressed high-tide salinities to values below those necessary for the experiment.

Following acclimation, oysters in each tank were exposed for 60 d to one of the several treatments described in Table 1. Tests were conducted from August through October 1982.

Exposure Conditions

Low salinity sea water (treatment E) was prepared by diluting incoming ambient sea water with groundwater from a shallow well. Normal ambient salinity at high tides ranged from 21 to 25°/₀₀. Low salinity water of 8-10°/₀₀ was desired, which required approximately a 40%:60% dilution of sea water:groundwater. Salinity averaged 9.2°/₀₀ in the low salinity control sea water. Ambient temperatures of the resulting low salinity sea water ($\bar{x}$ = 23.9°C) were not significantly lower than ambient sea water temperatures ($\bar{x}$ = 24.6°C).

Preparations of chlorinated sea water were made at both low (8-10°/₀₀, $\bar{x}$ = 9.3) and high (21-25°/₀₀, $\bar{x}$ = 22.8) salinities by daily addition of NaOCl (Fisher Reagent Grade, 5%) to each tank at appropriate concentrations and dilutions. Nominal chlorine concentrations were 1.00, 0.56, and 0.10 mg/L in both high salinity (treatments B, C, and D, respectively) and low salinity (treatments F, G, and H, respectively) sea water. Exposure temperatures averaged 24.1°C in high salinity, CPO exposures compared to 23.6°C in low salinity, CPO exposures.

Oysters were also exposed to ambient high salinity (21-25°/₀₀) sea water (treatment A), to serve as indicators of natural or control conditions.

A total of 120 oysters were placed in each test aquaria and were exposed to each treatment for 60 d (Table 1). Continuous aeration of each aquarium was provided by a piston pump (DP Model 440), while circulation was provided by a submersible pump (Little Giant) attached to a programmable timer to provide a realistic simulation of semidiurnal tidal currents in the Stono River estuary (5 h of flow followed by 1 h of slack water). Oysters fed on algae present in the sea water and were exposed to the ambient light:dark (14 h:10 h) cycle and sea water temperatures. Water changes were made and test aquaria were cleaned daily throughout the experiment.

Test Parameters Measured

During each experiment, the following measurements were used to assess the physiological effects of each treatment on adult oysters:

1. Survival and mortality percentages were recorded daily using three replicates per treatment.

2. Prior to and at the end of each exposure, shell height (cm) was measured, using Vernier calipers.

3. Chlorine-produced oxidant (CPO) concentrations (free plus combined chlorine, bromine, and other residual oxidants) were measured with a Fisher Titrator (Model 397). Phenylarsine oxide and pH 4 buffer were standard Fisher reagents. The 5% KI solution and chlorine (Fisher NaOCl analytical reagent, minimum 5% NaOCl) stock solutions were prepared with deionized water. Stock solutions were kept in lightproof containers to prevent photodecomposition. Measurements of CPO concentrations employed procedures outlined in Standard Methods (APHA, 1971). All water samples were taken directly from treatment aquaria at least every 2-3 d during each experiment. Agitation of samples was avoided to reduce the potential for error by photodecomposition or flashing-off of CPO (EPA, 1974). The mixing order of Crecelius *et al.* (1978) was used in each titration: addition of 2 ml of pH 4 buffer and 2 ml of 5% KI solution to the sampling cup followed by addition of the sea water sample. Values are reported in mg/L.

4. In each experiment, ambient sea water temperature and salinity were measured daily in the incoming sea water from the Stono River. A mercury thermometer and a refractometer (American Optics) were used to measure temperature and salinity, respectively. In addition, periodic measurements of dissolved oxygen (Yellow Springs Instrument Company) and

Table 1. Description of treatment codes and physical/chemical parameters measured in each of the various salinity and chlorination treatments used in this study. Statistical measurement included calculation of mean ($\bar{x}$) and standard error ($s/\sqrt{n}$). Single (*) or double (**) asterisks indicate measured CPO concentrations in each low salinity, CPO treatment were significantly ($p<0.05$) or very significantly ($p<0.01$) different from similar high salinity, CPO treatments.

Treatment Code	Treatment Description[1]	Measured CPO Concentration (mg/L)[2]		Exposure Salinity (ppt)[3]		Exposure Temperature (°C)[4]	
		$\bar{x}$	$s/\sqrt{n}$	$\bar{x}$	$s/\sqrt{n}$	$\bar{x}$	$s/\sqrt{n}$
A	High salinity control	0.00	0.000	22.6	0.42	24.2	0.40
B	High salinity, 1.00 mg/L NaOCl concentration	0.06	0.007	22.9	0.21	24.2	0.40
C	High salinity, 0.56 mg/L NaOCl concentration	0.03	0.005	22.7	0.04	24.0	0.39
D	High salinity, 0.10 mg/L NaOCl concentration	0.01	0.003	22.9	0.21	24.0	0.39
E	Low salinity control	0.00	0.000	9.2	0.18	23.9	0.38
F	Low salinity, 1.00 mg/L NaOCl concentration	0.11**	0.008	9.3	0.18	23.6	0.39
G	Low salinity, 0.56 mg/L NaOCl concentration	0.06**	0.007	9.6	0.59	23.6	0.39
H	Low salinity, 0.10 mg/L NaOCl concentration	0.02*	0.004	9.0	0.17	23.6	0.39

1 = NaOCl refers to nominal NaOCl concentrations added.
2 = Sample size (n) ranged from 36 to 38.
3 = n ranged from 57 to 58.
4 = n ranged from 51 to 52.

pH (Corning pH meter) were made. Dissolved oxygen measurements showed that oxygen concentrations were always at saturation. The pH ranged from 7.40 to 8.20 during the study.

5. Fecal production or biodeposition rate measurements were made every 2-3 d in all treatments by observing the number of oysters producing feces per day.

6. Perkinsus marinus densities in oyster tissues were determined using methods described by Ray (1952, 1966), Quick and Mackin (1971), and Farley (1965), but modified slightly. Fifteen adult oysters were removed from each test aquaria on days 0, 15, 30, and 60 of the experiment. A total of 490 oysters were examined for P. marinus densities. Each oyster was scrubbed thoroughly and then opened. The gill tissues and gonads were removed for tissue respiration and gonadal index measurements. The rectum and labial palps were removed, weighed, and then placed into culture tubes (17 x 125 mm with screw caps) containing 10 ml of fluid thioglycollate medium which had been fortified with myostatin (E.R. Squibb and Sons, Inc., NDC0003-0543-20 and M175C/59320) and chlormphenicol (Park-Davis M379). Innoculated test tubes were incubated in the dark at 25°C for a minimum of 7 d to a maximum of 18 d. Tissues were checked after 4, 10, and 18 d of incubation for the presence of hypnospores.

After incubation, each piece of tissue was removed from the culture media, blotted to remove excess liquid, and was placed on a labeled microscope slide. Several drops of Lugol's aqueous iodine potassium iodide solution was then added to stain each tissue. After 10 min, a cover glass was pressed onto the preparation, and examinations were then made at 50x and 100x magnifications using an Olympus light microscope. Tissue explant size and slide preparation thickness were kept as constant as possible to minimize error in infection density measurements.

Incidence of infection and infection density were determined by counting the number of hypnospores in either a 5-mm (50x) or 2-mm (100x) field of view. Three replicate counts of the most heavily infected area of each tissue were made to provide statistical quantification of infection density. Based upon the average infection density, oysters were then classed according to the method used by Ray (1953, 1954):

a. negative infection (code #0) with no hypnospores present;
b. Very light infection (1) with a density of 1-5 hypnospores/5-mm field;

c. Light infection (2) with a density of 6-10 hypnospores/5-mm field;
d. Lightly moderate infection (3) with a density of 11-30 hypnospores/ 5-mm field;
e. Moderate infection (4) with a density of 31-300 hypnospores/5-mm field;
f. Moderately heavy infection (5) with a density of 301-1000 hypnospores/5-mm field;
g. Heavy infection (6) with a density of 1001-3000 hypnospores/5-mm field; and
h. Very heavy infection (7) with a density of >3001 hypnospores/5-mm field.

Moderate (4) to very heavy (7) infection densities have generally been correlated with parasite infection densities which will cause immediate oyster death.

Infection density code numbers were averaged for each group of oysters to produce the weighted incidence of infection, which is a value reflecting both the proportion of the population infected and the intensity of infection. In addition, calculations of incidence of infection were made for each group of oysters from each of the various treatments.

Statistical Treatment

Statistical analysis of each experimental measurement in oysters from each treatment included determination of mean, standard deviation, and standard error. T-tests (parametric) and Wilcoxon T-tests (nonparametric) were used to determine if measured differences between various treatments were statistically significant. The 95% confidence level ($p<0.05$) was the minimum statistical significance accepted.

RESULTS AND THEIR INTERPRETATION

Salinity Effects

Results indicating the effect of high and low salinity exposure on oyster mortalities associated with infections of *P. marinus* are depicted in Figure 1A. Oyster mortalities were virtually the same in high and low salinity during the first 33 d of exposure (16% vs. 18% mortality, respectively); however, after 60 d of exposure, oyster mortalities were significantly ($p<0.05$) reduced in the low salinity treatment by an average of 25% ($\overline{x}$ = 35%) when compared to high salinity controls ($\overline{x}$ = 46.7%).

Measurement of fecal production or biodeposition rates are listed in Table 2. Fecal production was significantly ($p \leq 0.01$) lower in low salinity controls ($\bar{x}$ = 64.7%) when compared to high salinity controls ($\bar{x}$ = 83.4%). The lower rate of fecal production in the low salinity oysters was probably related to the lower phytoplankton density (food availability) in the low salinity treatment.

Table 2. Summary of fecal production values measured in oysters exposed to various salinity/chlorination conditions. Note the significant ($p \leq 0.01$) reduction in fecal production in low salinity and low salinity, CPO-exposed oysters. Statistical parameters included calculations of mean ($\bar{x}$), standard error ($s/\sqrt{n}$), sample size (n), and Students t-value. Asterisks (**) indicate sample means were significantly ($p \leq 0.01$) different.

Treatment	Nominal NaOCl Concentration (mg/L)	Fecal Production (%)				
		Range	$\bar{x}$	$s/\sqrt{n}$	n	t-Value
A	0.00	60-100	83.4	1.54	41	
B	1.00	20-100	79.2	1.56	40	
C	0.56	48-100	80.3	2.08	40	
D	0.10	64-100	82.5	1.65	39	
E	0.00	4-100	64.7**	3.61	38	-4.76[a]
F	1.00	16-92	65.5**	2.61	38	-4.51[b]
G	0.56	12-92	72.5**	2.28	39	-2.53[c]
H	0.10	12-92	66.9**	3.19	39	4.34[d]

[a] t-value calculated from a comparison of treatments A and E.
[b] t-value calculated from a comparison of treatments B and F.
[c] t-value calculated from a comparison of treatments C and G.
[d] t-value calculated from a comparison of treatments D and H.

Gill tissue respiration values are depicted graphically in Figure 1B. Mean gill tissue respiration rates declined throughout the study in both high ($\bar{x}$ = 1280.6 to 1201.5 µl O_2/g/h) and low salinity ($\bar{x}$ = 1280.6 to 1201.3 µl O_2/g/h) controls in response to declining sea water temperatures (28-20°C) measured in the study. There were no significant differences in respiration rates measured among oysters in comparison of high and low salinity controls.

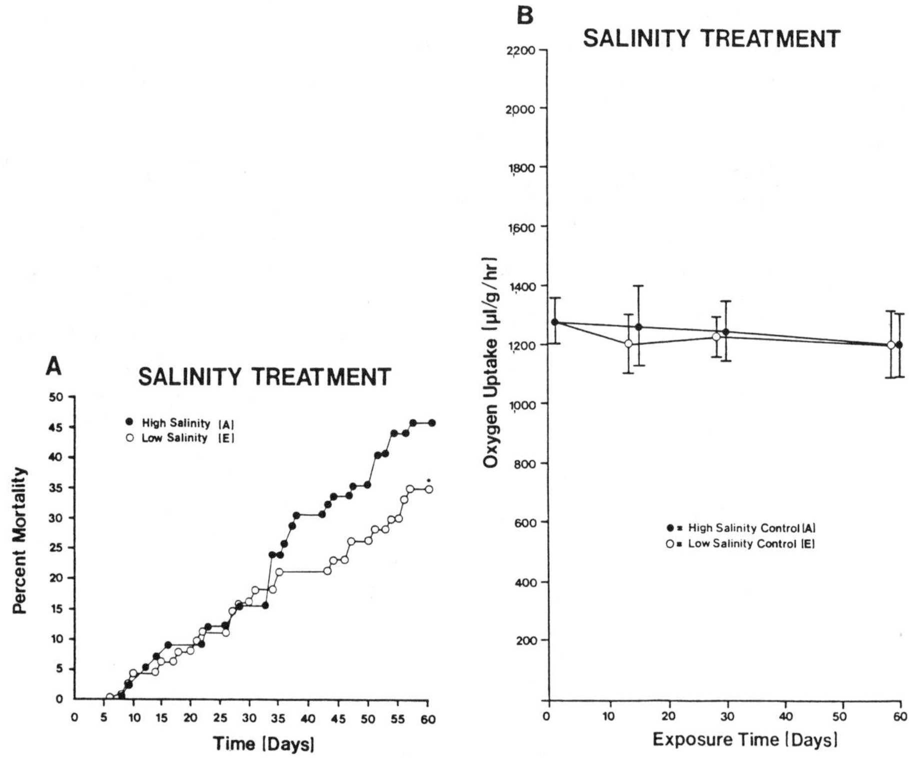

Figure 1A. Mortality data for oysters exposed to high (A) and low (E) salinity treatments. Note that mortality differences between the two treatments were only significant after 30 days of exposure when temperatures fell below 25°C. Asterisk (*) indicates that 60-day mortality results were significantly ($p<0.05$) different. Standard error values were less than ± 4% in both treatments.

1B. Mean gill tissue respiration rates (µl O_2/g/h) measured in oysters exposed to high (A) and low (E) salinity treatments. Note the similarity in respiration rates among high and low salinity oysters throughout the entire 60 days of exposure. Bars (I) represent sample standard error.

Hypnospore densities of *P. marinus* in rectal and labial palp tissues of surviving oysters are listed in Table 3. The incidence of infection was 100% for both high and low salinity control oysters. Hypnospore densities in rectal tissues increased throughout the 60 d of exposure in both high salinity controls ($\bar{x}$ = 66.23 to 670.18 hypnospores per 5-mm field) and low salinity controls ($\bar{x}$ = 75.8 to 1333.3 hypnospores per 5-mm field). Statistical analysis indicated that hypnospore densities were not significantly different between rectal and labial palp tissues in oysters exposed to either high or low salinity. Additionally, hypnospore densities in rectal tissues were not significantly different in comparisons between high and low salinity treatments. Also, hypnospore densities in labial palp tissues were not significantly different between high and low salinity controls.

Weighted indices of infection-intensity calculations for *P. marinus* in rectal and labial palp tissues are listed in Table 4. Initially, low salinity exposure significantly ($p<0.01$) lowered infection intensities in both rectal and labial palp tissues as hypnospore densities were substantially reduced during the first 15 d of exposure. However, after 30 d of exposure and continuing through 60 d of exposure, infection intensities were practically the same for low and high salinity controls in both rectal and labial palp tissues.

Chlorination Treatment

Nominal NaOCl and measured CPO concentrations are listed in Table 1 for both high and low salinity exposures. Measured CPO concentration for each selected nominal NaOCl concentration varied significantly in both high and low salinity, chlorinated sea water combinations. Further analysis of these results indicated that measured CPO concentrations for each selected nominal NaOCl concentration were significantly ($p<0.01$-0.05) higher (45%-50%) in low salinity, chlorinated sea water exposures (F, G, H) when compared with high salinity, chlorinated sea water exposures (B, C, D). Higher CPO concentrations in the low salinity sea water exposures would be expected due to the lower chlorine demand (produced by the presence of all forms of dissolved and suspended organic matter) exerted by the low salinity sea water (as it was diluted with groundwater which reduced phytoplankton densities by 66%).

Results indicating the effect of chlorination on oyster mortalities associated with infections of *P. marinus* are

Table 3. Mean rectal and labial palp hypnospore densities in oysters exposed to various chlorination/salinity conditions. Hypnospore densities were not significantly different in comparison of high and low salinity, CPO-exposed oysters or in comparisons of controls. Statistical parameters included calculations of mean ($\bar{x}$), standard error ($s/\sqrt{n}$), sample size (n), and Students t-value.

Treat-ment	Exposure Time (days)	Exposure Temperature (°C)	Rectum Hypnospore Density[a]					Labial Palp Hypnospore Density[a]				
			$\bar{x}$	$s/\sqrt{n}$	n	t-value[1]	t-value[2]	$\bar{x}$	$s/\sqrt{n}$	n	t-value[1]	t-value[2]
A	0	28	77.4	66.23	14			75.8	64.89	15		
A	15	27	61.1	34.74	14			37.5	25.56	15		
B	15	27	2021.4	1065.53	15		+1.84	2007.7	1068.30	15		+1.84
C	15	27	1585.2	939.67	15		+1.62	1495.1	916.13	15		+1.59
D	15	27	81.6	70.77	15		+0.26	62.5	53.88	15		+0.42
E	15	27	3.3	1.41	15	-1.66		2.9	1.24	15	-1.35	
F	15	27	1058.4	559.58	15	+0.80	+1.89	1225.2	668.95	15	-0.62	+1.83
G	15	27	2122.9	1,131.79	15	+0.37	+1.87	1729.9	998.62	15	+0.17	+1.73
H	15	27	39.5	22.55	15	-0.57	+1.59	32.9	23.04	15	-0.51	+1.30
A	30	24	703.3	692.84	15			753.9	691.70	15		
B	30	24	27.7	7.41	15		-0.98	29.1	15.75	15		-1.05
C	30	24	712.8	697.73	15		+0.01	747.8	697.41	15		-0.01
D	30	24	771.0	713.80	15	+0.70		142.2	85.31	15	-0.88	
E	30	24	14.1	9.00	15		-0.99	10.5	6.07	15		-1.07
F	30	24	1009.5	726.23	15	+1.35	+0.23	997.7	726.04	15	+1.33	+1.17
G	30	24	795.6	759.27	15	+0.08	+0.02	799.9	76.85	15	+0.05	+0.84
H	30	24	155.3	90.69	15	+1.55	-0.86	130.4	64.92	15	+1.84	-0.11
A	60	20	765.0	670.18	15			752.9	658.44	15		
B	60	20	1578.1	961.99	15		+0.69	1508.2	952.28	15		+0.65
C	60	20	546.6	538.36	15		-0.25	696.5	684.71	15		-0.06
D	60	20	800.1	698.15	15		+0.04	809.1	714.65	15		+0.06
E	60	20	702.2	631.14	15	-0.07		1333.3	902.11	15	-0.52	
F	60	20	2141.2	1100.10	15	+0.39		1792.7	945.77	15	+0.21	+0.35
G	60	20	802.9	706.70	15	-0.29	+0.11	659.2	538.94	15	-0.04	-0.64
H	60	20	746.4	724.33	15	-0.05	+0.05	756.1	742.73	15	-0.05	-0.49

[a]Hypnospore per 5-mm field.
[1]Comparison of high salinity, CPO treatments with similar low salinity, CPO treatments (B vs. F, C vs. G, and D vs. H) and comparisons of high and low salinity controls (A vs. E).
[2]Comparison of high salinity controls (A) with high salinity, CPO-exposed oysters (B, C, D) and comparisons of low salinity controls (E) with low salinity, CPO-exposed oysters (F, G, H).

Table 4. Mean rectal and labial palps weighted indices of infection intensities (0-7) measured in oysters exposed to various chlorination/salinity conditions. No significant differences in infection intensities were observed in comparisons of high and low salinity, CPO-exposed oysters, but differences between high and low salinity controls and CPO-exposed oysters were observed. Statistical parameters included calculations of mean ($\bar{x}$), standard error ($s/\sqrt{n}$), sample size (n), and Students t-value. Asterisks (*) (**) indicate reported t-values were significant ($p<0.05$ and $p\leq0.01$, respectively).

Treatment	Exposure Time (days)	Exposure Temperature (°C)	Rectum Hypnospore Density[a]					Labial Palp Hypnospore Density[a]				
			$\bar{x}$	$s/\sqrt{n}$	n	t-value[1]	t-value[2]	$\bar{x}$	$s/\sqrt{n}$	n	t-value[1]	t-value[2]
A	0	28	1.73	0.36	15			1.60	0.38	15		
A	15	27	3.60	0.30	15			2.13	0.35	15		
B	15	27	3.00	0.59	15		+0.56	2.87	0.62	15		+1.04
C	15	27	3.33	0.55	15		+1.07	2.80	0.56	15		+1.01
D	15	27	1.13	0.22	15	-3.22**		1.20	0.14	15		
E	15	27	2.07	0.37	15		-0.97	1.87	0.35	15		-0.53
F	15	27	3.00	0.63	15	0.00	+2.80**	3.13	0.61	15	+0.30	+3.08**
G	15	27	2.60	0.62	15	-0.88	+2.23*	2.40	0.63	15	-0.47	+1.86
H	15	27	2.33	0.39	15	+0.48	+2.68*	2.00	0.34	15	+0.27	+2.18*
A	30	24	2.33	0.41	15			2.27	0.47	15		
B	30	24	2.73	0.27	15		+0.81	2.27	0.32	15		0.00
C	30	24	2.47	0.44	15		+0.23	2.20	0.46	15		-0.11
D	30	24	1.67	0.25	15		-1.23	2.20	0.46	15		-1.65
E	30	24	2.40	0.50	15	+0.11		2.33	0.51	15		
F	30	24	2.07	0.56	15	-1.06	-0.44	2.20	0.55	15	-0.11	-0.17
G	30	24	2.00	0.47	15	-0.73	-0.58	2.00	0.43	15	-0.32	-0.49
H	30	24	2.27	0.44	15	+1.19	-0.20	2.20	0.46	15	+1.54	-0.19
A	60	20	2.47	0.51				2.40	0.50	15		
B	60	20	3.47	0.74	15		+1.11	3.60	0.48	15		+1.73
C	60	20	2.00	0.45	15		-0.69	2.00	0.45	15		-0.59
D	60	20	2.73	0.49	15		+0.37	2.53	0.46	15		+0.19
E	60	20	2.27	0.49	15	-0.40		2.40	0.55	15		
F	60	20	3.40	0.55	15	-0.08	+1.53	3.07	0.56	15	-0.72	+0.85
G	60	20	2.53	0.51	15	-0.78	+0.37	2.60	0.49	15	+0.90	+0.27
H	60	20	2.33	0.44	15	-0.61	+0.09	2.07	0.45	15	-0.71	-0.46

[a]Hypnospore per 5-mm field.

[1]Comparison of high salinity, CPO treatments with similar low salinity, CPO treatments (B vs. F, C vs. G, and D vs. H) and comparisons of high and low salinity controls (A vs. E).

[2]Comparison of high salinity controls (A) with high salinity, CPO-exposed oysters (B, C, D) and comparisons of low salinity controls (E) with low salinity, CPO-exposed oysters (F, G, H).

depicted in Figure 2A. Oyster mortalities were significantly ($p<0.05$) lower in each low salinity, chlorination treatment when compared with equivalent high salinity, chlorination treatments. Mean oyster mortalities averaged 22.2% (ranging from 20 to 25%) in the various low salinity, chlorination exposures (F, G, H) compared to an average mortality rate of 51.1% (ranging from 41.7 to 56.7%) in the various high salinity, chlorination treatments (B, C, D).

Comparisons of mortality rates in the various high salinity, chlorination exposures (B, C, D) with high and low salinity controls (A and E, respectively) indicated the mean mortality rate in the nominal 1.00 mg/L, high salinity treatment (B = 56.7%) was significantly ($p<0.05$) higher than the mean mortality rates recorded in high and low salinity control oysters (A = 46.7% and E = 35%, respectively). The mean mortality rates in the nominal 0.56 mg/L and 0.10 mg/L, high salinity treatments (C = 45% and D = 41.7%, respectively) were slightly lower but not significantly different from the mortality rate for high salinity, control oysters (A = 46.7%). The mean mortality rates in the nominal 0.56 mg/L and 0.10 mg/L, high-salinity treatments (C = 45% and D = 41.7%, respectively) were significantly ($p<0.05$) higher than the mortality rate for low salinity, control oysters (E = 35%).

Comparisons of mortality rates in the various low salinity, chlorination exposures (F, G, H) with high and low salinity controls (A and E, respectively) indicated that the mean mortality rates were significantly ($p<0.05$) lower in each of the low salinity, chlorination treatments (F = 25%, G = 21.7%, and H = 20%) when compared to either the high salinity (A = 46.7%) or low salinity (E = 35%) controls.

Mortality was lowest (averaging 20%) in the 0.10 mg/L nominal NaOCl concentration, low salinity treatment (H). This represented a 57% reduction in the mortality rate measured in high salinity control oysters (A) and a 43% reduction in the mortality for low salinity control oysters (E).

Measurements of fecal production or biodeposition rates are listed in Table 2. Fecal production rates were significantly ($p<0.01$) lower in each low salinity, chlorination exposure when compared with equivalent high salinity, chlorination exposures. Mean fecal production rates averaged 68.3% (ranging from 65.6% to 72.5%) in the various low salinity, chlorination exposures (F, G, H) compared to an average rate of 80.7% (ranging from 79.2 to 82.5%) in the different high salinity, chlorination exposures (B, C, D). These results indicated that oyster fecal production in chlorinated, high salinity sea water was significantly ($p<0.05-0.01$) higher than in chlorinated, low salinity sea water. Further comparisons with high and low salinity, control oysters indicated that reduced salinity (or dilution), rather than chlorination, had a more profound effect on fecal production, probably related to the lower phytoplankton densities (food availability) in the low salinity, chlorination exposures.

Gill tissue respiration measurements for oysters exposed to various CPO concentration/salinity exposure regimes are depicted in Figure 2B. Gill tissue respiration rates in both high and low salinity, chlorination exposures initially increased from day 0 to day 15, then gradually declined from day 15 to day 60. Gill tissue respiration rates in high salinity, CPO-exposed oysters (B, C, D) generally were not significantly different from values measured in the low salinity, CPO exposures (F, G, H). The only exception to this trend occurred at days 30 and 60 in the 1.00 mg/L nominal NaOCl concentration, high salinity treatment values (B = 1568.7 and 1598.3 µl O_2/g/h, respectively) which were significantly ($p<0.01$) higher than corresponding low salinity, chlorination treatment values (F = 1110.5 and 1179.2 µl O_2/g/h, respectively) and high (A) and low (E) salinity control values.

Comparisons of high (A) and low salinity (E) controls with the other high salinity, chlorination exposures (C, D) indicated that gill tissue respiration rates generally were not significantly different between controls and high salinity, CPO exposures. Similar comparisons of different low salinity, CPO-exposed oysters (F, G, H) with high (A) and low (E) salinity controls indicated that generally gill tissue respiration rates of controls were not significantly different from low salinity, CPO exposures.

Hypnospore densities of *P. marinus* in rectal and labial palp tissues are listed in Table 3. The incidence of infection was 100% for both high salinity (B, C, D) and low salinity (F, G, H), CPO-exposed oysters. Hypnospore densities in rectal and labial palp tissues were variable but generally increased in both rectum (in 83% of the samples) and labial palp tissues (in 72% of the samples) throughout the 60 d of exposure in both high and low salinity, CPO-exposed oysters. Statistical analysis of high salinity (B, C, D) and low salinity (F, G, H), CPO-exposed oysters indicated that there were no significant differences in comparisons of rectal tissue hypnospore densities. Similarly, there were no significant differences in hypnospore

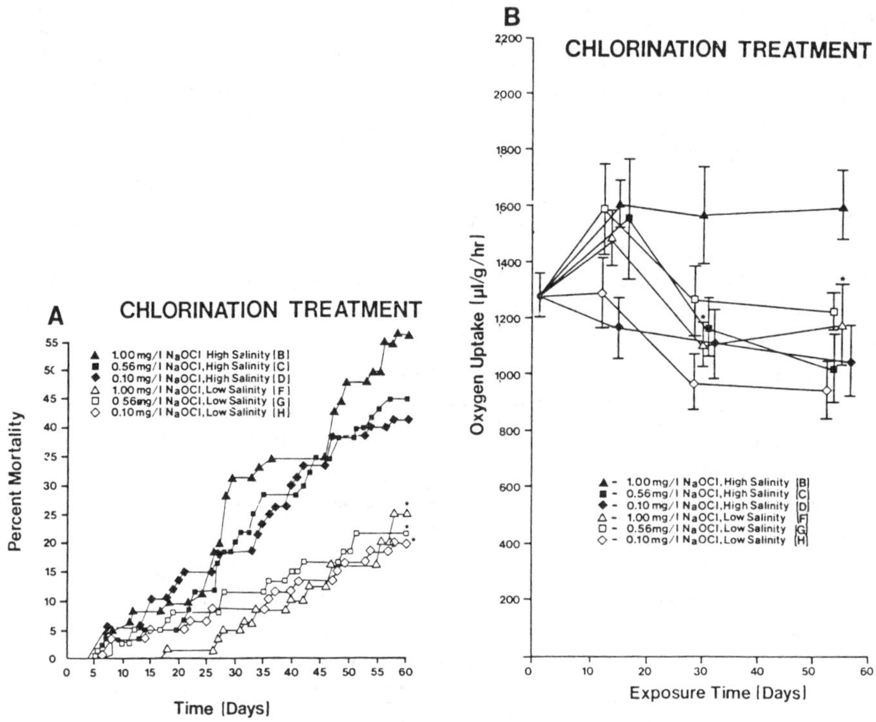

Figure 2A. Mortality data for oysters exposed to various high salinity (B, C, D) and low salinity (F, G, H), chlorination treatments. Note the significantly ($p \leq 0.05$) higher mortality in the high salinity, CPO-exposed oysters. Asterisk (*) indicates that 60-day mortality results were significantly ($p \leq 0.05$) different. Standard error values were less than ± 4% in all treatments throughout the 60 days of exposure.

2B. Mean gill tissue respiration rates (µl O_2/g/h) measured in adult oysters exposed to various high salinity (B, C, D) and low salinity (F, G, H), chlorination treatments for 60 days. Gill tissue respiration rates generally declined throughout the 60 days of exposure as a result of declining sea water temperatures. Note the similarity in gill tissue respiration rates between high and low salinity, CPO-exposed oysters with the exception of Treatment B at exposure days 30 and 60. Asterisk (*) indicates that 30- and 60-day gill tissue respiration rates in Treatment B were significantly ($p \leq 0.01$) higher than in Treatment F. Bars (I) represent sample standard error.

densities in labial palp tissue, between high and low salinity, chlorination treatments. Comparisons within each salinity/chlorination treatment indicated that rectal hypnospore densities were generally greater than labial palp tissue densities. Additional comparisons of high salinity (B, C, D) and low salinity (F, G, H), CPO-exposed oysters with high (A) and low salinity (E) controls indicated there were no significant differences between controls and the various chlorination treatments in hypnospore densities of rectal and labial palp tissues.

The weighted index of infection-intensity calculations for *P. marinus* in rectal and labial palp tissues are listed in Table 4. Infection intensity in both rectal and labial palp tissues generally increased (in 90% of the samples) throughout the 60 d of exposure of high salinity (B, C, D) and low salinity (F, G, H), CPO-exposed oysters. The general trend observed for most salinity/chlorination treatments was an initial increase in infection intensity (for both rectum and labial palps) from day 0 to day 15, followed by a slight decrease from day 15 to day 30, with a final increase being observed from day 31 to day 60.

Statistical analysis of high salinity (B, C, D) and low salinity (F, G, H), CPO-exposed oysters indicated there were no significant differences in comparison of weighted index of infection intensities among rectal tissues. Similarly, there were no significant differences in comparisons of the weighted index of infection intensities among labial palp tissues.

Comparisons of high salinity (B, C, D) and low salinity (F, G, H), CPO-exposed oysters with high (A) and low salinity (E) controls showed no significant differences in rectal infection intensities between controls and the various salinity/chlorination treatments. The same results were obtained for labial palp tissue.

DISCUSSION

Results from this study have generally indicated the significant interactive effects of salinity and CPO exposure in affecting oyster mortalities resulting from *P. marinus* infection.

Comparison of control oysters in high and low salinity exposures generally indicated that low salinity conditions resulted in exposure of oysters to water which was reduced in both salt content (major ions) and phytoplankton density (as food availability was reduced by 66%). Biodeposition rates (or feeding rates) were significantly reduced in low salinity exposed oysters, probably related to less food availability. Galtsoff (1964) reported similar reduced feeding rates in oysters exposed to low salinity. Exposure to low salinity conditions and accompanying reduced fecal production rates did not significantly affect oyster physiology as measurements of condition index, gonadal index, and gill tissue respiration rates were not significantly different (and were generally identical) throughout the 60 d of exposure (Scott, unpubl. data).

Analysis of mortality data from high and low salinity control conditions indicated that low salinity exposure significantly increased oyster survivability of *P. marinus* infections by 25%. This significant increase in survival occurred in low salinity control oysters even though there were generally no statistically significant differences in the incidence of infection, hypnospore density, and weighted index of infection intensity in comparisons of high and low salinity control oysters. Analysis of the weighted index of infection-intensity data suggested that low salinity exposure initially decreased infection intensity (at day 15) but that by days 30 and 60, infection intensities were equal for both high and low salinities. Earlier studies (Mackin & Wray, 1952; Mackin, 1956) have indicated a similar delay in hypnospore development from low salinity exposure.

While infection-intensity differences in comparisons of high and low salinity control oysters were only slight, actual differences may have been obscured by the higher mortality rate in high salinity control oysters, since the most heavily infected oysters were presumably dying first. This can be presumed since reported measurements were not made of gapers but rather of living specimens. The death of the most diseased individuals would result in selection of more resistant and/or less infected oysters which would tend to bias infection intensities downward to much lower levels than if all heavily diseased or dead individuals were included. Increased mortality among various treatments was therefore considered indicative of higher infection rates as all autopsied dead oysters had infection intensities of 7 (Scott, unpubl. data).

The mode of action for low salinity exposure in significantly reducing oyster mortalities has been reviewed by several authors (Ray, 1954; Andrews & Hewatt, 1957) who generally concluded that low salinity exposure must enhance survival due to possible reductions in the number of waterborne zoospores which would tend to limit and reduce both the incidence and intensity of infection. However, the fact that all oysters in this study were infected with *P. marinus* at day 0 and since the infection spreads only from the initial site of infection suggests that differences in survival were not related to differences in zoospore density in sea water, but rather to physiological differences in oysters exposed to different salinity.

Results from chlorination (CPO) exposure treatment indicated that the combination of low salinity and CPO exposure significantly ($p<0.05$) reduced oyster mortalities by 47-57% below high salinity control levels at all chlorine concentrations tested (0.10-1.00 mg/L nominal NaOCl concentrations; measured CPO concentrations of 0.02-0.11 mg/L). The combination of high salinity and CPO exposure increased mortality by 21% above high salinity control levels at high chlorination doses (1.00 mg/L nominal NaOCl concentration; measured CPO concentration of 0.06 mg/L), while in lower CPO concentrations mortality was equal to or was slightly lower than high salinity control mortality rates. Earlier studies (Scott, 1979; Scott & Vernberg, 1979) reported similar increased mortality 14% above control values in oysters exposed to a 1.00 mg/L nominal NaOCl concentration in high salinity waters during the same time period of the year.

Additional comparisons also indicated that the combination of low salinity and CPO exposure significantly reduced oyster mortality below low salinity control values.

Fecal production measurements indicated increased feeding rates in high salinity, CPO-exposed oysters (79.2-82.5%) when compared to low salinity, CPO-exposed oysters (65.5-72.5%). Fecal production rates in high and low salinity, CPO-exposed oysters were not significantly different from rates for each respective high and low salinity control. Decreased fecal production was considered attributable to low salinity effects rather than CPO exposure. These results differ significantly from earlier results obtained by Scott (1979), Scott and Middaugh (1978), and Scott and Vernberg (1979), who reported that CPO exposure caused significant reductions in fecal production rates at low chlorine concentrations (nominal 1.00 mg/L). However, these earlier studies were continuous-exposure, flow-through experiments in comparison to the intermittent (8-12 h of CPO exposure/d) conditions used in this study.

Comparisons of gill tissue respiration rates in high and low salinity, CPO-exposed oysters generally indicated that respiration rates were slightly lower in low salinity, CPO-exposed oysters. Gill respiration rates were only significantly ($p<0.01$) increased in oysters in the 1.00 mg/L nominal NaOCl concentration, high salinity exposure. Scott (1979), Scott and Vernberg (1979), and Scott *et al*. (1980) have reported similar increased gill tissue respiration rates in oysters chronically (60 d) exposed to 1.00 mg/L nominal NaOCl concentration under similar exposure conditions, which resulted from increased anaerobic metabolism by oysters to survive CPO exposure.

Results from this study suggested that CPO exposure at similar chlorine concentrations (1.00 mg/L nominal NaOCl) increased gill tissue respiration rates, but only under high salinity conditions. Higher gill tissue respiration rates in these oysters resulted from increased reliance upon gonadal glycogen reserves to survive CPO exposure under high salinity conditions. Absence of these effects in lower CPO concentrations under high salinity exposure and in low salinity, CPO exposures may imply that CPO exposure must interact with factors representative of normal high salinity exposure conditions per se, such as increased *P. marinus* infection intensity (indicated by higher mortality), to cause increased stress in these oysters. Such an interaction would result in a more rapid depletion of stored gonadal glycogen reserves due to increased metabolic utilization of glycogen by the oyster to survive CPO exposure and by the parasite as the infection intensifies and spreads. There was a rapid decline of gonadal indices in high salinity, CPO-exposed oysters in this study which may be indicative of this effect (Scott, unpubl. data). The rapid depletion of gonadal glycogen stores resulting from this combined host/parasite effect would probably result in earlier utilization of protein stores (amino acids) as glycogen levels declined. This would signify utilization of different metabolic pathways both by the host (oyster) and parasite and may result in an ultimate buildup of different secondary metabolites which may adversely affect the oyster. This increased mortality rate observed in 1.00 mg/L nominal NaOCl concentration, high salinity exposed oysters was probably indicative of this combined toxicant/host/parasite effect.

Consideration must be given to physiological and biochemical differences between high and low salinity, control oysters. Oysters are euryhaline organisms surviving wide ranges of salinity as low as 5-7°/oo (Galtsoff, 1964; Vernberg & Vernberg, 1972). Additionally, oysters are considered osmoconformers which prefer salinities of 11-30°/oo (Galtsoff, 1964; Dupuy *et al*., 1977). Osmoregulation in low salinity sea water appears to result from volume regulation, the result of intracellular solute loss (Pierce, 1971). The source of these intracellular solutes has been reported to be in the intracellular, free amino acid pool. Decreased levels of taurine, alanine, glycine, and proline have been reported in intertidal bivalve species, while reduced levels of taurine, alanine, and glycine have been reported in subtidal bivalve species (Pierce, 1971). Under low salinity exposure, free amino acid levels may decline by as much as 84%, particularly ninhydrin positive (i.e., glycine and alanine) substances (Lange, 1970; Vikar & Webb, 1970). Low salinity exposure would significantly reduce the amount of intracellular, free amino acids in oysters during this study.

This reduction of free amino acid levels in low salinity control oysters may, perhaps, explain the high survival observed under low salinity exposure. Low salinity oysters may have increased survival due to the reduced availability of protein or amino acids as food substrate for *P. marinus*, particularly as the infection intensifies to heavy intensities.

Additional consideration must be given to results from Soniat and Koenig (1982) who found that free amino acid levels in oysters declined with increasing levels of *P. marinus* infection. This suggests increased utilization of free amino acids by both the host (oyster) and parasite during periods of infection. Soniat and Koenig (1982) also reported increased levels of taurine and aspartic acid with increased infection intensity, although total free amino acid levels declined with increasing infection intensity. Similarly, Lange (1963) reported increased levels of free amino acids with increased salinity in *Mytilus edulis*, with taurine having the most dramatic increase with increased salinity. Giles (1972) reported a 55% decrease in taurine and an 82% decline in aspartic acid levels in *Mytilus edulis* exposed to low salinity sea water. When these findings are considered, our data may suggest that the higher rates of survival measured in low salinity control and low salinity, CPO-exposed oysters may have resulted from decreased levels of taurine and aspartic acid resulting from osmoregulation at low salinity conditions.

Exposure of oysters to low salinity and low salinity, CPO conditions probably resulted in oysters which had lowered levels of both glycogen and free amino acids. This would initially tend to retard secondary spread and intensification of the disease. The higher mortality rate observed in high salinity controls and high salinity, CPO-exposed oysters would suggest that the lowered levels of free amino acids in low salinity control and low salinity, CPO-exposed oysters offered some degree of protection to the oyster by preventing the disease from becoming toxic. What is suggested by these data is that the lowered free amino acid levels in low salinity control oysters and low salinity, CPO-exposed oysters may have resulted in less available nitrogen-containing molecules for *P. marinus* to utilize. Further research in this area is needed to better understand the exact mechanism of host-parasite interactions to chlorine as well as other environmental pollutants.

LITERATURE CITED

APHA (American Public Health Association). 1971. Standard Methods for Examination of Water and Waste Water, 14th Ed. American Water Works Association, Water Pollution Control Federation, New York. 874 pp.

Andrews, J.D. and W.G. Hewatt, 1957. Temperature control experiments on the fungus disease, *Dermocystidium marinum*, of oysters. Virginia Fisheries Laboratory 62: 129-133.

Block, R.N., D.T. Burton, S.R. Gullans, and L.B. Richardson. 1977. Respiratory and osmoregulatory responses of white perch (*Morone americana*) exposed to chlorine and ozone in estuarine waters. pp. 351-360. *In*: Water Chlorination: Environmental Impact and Health Effects, Vol. 2. R.L. Jolley, H. Gorchev, and D.N. Hamilton, Jr. (eds.). Ann Arbor Press, MI.

Couch, J.A. 1978. Diseases, parasites, and toxic responses of commercial Penaeid shrimps of the Gulf of Mexico and south Atlantic coasts of North America. Fish. Bull. 76(1): 1-44.

Crecelius, E.A., G. Roesijadi, and T.O. Thatcher. 1978. Correspondence. Environ. Sci. Technol. 12: 1088.

Dupuy, J.L., N.T. Windsor, and C.E. Sutton. 1977. Manual for Design and Operation of an Oyster Seed Hatchery for the American Oyster, *Crassostrea virginica*. Appl. Mar. Sci. and Ocean Eng., Virginia Institute of Marine Science, Special Report No. 142. Gloucester Point, VA. 104 pp.

EPA (Environmental Protection Agency). 1974. Handbook of Analytical Control in Water and Waste Water Laboratories. Anal. Qual. Control. Lab., Cincinnati, OH. 45 pp.

Farley, C.A. 1965. Acid-fast staining of haplosporidian spores in relation to oyster pathology. J. Invert. Pathol. 7(2): 144-147.

Galtsoff, P.S. 1964. The American Oyster, *Crassostrea virginica* (Gmelin). U.S. Fish and Wild. Serv., Fish. Bull. 64: 1-480.

Giles, R. 1972. Osmoregulation in three molluscs: *Acanthochitona discrepans* (Brown), *Glycymeris glycymeris* (L), and *Mytilus edulis* (L). Biol. Bull. 142: 25-35.

Hall, L.W., Jr., G.R. Helz, and D.T. Burton. 1981. Power Plant Chlorination: A Biological and Chemical Assessment. Ann Arbor Science Publishers, Ann Arbor, MI. 237 pp.

Lange, R. 1963. The osmotic function of free amino acids and taurine in the mussel, *Mytilus edulis*. Comp. Biochem. Physiol. 10: 173-179.

Lange, R. 1970. Isosmotic intracellular regulation and euryhalinity in marine bivalves. J. Exp. Mar. Biol. Ecol. 5: 170-179.

Mackin, J.G. 1956. *Dermosystidium marinum* and salinity. Proc. Natl. Shellfish Assoc. 46: 116-128.

Mackin, J.G. 1962. Oyster disease caused by *Dermocystidium marinum* and other microorganisms in Louisiana. Publ. Inst. Mar. Sci. 7: 132-229.

Mackin, J.G. and D.A. Wray. 1952. Report on the Second Study of Mortality of Oysters in Barataria Bay, Louisiana, and Adjacent Areas. Texas A & M Res. Proj. No. 9, College Station, TX. 40 pp.

Perkins, F.P. 1966. Life history studies of *Dermocystidium marinum*, an oyster pathogen. Ph.D. Dissertation. Florida State Univ., Tallahassee. Dissertation Abstracts, 66-9078, 272 pp.

Pierce, S.K., Jr. 1971. A source of volume regulation in marine mussels. Comp. Biochem. Physiol. 38A: 619-635.

Quick, J.A. and J.G. Mackin. 1971. Oyster parasitism by *Labyrinthomyxa marina* in Florida. Florida Department Natural Resources Marine Laboratory, Professional Paper Series 13. 55 pp.

Ray, S.M. 1952. A culture technique for the diagnosis of infections with *Dermocystidium marinum* (Mackin, Owen, and Collier) in oysters. Science 116(3014): 360-361.

Ray, S.M. 1953. Studies on the occurrence of *Dermocystidium marinum* in young oysters. Proc. Natl. Shellfish Assoc. 43: 80-88.

Ray, S.M. 1954. Biological Studies of *Dermocystidium marinum*, a Fungus Parasite of Oysters. Rice Inst.

Pam., Spec. Issue (Nov. 1954), Monographs in Biology. 114 pp.

Ray, S.M. 1966. A review of the culture method for detecting *Dermocystidium marinum*, with suggested modifications and precautions. Proc. Natl. Shellfish Assoc. 54: 55-69.

Ray, S.M., J.G. Mackin, and J.L. Boswell. 1953. Quantitative measurement of the effect on oysters of the disease caused by *Dermocystidium marinum*. Bull. Mar. Sci. Gulf and Carib. 3: 6-33.

Roberts, M.H., Jr. and B.B. Casey. In press. Depression of larval growth and metamorphosis of oysters (*Crassostrea virginica*) exposed to chlorinated sewage. *In*: Water Chlorination: Environmental Impact and Health Effects, Vol. 5. R.L. Jolley, W.A. Brungs, J.A. Cotruro, R.B. Cummings, J.S. Mattice, and V.A. Jacobs (eds.). Ann Arbor Press, Ann Arbor, MI.

Rosenberg, W.H., J.C. Rhoderick, R.M. Block, V.S. Kennedy, S.R. Gullans, S.M. Vreenegoor, A. Rosenkranz, and C. Collette. 1980. Effects of the chlorine produced oxidants on survival of larvae of the oyster, *Crassostrea virginica*. Mar. Ecol. Prog. Ser. 3: 91-96.

Scott, G.I. 1979. The effects of seasonal chronic chlorination on the growth, survival, and physiology of the American oyster, *Crassostrea virginica* (Gmelin). Ph.D. Dissertation. Univ. of South Carolina, Columbia. 169 pp.

Scott, G.I. and D.P. Middaugh. 1978. Seasonal chronic toxicity of chlorination to the American oyster, *Crassostrea virginica*. pp. 311-328. *In*: Water Chlorination: Environmental Impact and Health Effects, Vol. 2. R.L. Jolley, H. Gorcher, and D.H. Hamilton, Jr. (eds.). Ann Arbor Press, Ann Arbor, MI.

Scott, G.I. and W.B. Vernberg. 1979. Seasonal effects of chlorine-produced oxidants on the growth, survival, and physiology of the American oyster, *Crassostrea virginica* (Gmelin). pp. 415-535. *In*: Marine Pollution: Functional Responses. W.B. Vernberg, F.P. Thurberg, A. Calabrese, and F.J. Vernberg (eds.). Academic Press, New York.

Scott, G.I., D.P. Middaugh, A.M. Crane, N.H. McGlothlin, and N. Watabe. 1980. Physiological effects of chlorine-produced oxidants and uptake of chlorination by-products in the American oyster, *Crassostrea virginica* (Gmelin). pp. 501-516. *In*: Water Chlorination: Environmental Impact and Health Effects, Vol. 3.

R.L. Jolley, W.A. Brungs, and R.B. Cumming (eds.). Ann Arbor Press, Ann Arbor, MI.

Scott, G.I. 1983. Physiological effects of chlorine-produced oxidants, dechlorinated effluents, and trihalomethanes on marine invertebrates. pp. 827-842. In: Water Chlorination: Environmental Impact and Health Effects. R.L. Jolley, et al. (eds.). Ann Arbor Press, Ann Arbor, MI.

Sinderman, C.J. 1978. Pollution-associated diseases and abnormalities of fish and shellfish: A review. Fish. Bull. 76(4): 717-749.

Soniat, T.M. and M.L. Koenig. 1982. The effects of parasitism by *Perkinsus marinus* on the free amino acid composition of *Crassostrea virginica* mantle tissue. J. Shellfish Res. 2(1): 25-28.

Vernberg, F.J. and W.B. Vernberg. 1972. Environmental Physiology of Marine Animals. Springer-Verlag Press, New York. 346 pp.

Vikar, R.A. and K.L. Webb. 1970. Amino acids in the clam, *Mya*. Comp. Biochem. Physiol. 32: 775-783.

White, G.C. 1972. Handbook of Chlorination. Van Nostrand Rheinhold Co., New York. 744 pp.

EFFECTS OF EMBRYONIC EXPOSURE TO SUBLETHAL CONCENTRATIONS OF DIMILIN® ON THE PHOTOBEHAVIOR OF GRASS SHRIMP LARVAE

J.E. Wilson, R.B. Forward, Jr., and J.D. Costlow

SUMMARY

The photobehavior of grass shrimp, Palaemonetes pugio, larvae hatched from embryos exposed to sublethal concentrations of Dimilin® was investigated using a static bioassay. The sign of phototaxis of untreated and treated light-adapted Stage I larvae varied significantly ($p \leqq 0.001$) with stimulation intensity.

Larvae hatched from untreated embryos exhibited strong positive phototaxis at high light intensities (3.0×10^{-1} and 10^{-2} W/m^2). At lower light intensities (3.0×10^{-3} - 10^{-5} Wm^2) they became strongly photonegative. On the other hand, larvae hatched from embryos exposed to a single dose of Dimilin® (0.3 - 1.0 ppb) generally showed strong negative responses to both high and low light intensities, and a concomitant loss of positive phototaxis. The minimum test concentration of Dimilin® to alter the measured responses was 0.3 ppb. Depression in positive phototaxis and elevation in negative phototaxis were most pronounced at 0.5 ppb. At 2.5 and 5.0 ppb (where morphological abnormalities were observed to be most severe) both positive and negative phototaxis were severely suppressed.

Reversals in phototactic responses were also observed when embryos at different stages of development were exposed to 0.5 ppb Dimilin® for 4 d. The degree of alteration of phototaxis depended on which embryonic stage was exposed. Reversal of phototactic response was most pronounced upon exposure of stage M2, i.e., at commencement of body segmentation. For all stages, the lowest test concentration to affect phototaxis was 0.3 ppb. The possible ecological consequences of these findings are discussed. Phototaxis of larvae hatched from treated embryos may be useful as a rapid, sensitive indicator of adverse effects of chemical pollutants such as Dimilin® in estuaries.

INTRODUCTION

Since the introduction of insect growth regulators and juvenile hormone mimics and their analogues as potentially useful chemicals for controlling various insect pests they have been the subject of much research. These compounds are particularly promising as pesticides because they are usually nonpersistent and therefore do not bioaccumulate in the food chain. Also their effect on target species are morphogenetic rather than directly toxic (Christiansen *et al*., 1977).

Dimilin® is an example of such insect growth regulators that has received much attention in recent years because of its unique mode of action and its effectiveness against a wide variety of arthropods. According to Post *et al*. (1974) the most conspicuous feature in the toxicity of this compound is the abortion of moult, *i.e.*, it interferes with cuticle formation. Histological examination by Mulder and Gijswijt (1973) revealed that the endocuticle was replaced by "a mass of granular particles," so that the newly formed cuticle consisted only of epicuticle and exocuticle improperly attached to the epidermis. As a result the new cuticle cannot resist the muscular contraction during molting and the larvae succeed only partially, or not at all, in casting off their exuviae (Mulder & Gijswijt, 1973; Christiansen *et al*., 1978; Christiansen & Costlow, 1982).

Much attention has been given to acute toxic effects of Dimilin® on both target and non-target arthropods. Most of these studies are on freshwater arthropods. Among the studies on the toxic effects of Dimilin® on marine and estuarine crustaceans are Cunningham (1976); Christiansen *et al*. (1978); Costlow (1979); Nimmo *et al*. (1979); Nimmo *et al*. (1980); and Tester and Costlow (1981). All of the studies of marine and estuarine crustaceans have demonstrated that at concentrations of 1-10 ppb, Dimilin® is lethal. Their hatching success and larval development, but not fecundity, are affected in varying degrees. However, there has been a growing awareness among many investigators that determination of acute toxicity is not sufficient; and there must be more concern with sublethal effects. The literature of such studies for Dimilin® is, however, sparse. Tester and Costlow (1981) reported that *Acartia tonsa* held in 1 and 10 ppb Dimilin® for 36 h failed to produce viable nauplii even after they had been placed in clean sea water. Similarly, Nimmo *et al*. (1980) observed that chronic exposure of *Mysidopsis bahia* to 0.075 ppb Dimilin® reduced the reproductive success of both parents and progeny even after they were transferred to uncontaminated sea water.

By studying the behavioral modifications of both larvae and adult crustaceans, the sensitivity of sublethal toxicity tests have been increased. Both phototaxis and swimming speeds have been used as behavioral indicators of sublethal effects of various pollutants. Forward and Costlow (1976) continuously exposed *Rhithropanopeus harrisii* larvae to the insect juvenile hormone mimc (MON-0585) at the optimum temperature and salinity for development. They observed that concentrations of 0.1 ppm and higher caused significantly (with reference to the controls) faster swimming in Stage II larvae; Stages III and IV were unaffected. The lowest concentration that caused altered phototaxis (depression of both positive and negative phototaxis but not reversal of phototaxis) was 0.1 ppm for Stage II larvae; Stages III and IV were unaffected. In a similar study with three insect growth regulators (Methroprene:ZR-515; Hydroprene:ZR-512; and Dimilin®:TH-6040), Forward and Costlow (1978) demonstrated that Dimilin® was the most potent. Swimming speeds were significantly increased in Stages I, II, and III zoea at a concentration of 0.3 ppb, while phototaxis was altered at 0.1 ppb. Other studies employing behavioral modifications as indicators of sublethal toxicity include those of Bigford (1977) for oil on *Cancer irroratus* larvae; Lang *et al*. (1980) for copper on barnacle nauplii; and Moyer and Barthalmus (1979) for the herbicide, Weeder-64 on adults and larval stages of *Palaemonetes pugio*.

In all these studies, the animals were directly exposed to the toxicants for different lengths of time prior to monitoring photobehavior. The possible effects of short-term exposure of embryos to sublethal doses of toxicants on larval photobehavior remains unreported. The present study was therefore designed in an attempt to address this issue. We measured the photoresponse of one-day old larvae from embryos that were exposed at different embryonic stages to various sublethal concentrations of Dimilin®. The results indicate that Dimilin® exposure affects larval photobehavior, and sensitivity varies with the stage at which the embryos were exposed.

MATERIALS AND METHODS

Laboratory-raised (winter breeding) ovigerous females of the grass shrimp, *Palaemonetes pugio*, were obtained from a laboratory "habitat." The habitat consisted of a fiberglass tank (500 L) fitted with an undergravel filter and a water heater. Illumination was by two 15-watt daylight fluorescent lamps. Photoperiod (14:10 L:D) and temperature were controlled by timers. The whole set-up was covered with black plastic. Laboratory animals were used in this study because they could be induced to breed during the winter, and because they were relatively homogeneous and gave less variable results than field animals. The wettable powder formulation of Dimilin® (WP-25%) is the "practical or field formulation," hence it was used in this study so that the results will be more meaningful ecologically.

The females were sorted according to stage of embryonic development (based on yolk consumption and eye development) and placed in 20-cm diameter finger bowls containing varying concentrations of the wettable powder-25% formulation of Dimilin® for 4 d (single dose). The test concentrations were 0, 0.3, 0.5, 1.0, 2.5, and 5.0 ppb (μg/L) made up in 20‰ sea water. Shrimp were maintained at a density of five per liter of solution. Following the 4-d exposure to Dimilin® solutions, the animals were transferred into clean sea water (20‰) which was changed daily until the eggs hatched. All the animals were kept in temperature controlled cabinets set at 25°C and a 12:12 L:D cycle. Animals were fed freshly hatched *Artemia salina* nauplii daily, except just prior to hatching of the eggs, when they were fed twice a day to reduce the incidence of parental cannibalism.

First larval stage (Stage I zoea) were reared in 9-cm finger bowls containing 20‰ filtered sea water for 24 h at 20°C and a 12:12 L:D cycle. They were fed freshly hatched *Artemia* nauplii daily. Only one-day old larvae were used for the experiments. To avoid complications due to biological rhythms in behavior, all experiments were conducted between midnight and 3 am. This time was chosen because it ensured that the larvae tested were exactly one day old and because preliminary data suggested that the animals were maximally responsive to light at this time. The larvae were fed throughout the phototaxis experiments to reduce the possibility of altered phototaxis due to starvation (Cronin & Forward, 1980; Lang *et al.*, 1980).

Phototaxis

Phototaxis was measured upon light stimulation in the horizontal plane in order to avoid complications due to geotaxis. The stimulus source was a slide projector fitted with a 300W incandescent bulb, the light from which was interference-filtered to 500 nm (7nm half bandwidth). This wavelength was chosen because it is the spectral sensitivity maximum for *P. pugio* (Douglass, unpubl. data) and *P. vulgaris* (White, 1942). The light intensity was controlled by neutral density filters (Ditric Optics, Inc.). The experimental chamber was a lucite trough (14.9 x 3.8 x 3.5 cm). The chamber was divided into five equal sections by lucite partitions which could be raised and lowered simultaneously.

The experimental protocol was a follows:

1. Animals were light-adapted for 4-6 h to a light intensity of 12.53 W/m^2 using incandescent light from a Tensor lamp, supplemented with room lights.
2. The trough was filled with 110 ml filtered sea water ($20^{o}/_{oo}$) to a depth of 2.5 cm and placed on a stage in front of the light source.
3. Ten to 15 light-adapted larvae were placed in the central section of the trough and all lights turned off for 30 sec.
4. After 30 sec in darkness, the partitions were gently raised and the stimulus light turned on simultaneously. Larvae were stimulated for 60 sec after which the partitions were lowered, and the stimulus light turned off.
5. The number of larvae in each section of the trough was recorded and the animals discarded. Another set of 10 larvae were then introduced. The process was repeated at least three times for each stimulus condition using new larvae each time.
6. Neutral density filters were changed and phototaxis was determined at a different light intensity.

Six to 9 different light intensities were tested. A 'dark control' was run for each batch of larvae, following the protocol outlined above, except that the stimulus light was not turned on.

Positive phototaxis was defined as the movement towards the light source and negative phototoxis as the movement away from the light source. The proportion of animals in the two compartments closest to the light source was regarded as showing positive phototaxis; and those in the two compartments farthest from the light source as negatively phototactic. The mean ($\bar{x}$) percentage positive and negative response and their standard errors (SE) were calculated at each light intensity. For statistical analysis, the percentages were first arc sine transformed. Chi square tests and analysis of variance were performed on the results as outlined by Sokal and Rohlf (1981). The level of significance was set at $p \leq 0.05$ for all tests except where otherwise mentioned.

RESULTS

In order to determine whether the duration of incubation of eggs in the culture cabinets had any significant effect on phototaxis, the response patterns of light-adapted Stage I larvae hatched from untreated embryos was compared (Fig. 1). Qualitatively the patterns are almost identical, irrespective of the embryonic stage at which laboratory incubation started. As a result of this similarity in response patterns, all the data for sea water control Stage I larvae were pooled and the values presented in Figure 2 as the "main control." Positive and negative phototaxis are significantly affected by light intensity ($p \leq 0.001$, ANOVA). Larvae exhibit strong positive phototaxis at high stimulation intensities (3.0×10^{-1} and 10^{-2} W/m^2) and became strongly photonegative at lower stimulation intensities (between 10^{-3} and 10^{-5} W/m^2; Fig. 2). The various periods of incubation of the embryos in both sea water and Dimilin®, starting at different stages of embryogenesis, is shown in Figure 3. It is evident from this graph that the incubation period in sea water following Dimilin® treatment decreases as the embryonic stage at pre-exposure progresses, but the total exposure time to clean sea water was the same for each treatment.

Effect of exposure to Dimilin®

The intensity versus response curves (both positive and negative phototaxis) for Stage I larvae hatched from embryos at the M2 stage (commencement of body segmentation) treated with different concentrations of Dimilin® are shown in Figure 4. There is a distinct difference in the pattern of phototaxis for larvae from treated and untreated embryos. While untreated larvae are positively phototactic at high light intensities (10^{-1} and 10^{-2} W/m^2), Dimilin®-treated larvae become increasingly photonegative at these inten-

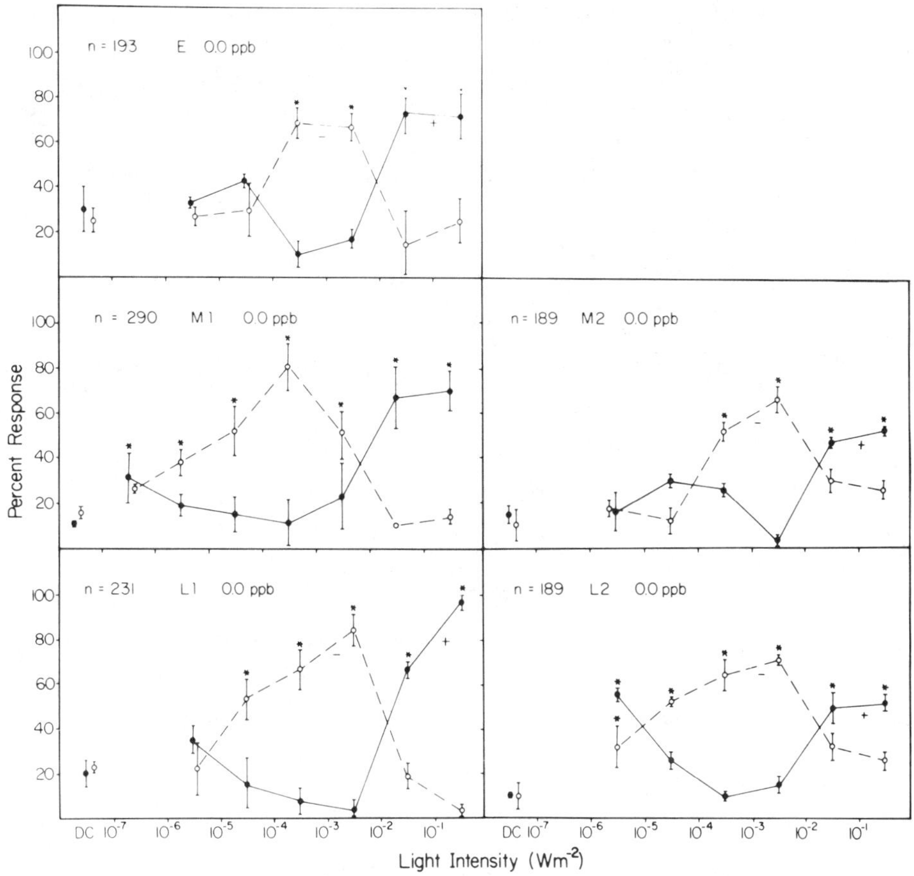

Fig. 1. Percent response (ordinate) of positive phototaxis (closed circles, solid lines) and negative phototaxis (open circles, dashed lines) upon stimulation with intensities of 500 nm light (abscissa) by Stage I larvae hatched from untreated embryos. Embryos were incubated in $20^{o}/_{oo}$ filtered sea water starting at different stages of development, Viz : E, blastoderm stage; M1, gastrula stage; M2, commencement of body segmentation; L1, embryonic eye stage; and L2, organogeny and body segmentation completed. Dark control (D C) represents response due to random swimming in darkness. Mean values and their respective standard errors are shown. Total sample sizes (n) are given for each set of experiments. Asterisk indicates response level greater than the appropriate dark control at $p \leq 0.05$.

sities as the exposure concentration is increased. For 0.3 ppb, increased negative phototaxis is only observed at 10^{-2} W/m^2; however, at concentrations of 0.5 and 1.0 ppb, photonegativity is observed at both 10^{-2} and 10^{-1} W/m^2. The animals become increasingly photonegative, and as a result the positive response is no longer evident. Larvae from 2.5 and 5.0 ppb treated embryos exhibit almost negligible photoresponse, both positive and negative phototaxis being severely suppressed.

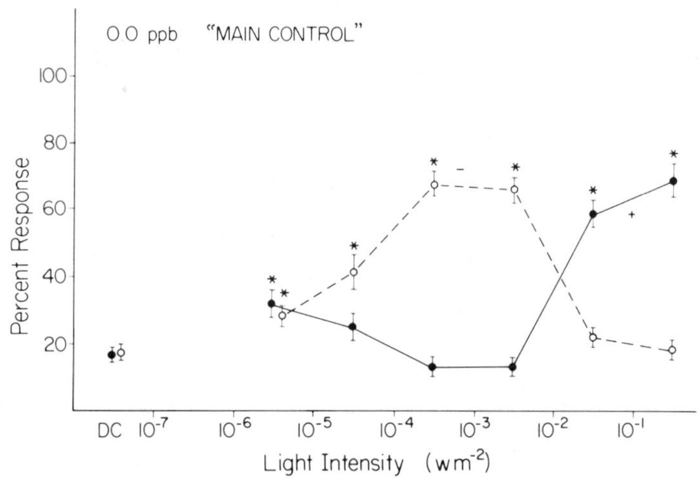

Fig. 2. The percentage phototactic response of light-adapted Stage I larvae of *P. pugio* as a function of light intensity at a wavelength of 500 nm. Data points represent the means ± SE for positive phototaxis (closed circles, solid lines) and negative phototaxis (open circle, dashed lines). D C indicates dark control. This graph is designated "main control" and represents the mean percent response of all larvae hatched from untreated embryos, irrespective of the embryonic stage at which incubation in culture cabinet started. Average sample size at each light intensity = 161. Asterisks indicate values that are significantly different ($p \leq 0.05$) from the dark controls.

The dose-dependent effect of embryonic exposure to Dimilin® on the photoresponse of the larvae is clearly seen when the percentage negative response at 10^{-3} and positive response at 10^{-1} W/m^2 are plotted against exposure concentration (Fig. 5). These light intensities were chosen because they always evoked maximum negative and positive responses in untreated larvae (Figs. 1 & 2). For larvae hatched from embryos treated at stage M2, there is a steady rise (significant at $p \leqq 0.05$, ANOVA then sum of squares simultaneous test procedure) in negative phototaxis from 50% at 0.0 ppb to a peak (90%) at 0.5 ppb; after which the response declines to about 75% at 1.0 ppb. At 2.5 and 5.0 ppb there is essentially no photoresponse since the percent response is less than that observed during random swimming in the dark (Fig. 5). Positive phototaxis follows a different pattern, in that there is a non-significant increase ($p = 0.05$ sum of squares simultaneous test procedure) from about 50% at 0.0 ppb to 60% at 0.3 ppb; thereafter the response decreased significantly to less than 20% at concentrations $<$ 0.5 ppb.

Effect of Dimilin® on Different Embryonic Stages

The intensity versus response curves for larvae from embryos that were exposed to 0.5 ppb Dimilin® (starting at different embryonic stages) are shown in Figure 6. This concentration was used because it has a pronounced effect upon phototaxis. With the exception of Stage LI, the general shape of the curves are similar. All show a broader-than-normal (with reference to untreated larvae; Fig. 1) range of stimulation intensities that evoke maximum negative response. However, the details of the curves vary somewhat with the embryonic stage at exposure. For embryos less than 6-h old, both positive and negative phototaxis were only slightly affected at an intensity of 10^{-1} W/m^2. At 10^{-2} W/m^2 there is a loss of positive phototaxis and enhancement of negative phototaxis. With stages E, M1, and M2 embryos, there is a reversal in the sign of phototaxis from positive to negative at stimulation intensities 10^{-1} and 10^{-2} W/m^2.

The extent to which positive phototaxis is suppressed and negative phototaxis is enhanced appears to vary with the embryonic stage at exposure (Fig. 7). If there is no effect due to embryonic stage at exposure, the data points should all fall within a narrow range (as those for the respective untreated larvae) between 50 and 80%. But a distinct pattern emerges, where the photoresponses are most severely affected at stage M2 (Fig. 7). Except for 6 h embryos, there was a significant ($p<0.001$) decrease in positive phototaxis in treated larvae compared to their respective controls. Positive phototaxis decreased steadily from 6 h embryos to M2, which exhibited the most significant ($p \leqq 0.05$) decrease when treated embryos are compared with each other. With negative phototaxis, the trend is more variable. Nevertheless except for 6 h and L1, negative phototaxis was higher in treated embryos than in the untreated larvae. The increase in photonegativity at M1 and M2 was significant ($p \leq 0.05$), but at E it was not. At Stage L1 a decrease in negative phototaxis is observed.

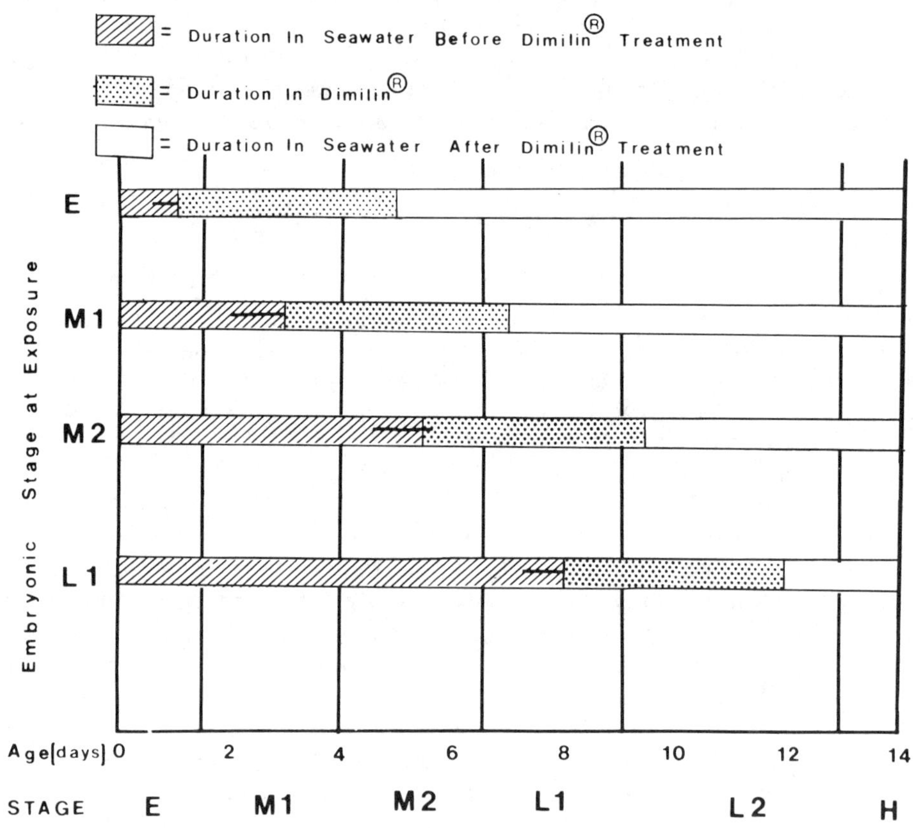

Fig. 3. Palaemonetes pugio. Incubation periods of embryos in sea water and Dimilin®. Striped = duration in sea water prior to Dimilin® treatment; stippled = duration in Dimilin®, unshaded = duration in sea water after Dimilin® treatment. E, M1, M2, and L1 are the different embryonic stages used; abbreviations are as for Figure 1. H = hatch (time of larval release).

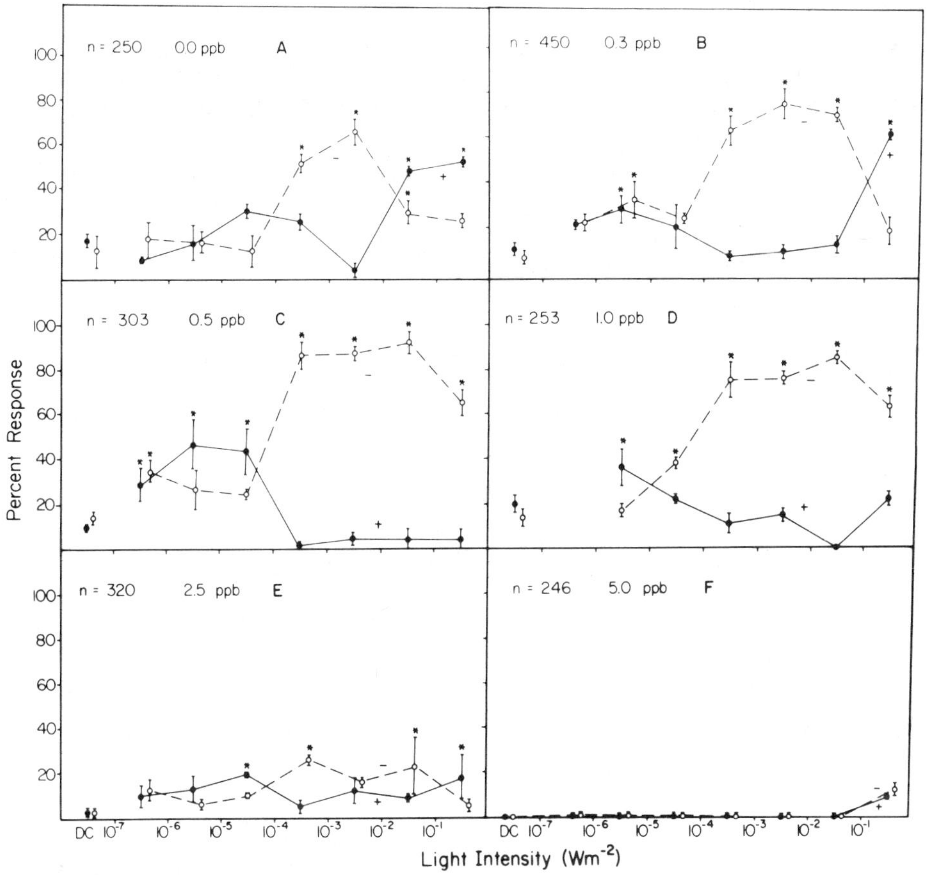

Fig. 4. Intensity versus response curves for positive phototaxis (closed circle, solid lines) and negative phototaxis (open circles, dashed lines) for light-adapted Stage I larvae hatched from embryos treated with various concentrations of Dimilin® for 4 d (A = 0.0 ppb [untreated], B = 0.3 ppb; C = 0.5 ppb; D = 1.0 ppb; E = 2.5 ppb; F = 5.0 ppb). All the embryos were in stage M2 when exposure to Dimilin® started. Data points are means ± SE. Dark control is represented by D C. Total sample size (n) for each experiment is given. Asterisk indicates a response level significantly ($p \leqq 0.05$) greater than the appropriate dark control.

DISCUSSION

Both phototaxis and swimming speeds have been used as indicators of sublethal effects of various pollutants; e.g., effects of oil on *Cancer irroratus* larvae (Bigford, 1977);

effects of insect juvenile hormone mimic (MON-0585) on Rhithropanopeus harrisii larvae (Forward & Costlow, 1976); effects of three insect growth regulators (including Dimilin®) on R. harrisii larvae (Forward & Costlow, 1978); the effects of copper and cadmium on the behavior and development of barnacle larvae (Lang et al., 1981); and effects of the herbicide Weeder-64 on adult and larval Palaemonetes pugio (Moyer & Barthalmus, 1979).

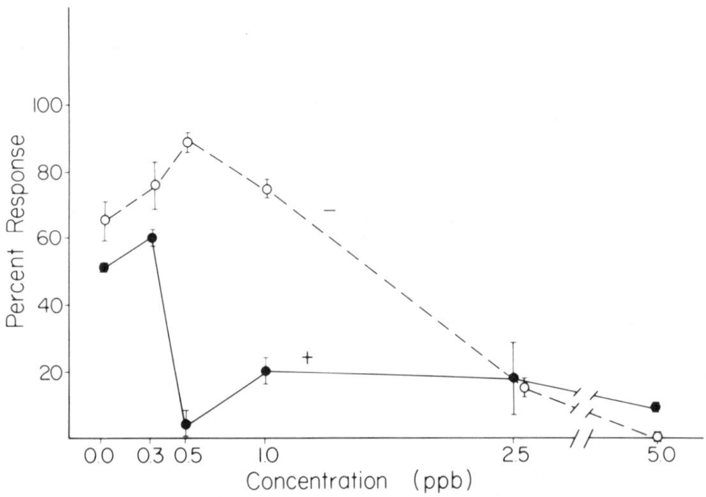

Fig. 5. Level of response at two light intensities which evoke greatest negative phototaxis (10^{-3} W/m^2) and positive phototaxis (10^{-1} W/m^2) in untreated light-adapted Stage I larvae. Graphs show variations in response level (percent) with Dimilin® concentration (ppb). Experiments were conducted with embryos in stage M2 (with body segmentation in progress). Open circles, dashed lines are for negative phototaxis and closed circles, solid lines for positive phototaxis. Data points are means ± SE.

In the present study, the effects of Dimilin® upon phototaxis of larvae of P. pugio were studied. In larvae hatched from untreated embryos, the phototactic pattern was positive response at high stimulation intensities (10^{-1} - 10^{-2} W/m^2) and negative phototaxis at lower intensities (10^{-3} - 10^{-5} W/m^2) (Fig. 1). The response maxima for negative and positive phototaxis were between 10^{-3} - 10^{-4} and 10^{-2} - 10^{-1} W/m^2, respectively. This phototactic pattern is changed upon Dimilin® treatment. The most obvious effect upon exposure of the M2 stage is a depression in positive phototaxis with a concomitant increase in negative phototaxis (Fig. 4). At 0.3 ppb, there is a change in sign of phototaxis from positive to negative at only 10^{-2} W/m^2 stimulation intensity. At 0.5 and 1.0 ppb, photoreversal (positive to negative) is observed at both 10^{-2} and 10^{-1} W/m^2. However, at 2.5 ppb both positive and negative phototaxis are suppressed (response maxima<25%), despite the fact that photoreversal is present at 10^{-2} W/m^2. At 5.0 ppb, there is practically no photoresponse (Fig. 4). It should be noted that larvae hatched from embryos treated with 2.5 and 5.0 ppb Dimilin® always show severe morphological abnormalities which resulted in impaired swimming (unpubl. data).

The depression in positive phototaxis and elevation in negative phototaxis are most pronounced at 0.5 ppb (Fig. 5). The general trend observed in this curve is for negative phototaxis to increase gradually with increase in concentration from 50% at 0.0 ppb to 90% at 0.5 ppb. Thereafter the response decreases steadily to 25% at 2.5 ppb and then to 0.3% at 5.0 ppb. Trends opposite to that outlined above were observed for positive phototaxis, i.e., a decrease in response with increase in concentration from 0.0 ppb up to 0.5 ppb where a sharp inflection is seen followed by a slight rise at 2.5 ppb then decreasing to 10% at 5.0 ppb. The minimum test concentration which altered the measured responses was 0.3 ppb.

These observations, especially for negative phototaxis, together with those of Forward and Costlow (1976, 1978) and Lang et al. (1980, 1981), clearly show that toxicant effects on crustacean behavior follow a predictable pattern. There is usually an elevation in behavior at the lowest concentration causing an effect, followed by suppression at the highest concentration.

Apart from the work of Lang et al. (1980), there has not been any report of a reversal in phototactic sign in crustacean larvae due to toxicant exposure. As in the present study, Lang et al. (1980) observed that 150 ppb copper caused reversal of phototactic sign from positive to negative in barnacle nauplii. At lower concentrations only a decrease in photopositivity was observed. Contrary to these findings, the present study shows that at the high toxicant concentration there is only a decrease in the responses (both positive and negative phototaxis); but at lower concentration there is reversal in phototactic sign. The observed differences between the two studies may be the result of using different toxicants and experimental animals. Also, in the present study the animals were exposed during embryonic development; while in all of the above-mentioned reports, the larvae were directly exposed to the toxicants. It is interesting to note, however, that Forward and Costlow (1979) observed no photoreversal in Stage I, II, and III zoeae of *R. harrisii* after a 24-h direct exposure to sublethal concentrations of Dimilin®. There are two possible reasons for their failure to observe a change in the sign of phototaxis using the same pesticide: (1) they did not monitor the photoresponse of treated larvae over a wide range of stimulation intensities as was done in this study, and (2) they exposed larvae which may be less sensitive than the embryos.

The alterations in photoresponses also varied with the embryonic stage at which exposure to Dimilin® commenced (Figs. 6 & 7). Stage M2 embryos appear to be the most vulnerable to Dimilin®, since photoreversal and suppression of positive phototaxis were most pronounced at this stage. Embryos exposed at an early stage of development perhaps had longer periods (ca. 10 d) in clean sea water (Fig. 3) for depuration prior to hatching. Therefore, the effects may have been reversed. Embryos treated late in their development (Stage L1) may have developed past the "stage of greatest vulnerability." Whether the observed effects were due to depuration time or stage susceptibility is not clear.

The probable ecological consequences of the observed alterations in phototaxis reported here have not yet been investigated. Nevertheless, from the available data, certain inferences can be made. Both phototaxis (and swimming speed) play vital roles in the maintenance of proper depth in the water column by crustacean and other planktonic larvae (Sulkin, 1975, 1984; Forward, 1976a). Appropriate water column position can be crucial for larval retention in an estuary (Miller, 1980) and also for their horizontal migrations (Sandifer, 1975). Phototaxis may also be involved in predator avoidance (Forward, 1976b; Forward & Cronin, 1978) and avoidance of adverse salinity (Latz & Forward, 1977). Farr (1977) has demonstrated that for adult grass shrimp exposed to sublethal doses of Parathion, their antipredatory behavior was affected because of increased hyperactivity and decreased swimming stamina. It is to be expected therefore, that any significant disruption of normal phototactic responses of the larvae will reduce their survival potential. For example, in the present study where there is a change in sign from positive to negative phototaxis at high light-intensities, larvae that should normally be found at or close to the surface during the day may be distributed close to the bottom of the water column. This may result in the transportation of these larvae into areas not suitable for their development, metamorphosis, and settlement.

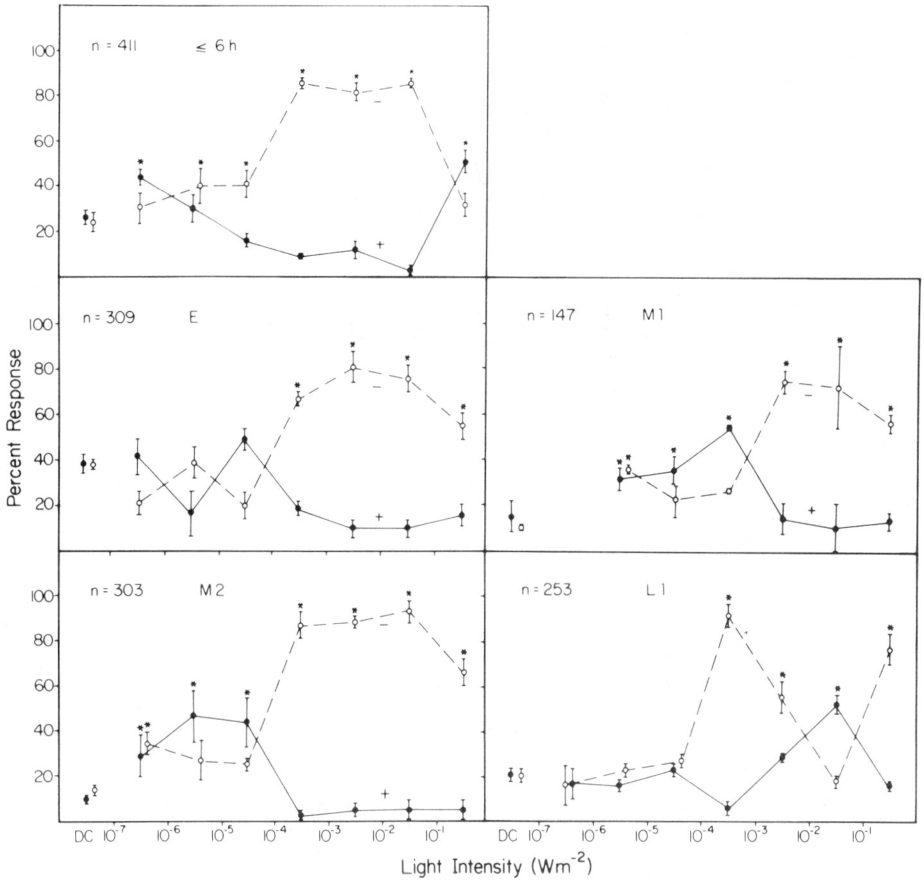

Fig. 6. Intensity versus response curves for all light-adapted Stage I larvae hatched from embryos that were exposed to 0.5 ppb Dimilin® for 4 d at different stages of embryonic development (≦ 6 h = cleavage in progress; E = blastoderm stage; M1 = gastrula stage; M2 = commencement of body segmentation; L1 = embryonic eye stage). Open circles, dashed lines represent negative phototaxis and closed circles, solid lines represent positive phototaxis. Data points are means ± SE. The sample size (n) at each stage is given. Asterisk indicates a response level greater than the appropriate dark controls at $p \leqq 0.05$.

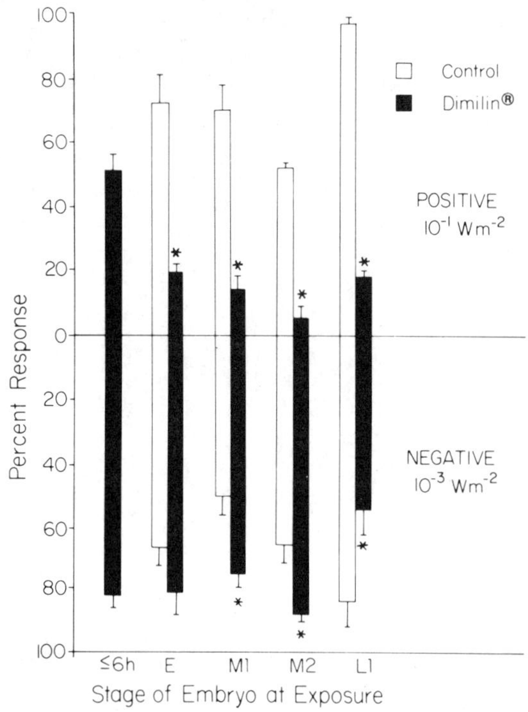

Fig. 7. Mean (± SE) percent response of light-adapted Stage I larvae hatched embryos exposed to 0.5 ppb Dimilin® starting at different stages of embryonic development. Stimulus light intensities which evoke maximal phototactic response in untreated larvae were used. These light intensities were 10^{-3} W/m^2 for negative response and 10^{-1} W/m^2 for positive response. Symbols are the same as for Figure 1. Vertical lines represent the standard error of the mean. Asterisks denote values significantly different from controls at $p \leqq 0.05$.

It should be noted that the present study was carried out at optimal salinity and temperature for the species tested. In the estuary where these conditions fluctuate daily, it is possible that the observed effects may be further amplified. For example, Vernberg *et al.* (1973) reported a decrease in positive phototaxis in *Uca pugilator* larvae exposed to suboptimal temperatures for 24 h, and the addition of 1.8 ppb mercury decreased positive phototaxis even further (by 20%). Similarly, Latz and Forward (1977) observed that when *R. harrisii* larvae were exposed to low salinity sea water there was a suppression of positive phototaxis and an increase in photonegativity. A similar response was observed by Lyon (1906) for *Palaemonetes* larvae.

In some of the above reports, alterations in phototaxis due to temperature and salinity changes were only temporary, with the larvae regaining their positive response in less than 10 min after a salinity decrease (Lyon, 1906; Latz & Forward, 1977). Such rapid reestablishment of normal phototactic responses were not observed in the present study. However it is possible that *P. pugio* may regain normal photoresponsiveness if reared in clean water to later larval stages.

In conclusion, both the level of the photoresponse and the sign of phototaxis are altered in light-adapted Stage I larvae after 4-d exposure of the embryos to sublethal concentrations of Dimilin®. These alterations in photoresponses varied with the concentration of Dimilin® and the embryonic stage at exposure. The present findings confirm those of earlier workers that photobehavior is a sensitive indicator of sublethal toxicity in invertebrates. Work is in progress to study the possible alterations of spectral sensitivity in larvae due to embryonic exposure to Dimilin®. These data should help us to obtain a better understanding of how embryonic exposure to Dimilin® and other insect growth regulators affects the photophysiology of larval crustaceans.

ACKNOWLEDGMENTS

This work was funded by a Marine Biomedical Center Grant No. E S 0 908 from the National Institute of Environmental Health Sciences and a Department of Energy Grant No. D E AS05-76E04377 to Dr. J.D. Costlow. In addition J. Wilson was supported by an AFGRAD Fellowship from the African American Institute. We gratefully acknowledge the technical assistance of M. Forward, S. Taylor, and A. Wilson. Also we thank the Thompson-Harward Chemical Company, Kansas City, Kansas, for providing the Dimilin®.

LITERATURE CITED

Bigford, T.E. 1977. Effects of oil on behavioral responses to light and gravity in larvae of the rock crab *Cancer irroratus*. Mar. Biol. 45: 137-148.

Christiansen, M.E. and J.D. Costlow, Jr. 1982. Ultrastructural study of the exoskeleton of the estuarine crab

Rhithropanopeus harrisii: Effect of the insect growth regulator Dimilin (Diflubenzuron) on the formation of the larval cuticle. Mar. Biol. 66: 217-266.

Christiansen, M.E., J.D. Costlow, Jr., and R.J. Monroe. 1977. Effects of the juvenile hormone mimic ZR-515 (Altosid) on larval development of the mub-crab *Rhithropanopeus harrisii* in various salinities and cyclic temperatures. Mar. Biol. 39: 267-279.

Christiansen, M.E., J.D. Costlow, Jr., and R.J. Monroe. 1978. Effects of the insect growth regulator Dimilin (T H 6040) on larval development of two estuarine crabs. Mar. Biol. 50: 29-36.

Costlow, J.D. 1979. Effect of Dimilin on development of larvae of the stone crab *Menippe mercenaria*, and the blue crab, *Callinectes sapidus*. pp. 355-363. *In*: Marine Pollution: Functional Responses. W.B. Vernberg, F.P. Thurberg, A. Calabrese, and F.J. Vernberg (eds.). Academic Press, New York.

Cronin, T.W. and R.B. Forward, Jr. 1980. The effects of starvation on phototaxis and swimming of the larvae of the *Rhithropanopeus harrisii*. Biol. Bull. 158: 283-294.

Cunningham, P.A. 1976. Effects of Dimilin (T H 6040) on reproduction of the brine shrimp, *Artemia salina*. Environ. Entomol. 5: 701-706.

Farr, J.A. 1977. Impairment of antipredator behavior in *Palaemonetes pugio* by exposure to sublethal doses of parathion. Trans. Amer. Fish. Soc. 106: 287-290.

Forward, R.B., Jr. 1976a. Light and diurnal vertical migration: Photobehavior and photophysiology of plankton. pp. 157-209. *In*: Photochemical and Photobiological Reviews. Vol *1*. K.C. Smith (ed.). Plenum Publishing Corporation, New York.

Forward, R.B., Jr. 1976b. A shadow response in a larval crustacean. Biol. Bull. 151: 126-140.

Forward, R.B., Jr. and J.D. Costlow, Jr. 1976. Crustacean larval behavior as an indicator of sublethal effects of an insect juvenile hormone mimic. pp. 279-289. *In*: Estuarine Processes, Vol. 1. M. Wiley (ed.). Academic Press Inc., New York.

Forward, R.B., Jr. and J.D. Costlow, Jr. 1978. Sublethal effects of insect growth regulators upon crab larval behavior. Water, Air and Soil Pollution 9: 238-277.

Forward, R.B., Jr. and T.W. Cronin. 1978. Crustacean larval phototaxis: Possible functional significance. pp. 253-261. *In*: Proc. 12th European Symposium in Marine

Biology. D.S. McLusky and A.J. Berry (eds.). Pergamon Press, New York.

Lang, W.H., R.B. Forward, Jr., D.C. Miller, and M. Marcy. 1980. Acute toxicity and sublethal behavioral effects of copper on barnacle nauplii (Balanus improvisus). Mar. Biol. 58: 139-145.

Lang, W.H., D.C. Miller, P.J. Ritacco, and M. Marcy. 1981. The effects of copper and cadmium on the behavoir and development of barnacle larvae. pp. 165-203. In: Biological Monitoring of Marine Pollutants. F.J. Vernberg, A. Calabrese, F.P. Thurberg, and W.B. Vernberg (eds.). Academic Press, New York.

Latz, M.I. and R.B. Forward, Jr. 1977. The effects of salinity upon phototaxis and geotaxis in larval crustacean. Biol. Bull. 153: 163-179.

Lyon, E.P. 1906. Note on the heliotropism of Palaemonetes larvae. Biol. Bull. 12: 23-25.

Miller, D.C. 1980. Some applications of locomotion response in pollution effects monitoring. Rapp. P. -v. Reun. Cons. Int. Explor. Mer 197: 154-161.

Moyer, J.C. and T. Barthalmus. 1979. Phototactic behavior: An index for subacute effects of the herbicide 2,4-dichlorophenoxyacetic acid in estuarine grass shrimp. Neurotoxicology 1: 105-123.

Mulder, R. and M.J. Gijswijt. 1973. The laboratory evaluation of two promising new insecticides which interfere with cuticle deposition. Pesticide Sci. 4: 737-745.

Nimmo, D.R., T.L. Hamaker, J.C. Moore, and R.A. Wood. 1980. Acute and chronic effects of Dimilin® on survival and reproduction of Mysidopsis bahia. pp. 366-376. In: Aquatic Toxicology. J.G. Eaton, P.R. Parrish, and A.C. Hendricks (eds.). American Society for Testing and Materials, Philadelphia.

Nimmo, D.L., T.L. Hamaker, J.C. Moore, and A.C. Summers. 1979. Effect of diflubenzuron on an estuarine crustacean. Bull. Envir. Contam. Toxicol. 22: 767-770.

Post, L.C., B.J. De Jong, and W.R. Vincent. 1974. 1-(2,6 disubstituted benzoyl)-3-phenylurea insecticides: Inhibition of chiton synthesis. Pest. Biochem. Physiol. 4: 473-490.

Sandifer, P.A. 1975. The role of pelagic larvae in recruitment to populations of adult decapod crustaceans in the York River estuary and adjacent lower Chesapeake Bay. Est. Coastal Mar. Sci. 3: 269-279.

Sokal, R.R. and F.J. Rohlf. 1981. Biometry. 2nd Edition. Freeman and Company, San Francisco. 859 pp.

Sulkin, S.D. 1975. The influence of light in the depth regulation of crab larvae. Biol. Bull. 148: 333-343.

Sulkin, S.D. 1984. Behavioral basis of depth regulation in the larvae of brachyuran crabs. Mar. Ecol. Prog. Ser. 15: 181-205.

Tester, P.A. and J.D. Costlow, Jr. 1981. Effect of insect growth regulator Dimilin® (T H 6040) on fecundity and egg viability of the marine copepod *Acartia tonsa*. Mar. Ecol. Prog. Ser. 5: 297-302.

Vernberg, W.B., P.J. DeCoursey, and W.J. Padgett. 1973. Synergistic effects of environmental variables on larvae of *Uca pugilator*. Mar. Biol. 22: 307-312.

White, G.M. 1942. Reactions of the larvae of the shrimp, *Palaemonetes vulgarus* and the squid, *Loligo pealii*, to monochromatic light. Biol. Bull. 47: 265-273.

ASSOCIATIONS BETWEEN PHYSIOLOGICAL ALTERATIONS AND POPULATION CHANGES IN AN ESTUARINE MYSID DURING CHRONIC EXPOSURE TO A PESTICIDE

Charles L. McKenney, Jr.

SUMMARY

A number of vital life processes of an estuarine mysid (*Mysidopsis bahia*) were examined throughout its life cycle during exposure to the thiocarbamate herbicide, thiobencarb. Initial exposure of juvenile mysids to thiobencarb resulted in elevated respiration rates. Concentrations of thiobencarb, that produced significant reductions in population survival through a complete life cycle (approximately 24 d), significantly stimulated respiration rates of juveniles after only 4-d exposure. Increased metabolic demands with sublethal thiobencarb exposure reduced the amount of assimilated energy available for production of new tissue by juvenile mysids, resulting in retarded juvenile growth rates. Modifications in the energy metabolism of individual mysids exposed to thiobencarb were related to decreases in total young production of discrete populations. Greater usage of proteinaceous substrates for energy metabolism during the maturation of *M. bahia* was altered by exposure to high sublethal concentrations of thiobencarb. Higher O:N ratios during the maturation of these thiobencarb-exposed mysids suggest a greater reliance on the more energy-rich lipid substrates to support the elevated rates of oxidative metabolism, resulting in less lipid material being available for gamete production.

INTRODUCTION

To adequately assess the biotic hazard of contaminants entering the marine environment, suitable criteria must be determined for detecting the ecological damage resulting from long-term exposure to low concentrations of contaminants. While it is generally recognized that lower levels of biological organization (cellular and organismal) respond to environmental stress long before higher levels (community and ecosystem), the ecological significance of biotic responses at lower levels may be difficult to interpret (Waldichuk, 1979; Bayne *et al*., 1980; Kinne, 1980; Sastry & Miller, 1981). However, for a number of years ecologists have examined the productive processes of communities and ecosystems by studying the underlying physiological patterns of response of a species throughout its life cycle to environmental variables (Mann, 1969; Winberg, 1971; Steele, 1973, 1974; Grodzinski *et al*., 1975). Application of these concepts of ecological energetics to the field of aquatic toxicology may provide valuable "tools" for use both in the laboratory as short-term predictors of chronic effects and in the field for monitoring long-term ecological effects of reduced water quality.

In the last decade, a mysid shrimp (*Mysidopsis bahia*) has been shown to be very sensitive to a variety of toxic substances when compared with other marine species (for a review see Nimmo & Hamaker, 1982). The responses of this zooplankter to chronic exposure to microcontaminants (including increased mortality in the population, retarded growth, and diminished reproductive success) have direct effects on the population production of the species, which serves as an important link in the marine food chain between primary producers and fish. Furthermore, a recent study (McKenney, 1982) demonstrated the interrelationships between energy metabolism, growth dynamics, and reproduction of *M. bahia* exposed through an entire life cycle to the chlorinated organic pesticide, endrin. Increased metabolic demands for maintenance during endrin exposure reduced the growth efficiency of juveniles, resulting in retarded somatic production rates. At the same sublethal concentrations of endrin, increased lipid catabolism of maturing juveniles led to reduced reproductive capacity.

The objective of this study was to gain further understanding of the ecological consequences of physiological effects of toxic organics on a sensitive estuarine zooplankton species. Dose-response relationships of individual physiological functions were examined for various life stages of *M. bahia* exposed throughout its life cycle to the organic pesticide, thiobencarb, and these physiological responses were compared with production processes of the population. Individual physiological responses were combined into indices of bioenergetic and metabolic significance for the various life stages and their dose-response relationships evaluated on the basis of their ecological significance.

MATERIALS AND METHODS

Mysidopsis bahia was exposed to thiobencarb (S-[4-chlorophenyl] methyl diethylcarbamothionate) through an entire life cycle using a proportional diluter as described by Schoor and McKenney (1983). Every 3 min the diluter produced one liter of filtered sea water for each of six thiobencarb exposure concentrations and a carrier control, each containing an equal portion of triethylene glycol. Each liter was delivered to a glass aquarium divided into three replicate chambers. Sea water in the exposure system was maintained at 20 ± 2 $^{o}/_{oo}$ salinity and 25 ± 1°C.

Throughout the course of the study, thiobencarb concentrations for each exposure condition were determined weekly by gas chromatography in accordance with procedures described by Schimmel *et al.* (1983). The lower limit of detection was 0.50 µg/L, and recovery from fortified samples was 96%. Mean measured concentrations of thiobencarb and their standard deviations for all exposure conditions were <0.50 (carrier control), 22 ± 6, 35 ± 10, 76 ± 35, 181 ± 56, 345 ± 140, and 734 ± 48 µg/L.

Newly released juvenile *M. bahia* (<24 h old) were obtained from ovigerous females maintained in static, recirculating cultures at the same salinity/ temperature conditions as the exposure system. For the survival and reproductive portions of the study, 15 juveniles were pipetted into Nitex® baskets in each of the triplicate chambers. Additional groups of mysids were placed in separate baskets at each exposure concentration and subsampled for metabolic and growth measurements. Throughout the study, all groups of mysids were fed an abundance of freshly hatched *Artemia* nauplii daily.

Survival observations were made at each exposure concentration on four life stages: early juveniles (Day 1), juveniles (Day 4), advanced juveniles (Day 10), and young adults (Day 16). In addition, ovigerous females were isolated in pairs with mature males within Nitex® brood cups. Daily records were maintained for the day of release of the first brood and number of young released in the first brood by isolated females.

On days representative of the various mysid life stages, respiration rates and ammonia excretion rates were determined for seven individuals of each life stage per exposure condition in accordance with methods described by McKenney (1982). To reduce variability in these physiological responses due to diurnal rhythms, respiration and excretion measurements were taken for the various life stages at the same time of the day. Even though mysids were last fed *Artemia* approximately 18 h prior to these metabolic measurements, food remained available to them until their isolation period in the respirometers. O:N ratios were calculated from the atoms of oxygen consumed to atoms of nitrogen excreted. In addition, dry weights of individual mysids from each life stage and thiobencarb concentration were converted into values of daily weight gain (μg dry weight/day), and the caloric equivalents of these growth rates and the respiration rates were used to produce bioenergetic terms of production (P) and daily maintenance costs (R) as described by McKenney (1982). These values were used to derive net growth efficiency values (K_2) (Winberg, 1971) for each mysid stage under each thiobencarb exposure condition by the formula:

$$K_2 = \frac{P}{(P+R)} \qquad (1)$$

Differences in biological responses between thiobencarb concentrations were analyzed by analysis of variance, using Dunnett's procedure for multiple comparisons of treatment means to the control mean (Zar, 1974). Relationships between biological responses and thiobencarb exposure concentrations were examined by linear regression analysis, while more complex relationships between biological responses and thiobencarb exposure concentration for the various mysid life stages were examined by multiple regression analysis (McNeil *et al.*, 1975; Box *et al.*, 1978).

The most complex model was a two-factor, second-order polynomial equation, in which the entire data set for a

---

®Trademark; Tobler, Ernst, and Trabor, Inc., Murray St., New York, NY. Mention of commercial products does not imply endorsement by the U.S. Environmental Protection Agency.

particular biological response was regressed on the quadratic function of day in life cycle and thiobencarb concentration:

$$\hat{Y} = b^0x^0 + b^1\text{CONC} + b^{11}\text{CONC}^2 + b^2\text{DAY} + b^{22}\text{DAY}^2 + b_{12}(\text{CONC} \times \text{DAY}) \quad (2)$$

where $\hat{Y}$ = the estimated biological response; x^0 = the intercept; CONC and DAY = the linear effect of thiobencarb concentration and day in life cycle, respectively; CONC^2 and DAY^2 = the quadratic effect of thiobencarb concentration and day in life cycle, respectively; CONC x DAY = the linear by linear interaction between thiobencarb concentration and day in life cycle; and b_0, b_1, etc. = the regression coefficients (Meyers, 1971, Alderdice, 1972). If this model best explained the data, response surface curves were generated at a range of levels of these two factors by the General Linear Model (GLM) and PLOT procedures available in the SAS statistical package (Barr *et al*., 1976).

RESULTS

Exposure to a range of thiobencarb concentrations from 35 to 735 µg/L produced differential survival rates in populations of *Mysidopsis bahia* through an entire life cycle. Four days of exposure to concentrations of 345 and 735 µg thiobencarb/L resulted in significant ($p<0.05$) mortality (23 and 100%, respectively) in the mysid populations. No other concentrations produced significant mortality with continued exposure through the juvenile phase of the life cycle (approximately 2 wk). Survival of *M. bahia* exposed through an entire life cycle (24 d) was linearly related to exposure concentration of thiobencarb (Fig. 1A) and measured thiobencarb concentrations above 76 µg/L resulted in significant reduction in survival.

Production of young in the first brood by isolated females was linearly reduced by thiobencarb in such a manner that no young were produced in 345 µg thiobencarb/L (Fig. 1B). Total young production in the first brood of the population, however, was related to thiobencarb concentration in a curvilinear fashion (Fig. 1C). Significantly fewer young were produced by populations exposed to thiobencarb concentrations of 35 µg/L and higher. This inhibition in reproductive success of the population was maximum at concentrations higher than 181 µg thiobencarb/L.

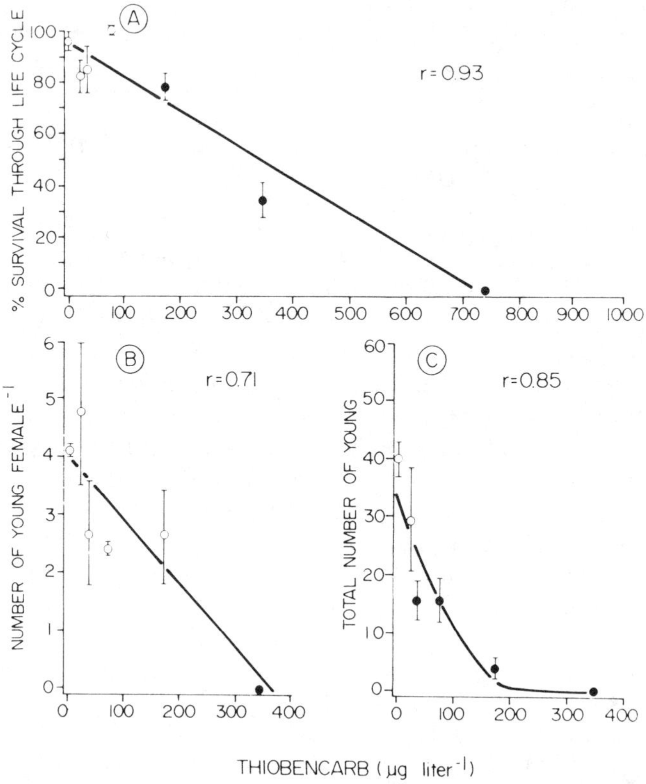

Fig. 1. (A) Survival percentage, (B) young production per female, and (C) total production of young by a Mysidopsis bahia population as a function of thiobencarb exposure through a complete life cycle. Values represent mean ± SE. Darkened circles are significantly different from control at 0.05 probability level.

Since a two-way analysis of variance demonstrated that mysid respiration was significantly ($p = 0.0001$) influenced by an interaction between life stage and thiobencarb concentration, respiration rates for individual life stages were regressed onto thiobencarb concentration (Fig. 2). As early as 24 h after exposure, mysid respiration was significantly ($p<0.05$) elevated by progressively higher concentrations of thiobencarb. As shown by the changing slopes and correlation coefficients (r-values) of the regressions, the linearity of this response increased with continued exposure through the juvenile phase of the life cycle. As juveniles matured and weight-specific respiration rate decreased, a quadratic regression curving downward at the lower concentrations best explained the respiration response of adult mysids to thiobencarb exposure.

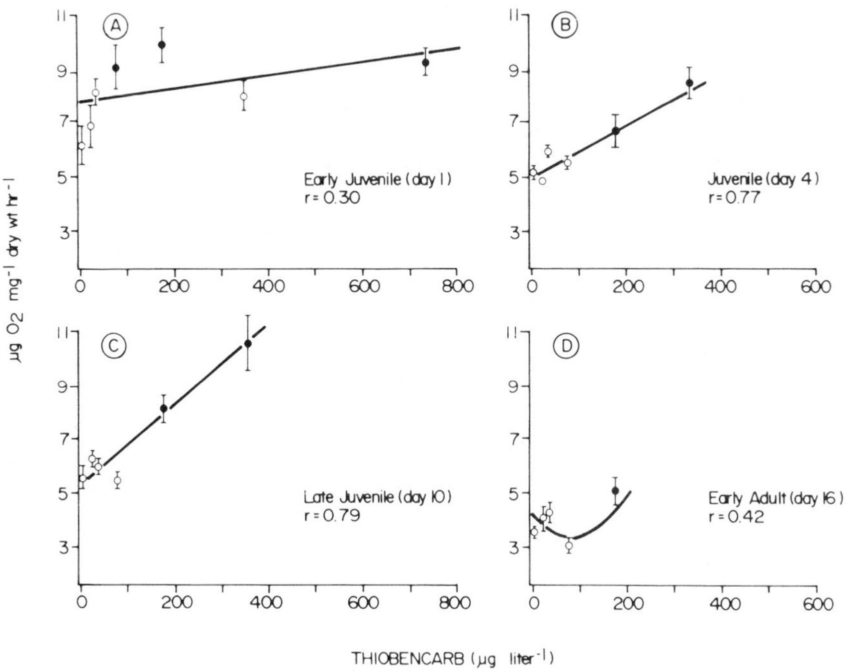

Fig. 2. Respiration rates of various life stages of Mysidopsis bahia as a function of continuous thiobencarb exposure through a complete life cycle. Values represent means ± SE. Darkened circles are significantly different from control at 0.05 probability level.

To best illustrate the influence of thiobencarb on respiration of M. bahia throughout its life cycle, the entire data set for respiration was regressed on the quadratic function of day in life cycle and thiobencarb concentration as described in Equation 2 ($r = 0.73$). The coefficients of this function (Table 1) were used to develop the response surface shown in Figure 3. The response curves depict the dominant influence of the linear interaction between day in life cycle and thiobencarb concentration and the significant main effect of each factor acting individually on mysid respiration. Even though weight-specific respiration rates of juvenile mysids decreased as they developed, the stimulation of mysid respiration by thiobencarb exposure became exaggerated with prolonged exposure.

Table 1. Coefficients for models describing the physiological responses of *Mysidopsis bahia* to thiobencarb exposure throughout a complete life cycle. Abbreviations of model terms are those described for Equation 2.

Physiological Response	Model Term	Coefficient
Oxygen consumption ($\mu g\ O_2$/mg dry wt/h)	Intercept	7.045042
	CONC	0.006978
	$CONC^2$	-0.000006
	DAY	-0.171730
	DAY^2	-0.004321
	CONC x DAY	0.000737
O:N ratio	Intercept	57.052402
	CONC	-0.084752
	$CONC^2$	0.000053
	DAY	-3.316112
	DAY^2	0.097934
	CONC x DAY	0.005928
Growth (µg dry wt/d)	Intercept	-8.625074
	CONC	-0.139513
	$CONC^2$	0.000330
	DAY	9.666398
	DAY^2	-0.262703
	CONC x DAY	-0.013189
Net growth efficiency (K_2 values)	Intercept	89.174004
	CONC	-0.096446
	$CONC^2$	-0.000162
	DAY	-2.709149
	DAY^2	0.088125
	CONC x DAY	0.006211

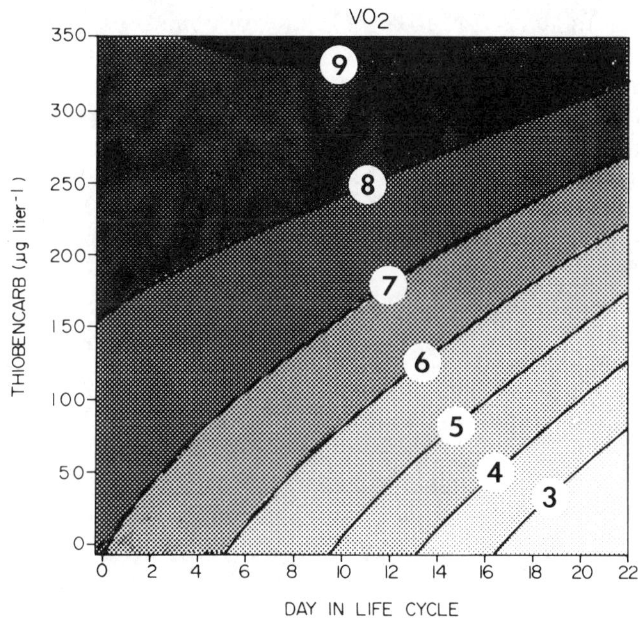

Fig. 3. Response surface of respiration rates ($\mu g O_2$/mg dry wt/h) of Mysidopsis bahia as influenced by continuous thiobencarb exposure through a complete life cycle.

Because the two-way analysis of variance of the ammonia excretion data showed a highly significant ($p = 0.0002$) interaction between life stage of exposed mysids and exposure concentration, ammonia excretion rates of individual life stages were regressed against thiobencarb concentration (Fig. 4). After 24 h of exposure, ammonia excretion rates of juveniles were not significantly elevated by thiobencarb exposure. Quadratic regressions best fit the excretion data of older juveniles, for which lower thiobencarb concentrations had no effect and exposure to a concentration of 345 μg/L significantly ($p<0.05$) increased ammonia excretion. As juveniles matured, ammonia excretion rates were reduced with exposure to lower thiobencarb concentrations, a significant reduction occurring at 75 μg/L.

Even though a two-way analysis of variance of log-transformed O:N ratios for thiobencarb-exposed mysids demonstrated a highly significant ($p = 0.0001$) influence by thiobencarb and life stage individually and a less significant ($p = 0.08$) influence by the interaction of life stage and thiobencarb, linearity of O:N ratios to thiobencarb concentration was poor (Table 2). Nevertheless, O:N ratios of juveniles exposed for at least 4 d to 345 µg thiobencarb/L were significantly ($p<0.05$) reduced. O:N ratios of adults were significantly lower than those of juveniles. However, these adult O:N ratios were significantly elevated in thiobencarb concentrations above 35 µg/L.

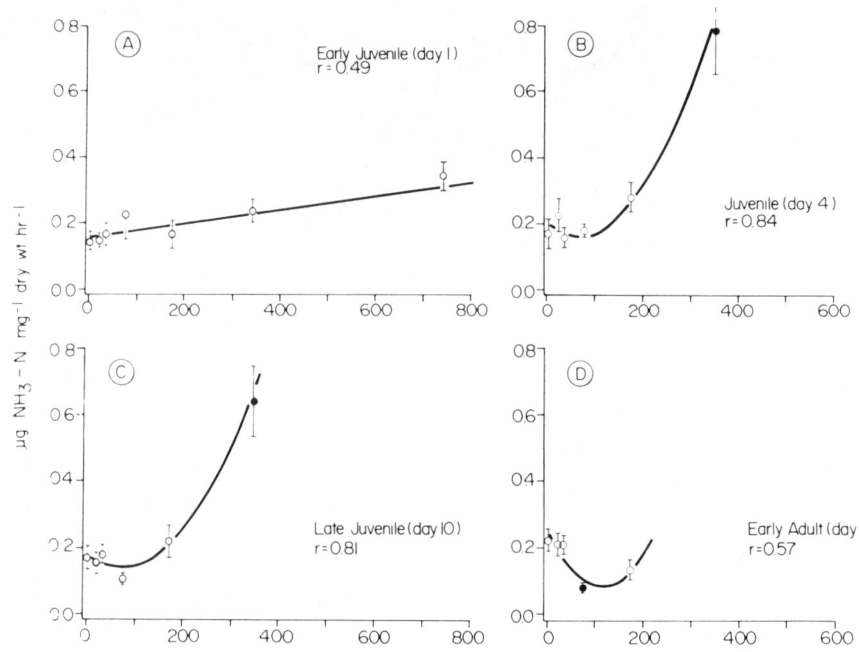

Fig. 4. Ammonia excretion rates of various life stages of Mysidopsis bahia as a function of continuous thiobencarb exposure through a complete life cycle. Values represent means ± SE. Darkened circles are significantly different from control at 0.05 probability level.

These relationships between O:N ratios of different mysid life stages and thiobencarb exposure are best illustrated by the response surface of O:N ratios (Fig. 5) produced from coefficients of the quadratic function of day in life cycle and thiobencarb exposure concentration ($r = 0.43$) (Table 1). This surface depicts the depressed O:N ratios of exposed mysid juveniles, the reduction of O:N ratios as juveniles mature, and the elevation of adult O:N ratios with thiobencarb exposure.

Table 2. Influence of thiobencarb exposure on O:N ratio of different life stages of *Mysidopsis bahia* continually exposed through an entire life cycle. Values are the means and standard errors.

Thiobencarb Concentration (μg/L)	O:N Ratio			
	Early Juvenile	Juvenile	Late Juvenile	Early Adult
<0.50 (control)	42 ± 13	35 ± 8	36 ± 10	15 ± 2
22	55 ± 12	30 ± 10	53 ± 15	20 ± 4
35	50 ± 8	41 ± 5	34 ± 5	20 ± 3
76	62 ± 17	31 ± 3	48 ± 7	*36 ± 6
181	72 ± 20	23 ± 3	43 ± 11	*48 ± 20
345	36 ± 5	*10 ± 2	*17 ± 3	-
734	30 ± 5	-	-	-

* Significantly different from the control at 0.05 probability level.

Because a two-way analysis of variance showed a highly significant ($p = 0.0001$) interaction between growth of particular life stages and thiobencarb concentration, daily growth rates of mysids were regressed across thiobencarb concentrations for each phase in the life cycle (Fig. 6). For all life stages, a quadratic regression best fit the growth data, indicating the curvilinear nature of the response to thiobencarb concentration. Exposure concentrations from 22 to 345 μg thiobencarb/L significantly ($p<0.05$) reduced growth rates of juvenile mysids. Retardation of growth rates by thiobencarb exposure was most severe during the most rapidly growing period in the life cycle (day 4 to day 10). As juveniles approached maturity, growth rates of unexposed mysids were reduced. However, growth of juveniles, that was previously retarded in low sublethal concentrations of thiobencarb, was significantly enhanced at these same concentrations as they matured into adults.

Net growth efficiency values (K_2 values) of mysids were significantly ($p = 0.0001$) influenced by an interaction between life stage and thiobencarb exposure concentration. Regressions of K_2 values during different life stages demonstrate these relationships (Fig. 7). Growth efficiency of mysids during the first 4 d of their life cycle was significantly ($p<0.05$) reduced by thiobencarb exposure (≦22 μg/L) in a highly significant linear fashion ($r = 0.98$). The sensitivity of this response was less as juveniles continued to grow. As mysids matured, growth efficiency was significantly increased in sublethal thiobencarb concentrations that had previously reduced efficiency of juveniles.

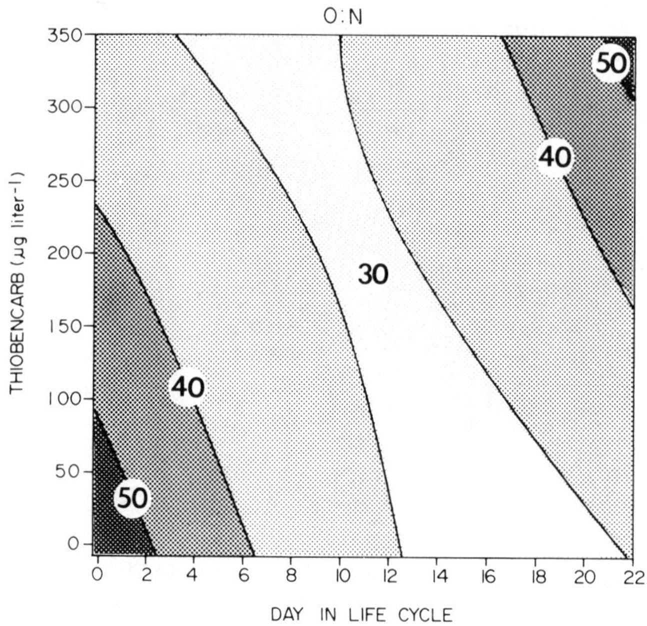

Fig. 5. Response surface of O:N ratios of *Mysidopsis bahia* as influenced by continuous thiobencarb exposure through a complete life cycle.

The changing relationships between net growth efficiency of mysids and thiobencarb exposure during a complete life cycle of *M. bahia* are best depicted by the response surface (Fig. 8) produced from coefficients of the regression of all K_2 values onto the quadratic function of day in life cycle and thiobencarb concentration ($r = 0.86$) (Table 1). This surface illustrates the direct inhibition of growth efficiency during the early juvenile period, the reduced growth efficiency of mysids as they mature, and the diminished influence of thiobencarb on growth efficiency as mysids mature.

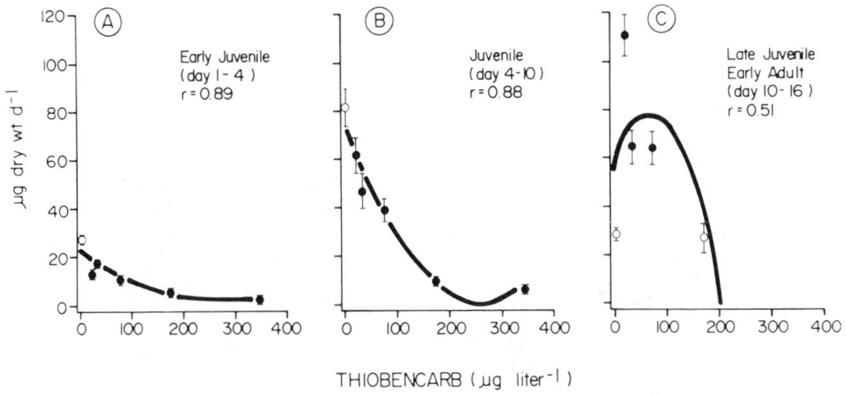

Fig. 6. Growth rates of various life stages of *Mysidopsis bahia* as a function of continuous thiobencarb exposure through a complete life cycle. Values represent means ± SE. Darkened circles are significantly different from control at 0.05 probability level.

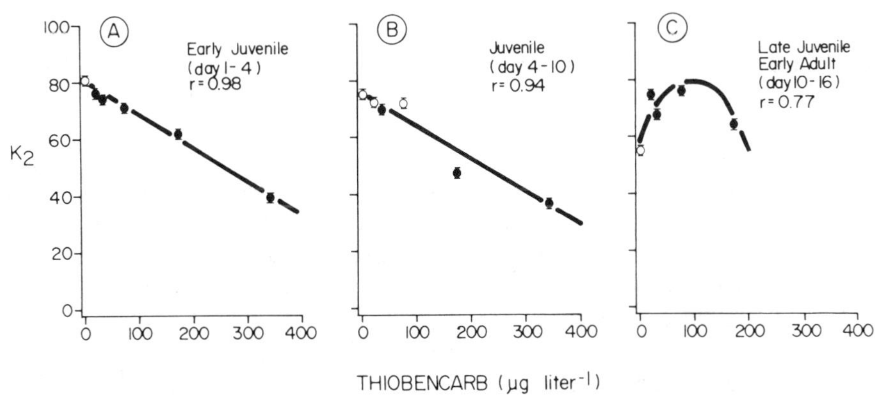

Fig. 7. Net growth efficiencies (K_2 values) of various life stages of *Mysidopsis bahia* as a function of continuous thiobencarb exposure through a complete life cycle. Values represent means ± SE. Darkened circles are significantly different from control at 0.05 probability level.

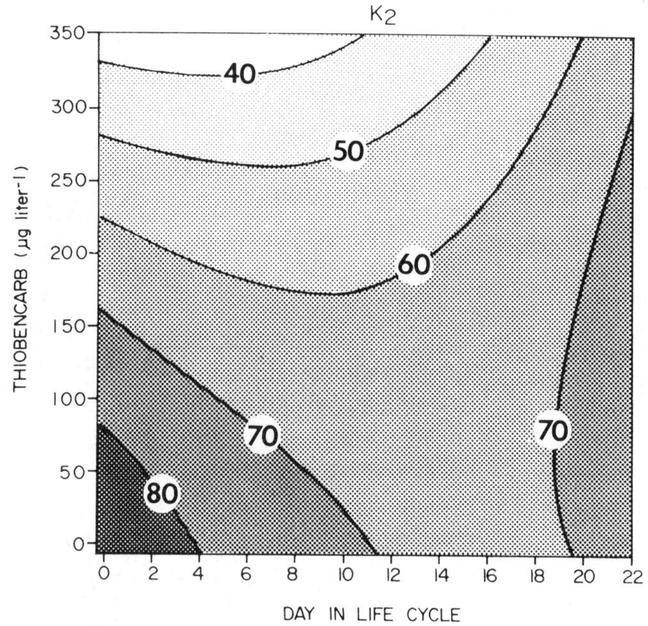

Fig. 8. Response surface of net growth efficiencies (K_2 values) of Mysidopsis bahia as influence by continuous thiobencarb exposure through a complete life cycle.

DISCUSSION

Chronic exposure of Mysidopsis bahia through an entire life cycle to the organic pesticide, thiobencarb, produced lower survival rates and inhibited the reproductive success of the population at sublethal concentrations. Production of young by individual females was inversely and linearly related to exposure concentration. The inherently high variability in this response, however, resulted in it being no more sensitive statistically to thiobencarb exposure than lethality. Nevertheless, total production of young by the population, representing a summation of the effects of these reduced survival and individual fecundity rates on the reproductive success of the population, was inhibited by concentrations an order of magnitude below those concentrations which were acutely lethal for this species (96-h LC50 = 330 µg thiobencarb/L) (Schimmel et al., 1983). These findings are in agreement with previous studies that have demonstrated inhibited reproduction to be the most sensitive sublethal, population response of this species to chronic pesticide exposure (McKenney, 1982 and references therein).

Metabolic alterations in juvenile *M. bahia* exposed to thiobencarb preceded population-level responses. Respiration rates of juvenile mysids were significantly elevated by acute exposure to the same concentrations of thiobencarb that ultimately led to significant reductions in population survival rates with chronic exposure through an entire life cycle. Thiobencarb stimulation of mysid respiration may be a result of functional disruptions of cytochrome oxidases, as has been described with other thiocarbamates (Ashton & Crafts, 1981), or a result of oxidative metabolism by the microsomal mixed-function oxidases that degrade adsorbed thiobencarb (Hubbell & Casida, 1977; Casida, 1978; Lee, 1981). Regardless of the mechanism, higher respiration rates of juvenile mysids might be useful in the laboratory and, given adequate baseline information, in the field to predict contamination that would produce lowered survival rates of the population with chronic exposure. Since similar correlations were reported between juvenile respiration and population survival for *M. bahia* exposed to an organochlorine pesticide (endrin) (McKenney, 1982), the utility of this response may apply to many classes of toxic organics. Furthermore, results of a study by Vargo (1981) suggest that metabolic alterations in marine zooplankton may have implications for long-term ecological disruptions at even higher levels of biological organization. Altered patterns of respiration and excretion of individual zooplankters (of separate species) immediately after exposure to low concentrations of oil in an enclosed microcosm led to structural changes in the community after 6-wk exposure.

O:N ratios for the various life stages of *M. bahia* indicated a shift in the energy metabolism toward greater catabolism of protein as juveniles matured. Increased catabolism of proteins may reflect metabolic shifts of lipids from use as an energy source to mobilization and transfer to developing gonads as a prerequisite for successful gametogenesis. O:N ratios obtained for *M. bahia* adults were similar to those reported for the adults of several other marine mysids (Chin, 1974; Gaudy *et al.*, 1980). In addition, similar shifts in energy metabolism previously have been reported for maturing mysids (McKenney, 1982) and for reproductively active copepods (Conover & Corner, 1968; Bamstedt & Matthews, 1975). Mobilization, storage, and utilization of lipids during reproductive preparations have been reported for other marine zooplankton (Lawrence, 1976; Gatten *et al.*, 1979).

Modifications in the energy metabolism of individual mysids exposed to thiobencarb were related to reproductive inhibitions, with long-term implications on the ecological fitness of the mysid population. Ammonia excretion rates, although stimulated concurrently with metabolic rates at chronically lethal thiobencarb concentrations, were neither as sensitive to thiobencarb exposure nor as meaningful in biological interpretation as when combined with oxygen consumption rates into an O:N ratio. An apparent changing pattern of substrate utilization during the maturation of *M. bahia* was altered by exposure to high sublethal concentrations of thiobencarb. Higher O:N ratios during the maturation of these thiobencarb-exposed mysids suggest a greater reliance on lipid material as a substrate for oxidative metabolism, which would result in less lipid material being available for reproductive preparations. Similar relationships existed between energy metabolism and reduced reproductive success of *M. bahia* exposed to endrin (McKenney, 1982).

It should be noted that both ammonia excretion rates and O:N ratios of mysids were more variable than respiration rates, and neither parameter was as sensitive to thiobencarb exposure as respiration rates. Assurance of the nutritional state of each organism by feeding immediately prior to metabolic measurements might stabilize the observed variation in these measurements and strengthen their dose-response patterns. Similar coordination of feeding and metabolic measurements could result in higher respiration rates reflective of maximum early post-absorptive metabolism (J. Capuzzo, pers. comm.). Regardless of these potential sources of error in absolute values, the relative influence of thiobencarb exposure on the bioenergetics of *M. bahia* remains the same.

Retarded growth rates of juvenile mysids exposed to thiobencarb were accompanied by decreasing net growth efficiencies, suggesting that the increased metabolic demands of exposed mysids, reflected in the elevated respiration rates, allowed for less assimilated energy to be available for somatic production. Similarly, elevated respiration rates of juvenile *M. bahia* exposed to endrin resulted in retarded growth rates (McKenney, 1982).

Other marine zooplankton have shown disruptions in bioenergetic processes following sublethal exposure to toxic organics. Acute exposure to oil reduced ingestion rates and elevated respiration rates of adult planktonic shrimp (Lee *et al.*, 1978), but caused significant reductions in the respiration and excretion rates of larval lobsters (Capuzzo & Lancaster, 1982). Juvenile mysids (*M. almyra*), exposed to a drilling mud-sea water mixture for a week, displayed higher rates of ingestion and respiration, accompanied by reduced growth rates and lowered growth efficiencies (Carr *et al.*, 1980). Crab larvae reared in oil-treated sea water exhibited reduced rates of ingestion and growth, while respiration rates were increased (Johns & Pechenik, 1980). None of these studies, however, examined bioenergetic processes of the various life stages exposed throughout an entire life cycle; therefore, precluding the establishment of dose-response relationships for these physiological responses and their comparison with long-term alterations in population responses of ecological significance (*i.e.*, survival, growth, and reproduction).

Dose-response relationships for different physiological responses of pesticide-exposed mysids varied with life stage. The more linear relationships between respiration and thiobencarb concentration observed in juvenile mysids resulted in lower net growth efficiencies and retarded growth rates of juveniles. The absence of an inhibitory effect by thiobencarb on growth efficiencies of maturing juveniles, combined with reproductive inhibitions at these same concentrations, imply that as juveniles mature the physiological effects of thiobencarb exposure influence reproductive processes more than growth. Furthermore, thiobencarb exposure affected substrate utilization patterns of maturing juveniles more than those of earlier juveniles. These changing patterns were correlated with lower reproduction rates, suggesting that higher lipid catabolism of mysids maturing in sublethal thiobencarb concentrations may be the mechanism responsible for inhibited reproductive capacities of adult mysids. Additional research involving detailed analysis of biochemical composition could provide supportive information in this area.

Alterations in environmental quality which ultimately modify the normal structural and functional properties of an ecosystem are a product of population and community changes that result from initial damage at the organismic and suborganismic level (NAS, 1971; Sprague, 1971; Waldichuk, 1979; Kinne, 1980; Sastry & Miller, 1981). Environmental modifications of physiological activity of individuals may reflect distortions in ecological mechanisms, such as secondary production. Lower secondary production in a crustacean population, as indicated in this study by retarded mysid growth rates and inhibited reproduction rates, would alter the energy-flow patterns between connected trophic levels in the ecosystem. Although thiobencarb is directly toxic to fish in low mg/L concentrations (Ochiai & Kubota, 1978; Johnson & Finley, 1980; Metelev & Brichko, 1980; Schimmel et al., 1983), the reduced secondary production of mysid populations with exposure to low µg/L concentrations of thiobencarb could indirectly reduce fish populations that depend upon mysids as a food source (Darnell, 1958; Odum, 1971; Carr & Adams, 1973; Adams, 1976; Livingston, 1980) and disrupt the delicate balance of marine food webs.

From the responses of M. bahia to chronic thiobencarb exposure through an entire life cycle, it may be concluded that metabolic dysfunctions in individual mysids preceded population responses important in determining community tropic patterns. Lower secondary production rates of the mysid populations, manifested by retarded growth rates in juveniles and inhibited reproductive performance in adults, were preceded by modifications in the energy metabolism of individual mysids. Integration of several individual functional responses into physiological and bioenergetic indices proved to be more sensitive than individual physiological measurements for assessing environmentally-induced alterations in biological fitness. Short-term measurements of altered metabolic patterns in contaminated zooplankton, therefore, may offer the potential for monitoring and predicting ecological disruptions in the marine ecosystem at higher levels of biological organization (i.e., the population or community level) (Fig. 9).

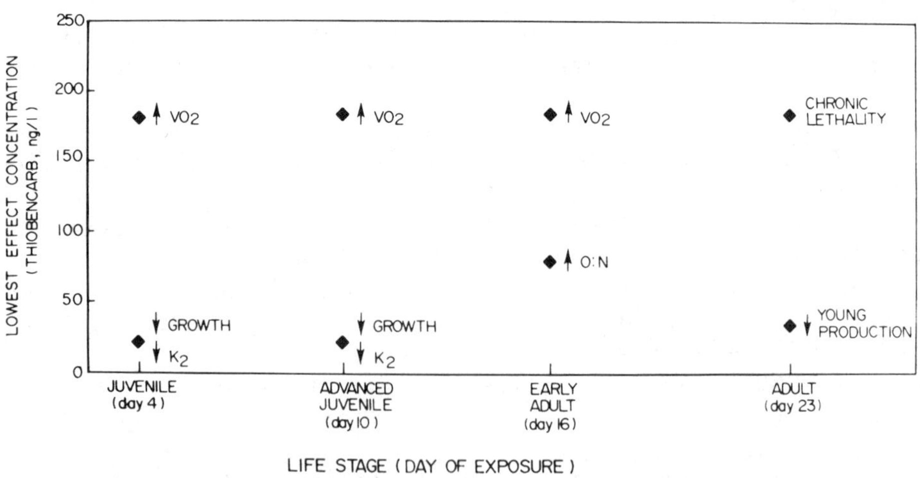

Fig. 9. Summary of the effects of thiobencarb on Mysidopsis bahia exposed through an entire life cycle.

ACKNOWLEDGMENTS

The valuable assistance of Ed Matthews, Debra Lawrence, Neil Santarella, and Meta Leckband during portions of this study is greatly acknowledged. Special thanks are extended to Jim Moore and Johnnie Knight for their quantitative analyses of thiobencarb.

LITERATURE CITED

Adams, S.M. 1976. Feeding ecology of eel grass fish communities. Trans. Amer. Fish. Soc. 105: 514-519.

Alderdice, D.F. 1972. Factor combinations: Responses of marine poikilotherms to environmental factors acting in concert. pp. 1659-1772. In: Marine Ecology. Vol. 1. Environmental Factors. O. Kinne (ed.). Wiley-Interscience, London.

Ashton, F.M. and A.S. Crafts. 1981. Mode of Action of Herbicides, 2nd Ed. John Wiley and Sons, New York. 525 pp.

Bamstedt, U. and J.B.L. Matthews. 1975. Studies of the deep-water pelagic community of Korsfjorden, western Norway. The weight and biochemical composition of *Euchaeta norvegica* Boeck in relation to its life cycle. pp. 311-327. In: Proceedings of the 9th European Marine Biology Symposium. H. Barnes (ed.). Aberdeen University Press, Aberdeen.

Barr, A.J., J.H. Goodnight, J.P. Sall, and J.T. Helwig. 1976. A User's Guide to SAS 76. Statistical Analysis Systems Institute, Inc., Raleigh, NC. 329 pp.

Bayne, B.L., J. Anderson, D. Engel, E. Gilfillan, D. Hoss, R. Lloyd, and F.P. Thurberg. 1980. Physiological techniques for measuring the biological effects of pollution in the sea. Rapports et Proces-Verbaux des Reunions Conseil International pour l'Exploration de la Mer 179: 88-99.

Box, G.E.P., W.G. Hunter, and J.S. Hunter. 1978. Statistics for Experimenters. John Wiley and Sons, New York. 653 pp.

Capuzzo, J.M. and B.A. Lancaster. 1982. Physiological effects of petroleum hydrocarbons on larval lobsters (*Homarus americanus*): Hydrocarbon accumulation and interference with lipid metabolism. pp. 477-501. In: Physiological Mechanisms of Marine Pollutant Toxicity. W.B. Vernberg, A. Calabrese, F.P. Thurberg, and F.J. Vernberg (eds.). Academic Press, New York.

Carr. R.S., L.A. Reitsema, and J.M. Neff. 1980. Influence of a used chrome lignosulfonate drilling mud on the survival, respiration, feeding activity and net growth efficiency of the opossum shrimp *Mysidopsis almyra*. pp. 944-963. *In*: Symposium/Research on Environmental Fate and Effects of Drilling Fluids and Cuttings. Courtesy Associates, Washington, D.C.

Carr, W.E.S. and C.A. Adams. 1973. Food habits of juvenile marine fishes occupying seagrass beds in the estuarine zone near Crystal River, Florida. Trans. Amer. Fish. Soc. 102: 511-540.

Casida, J.E. 1978. Metabolism of pyrethroid insecticides and thiocarbamate herbicides. pp. 163-170. *In*: Symposium on Environmental Transport and Transformation of Pesticides. D.W. Duttweiler (ed.). U.S. Environmental Protection Agency. Athens, GA.

Chin, P. 1974. Acclimation and metabolic attitude of *Neomysis awatschensis*. Publications of the Marine Laboratory in Busan Fisheries College. 7: 1-20.

Conover, R.J. and E.D.S. Corner. 1968. Respiration and nitrogen excretion by some marine zooplankton in relation to their life cycle. J. Mar. Biol. Assoc. U.K. 48: 49-75.

Darnell, R.M. 1958. Food habits of fishes and larger invertebrates of Lake Pontchartrain, Louisiana. Contrib. Mar. Sci. 5: 353-416.

Gatten, R.R., E.D.S. Corner, C.C. Kilvington, and J.R. Sargent. 1979. A seasonal survey of the lipids in *Calanus helgolandicus* Claus from the English Channel. pp. 275-284. *In*: Cyclic Phenomena in Marine Plants and Animals. E. Naylor and R.G. Hartnoll (eds.). Pergamon Press, Oxford.

Gaudy, R., J.P. Guerin, and M. Pagano. 1980. Ecophysiologie comparee des mysidaces *Hemimysis speluncola* Ledoyer (cavernicole) et *Leptomysis lingvura* G.O. Sars (noncavernicole). Respiration et excretion. J. Exp. Mar. Biol. Ecol. 44: 29-46.

Grodzinski, W., R.Z. Klekowski, and A. Duncan. 1975. Methods for Ecological Bioenergetics. Blackwell Scientific Publications, Oxford. 367 pp.

Hubbell, J.P. and J.E. Casida. 1977. Metabolic fate of the N, N-dialkylcarbamoyl moiety of thiocarbamate herbicides in rats and corn. J. Agri. Food Chem. 25: 404-413.

Johns, D.M. and J.A. Pechenik. 1980. Influence of the water-accommodated fraction of no. 2 fuel oil on energetics of Cancer irroratus larvae. Mar. Biol. 55: 247-254.

Johnson, W.W. and M.T. Finley. 1980. Handbook of Acute Toxicity of Chemicals to Fish and Aquatic Invertebrates. Resource Publication 137. U.S. Fish and Wildlife Service. Washington, D.C. 98 pp.

Kinne. O. 1980. Summary of symposium papers and conclusion. In: 14th European Marine Biology Symposium: Protection of Life in the Sea. O. Kinne and H.-P. Bulnhiem (eds.). Helgol. wiss. Meeres. 33: 732-761.

Lawrence, J.M. 1976. Patterns of lipid storage in postmetamorphic marine invertebrates. Amer. Zool. 16: 747-762.

Lee, R.F. 1981. Mixed function oxygenases (MFO) in marine invertebrates. Mar. Biol. Lett. 2: 87-105.

Lee, W.Y., K. Winters, and J.A.C. Nicol. 1978. The biological effects of the water-soluble fractions of a no. 2 fuel oil on the planktonic shrimp, Lucifer faxoni. Environ. Poll. 15: 167-183.

Livingston, R.J. 1980. Ontogenetic trophic relationships and stress in a coastal seagrass system in Florida. pp. 423-435. In: Estuarine Perspectives. V.S. Kennedy (ed.). Academic Press, New York.

Mann, K.H. 1969. The dynamics of aquatic ecosystems. pp. 1-81. In: Advances in Ecological Research. J.B. Cragg (ed.). Academic Press, London.

McKenney, C.L., Jr. 1982. Interrelationships between energy metabolism, growth dynamics, and reproduction during the life cycle of Mysidopsis bahia as influenced by sublethal endrin exposure. pp. 447-476. In: Physiological Mechanisms of Marine Pollutant Toxicity. W.B. Vernberg, A. Calabrese, F.P. Thurberg, and F.J. Vernberg (eds.). Academic Press, New York.

McNeil, K.A., F.J. Kelley, and J.T. McNeil. 1975. Testing Research Hypothesis Using Multiple Linear Regression. Southern Illinois University Press, Carbondale and Edwardsville. 587 pp.

Metelev, V.V. and V.F. Brichko. 1980. Effect of biothiocarb on fish. Veterinariya (Moscow) 12: 58-60.

Meyers, R.H. 1971. Response Surface Methodology. Allyn Bacon, Boston. 246 pp.

Nimmo, D.R. and T.L. Hamaker. 1982. Mysids in toxicity testing - A review. Hydrobiologia 93: 171-178.

National Academy of Science. 1971. Marine Environmental Quality. National Academy of Science, Washington, D.C. 107 pp.

Ochiai, T. and S.S. Kubota. 1978. Studies on toxic effects of pesticides on fish. II. Toxic effects of the herbicide, benthiocarb, on carp. Bull. Fac. Fish., Mie University 5: 155-163.

Odum, W.E. 1971. Pathways of Energy Flow in a South Florida Estuary. University of Miami Sea Grant Program, Miami, FL. 158 pp.

Sastry, A.N. and D.C. Miller. 1981. Application of biochemical and physiological responses to water quality monitoring. pp. 265-294. In: Biological Monitoring of Marine Pollutants. F.J. Vernberg, A. Calabrese, F.P. Thurberg, and W.B. Vernberg (eds.). Academic Press, New York.

Schimmel, S.C., R.L. Garnas, J.M. Patrick, Jr., and J.C. Moore. 1983. Acute toxicity, bioconcentration, and persistence of AC 222,705, benthiocarb, chlorpyrifos, fenvalerate, methyl parathion, and permethrin in the estuarine environment. J. Agri. Food Chem. 31: 104-113.

Schoor, W.P. and C.L. McKenney, Jr. 1983. Determination of fenvalerate in flowing-seawater exposure studies. Bull. Environ. Contam. Toxicol. 30: 84-92.

Sprague, J.B. 1971. Measurement of pollutant toxicity to fish. III. Sublethal effects and "safe" concentrations. Water Res. 5:245-266.

Steele, J.H. 1973. Marine Food Chains. Oliver and Boyd, Edinburgh. 552 pp.

Steele, J.H. 1974. The Structure of Marine Ecosystems. Harvard University Press, Cambridge, MA. 128 pp.

Vargo, S.L. 1981. The effects of chronic low concentrations of no. 2 fuel oil on the physiology of a temperate estuarine zooplankton community in the MERL microcosm. pp. 295-322. In: Biological Monitoring of Marine Pollutants. F.J. Vernberg, A. Calabrese, F.P. Thurberg, and W.B. Vernberg (eds.). Academic Press, New York.

Waldichuk, M. 1979. Review of the problem. pp. 1-26. In: The Assessment of Sublethal Effects of Pollutants in the Sea. H.A. Cole (ed.). The Royal Society, London.

Winberg, G.G. 1971. Methods for the Estimation of Production of Aquatic Animals. Academic Press, New York. 175 pp.

Zar, J.H. 1974. Biostatistical Analysis. Prentice-Hall. Englewood Cliffs, New Jersey. 620 pp.

THE EFFECT OF THE PCB, AROCLOR 1254, AND ETHANOL ON THE HUMORAL IMMUNE RESPONSE OF A MARINE TELEOST TO A SLUDGE BACTERIAL ISOLATE OF *E. COLI*

Joanne S. Stolen

SUMMARY

PCBs are a persistant environmental threat. Fishes exposed to PCBs may appear healthy, but if their immune system is compromised by this exposure, their resistance to disease could be significantly lowered. These studies show that exposure by feeding and injection of PCBs dissolved in peanut oil only slightly suppresses and delays the humoral immune response in summer flounder (*Paralichthys dentatus*) to *Escherichia coli*. Fish pretreated with PCBs in ethanol or ethanol alone showed no response to *E. coli*. This dramatic suppression was due to the ethanol. A second injection of *E. coli* was given to the ethanol-PCB groups 5 wk after the first injection. After one week, antibodies were detectable, but those fish given PCBs showed a slight delay in appearance of circulating antibody levels when compared with the groups pretreated with ethanol alone, indicating a possible effect of PCBs in delaying the immune response. Fish species vary with respect to the accumulation of PCBs. This fish species does not bioaccumulate high levels of PCBs. This might be an explanation for the marginal effect of PCBs on the immune response.

INTRODUCTION

Polychlorinated biphenyls (PCBs) are a group of synthetic compounds that are extremely resistant to thermal and chemical degradation. Because of their unique characteristics they were used freely (since the 1980's) for industrial applications. In the United States, the most common PCB is Aroclor, a trade name followed by a four digit number to designate the mixture of compounds (*e.g.*, Aroclor 1254). PCBs first made headlines when Japanese, who ate PCB-tainted rice oil, became ill with skin lesions and damage to liver and nerves. PCBs were also detected in fishes (Jensen, 1966). Further tests for PCBs found the contaminant to be detected throughout the food chain of living freshwater, estuarine, and marine species (Risebrough *et al.*, 1968). Although PCBs have been banned since 1976, they still impose a significant environmental threat. Rivers and coastal sediments serve as reservoirs of PCBs. They are absorbed by all organisms in the food chain and concentrate in adipose tissues over a period of time even if actual exposure levels are low.

Fish exposed to PCBs were found to have higher incidences of epidermal lesions (Perkins *et al.*, 1972), fin erosion (Mearns & Sherwood, 1974; McDermott-Ehrlich *et al.*, 1977), blood anemia and altered cholesterol metabolism (Johansson *et al.*, 1972).

PCBs have been found to suppress the immune defense system in various species of mammals (Friend & Grainer, 1970; Vos & de Roij, 1972; Vos & Driel-Grootenhuis, 1972). Those species exposed to PCBs might appear healthy, yet the immune system has been compromised and disease might eliminate a significantly larger segment of the population than would otherwise succumb.

This is a study of the effect of the PCB mixture Aroclor 1254 with an inflammatory agent (ethanol) used as a diluent, on the immunity of summer flounder (*P. dentatus*) to the bacterium *E. coli* isolated from the municipal sewage dumpsite in the New York Bight.

MATERIALS AND METHODS

Fish Summer flounder 30-45 cm in length were caught by otter trawl in Sandy Hook Bay and maintained at a water temperature of 18-20°C in 250-gallon flow-through salt water tanks. All fish were acclimated at least two weeks prior to any experimental procedure.

PCB The PCB, Aroclor 1254, was dissolved in either ethanol or peanut oil to a concentration of 10 mg/ml. Fish were injected with the PCB in ethanol intraperitoneally (IP). A total concentration of 5 µg/g body weight was given per fish. Fish were given an average of 0.35 ml total volume of PCB in diluent. Another group of fish were exposed to PCB in peanut oil, half of the dosage (2.5 μg/g body weight) was IP-injected and the other half was given orally in gelatin capsules.

Antigen Fish were injected IP with 0.15 ml of a 20% suspension of formalin-killed bacteria in Freund's complete adjuvant (FCA). *E. coli* was isolated from sewage sludge as previously described (Stolen *et al.*, 1983).

Antibody Fish were bled weekly for serum by puncture of the caudal vein. Microtiter agglutination tests were performed to quantitate the antibody response to *E. coli* (Stolen *et al.*, 1983). Titers were expressed as $\log_2$ dilutions.

The experimental protocol is shown in Table 1.

Table 1. Experimental Protocol.

Group	Number of Fish	Treatment on Day 0	7	14	49
1	8	-	*E. coli*	-	-
2	8	ETOH*	ETOH	*E. coli*	*E. coli*
3	8	PCB (ETOH)	*E. coli*	-	*E. coli*
4	8	PCB (ETOH)	-	*E. coli*	*E. coli*
5	8	PCB (ETOH)	PCB (ETOH)	*E. coli*	*E. coli*
6	8	PCB (peanut oil)	*E. coli*	-	-

* 0.3-0.4 ml ethanol per fish

RESULTS

Aroclor 1254 dissolved in ethanol

Fish injected with either Aroclor 1254 dissolved in ethanol or ethanol alone showed total suppression of the antibody response to *E. coli* as compared with fish injected with just *E. coli* (Fig. 1). When this group of fish was given a second injection of *E. coli* 5 wk later, titers were roughly equivalent to a primary response in untreated fish by 28 d. Lower titers were evident with PCB-injected fish at 7 and 14 days as compared with fish pretreated with just the diluent ethanol (Fig. 2). At day 21, the group injected with PCB two weeks before the first injection of *E. coli*, showed equivalent titers to ethanol-injected fish, while Groups 2 and 4 were still suppressed. At day 28 all titers were roughly equivalent.

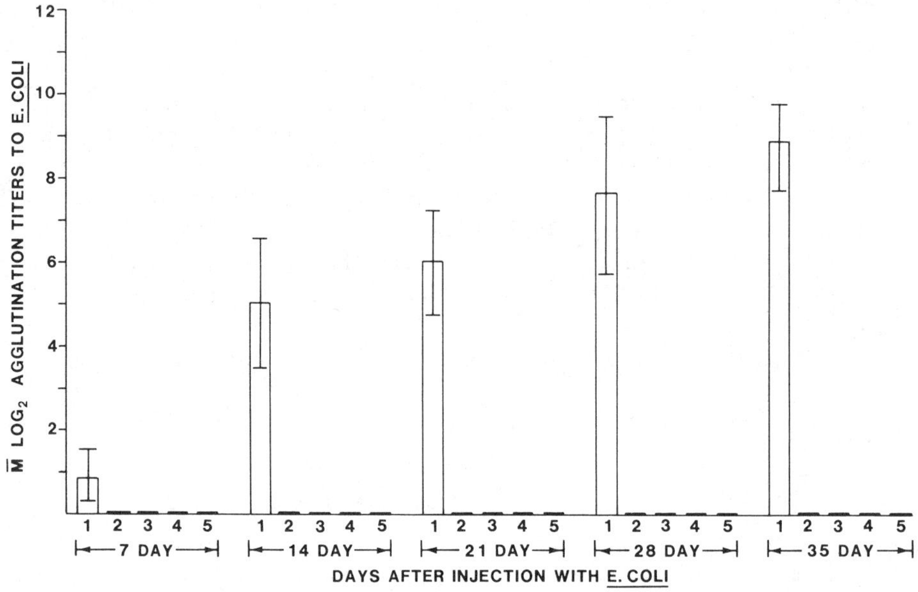

Fig. 1. The effect of Aroclor 1254 on the immune response of summer flounder. For experimental groups of 8 fish/group see Table 1. Vertical lines indicate standard deviation.

Aroclor in peanut oil

Fish given an oral dosage and an IP injection of PCB in peanut oil showed suppressed antibody titers to *E. coli* (Fig. 3) when compared with untreated fish.

DISCUSSION

It has been well recognized that stress and resistance to disease are related and that stress causes many changes in the physiological systems of the body including the immune system. Direct correlation between disease states and the presence of pollutants have been determined by numerous investigators. The effect of toxic agents on the immune system has been extensively reviewed (Zeeman & Brindley, 1981).

Studies on a number of mammalian species have shown PCBs to be a suppressor of the immune system. Silkworth and Grabstein (1982) found that in mice the immunotoxic potential of PCB isomers were dependent on conformational planarity. Thomas and Hinsdill (1980) found lower contact sensitivity in PCB-treated rabbits. Vos and de Roij (1972) saw immunosuppression in guinea pigs to tetanus toxoid; both the humoral and cellular response were affected.

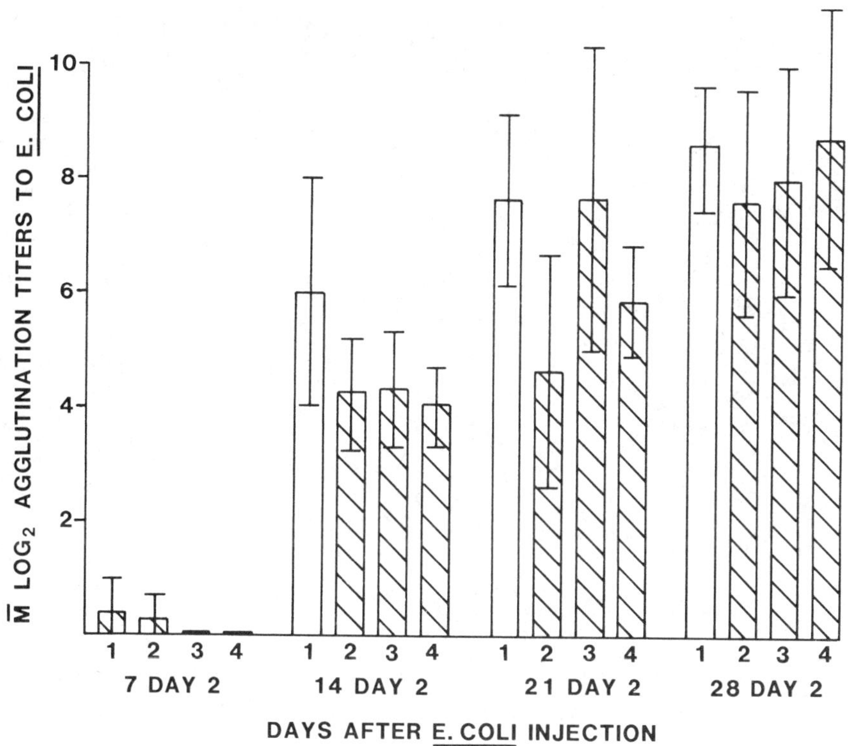

Fig. 2. The effect of Aroclor 1254 on a second injection of *E. coli* in summer flounder. Fish were given a second injection of *E. coli* 35 d after the first. Group 1 represents two injections of ethanol; Group 2 and 3 received one injection of PCB in ethanol; in Group 3 there was a lag of an additional one week delay before the first injection of antigen; Group 4 received two injections of PCB in ethanol. Vertical lines indicate standard deviation.

Numerous investigators have linked PCBs with decreased disease resistance of fish. We have shown in these studies that humoral antibody formation in summer flounder can be slightly suppressed or delayed by pretreatment with PCBs diluted by peanut oil. We also showed an example of extreme suppression of antibody responses using ethanol which induces an inflammatory response. Agents such as turpentine have been shown to be inflammatory and induce stress on the immune system (White et al., 1981). Fish injected with ethanol showed a complete lack of immune response to E. coli. After 5 wk the effect of ethanol appears to have disappeared, while the effect of PCBs seems to have persisted for a longer period of time.

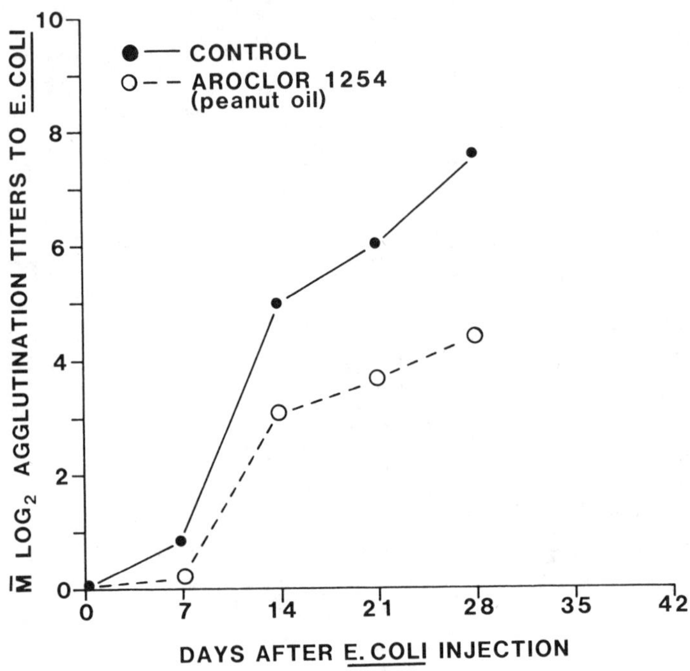

Fig. 3. The effect of PCBs on the immune response of summer flounder. Aroclor 1254, dissolved in peanut oil, was given orally and by IP injection to a final concentration of 5 μg/g body weight to 8 fish one week before injection of E. coli (0----0). Unpretreated controls were given the same dose of antigen (· —— ·). Numbers in parenthesis indicate standard deviation. $p<0.05$.

There is a wide variation in the bioaccumulation of PCBs in different species of fish. The summer flounder accumulates far less PCB than certain other species, such as striped bass and bluefish (Belton et al., 1982). Future experiments should involve a comparison of the immunotoxicity of PCBs in different species of marine teleosts.

ACKNOWLEDGMENTS

The research was supported by contract no. NO-80-FAC-0050, U.S. Department of Commerce, National Marine Fisheries Service to Drew University. My thanks to S. Draxler and W. Kaplan for their technical help; M. Cox for the illustrations; and Drs. Carl Sindermann and John Pearce NMFS, NEFC, Sandy Hook Laboratory for their support.

LITERATURE CITED

Belton, T., B. Rupple, K. Lockwood, and M. Boriek. 1982. PCB's in selected finfish. N.J. Dept. Env. Protect. Off. Sci. & Res. 35 pp.

Friend, M. and D.O. Trainer. 1970. Polychlorinated biphenyl: Interactions with duck hepatitis virus. Science 170: 314-316.

Jensen, S. 1966. Report of a new chemical hazard. New Sci. 32: 612.

Johansson, N., A. Larssen, and K. Lewander. 1972. Metabolic effects of PCB (polychlorinated biphenyls) on the brown trout (*Salmo trutta*). Comp. Gen. Pharmac. 3: 310-314.

McDermott-Ehrlich, D.J., M.J. Sherwood, T.C. Heesen, D.R. Young, and A.J. Mearns. 1977. Chlorinated hydrocarbons in Dover sole *Microstomus pacificus*: Local migrations and fin erosion. Fish. Bull., U.S. 74: 513-517.

Mearns, A. and M. Sherwood. 1974. Environmental aspects of fin erosion and tumors in southern California Dover sole. Trans. Amer. Fish. Soc. 4: 799-810.

Perkins, E.J., J.R.S. Gilchrist, and O.T. Abbott. 1972. Incidence of epidermal lesions in fish of the northeast Irish Sea area. Nature 238: 101-103.

Risebrough, R.W., P. Rieche, D.B. Peakall, S.G. Herman, and M.N. Kirven. 1968. Polychlorinated biphenyls in the global ecosystem. Nature 220: 1098-1102.

Silkworth, J.B. and E.M. Grabstein. 1982. Polychlorinated biphenyl immunotoxicity: Dependence on isomer planarity and the Ah gene complex. Toxicol. Appl. Pharmacol. 65: 109-115.

Stolen, J.S., V. Kasper, V. Lipcon, J.J. Nagle, and W.N. Adams. 1983. Monitoring environmental pollution in marine fishes by immunological techniques. Biotechn. 1: 66-68.

Thomas, P.T. and R.D. Hinsdill. 1980. Perinatal polychlorinated biphenyl exposure and its effect on the immune system of young rabbits. Drug. Chem. Toxicol. 3: 173-184.

Vos, J.C. and de Roij, T. 1972. Immunosuppressive activity of a polychlorinated biphenyl preparation on the humoral immune response in guinea pigs. Toxicol. Appl. Pharmacol. 21: 549-555.

Vos, J.G. and L. Van Driel-Grootenhuis. 1972. PCB-induced suppression of the humoral and cell-mediated immunity in guinea pigs. Sci. Total Environ. 1: 289-302.

White, A., T.C. Flètcher, M.B. Pepys, and B.A. Baldo. 1981. The effect of inflammatory agents on C-reactive protein and serum Amyloid P-component levels in plaice (*Pleuronectes platessa* L.) serum. Comp. Biochem. Physiol. 69: 325-329.

Zeeman, M.G. and W.A. Brindley. 1981. Effects of toxic agents upon fish immune systems: A review. pp. 1-60. *In*: Immunological Considerations in Toxicology 11. R.P. Sharma (ed.). CRC Press, Boca Raton, FL.

BIOENERGETICS AND SURVIVAL OF THE MARINE MUSSEL, *MYTILUS EDULIS L.*, DURING LONG-TERM EXPOSURE TO THE WATER-SOLUBLE FRACTION OF COOK INLET CRUDE OIL

William B. Stickle, Jr., Stanley D. Rice, Catherine Villars, and William Metcalf

SUMMARY

The sensitivity of *Mytilus edulis* on exposure to the WSF of Cook Inlet crude oil increased with duration of exposure. Scope for growth declined in an inverse manner with WSF concentration to zero at 1163 ppb aromatic hydrocarbons, this value represents 82% of the 28-d LC50. Therefore, *M. edulis* is not a particularly sensitive species to utilize for the study of sublethal stress during WSF exposures over long-term (one month) exposure periods. The primary determinant of a decline in scope with increased concentration of WSFwas a decline in algal consumption. Scope for growth regressed in an inverse manner against total tissue aromatic hydrocarbons at WSF concentrations up to 1083 ppb over 28-d exposure, but no relationship existed between scope for growth and total tissue concentration of aliphatic hydrocarbons. At a higher WSF aromatic hydrocarbon concentration (1083 ppb) the uptake of aromatic hydrocarbons was probably reduced because of a reduced algal clearance (= feeding) rate. Oxygen:nitrogen ratios were low indicating protein catabolism, and did not decline with increased WSF exposure; however, mussels were in the postspawning phase of their reproductive cycle.

INTRODUCTION

The sessile, filter-feeding bivalve mollusc *Mytilus edulis* has been employed on a world-wide scale as a ubiquitous species suitable for the chemical monitoring of pollutants, including petroleum hydrocarbons, in coastal and estuarine environments (Goldberg, 1975; Goldberg *et al*., 1978; Phillips, 1980). Marine molluscs have been shown to accumulate petroleum hydrocarbons to a marked degree over sea water levels (Farrington & Quinn, 1973; Fossato & Siviero, 1974; Di Salvo *et al*., 1975; Ehrhardt & Heinemann, 1975; Gilfillan *et al*., 1977; Widdows *et al*., 1982; Stickle *et al*., 1984).

Although there is a plethora of investigations that use mussels and other molluscs to monitor the chemical accumulation of petroleum hydrocarbons, there are relatively few studies that have determined the physiological responses of these molluscs to petroleum hydrocarbons. Widdows *et al*. (1982) exposed *M. edulis* to low concentrations (7-68 μg/L) of the water-accommodated fraction of North Sea crude oil for up to five months and found the stress indices scope for growth and lysosomal latency to be negatively correlated with tissue aromatic hydrocarbons. Stickle *et al*. (1984) reported similar findings for scope for growth when the carnivorous marine prosobranch *Thais lima* was exposed to the water-soluble fraction (WSF) of Cook Inlet crude oil for 4 wk. While many physiological measurements are possible, the determination of scope for growth over a period of exposure time will integrate many of the affected physiological responses.

In this study we investigated the effects of the WSF of Cook Inlet crude oil on the bioenergetics (scope for growth and O:N ratio), soft tissue aromatic and aliphatic hydrocarbon concentration, and survival of *M. edulis* during 28-d exposure. The main objectives of our experimental design were (1) determination of the relationship between the 28-d LC50 WSF concentration of Cook Inlet crude oil and that concentration which significantly reduces the bioenergetics of the mussels and, (2) examination of the relationship between the body burden of aromatic and aliphatic hydrocarbons and altered bioenergetics of the mussels.

MATERIALS AND METHODS

Animal Collection and Maintenance

Mytilus edulis were collected on June 18, 1981 from Lena Cove near Juneau, Alaska, taken to the Auke Bay Laboratory and sorted to a length range of 58-71 mm. Mussels were held for 7 d prior to the experiment in a tank through which sea water (10°C and 30°/$_{oo}$ S) flowed at a rate of 50 L/min. For the tolerance bioassay, mussels were divided into eight groups of 12 mussels per tank. For the bioenergetic study, six groups of mussels were placed in 50-L exposure tanks. Individual mussels within each group were numbered with non-toxic acrylic paint. For the uptake study, an additional 40 mussels were placed into each of the eight dosing chambers. Ten mussels were removed from each dose and frozen for future aromatic and aliphatic hydrocarbon analyses on days 7, 14, 21, and 28 of the experimental period. An inflow concentration of 1.2 x 10^4 cells/ml of *Phaeodactylum tricornutum* was continuously added to each tank throughout the experimental periods.

Toxicant Dosing, Bioassays, and Aromatic Hydrocarbon Uptake

The water-soluble fraction (WSF) of Cook Inlet crude oil was generated in a flow-through dilution apparatus described by Moles *et al*. (in press). The WSF effluent stock was diluted to produce seven exposure doses plus one control. Average aromatic hydrocarbon concentrations ($\bar{x}$ ± 95% C.I.) of 0, 93 ± 18, 207 ± 20, 381 ± 51, 604 ± 122, 1083 ± 111, 1306 ± 145, and 2100 ± 310 ppb (21 d only) were produced during the 28-d experiment. The eight 50-L exposure tanks each had flow-through water rates of 400 ml/min.

The LC50 of mussels was determined after 7-, 14-, 21-, and 28-d exposure to the WSF of Cook Inlet crude oil. Logit analysis, requiring at least two partial kills, was employed for the determination of LC50s (Silverstone, 1957).

The concentration of the water-soluble fraction stock was analyzed daily by gas chromatography. We estimated the concentration of total aromatic hydrocarbons in the WSF, which are the most toxic components of the WSF of Cook Inlet crude oil (Rice *et al*., 1977), by measuring the concentrations of the ten most common aromatic hydrocarbons (Benzene; toluene; o-, m- and p-xylenes; ethylbenzene; naphthalene; 1-

and 2-methylnaphthalenes, and 2, 6-dimethylnaphthalenes). Dilutions of the stocks were measured daily by ultraviolet spectrophotometry (Neff & Anderson, 1975). The oil samples were extracted with *n*-hexane, and the ultraviolet absorption was measured at 220 nm.

Aromatic and aliphatic concentrations in mussel soft tissues were determined after 14- and 28-d exposure. About 10 g of wet tissue was digested with an internal standard in hydroxide-hexane, centrifuged, and the hexane fraction retained. The hexane was concentrated and passed through a silica gel column. The eluate was collected, concentrated, and analyzed by gas chromatography for the predominant aromatic and aliphatic hydrocarbons. Total aromatic hydrocarbons were calculated in this study as the sum of the peak areas in chromatographic regions surrounding the ten most common aromatic hydrocarbons in the WSF of Cook Inlet crude oil. The peak areas in the region of each standard were first added together, and all further calculations assumed a single peak of equivalent area to be that of the standard.

Bioenergetic Measurements

Two sublethal physiological stress indices, scope for growth and oxygen: nitrogen ratio (Bayne, 1975), were computed from the rate functions measured during the experiment. Scope for growth is based on the balanced energy equation of Winberg (1960):

$$C - F = Ab = R + U + P$$

where C = the food energy consumed, F = the energy lost as feces, Ab = energy absorbed from the food, R = energy loss by respiration, U = energy excreted, and P = energy incorporated into somatic growth and gamete production. Production, P, (or growth) is referred to as scope for growth because it is derived by subtracting energy respired and excreted from energy absorbed from the food; scope for growth is:

$$P = Ab - (R + U)$$

Scope for growth includes energy available for somatic growth and the production of gametes. Scope for growth may be positive or negative; mussels exhibiting a negative scope for growth will eventually die of starvation.

Another potentially useful bioenergetic stress index is the oxygen:nitrogen (O:N) ratio; the ratio in atomic equivalents of oxygen consumed to nitrogen excreted (Conover & Corner, 1968). High O:N ratios (>25-30) indicate the catabolism of primarily carbohydrate and/or lipid as substrates, whereas lower values indicate that protein is the catabolic substrate (Widdows, 1978a, b; Widdows et al., 1981, 1982; Stickle, 1984).

Rates of algal clearance, oxygen consumption, and ammonia excretion were determined for mussels which were placed in individual chambers of a flow-through system. Water was pumped via a submersible pump from the appropriate dosing compartment through a manifold, with excess water returned to that dosing compartment via a return tube. Fifteen side ports from the manifold were connected to flow-through animal chambers so that blank chambers were placed at both ends and at the center of the manifold. One hour was allowed between loading mussels and the initiation of respiration and ammonia excretion determinations to allow for mussel acclimation to the chambers. Flow rates were regulated at 3-7 ml/min by inserting various gauge needles into the end of tubes on the outflow side of the animal chambers. Following the determination of oxygen consumption and ammonia excretion rates, the flow rate through each chamber was increased to a rate between 50-70 ml/min by inserting a larger gauge needle into the end of the outflow tube of each mussel chamber. One hour was allowed for mussel acclimation to the increased flow rate before determining clearance rates.

Consumption, C, was measured by determining the average of three measurements (taken over a 1.5 h period) of the clearance rate of *Phaeodactylum tricornutum* by mussels. The cell density in the outflow from the three blank and twelve mussel chambers was determined with a model ZB Coulter Counter. The clearance rate in liters per day was calculated according to the equation generated by Coughlan (1969):

$$\frac{\text{Blank chamber cells/ml} - \text{animal chamber cells/ml}}{\text{animal chamber cells/ml}} \times \begin{matrix}\text{flow}\\ \text{rate}\\ \text{(L/24 h)}\end{matrix} = \begin{matrix}\text{clearance}\\ \text{rate}\\ \text{(L/24 h)}\end{matrix}$$

Clearance rate per day was converted to joules ingested per day by multiplying the liters cleared by the algal cell density (1.2×10^4 cells/ml = 0.84 mg/L) and the energy value of the algal cells was determined in a Phillipson microbomb calorimeter (16.43 joules/mg algal cells).

The absorbed ration, Ab, was determined by multiplying joules ingested per day by the absorption efficiency of the ingested ration. Absorption efficiency was calculated by the method of Conover (1966).

Energy losses due to respiration, R, were determined by measuring the oxygen consumption rate of mussels. A Radiometer oxygen electrode housed in a flow-through thermostated cell was used to determine the PO_2 of water collected anaerobically in 1-ml syringes from outflow tubes of blank and animal chambers. Oxygen consumption rates were calculated in ml O_2/h and converted to energy expenditure rates by multiplying by the oxycalorific value of 20.07 joules (= 4.8 calories) per ml of oxygen consumed at STP (Crisp, 1971).

Energy losses due to excretion, U, were determined as the energetic equivalent of ammonia lost from mussels to the ambient sea water by the phenol hypochlorite method of Solorzano (1969). Ammonia excretion rates, µM NH^{4+}/h, were multiplied by the energy equivalent of excreted ammonia of 3.779 joules (= 0.9019 calories)/µM NH^{4+} given by Elliott and Davison (1975).

Statistical Analyses

In the initial data analysis of clearance rate, oxygen consumption, and ammonia excretion, the relationship between rate functions and body weight were described for each experimental condition by the allometric equation: $Y = aX^b$, where Y = the physiological rate, X = the mussel dry tissue weight (in mg), and *b* and *a* are the slope and intercept of the $\log_{10}Y/\log_{10}X$ regression, respectively. The individual regression lines of each rate function were then compared by analysis of covariance to test variation among treatments, and Scheffe's Test was used to identify significant differences between those treatment means (Barr *et al*., 1979).

All rate function data are based on average daily values for mussels exposed to each concentration of the WSF of Cook Inlet crude oil. The average dry weight of all mussels used in our experiment was 726 mg, and ANOVA indicated that there was no significant variation in mussel dry weights among treatments. Furthermore, log-transformed regressions of 27 rate functions on dry weight treatment-day combinations each of oxygen consumption, ammonia excretion, O:N ratios, and algal clearance rate were only significant for 0, 2, 4, and 5 combinations respectively. Therefore all data are presented on the basis of a uniform dry weight mussel weighing 762 mg.

RESULTS

Tolerance

The LC50 of *M. edulis* was >2500 ppb aromatic hydrocarbons through 7-d exposure. The LC50 of mussels was 2510 ± 710 ppb (95% C.I.) aromatic hydrocarbons on day 14 of exposure; 1930 ± 480 ppb on day 21; and 1420 ± 300 ppb on day 28.

Energy Consumed and Absorbed, C and Ab

The mussels' algal consumption rate, C, (averaged over 28-d exposure) regressed significantly in a negative relationship on WSF aromatic hydrocarbon concentration according to the following equation: algal consumption rate (joules/day) = 495 - 0.277 ppb aromatic hydrocarbons ($p<0.01$; Fig. 1). Algal consumption rates of mussels exposed to 0, 93, and 207 ppb aromatic hydrocarbons over 28 d were significantly higher than the rate for mussels exposed to 1083 ppb. No significant differences existed in the algal consumption rate by mussels exposed to other WSF concentrations of aromatic hydrocarbons over 28 d (Table 1). The duration of exposure also had a significant effect on algal consumption rate ($p>0.01$; ANOVA), with the consumption rate being significantly lower on day 21 than on day 7 (Scheffe's test). Visual inspection of the data showed that energy consumption declined directly with the duration of exposure for mussels exposed to 604 and 1083 ppb aromatic hydrocarbons for 28 d.

Mussel absorption efficiency varied between 46 and 84% during 28-d exposure to 0, 93, 207, and 381 ppb aromatic hydrocarbons (Table 1). The absorption efficiency of mussels exposed to 604 ppb aromatic hydrocarbons declined to 43% on day 28, while the absorption efficiency of mussels exposed to 1083 ppb aromatic hydrocarbons declined to 48% on day 14 and 41% on day 21 and 28 of the experiment.

The energy absorption rate, Ab, over the 28-d exposure period regressed significantly in a negative relationship with respect to WSF aromatic hydrocarbon concentration according to the equation: algal absorption rate (joules/day) = 341 - 0.199 ppb aromatic hydrocarbons ($p<0.01$; Fig. 1). Absorption rate was significantly higher in mussels exposed to 0, 93, and 207 ppb aromatic hydrocarbons than in mussels exposed to 1083 ppb. The duration of exposure also exerted a significant effect on the algal absorption rate ($p>0.01$;

ANOVA), with the rate of energy absorption being significantly lower on day 21 than on day 7 across all treatments (Scheffe's test). Energy absorbed by mussels exposed to 604 and 1083 ppb aromatic hydrocarbons for 28 d declined directly with the duration of exposure (visual inspection).

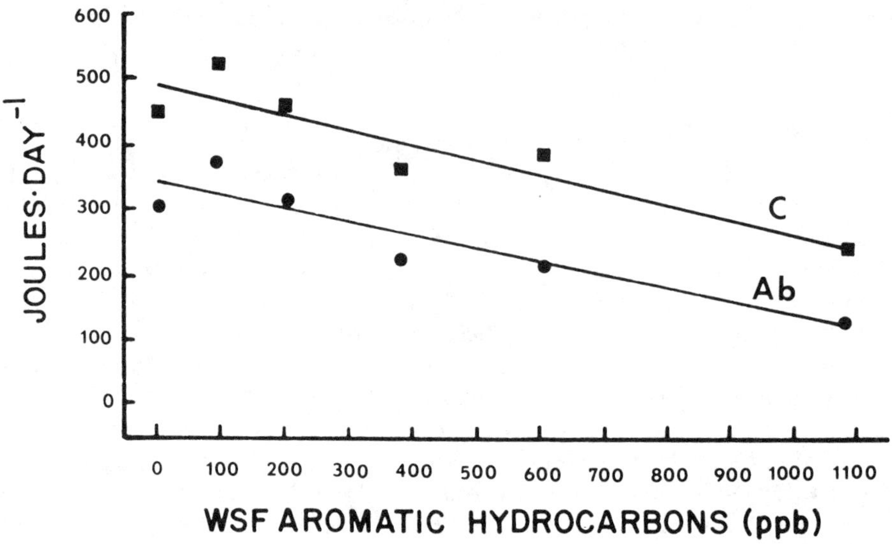

Fig. 1. Energy consumption (C) and absorption (Ab) rate (joules/standard mussel/day) of *M. edulis* over 28-d exposure to the WSF of Cook Inlet crude oil. WSF concentrations are given in ppb aromatic hydrocarbons. MSD is the minimum significant difference between treatment means as determined by Scheffe's test.

Maintenance Energy, (R + U)

The total energy expenditure rate of mussels did not regress in a significant manner against the WSF concentration of aromatic hydrocarbons over the 28-d experimental period ($p>0.05$) nor did the rate vary significantly as a function of exposure concentration or duration of exposure (Table 2).

Oxygen consumption accounted for an average of 92% of maintenance costs over all treatment combinations, and ammonia excretion accounted for 8%. Neither the concentration of aromatic hydrocarbons nor duration of exposure affected the percentage maintenance costs attributable to oxygen consumption and ammonia excretion (ANOVA; Table 2).

Table 1. Rates of algal consumption (C), absorption efficiency (AE in %), and algal absorption (Ab) are given in joules/day for standard weight (726 mg) *Mytilus edulis* exposed to the WSF of Cook Inlet crude oil for 28 d. WSF concentrations are given in ppb total aromatic hydrocarbons. Sample size is 10. Minimum significant difference (Scheffe's test) between means for day 7-28 average is 200 for C and 133 for Ab. *Mean values for the control treatment (0 ppb) include day 0 data.

Concentration (ppb)	Variable	Day 0	7	14	21	28	7-28 $\bar{x}$*
0	C	826	301	335	275	522	452
	AE	69	79	46	82	66	68
	Ab	566	239	165	225	343	308
93	C	905	544	502	473	578	521
	AE	68	64	68	80	74	71
	Ab	620	349	341	377	426	371
207	C	575	543	345	324	665	469
	AE	69	63	80	76	60	70
	Ab	395	342	277	246	400	272
381	C	585	518	452	200	285	364
	AE	69	56	68	52	73	64
	Ab	401	288	309	104	209	227
604	C	699	586	475	241	219	385
	AE	69	61	57	58	43	57
	Ab	480	355	272	140	93	218
1083	C	549	409	269	175	117	242
	AE	69	68	48	41	41	53
	Ab	376	279	130	72	48	132

Scope for Growth, C - F = Ab - (R + U)

Energy available for somatic growth and reproduction (or scope for growth) was affected by WSF concentrations, regressed significantly against the WSF aromatic hydrocarbon concentration in a negative manner, and fit the equation: Scope for growth (joules/day) = 270 - 0.232 ppb ($r = 0.9443$; $p<0.01$). Scope for growth was significantly higher in mussels exposed to 93 ppb aromatic hydrocarbons for 28 d than in those exposed to 604 and 1083 ppb (Fig. 2). Duration of exposure did not affect scope for growth values during the 28-d experimental period when analyses were compared for all treatments. However, visual inspection of scope for growth data showed a direct decline in scope for growth with increasing duration of exposure for mussels exposed to 604 and 1083 ppb aromatic hydrocarbons for 28 d.

Table 2. Rates of total energy expenditure (R + U) and percent energy expenditure due to oxygen consumption (% R), and ammonia excretion (% U), are given as joules/day and % for standard weight (726 mg) *Mytilus edulis* exposed to the WSF of Cook Inlet crude oil for 28 d. WSF concentrations are given in ppb total aromatic hydrocarbons. Sample size is 10. Minimum significant difference between means for day 7-28 average is 33 for (R + U), 3 for % R, and 3 for % U. *Mean values for the control treatment include day 0 data.

Concentration (ppb)	Variable	Day 0	7	14	21	28	7-28 $\bar{x}$*
0	(R + U)	110	77	65	82	91	85
	% R	95	91	89	89	91	91
	% U	5	9	11	11	9	9
93	(R + U)	89	84	66	78	81	78
	% R	87	92	91	93	92	92
	% U	13	8	9	7	8	8
207	(R + U)	99	91	56	95	85	82
	% R	84	88	95	96	94	93
	% U	16	12	5	4	6	7
381	(R + U)	54	59	61	73	74	67
	% R	92	91	92	93	95	93
	% U	8	9	8	7	5	7
604	(R + U)	84	51	84	77	70	71
	% R	83	90	93	90	91	91
	% U	17	10	7	10	9	9
1083	(R + U)	94	59	73	59	71	66
	% R	92	94	94	93	91	94
	% U	8	6	6	7	9	6

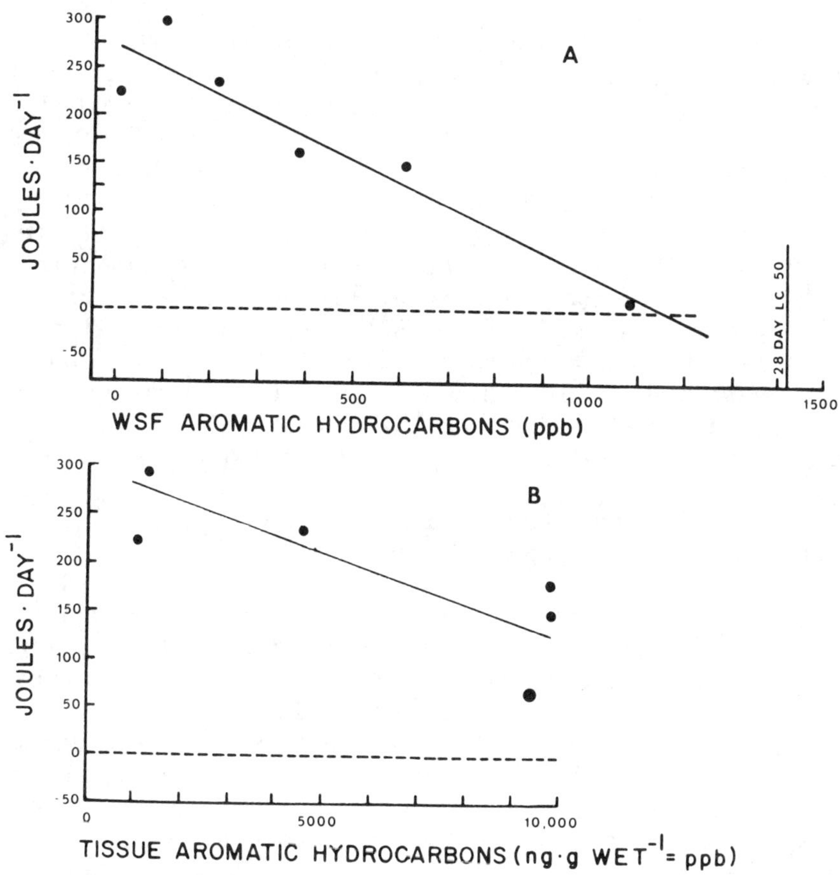

Fig. 2. Scope for growth (joules/standard mussel/day) of *M. edulis* related to 28-d exposure to the WSF of Cook Inlet crude oil (A) and to tissue aromatic hydrocarbon levels (B).

Hydrocarbon Body Burdens

Mussels bioaccumulated total aromatic and aliphatic hydrocarbons in a direct relationship with the WSF concentration. Total aromatic hydrocarbons in mussel soft tissue also increased in a direct manner with duration of exposure at WSF concentrations up to 604 ppb aromatic hydrocarbons (Fig. 3a). No difference existed in total aromatic hydrocarbons concentration of mussels exposed to an average of 1039 ppb aromatic hydrocarbons for 14 d and those exposed to 1083 ppb over 28 d. The contribution of the diaromatic hydrocarbons (naphthalene, 2-methylnaphthalene, 1-methylnaphthalene, 2,6-dimethylnaphthalene, and acenaphthene) to the total aromatic hydrocarbon body burden increased with duration of WSF exposure, 28 versus 14 d, in WSF concentrations up to 381-413 ppb, but did not vary with time of exposure at higher WSF concentrations.

Total aliphatic hydrocarbon body burdens increased linearly with the WSF concentration after 14-d exposure, but not after 28 d (Fig. 3b). Total tissue aliphatic body burden in mussels exposed to the WSF for 28 d had fallen to 43 and 80% of the aliphatic body burden after 14-d exposure in the 586-604 and 1039-1083 ppb aromatic hydrocarbon concentrations, suggesting the elimination of aliphatics from the mussel body between days 14 and 28 of the exposure period.

Relationship Between Scope for Growth and Hydrocarbon Body Burdens

Scope for growth of mussels exposed to WSF concentrations of the WSF of Cook Inlet crude oil regressed significantly in an inverse manner on tissue aromatic hydrocarbon body burden (Fig. 2b). The regression equation generated was scope for growth = 281 - 0.016 ppb tissue aromatic hydrocarbons ($r = -0.8359$, $p<0.05$, $n = 6$). Scope for growth did not regress in a significant manner against tissue aliphatic hydrocarbons or the sum of tissue aliphatic and aromatic hydrocarbons.

Oxygen Nitrogen (O:N) Ratios

No significant difference existed in the O:N ratios of mussels exposed to the WSF of Cook Inlet crude oil for 28 d. Average O:N values ranged from 20.0 for mussels exposed to 93 ppb aromatic hydrocarbons to 27.7 for those exposed to 207 ppb and indicated that protein was the principal substrate catabolized (Table 3). Duration of exposure did not effect O:N ratios of mussels (ANOVA).

These low O:N ratios occurred at the stage of the mussel reproductive cycle shortly after spawning. Mantle tissue of mussels collected for the experiment was partially empty indicating that they had recently spawned. In addition, hematoxylin- and eosin-stained sections of mantle tissue from several mussels (David Lowe, pers. comm.) verified that mussels were in the postspawning stage of the reproductive cycle in which only residual gametes remained.

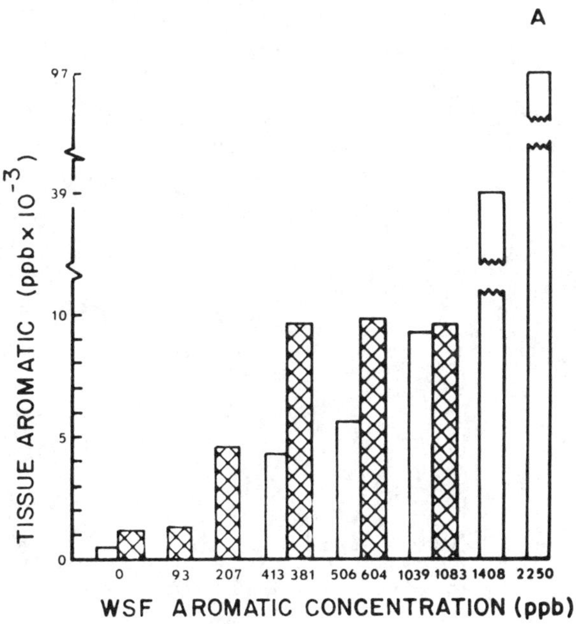

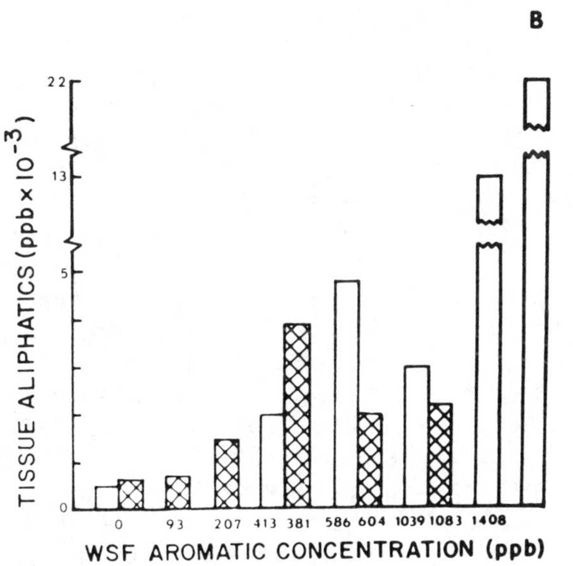

Fig. 3. Total aromatic (A) and aliphatic (B) petroleum hydrocarbon concentrations are given in ppb x 10^{-3} (nanograms/g wet weight) for mussels exposed to the WSF of Cook Inlet crude oil for 14 (clear bars), and 28 d (hatched bars). WSF concentrations on the x-axis are running mean values after 14- and 28-d exposure.

Table 3. Oxygen:nitrogen (O:N) ratios of standard weight (726 mg) *Mytilus edulis* exposed to the WSF of Cook Inlet crude oil for 28 d. WSF concentrations are given in ppb total aromatic hydrocarbons. Sample size is 10 and the minimum significant difference (Scheffe's test) between means of the day 7-28 average is 9.2.

Concentration (ppb)	Day 0	7	14	21	28	7-28 $\bar{x}$
0	32.7	18.6	14.9	13.6	20.6	20.1
93	23.3	21.6	14.4	24.6	19.4	20.0
207	17.8	15.2	30.3	37.3	30.1	27.7
381	17.1	25.6	19.4	28.2	33.0	26.5
604	14.6	21.9	26.7	17.8	19.8	21.6
1083	20.6	22.1	21.8	26.1	25.4	23.9

DISCUSSION

Tolerance of large (58-71 mm) *Mytilus edulis* to the WSF of Cook Inlet crude oil after 28-d exposure was lower than the tolerance of small mussels. Stickle *et al*. (1984) reported the 28-d LC50 of small mussels (15-30 mm) to be significantly higher than large mussels (this study) through 21-d exposure and declined to 1689 ± 42 ppb aromatic hydrocarbons on day 28, a value not significantly different from the 28-d LC50 value of 1420 ± 300 ppb for large mussels. Likewise, Rice (unpubl. data) determined the 28-d LC50 concentration of small (16-25 mm) *M. edulis* exposed simultaneously with the large mussels in this study to be >2500 ppb throughout the exposure period. Mussel size is obviously an important factor which affects their tolerance of aromatic hydrocarbons, probably because of the reproductive stage of the mussels. The large mussels had spawned, and were probably less capable of coping with simultaneous hydrocarbon stress.

Our results agree with those of Stickle *et al*. (1984) who emphasized that apparent differences in oil sensitivity between fish and invertebrates and between animals from different habitats diminish considerably if the exposure times for all species are increased. Earlier results from 96-h bioassays indicated that fish were affected more rapidly and were more sensitive than invertebrates. However, little mortality occurred in fish after the first few days of exposure. Pelagic animals were also more sensitive than intertidal animals (Rice *et al*., 1979).

Scope for growth of *M. edulis* declined inversely with WSF aromatic hydrocarbon concentration, but did not become negative in our study. An inverse relationship between scope for growth and WSF and/or body-burden pollutant concentration has been noted for *M. edulis* (Widdows *et al*., 1981, 1982) and *Thais lima* (Stickle *et al*., 1984). Regression of scope for growth on WSF aromatic hydrocarbon concentration in our study indicates that scope for growth does not decline to zero, the metabolic balance point between energy intake and losses, until a WSF concentration of 1163 ppb, which is 82% of the 28-d LC50 concentration. Scope for growth of *T. lima* declined to zero at a WSF concentration of 150 ppb representing 23% of the 28-d LC50 (Stickle *et al*., 1984). This relationship indicates that *T. lima* is physiologically stressed (Bayne, in press) at a much lower concentration relative to its LC50 concentration than *M. edulis*. Widdows *et al*. (1982) have shown that exposure of mussels to a water-accommodated fraction concentration of 30 ppb North Sea crude oil for 66-140 d caused a reduced scope for growth compared with control mussels, but which did not decline to zero. Consequently, although the scope for growth of *M. edulis* is sensitive to low-level aromatic hydrocarbon exposure over a long period of time (2-4 months), the time-course nature of changes in scope for growth of mussels indicates that *M. edulis* is not the species of choice for laboratory experiments relating oil effects to bioenergetics for experiments up to one month duration.

Energy intake (C) and absorption (Ab) rates were the only components of the energy budget of *M. edulis* to regress in a significant manner against the WSF concentration of aromatic hydrocarbons. Therefore, the inverse relationship between scope for growth and WSF aromatic hydrocarbon concentration was due to the effects of aromatic hydrocarbons on the mussels' algal clearance rate and absorption efficiency. Widdows *et al*. (1981) also found the rate of energy consumption (clearance rate) and absorption of *M. edulis*, which were transferred to a polluted location in Narragansett Bay for 30 d prior to the determination of scope for growth, declined by 44 and 24 joules/h (respectively) to 49 and 56% of the rates observed in mussels held at the control site. Stickle *et al*. (1984) also determined that energy intake (C) and absorption (Ab) were the primary driving factors affecting scope for growth of the gastropod *T. lima* exposed to the WSF of Cook Inlet crude oil for 28 d.

No relationship existed between total energy expenditure for maintenance (R + U) and aromatic hydrocarbon concentration in this study. Widdows *et al*. (1981) found energy expenditure for maintenance by *M. edulis* increased by only 4.1 joules/h (20% increase) in mussels held for 30 d at the most polluted site in their study as compared with the energy expenditure rate of control mussels.

The magnitude of the ratio of variation in energy consumption (C) to total maintenance costs (R + U) of invertebrates exposed to gradients of aromatic hydrocarbons, salinity, and temperature emphasizes the importance of the consumption component (C) to scope for growth variation. The C:(R + U) ratio of *M. edulis* was 5.0 (this study) and 10.7 (Widdows *et al*., 1981). Stickle (1984) reported this ratio to vary between 5.1 and 7.2 for five species of marine invertebrate carnivores studied. Consequently, more attention should be paid to the effects of pollutants on energy consumption and absorption than on energy expended for maintenance, if one goal of laboratory-based pollution studies is to predict ecological consequences of pollutant exposure.

A decline in oxygen-nitrogen (O:N) ratios with increased physiological stress is a generalized response pattern of *M. edulis* (Bayne, 1975; Bayne & Scullard, 1977; Widdows, 1978b; Widdows *et al*., 1981), but was not observed in our study. Oxygen:nitrogen ratios did not vary significantly as a function of WSF concentration or time of exposure, and low mean values between 20.0 and 27.7 indicate that protein was an important metabolic substrate being catabolized (Bayne, 1975; Stickle & Bayne, 1982). *M. edulis* used in our study were in the postspawning stage of their reproductive cycle. Widdows (1978b) and Widdows *et al*. (1981) have documented that mussels in this stage of the reproductive cycle have low O:N ratios; therefore, the O:N ratio is not a suitable stress index during the postspawning period of the reproductive cycle.

Tissue concentrations of total aromatic hydrocarbons were 2.1-5.8 times higher in *M. edulis* exposed in our study than in the study of Stickle *et al*. (1984). This discrepancy may be due to three causes: (1) the rate of aromatic hydrocarbon accumulation into tissues may be size dependent, (2) mussels exposed in this study were fed algae, whereas mussels used in the experiments of Stickle *et al*. (1984) were starved, and (3) unknown compounds which chromatographed near to each standard in this study were included under the peak area for the standard, whereas only the sum of the specific compounds were reported by Stickle *et al.* (1984). No data exist on the relationship between mussel size and petroleum hydrocarbon accumulation rate. Widdows *et al.* (1981) have shown that *M. edulis* exposed to very low concentrations (36 ppb) of the WAF (= WSF) of North Sea crude oil and fed algae simultaneously with the WAF for 12 h out of each day accumulated aromatic and aliphatic hydrocarbons at a rate which was 1.1-3.0 times faster than mussels exposed to the same concentration of WAF and algae during alternating 12-h periods of each day. Consequently, the simultaneous exposure of mussels to the WSF of oil and algae enhances the rate of aromatic hydrocarbon uptake.

M. edulis has been used in the mussel watch program because of its ability to bioaccumulate pollutants to concentrations considerably above sea water values (Goldberg, 1975; Goldberg *et al.*, 1978). Results obtained in this study, where *M. edulis* was exposed to the WSF of Cook Inlet crude oil for 14 or 28 d before body burden analysis, indicate that total aromatic hydrocarbons in mussel tissue increased linearly with WSF exposure and duration of exposure in WSF concentrations up to 604 ppb aromatics; at higher WSF concentrations a steady-state total aromatic hydrocarbon concentration had been reached after 14-d exposure. In contrast, total aliphatic body burdens increased linearly with WSF concentration after 14-d exposure, but not after 28 d. Indeed, our data suggest that aliphatic hydrocarbons are "eliminated" from tissues of *M. edulis* between day 14 and 28 of exposure to 580-1083 ppb aromatic concentrations in the WSF. Caution must be exercised in relating the WSF concentration of aliphatic hydrocarbons with body burden of this group of contaminants.

Increased body burden of petroleum hydrocarbons has recently been shown to correlate in a negative relationship with scope for growth of several species of molluscs (Widdows *et al.*, 1981, 1982; Stickle *et al.*, 1984). Our results indicate that scope for growth of *M. edulis* regressed in a negative relationship with total tissue aromatic hydrocarbon concentration of *M. edulis* in WSF exposures up to 1083 ppb. The total aromatic hydrocarbon concentration in the tissue of mussels exposed to 1083 ppb WSF for 28 d was slightly less (9362 ppb = ng/g wet wt) than that for mussels exposed to 604 ppb WSF (9815 ppb). No relationship existed between scope for growth of *M. edulis* and tissue body burden of aliphatic hydrocarbons. Widdows *et al*. (1981) have shown that aromatic hydrocarbon uptake is much higher in mussels feeding on oil-contaminated algae than when starved. We suggest that the lack of a tight relationship between scope for growth and total tissue aromatic hydrocarbon concentration of *M. edulis* exposed for 28 d to high (1083 ppb) WSF concentrations of aromatic hydrocarbons is due to a reduced feeding rate (Fig. 1) of mussels exposed to high WSF concentrations.

ACKNOWLEDGMENTS

This research was funded by the United States Office of Marine Pollution Assessment.

LITERATURE CITED

Barr, A.J., J.H. Goodnight, J.P. Sall, W.H. Blair, and D.M. Childo. 1979. SAS Users Guide, 1979 edition. Raleigh, North Carolina, SAS Institute, Inc.

Bayne, B.L. 1975. Aspects of physiological condition in *Mytilus edulis* L. with special reference to the effects of oxygen tension and salinity. pp. 213-238. *In*: Proc. 9th Europ. Mar. Biol. Symp. H. Barnes (ed.). Aberdeen University Press, Aberdeen.

Bayne, B.L. (in press). Responses to environmental stress: Tolerance, resistance and adaptation. *In*: Proc. 18th Europ. Mar. Biol. Symp. J.S. Gray (ed.). John Wiley and Sons, London.

Bayne, B.L. and C. Scullard. 1977. Rates of nitrogen excretion by species of *Mytilus* (Bivalvia:Mollusca). J. Mar. Biol. Assoc. U.K. 57: 355-369.

Conover, R.J. 1966. Assimilation of organic matter by zooplankton. Limnol. Oceanogr. 11: 388-354.

Conover, R.J. and E.D.S. Corner. 1968. Respiration and nitrogen excretion by some marine zooplankton in relation to their life cycles. J. Mar. Biol. Assoc. U.K. 48: 49-75.

Coughlan, J. 1969. The estimation of filtering rate from the clearance of suspensions. Mar. Biol. 2: 356-358.

Crisp, D.J. 1971. Energy flow measurement. pp. 197-1279. *In*: Methods for the Study of Marine Benthos. N.A. Holme and A.O. McIntyre (eds.). Blackwell Scientific Publications, Oxford.

DiSalvo, L.H., H.E. Guard, and L. Hunter. 1975. Tissue hydrocarbon burden of mussels as potential monitor of environmental hydrocarbon insult. Envir. Sci. Technol. 9: 247-251.

Ehrhardt, M. and J. Heinemann. 1975. Hydrocarbons in the blue mussels from the Kiel Bight. Envir. Pollut. 9: 263-282.

Elliott, J.M. and W. Davison. 1975. Energy equivalents of oxygen consumption in animal energetics. Oecologia 19: 195-201.

Farrington, J.W. and J.G. Quinn. 1973. Petroleum hydrocarbons in Narragansett Bay. I. Survey of hydrocarbons in sediments and clams (*Mercenaria mercenaria*). Est. Coastal Mar. Sci. 1: 71-79.

Fossato, V.U. and E. Siviero. 1974. Oil pollution monitoring in the Lagoon of Venice using the mussel *Mytilus galloprovincialis*. Mar. Biol. 25: 1-6.

Gilfillan, E.S., D.W. Mayo, D.S. Page, D. Donovan, and S. Hanson. 1977. Effects of varying concentrations of petroleum hydrocarbons in sediments on carbon flux in *Mya arenaria*. pp. 299-314. *In*: Physiological Responses of Marine Biota to Pollutants. F.J. Vernberg, A. Calabrese, F.P. Thurberg, and W.B. Vernberg (eds.). Academic Press, New York.

Goldberg, E.D. 1975. The mussel watch - A first step in global marine monitoring. Mar. Pollut. Bull. 6: 111.

Goldberg, E.D., V.T. Bowen, J.W. Farrington, G. Harvey, J.H. Martin, P.L. Parder, R.W. Resebrough, W. Robertson, E.D. Schneider, and E. Gamble. 1978. The mussel watch. Envir. Conserv. 5: 101-125.

Moles, A., S.D. Rice, and S.A. Andrews. (in press). Continuous-flow devices for exposing marine organisms to the water soluble fraction of crude oil and its components. Proc. Tenth Aquatic Toxicity Workshop, Halifax, Nova Scotia, Canada.

Neff, J.M. and J.W. Anderson. 1975. An ultraviolet spectrophotometric method for determination of naphthalene and alkylnaphthalenes in the tissues of oil-contaminated marine animals. Bull. Environ. Contam. Toxicol. 14: 122-128.

Phillips, D.J.H. 1980. Quantitative Aquatic Biological Indicators: Their Use to Monitor Trace Metal and Organochlorine Pollution. Applied Science, London. 488 pp.

Rice, S.D., A. Moles, T.L. Taylor, and J.F. Karinen. 1979. Sensitivity of 39 Alaskan marine species to Cook Inlet crude oil and no. 2 fuel oil. pp. 549-554. *In*: API, EPA, and USCG, 1979 Oil Spill Conference (Prevention, Behavior, Control, Cleanup), Proceedings. Washington, D.C. American Petroleum Institute.

Rice, S.D., J.W. Short, and J.F. Karinen. 1977. Comparative oil toxicity and animal sensitivity. pp. 78-94. *In*: Fate and Effects of Petroleum Hydrocarbons in Marine Ecosystems and Organisms. D.A. Wolfe (ed.). Pergamon Press, Oxford.

Silverstone, H. 1957. Estimating the logistic curve. J. Am. Stat. Assn. 52: 567-577.

Solorzano, L. 1969. Determination of ammonia in natural waters by the phenolhypochlorite method. Limnol. Oceanogr. 14: 799-801.

Stickle, W.B. (in press). Effects of environmental factor gradients on scope for growth in several species of carniverous marine invertebrates. Proc. 18th Europ. Mar. Biol. Symp. J.S. Gray (ed.). John Wiley and Sons, London.

Stickle, W.B. and B.L. Bayne. 1982. Effects of temperature and salinity on oxygen consumption and nitrogen excretion in *Thais* (*Nucella*) *lapillus* (L.). J. Exp. Mar. Biol. Ecol. 58: 1-17.

Stickle, W.B., S.D. Rice, and A. Moles. 1984. Bioenergetics and survival of the marine snail, *Thais lima*, during long-term oil exposure. Mar. Biol. 80: 281-289.

Widdows, J. 1978a. Combined effects of body size, food concentration, and season on the physiology of *Mytilus edulis*. J. Mar. Biol. Assoc. U.K. 58: 109-124.

Widdows, J. 1978b. Physiological indices of stress in *Mytilus edulis*. J. Mar. Biol. Assoc. U.K. 58: 125-142.

Widdows, J., T. Bakke, B.L. Bayne, P. Donkin, D.R. Livingstone, D.M. Lowe, M.N. Moore, S.V. Evans, and S.L. Moore. 1982. Responses of *Mytilus edulis* on exposure to the water-accommodated fraction of North Sea oil. Mar. Biol. 67: 15-31.

Widdows, J., D.K. Phelps, and W. Galloway. 1981. Measurement of physiological condition of mussels (*Mytilus edulis*) transplanted along a pollution gradient in Narragansett Bay, Rhode Island, USA. Mar. Env. Res. 4: 181-194.

Winberg, C.G. 1960. Rate of metabolism and food requirements of fishes. Fish. Res. Board Can., Translation Ser. 194: 202.

CONTRIBUTION OF PHENOL AND *P*-CRESOL TO THE TOXICITY OF CRUDE OIL TO PINK SALMON *(ONCORHYNCHUS GORBUSCHA)* FRY AND KELP SHRIMP *(EUALUS SUCKLEYI)*

Sid Korn, Stanley D. Rice,
D. Loren Cheatham, and Donald W. Brown

SUMMARY

Concentrations of phenolic compounds in the WSFs of Cook Inlet crude oil, Prudhoe Bay crude oil, and No. 2 fuel oil ranged from 0.040 to 0.088 ppm. Naphthols were not found. The concentration of phenol that killed 50% of the animals in 96 h (the 96-h LC50) was 3.7 ppm for pink salmon and 10.3 ppm for kelp shrimp; the LC50s for p-cresol were 3.4 ppm for pink salmon and 7.4 ppm for kelp shrimp. In 24 h, salmon and shrimp accumulated 11-22 times the initial exposure concentration of phenol and p-cresol. Both compounds were completely eliminated from pink salmon by day 7 or less. Shrimp depurated the compounds slower than pink salmon.

Phenol and p-cresol contributed less to the toxicity of the WSF than toluene and naphthalene. Although toluene is one-half as toxic as phenol, its concentration in the WSF is 50 times higher than the concentration of phenol. Naphthalene, which is 3-10 times more toxic than phenol, had concentrations in the WSF that were 2-7 times higher than the concentration of phenol. Furthermore, the phenols accumulate slowly and are eliminated rapidly. We therefore conclude that most of the toxicity of the WSF remains unaccounted for since phenols do not contribute substantially to the toxicity of the WSF of crude oil.

INTRODUCTION

The compounds that contribute the most to the toxicity of the water-soluble fraction (WSF) of crude oils are not necessarily the most common or most toxic of those found in the parent oils. Water-soluble fractions have a higher proportion of smaller, more water-soluble compounds than the parent oils.

There are no definitive studies that quantify the contributions of various compounds to the toxicity of oil WSFs because it is difficult to isolate the effect of one compound from the effects of other compounds in crude oil. Most research on petroleum toxicity has focused on low molecular weight aromatic hydrocarbons, such as substituted benzenes and naphthalenes--compounds that are more soluble than most of the other compounds in crude oil.

Mono- and dinuclear aromatic compounds are only slightly soluble in water but may be quite toxic. Anderson *et al*. (1974) and Rice *et al*. (1977) concluded that mono- and dinuclear aromatic hydrocarbons are the major contributors to petroleum toxicity. Caldwell *et al*. (1977), however, concluded that these aromatic hydrocarbons accounted for only 8.5% of the toxicity of oil to Dungeness crab (*Cancer magister*) larvae. Hydroxylated mono- and diaromatic compounds, such as phenol, cresol, and naphthol, are probably underrated as toxicants of crude oil (Clark & Brown, 1977), and phenolic compounds may be particularly important because they are much more water soluble than other hydrocarbons in crude oil and, thus, are easily extracted into the WSF.

For decades, phenolic compounds have been common industrial pollutants. They have been measured in effluents from petrochemical plants, refineries, and ballast-water treatment facilities (see Turnbull *et al*., 1954; Gould & Dorris, 1961; Rowe *et al*., 1983), and their toxicity, uptake, and depuration by fish and invertebrates has been measured by many researchers (reviews by Hodgins *et al*., 1977; Buikema *et al*., 1979; Jordan & Payne, 1980). The concentration of phenols present after oil spills or in the WSF of crude oil is unknown; the contribution of phenols to the toxicity of oil WSFs is also unknown. To assess the possible contribution of phenol and substituted phenols to the toxicity of petroleum WSF, we determined (1) the concentrations of phenolic compounds in the WSFs of two crude oils and No. 2 fuel oil, (2) the short-term toxicity of phenol and *p*-cresol to pink salmon (*Oncorhynchus gorbuscha*) and kelp shrimp (*Eualus suckleyi*), and (3) the accumulation and depuration of phenol and *p*-cresol by pink salmon and kelp shrimp. These results were then compared with the results of similar tests at our laboratory for toluene and naphthalene.

METHODS

Determination of Phenolic Compounds in the WSF of Crude Oils

Water-soluble fractions of Prudhoe Bay crude oil, Cook Inlet crude oil, and No. 2 fuel oil were prepared by the method of Anderson *et al*. (1974). Mixtures of 1% oil in sea water were mixed gently in 18-L glass jars with paint stirrers for 20 h, then the oil and aqueous phases were allowed to separate for 3 h before the aqueous phase was siphoned into clean 18-L jars. One-liter samples of each water-soluble fraction were collected 3, 24, and 96 h after the aqueous phase was siphoned, preserved with 10 ml concentrated sulfuric acid, and stored frozen at -10°C until extracted.

Alkanes, phenols, and other aromatic hydrocarbons were extracted from the 1-liter samples with 4 ml of 8:3 v/v dichloromethane/diethyl ether ($MeCl_2/Et_2O$) per 100 ml of sample and then with 2 ml of dichloromethane per 100 ml of sample. All glassware were rinsed with acid and solvent. The extract was concentrated, mixed with 1 ml of hexane, and separated on a column (10.5-mm i.d. column) containing 7 g SiO_2. The following eluting solvents were used: (1) Petroleum ether (PE) or pentane (15 ml), (2) 1:4 $MeCl_2$/PE (5 ml), (3) 4:6 $MeCl_2$/PE (25 ml), (4) $MeCl_2$ (15 ml), and (5) isopropanol (25 ml). Three fractions were collected that contained (1) alkanes (first 18 ml), (2) aromatic hydrocarbons (from the next 25 ml until the solvent front was 1 cm from the bottom of the SiO_2), and (3) phenols (next 6 ml).

The three fractions were analyzed with a Hewlett Packard 5840A gas chromatograph and Finnigan 3200 mass spectrometer with a 9500 model gas chromatographic data system and a Finnigan 6000 data system. (Reference to a trade name does not imply endorsement by National Marine Fisheries Service, NOAA.) The 30-cm by 0.25-mm i.d. Pyrex column was coated with FFAP, and pressure of the helium carrier gas was adjusted to a linear flow of 30 cm/s. The sample was injected with the split valve that was closed for the first 18 sec. At the time of injection, temperature of the oven was 50°C; after 5 min, the temperature was programmed to increase 4°C/min to 220°C. The detection limit (0.2 ppm) for each phenolic compound was determined by running a standard mixture of phenolic compounds through the entire analysis.

Determination of Acute Toxicity of Phenol and Cresol

Pink salmon fry were obtained from the Little Port Walter Hatchery, Alaska, and transferred to the Auke Bay Laboratory. Fry were fed Oregon Moist Pellets until they weighed 300 mg. Adult kelp shrimp (800 mg each) were collected in baited pots from Auke Bay, Alaska, and fed chopped Pacific herring (*Clupea harengus pallasi*).

The sensitivity of the salmon fry and shrimp to seven concentrations of phenol and *p*-cresol was determined using static 96-h tests (a dilution water control was also included). Test solutions were made from a stock solution of 100 ppm of phenol or *p*-cresol in sea water. Sea water from Auke Bay was used for all tests. Salinities averaged 32.3 ‰ and temperatures 6°C. Ten organisms were tested per concentration of toxicant, and tissue loading in the 18-L assay jars never exceeded 0.5 g/L. Test solutions were aerated. Low-intensity interior illumination was not controlled. Concentrations that killed half the animals (LC50) in 96 h were determined using probit analysis (Finney, 1971).

The stability of the toxicant concentrations was determined daily by ultraviolet spectrophotometry (absorbance peak for phenol, 268.5 nm; absorbance peak for *p*-cresol, 275 nm). Standard curves were prepared relating ultraviolet absorption to known concentrations of phenol or *p*-cresol dissolved in sea water.

Uptake and Depuration of Phenol and Cresol

The uptake and depuration rates of phenol and *p*-cresol were determined in adult female kelp shrimp and pink salmon fry. Shrimp or salmon were placed in 200-L covered Fiberglas tanks at a density of less than 0.7 g/L. Test solutions contained either a total of 1.5 ppm phenol including ring-labeled 14C phenol (Amersham Searle, specific activity 5.75 cpm/ng) or 1.5 ppm *p*-cresol including ring-labeled 14C *p*-cresol (Amersham Searle, specific activity 6.26 cpm/ng) and had specific activities ranging from 1.65 to 2.08 cpm/ng. Samples of the test solutions were analyzed with ultraviolet spectrophotometry and liquid scintillation radiometry at the beginning of the test and after 24 h.

Animals were kept in the test solutions for 24 h and then moved to clean, running sea water. Seven individuals of each species were individually analyzed at 3, 10, and 24 h during the exposure and at 3, 10, 24, 48, 96, and 168 h after the animals were placed in clean, flowing sea water.

Temperatures and salinities were the same as in the toxicity tests. The concentration of isotope in each animal was determined by oxidizing each animal in a Packard MD 306 sample oxidizer followed by analysis with a Beckman MD scintillation counter. Automatic external standardization was used to calculate counting efficiency of the scintillation counter, and all values were corrected accordingly.

RESULTS AND DISCUSSION

Presence of Phenol in Oil WSF

The total concentration of phenols (phenol; *o*-, *m*-, and *p*-cresol; and C_2- and C_3- substituted phenols) found in the water-soluble fraction of Cook Inlet crude oil, Prudhoe Bay crude oil, and No. 2 fuel oil (Table 1) ranged from 41 to 88 ppb and were apparently stable for 96 h. Increased concentrations of phenolic compounds would indicate biodegradation of aromatic hydrocarbons; decreased concentrations would indicate biodegradation of phenols. Naphthol and substituted dinuclear compounds were not detected (<0.2 ppb) in any WSF.

Table 1. Concentrations (ppm) of phenols in the water-soluble fraction of Prudhoe Bay crude oil, Cook Inlet crude oil, and No. 2 fuel oil (determined by gas chromatograph and mass spectrometric analyses).

Elapsed time (h)	Phenol	*p*-Cresol	C_2H_5 Phenols	C_3 Phenols	Total Phenols
		PRUDHOE BAY CRUDE OIL			
3	0.004	0.015	0.015	0.007	0.041
24	0.009	0.034	0.036	0.009	0.088
96	0.011	0.027	0.025	0.006	0.069
		COOK INLET CRUDE OIL			
3	0.007	0.020	0.013	0.0061	0.046
24	0.008	0.024	0.017	0.0069	0.056
96	0.004	0.047	0.013	0.0048	0.069
		NO. 2 FUEL OIL			
3	0.0009	0.016	0.031	0.015	0.063
24	0.0007	0.011	0.020	0.008	0.040
96	0.001	0.019	0.038	0.017	0.075

The total concentrations of phenols in all WSFs were similar, and only 1-5% of the total concentrations of aromatic hydrocarbons (Table 2); however, concentrations of specific phenols in the WSFs differed (Table 1). For example, the WSF of No. 2 fuel oil had lower concentrations of phenol and higher concentrations of C_3-phenols than the WSFs of crude oil. Concentrations of C_2-phenols and *p*-cresol in the WSF of No. 2 fuel oil were similar to concentrations of these compounds in the WSF of Prudhoe Bay crude oil (Table 1). Regardless of the oil type, the concentrations of C_2-phenols and cresols in the WSFs were usually higher than the concentrations of phenol or C_3-phenols.

Table 2. Concentration of phenols (range, ppm) in the water-soluble fraction of Cook Inlet crude oil, Prudhoe Bay crude oil, and No. 2 fuel oil. Concentrations of total aromatic hydrocarbons, toluene, and naphthalene are for a typical crude oil (mean of 20 values for Cook Inlet crude oil, 5 values for Prudhoe Bay crude oil, and 5 values for No. 2 fuel oil).

Oil	Total phenols (ppm)	Total[a] aromatic hydrocarbons (ppm)	Toluene (ppm)	Naphthalene (ppm)
Prudhoe Bay	0.040-0.088	4.820	2.000	0.084
Cook Inlet	0.046-0.069	6.610	2.500	0.150
No. 2 fuel oil	0.040-0.075	1.494	0.257	0.308

[a]Concentrations of total aromatic hydrocarbons from Rice *et al*. (1981).

Toxicity of Phenol and p-Cresol

Under the low light intensities of the 96-h toxicity tests, phenol remained within 5% of the initial concentration and *p*-cresol remained within 15% of the initial concentration; thus, static tests with these toxicants are appropriate. Continuous-flow toxicity tests should be used with unstable toxicants, such as toluene and naphthalene.

Para-cresol was more toxic to kelp shrimp than was phenol; however, pink salmon were more sensitive to both compounds than were the shrimp (Table 3). For example, the LC50 for pink salmon exposed to phenol was 3.7 ppm, and the 96-h LC50 for kelp shrimp exposed to phenol was 10.3 ppm.

The LC50s for phenol and *p*-cresol from this study are within the range of LC50s from previous studies (summarized in Buikema *et al*., 1979); however, the results of toxicity tests with phenol vary widely because of differences in methods, temperatures, and species tested. For example, LC50s of phenol in 87 short-term tests with invertebrates ranged from 2 to 1860 mg/L. In studies using fish, LC50s ranged from 4.2 to 34 mg/L phenol. In the test most comparable to our tests with pink salmon, the 24-h LC50 for rainbow trout (*Salmo gairdneri*) was 9-10 ppm (Herbert, 1962). Para-cresol has been used in fewer toxicity tests than phenol, and LC50s were 12-30 mg/L for *Daphnia magna* (two tests) and 19 mg/L with *Pimphales promelas* (Buikema *et al*., 1979).

Accumulation and Depuration

The tissue concentration of phenol and *p*-cresol in pink salmon and kelp shrimp after 24 h was 11-22 times the initial concentration in the test solutions. Apparently, neither compound reached equilibrium in the organisms (Fig. 1). Both species accumulated more *p*-cresol than phenol, and kelp shrimp accumulated significantly more of both compounds than pink salmon (Fig. 1).

Pink salmon depurated both toxicants significantly faster than kelp shrimp (Fig. 2): after 24 h in clean water, salmon retained only 21% of the accumulated phenol and 17% of the accumulated *p*-cresol; shrimp retained 64% of the accumulated phenol and 40% of the accumulated *p*-cresol. Pink salmon continued depurating the toxicants, and after 4 d, toxicants were undetectable. After the first 12 h, shrimp depurated the toxicants much more slowly than salmon, and the toxicants were undetectable after 14 d (Fig. 2).

The extent of bioaccumulation observed in our experiment was different from that observed with goldfish (*Carassius auratus*) exposed to phenol (Kobayashi *et al*., 1976) possibly because they used a different species and higher temperatures in their experiments, factors that could increase uptake and depuration rates. In our tests, pink salmon accumulated more phenol and retained it longer than the goldfish in their tests: the concentration of phenol in tissues of goldfish was only 1.2-2.3 times the exposure concentration, and the goldfish completed depuration of accumulated phenol within 2-4 h.

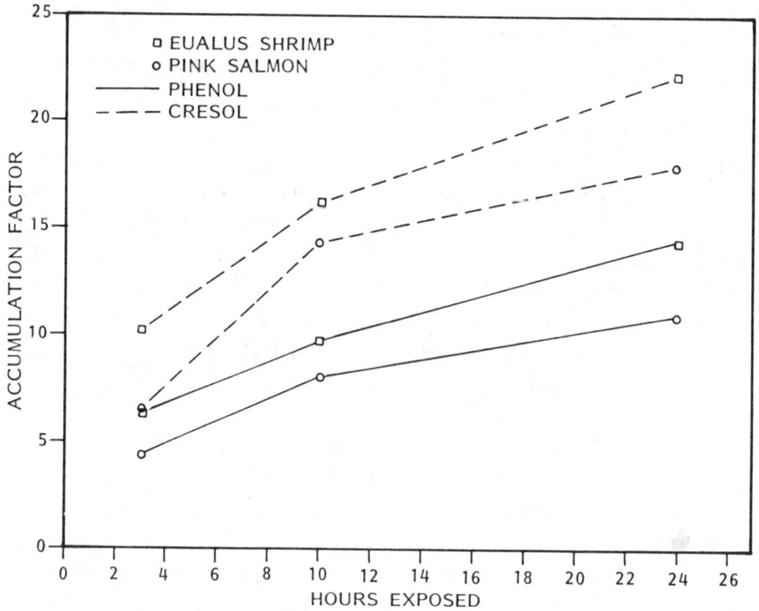

Fig. 1. Accumulation of phenol and *p*-cresol in pink salmon (*Oncorhynchus gorbuscha*) and kelp shrimp (*Eualus suckleyi*) exposed to 1.5 ppm of each toxicant. Accumulation factor is the mean tissue concentration divided by the mean exposure bath concentration.

Contribution of Phenols to the Toxicity of Oil WSF

The contribution of phenols to the short-term toxicity of WSFs of crude oil and No. 2 fuel oil can be inferred by comparing the presence of toxicity of phenols with the presence of toxicity of other important aromatic components in the WSFs. For these comparisons, we chose toluene and naphthalene because these compounds are representatives of mono- and diaromatic hydrocarbons that are thought to be major contributors to the toxicity of crude oil WSFs (Anderson *et al*., 1974; Rice *et al*., 1977). Furthermore, we have data from similar toxicity tests with pink salmon and kelp shrimp exposed to toluene and naphthalene (Korn *et al*., 1979).

Cresols, phenol, and other phenolic compounds probably contribute less to the short-term toxicity of the WSF than either toluene or naphthalene and much less than all the aromatic hydrocarbons in the WSF. Concentrations of *p*-cresol, phenol, and other phenolic compounds in WSFs of crude oil and No. 2 fuel oil are low (<0.1 ppm; Table 1), much lower than concentrations of total aromatic hydrocarbons or toluene, and slightly less than concentrations of naphthalene (Table 2). In addition, phenol and p-cresol are only slightly more toxic than toluene but less toxic than naphthalene (Table 3). When the concentrations of phenol, p-cresol, toluene, and naphthalene in the WSFs are divided by the LC50s for pink salmon and kelp shrimp, the percentage contributions of phenol or cresol to toxicity are quite low, between 0.078 and 1.4%; whereas the percents for toluene or naphthalene range from 11.5 to 33.1% (Table 3). Therefore, the potential contribution of toluene and naphthalene to the toxicity of WSFs is 10-100 times higher than that for phenol or cresol.

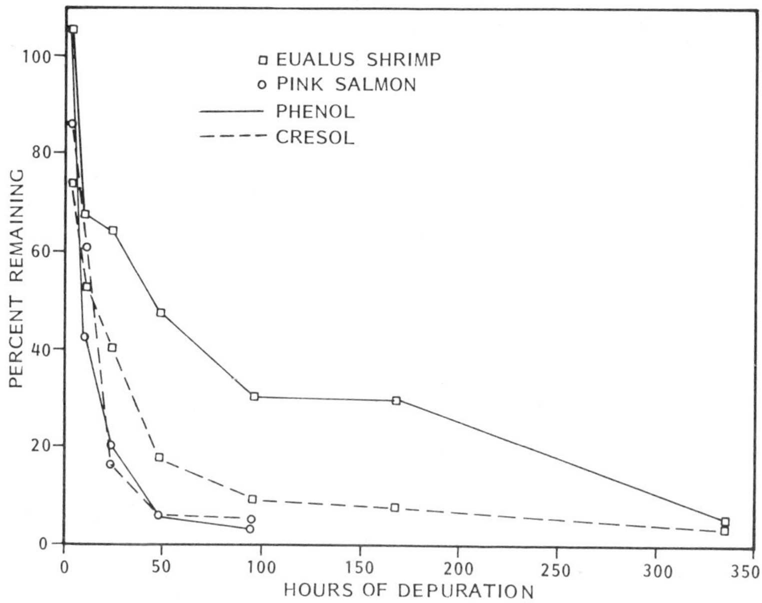

Fig. 2. Depuration of phenol and p-cresol in pink salmon and kelp shrimp that were placed in clean water after exposure to 1.5 ppm phenol or p-cresol for 24 h.

Comparing the accumulation of phenol and p-cresol to the accumulation of toluene and naphthalene further indicates the low contribution of phenols to oil toxicity. Pink salmon accumulated 10 times the initial concentration of phenol in the test solutions and 18 times the initial concentration of p-cresol. In contrast, this species accumulated 40 times the initial concentration of naphthalene (toluene and naphthalene data from Korn *et al*., 1981). Toluene accumulated to only 4 times the initial concentration of the test solutions; however, toluene is present in the WSF of oil at lower concentrations than phenol. Pink salmon accumulated less *p*-cresol than naphthalene (weighted by initial concentrations in test solutions); *p*-cresol is present at lower concentrations in the WSF than naphthalene.

Table 3. Short-term (96-h) toxicity of phenol, *p*-cresol, toluene, and naphthalene, and their concentration in the water-soluble fraction (WSF) of Cook Inlet crude oil. Concentrations are also expressed as a percentage of the LC50 for pink salmon and kelp shrimp. The LC50s for toluene and naphthalene are from Korn *et al*. (1979). The 95% confidence intervals are in parentheses.

Compound	LC50 (ppm)	Concentration in WSF (ppm)	Concentration in WSF % LC50
	Pink Salmon		
Phenol	3.73 (2.65 - 5.24)	0.008	0.22
p-cresol	3.36 (2.56 - 4.42)	0.047	1.4
Toluene	7.63 (6.86 - 8.48)	2.5	33.1
Naphthalene	1.84 (1.22 - 2.80)	0.15	11.5
	Kelp Shrimp		
Phenol	10.31 (9.51 - 11.1)	0.008	0.078
p-cresol	7.36 (6.23 - 8.69)	0.047	0.64
Toluene	20.2 (17.9 - 22.8)	2.5	12.4
Naphthalene	1.02 (0.77 - 1.34)	0.15	15.0

We conclude that phenol and *p*-cresol are unlikely to contribute substantially to the short-term toxicity of oil WSF because their presence is much lower than mono-aromatic hydrocarbons and their toxicity is much less than the diaromatic hydrocarbons. The concentration of phenol and *p*-cresol, as well as the concentration of all phenolic compounds in the oil WSFs, did not change during 96 h possibly because losses of phenolic compounds may have been offset by production of phenolic compounds from oxidized aromatic hydrocarbons. Phenolic compounds that are chronically discharged in aqueous effluents from oil refineries or processing plants may, however, lead to serious long-term effects.

Aromatic hydrocarbons are probably the major contributors to the toxicity of WSFs, even though previous work with mono- and dinuclear aromatic hydrocarbons (Caldwell *et al*., 1977) and the results of this study suggest that we have yet to account for much of the toxicity in the WSFs of crude oil and No. 2 fuel oil.

LITERATURE CITED

Anderson, J.W., J.M. Neff, B.A. Cox, and H.E. Hightower. 1974. Characteristics of dispersions and water-soluble extracts of crude and refined oils and their toxicity to estuarine crustaceans and fish. Mar. Biol. 27: 75-88.

Buikema, A.L., M.J. McGinness, and J.C. Cairns. 1979. Phenolics in aquatic ecosystems: A selected review of recent literature. Mar. Environ. Res. 2(2): 87-181.

Caldwell, R.S., E.M. Caldarone, and M.H. Mallon. 1977. Effects of a sea-water-soluble fraction of Cook Inlet crude oil and its major aromatic components on larval stages of the Dungeness crab, *Cancer magister* Dana. pp. 210-220. *In*: Fate and Effects of Petroleum Hydrocarbons in Marine Ecosystems and Organisms. Proceedings of a symposium. D.A. Wolfe (ed.). U.S. National Oceanic and Atmospheric Administration and Environmental Protection Agency. Pergamon Press, New York.

Clark, R.C. and D.W. Brown. 1977. Petroleum: Properties and analyses in biotic systems. pp. 1-75. *In*: Effects of Petroleum on Arctic and Subarctic Marine Environment and Organisms. Vol. 1. D.C. Malins (ed.). Academic Press, New York.

Finney, D.J. 1971. Probit Analysis. 3rd Edition. Cambridge Univ. Press, London. 333 pp.

Gould, W.R. and T.C. Dorris. 1961. Toxicity changes of stored oil refinery effluents. J. Water Pollut. Control Fed. 33: 1107-1110.

Herbert, D.W.M. 1962. The toxicity to rainbow trout of spent still liquors from the distillation of coal. Ann. Appl. Biol. 50: 755-777.

Hodgins, H.O., B.B. McCain, and J.W. Hawkes. 1977. Marine fish and invertebrate diseases, host disease resistance, and pathological effects of petroleum. pp. 97-173. *In*: Effects of Petroleum on Arctic and Subarctic Marine Environments and Organisms. Vol. II. D.C. Malins (ed.). Academic Press, New York.

Jordan, R.E. and J.R. Payne. 1980. Fate and weathering of petroleum spills in the marine environment, a literature review and symposium. Ann Arbor Science, Ann Arbor, Michigan. 174 pp.

Kobayashi, K., H. Akitake, and S. Kimura. 1976. Studies on the metabolism of chlorophenols in fish. VI. Turnover of absorbed phenol in goldfish. Bull. Jpn. Soc. Sci. Fish. 42(1): 45-50.

Korn, S., D.A. Moles, and S.D. Rice. 1979. Effects of temperature on the median tolerance limit of pink salmon and shrimp exposed to toluene, naphthalene, and Cook Inlet crude oil. Bull. Envir. Contam. Toxicol. 21: 521-525.

Korn, S., and S. Rice. 1981. Sensitivity to, and accumulation and depuration of, aromatic petroleum components by early life stages of coho salmon (*Oncorhynchus kisutch*). Rapp. P.-v. Reun. Cons. Int. Explor. Mer 178: 87-91.

Rice, S.D., S. Korn, C.C. Brodersen, S.A. Lindsay, and S.A. Andrews. 1981. Toxicity of ballast-water treatment effluent to marine organisms at Port Valdez, Alaska. pp. 55-61. *In*: Proceedings, 1981 Conference on Prevention and Control of Oil Spills. American Petroleum Institute. Washington, D.C.

Rice, S.D., J.W. Short, and J.F. Karinen. 1977. Comparative oil toxicity and comparative animal sensitivity. pp. 78-94. *In*: Fate and Effects of Petroleum Hydrocarbons in Marine Ecosystems and Organisms. D.A. Wolfe (ed.). Pergamon Press, Oxford.

Rowe, D.W., J.B. Sprague, and T.A. Heming. 1983. Sublethal effects of treated liquid effluent from a petroleum refinery. I. Chronic toxicity to flagfish. Aquat. Toxicol. 3: 149-159.

Turnbull, H.A., J.G. Demann, and R.F. Weston. 1954. Toxicity of various refinery materials to fresh water fish. Industrial Engineering Chem. 45: 324-333.

EFFECT OF OIL-CONTAMINATED PREY ON THE FEEDING AND GROWTH RATE OF PINK SALMON FRY *(ONCORHYNCHUS GORBUSCHA)*

Jack P. Schwartz

SUMMARY

Pink salmon fry fed 0.6, 3.2, and 6.5 ppm oil-contaminated prey (OCP) for 10 d ate less than fry fed uncontaminated prey. Fry fed 0.6 ppm OCP ate more than fry fed 6.5 ppm OCP. Fry feeding rates at each OCP dose level did not improve when the fish were fed uncontaminated prey for 5 d after exposure.

The growth rate of pink salmon fry exposed to OCP for 50 d decreased with increased dose levels of OCP. The wet weight of fry fed 0.6 ppm was significantly greater than fry fed 3.2 and 6.5 ppm OCP after 50 d. The decreases in fry feeding and growth rates were not in proportion to OCP hydrocarbon concentrations, but reductions in fry feeding rate correlated with reductions in growth rate. Reductions in the growth of fry exposed to the 96-h LC50 (0.7 ppm) of the water soluble fraction (WSF) of Cook Inlet crude oil were greater than the reductions in growth of fry fed OCP.

Pink salmon fry acquired bicyclic aromatic hydrocarbons from OCP during the first 12 h of exposure. However, hydrocarbon concentrations were below levels of detection in pink salmon fry tissues after 10-d exposure to OCP. In contrast, fry exposed to WSF rapidly accumulated mono- and bicyclic hydrocarbons within 24 h.

OCP were not acutely toxic to pink salmon fry. The fry were able to feed and grow at OCP hydrocarbon concentrations that were 5-10 times the 96-h LC50 for the WSF of Cook Inlet crude oil in sea water. However, reductions in the feeding rate of pink salmon fry both during and after exposure to

OCP, and reductions in fry growth rate from exposure to OCP, have the potential for reducing fry survival during early sea life.

INTRODUCTION

The effects of oil in sea water on reducing the feeding and growth rate of pink salmon fry, _Oncorhynchus gorbuscha_, have been documented (Rice _et al_., 1977; Moles & Rice, 1983). However, the effects of oil-contaminated prey (OCP) on feeding and growth rates of pink salmon fry are not known. Oil-contaminated zooplankton prey species can result from accidental oil spills (Conover, 1971; Grose & Mattson, 1977; Mackie _et al_., 1978), and could become a source of petroleum contamination for pink salmon fry. Although there is no information concerning the effects of OCP on feeding and growth rates of pink salmon fry, previous experiments with other salmonids exposed to oil in sea water indicate that oil can interfere with feeding behavior (Folmar _et al_., 1981; Folmar & Hodgins, 1982). Since exposure to oil in sea water can reduce prey consumption, exposure to OCP could also interfere with normal feeding.

Energy required for growth of pink salmon fry could become depleted during a critical stage in the fry's life history by prolonged exposure to OCP, which may result in reduced survival. For example, when Moles and Rice (1983) exposed pink salmon fry to sublethal concentrations of the water-soluble fraction (WSF) of Cook Inlet crude oil for 40 d, the fish ate very little food during the first several days of the experiment, and eventually utilized only 10% of their food ration. The growth rate of these fish was one-third the growth rate of control fry. The feeding behavior of juvenile coho salmon, _Oncorhynchus kisutch_, was impaired by exposure to sublethal concentrations of toluene and naphthalene, and decreased food intake was attributed to reducing the growth rate of these fish (Moles _et al_., 1981). It is not known if exposure to oil-contaminated prey will effect the growth of pink salmon fry, but previous studies with juvenile salmonids exposed to petroleum components in sea water indicate that fry growth can be reduced through reductions in food intake. Parker (1971) has shown that reductions in the size of juvenile salmon can reduce their survival during early sea life. Therefore, reductions in feeding and growth rate caused by exposure to OCP could reduce the survival of pink salmon fry.

The objective of this study was to determine the effects of oil-contaminated prey on the feeding and growth rate of pink salmon fry. Petroleum hydrocarbons acquired from the contaminated prey by the fry were also measured. The results of this experiment were used to determine the potential effect of oil-contaminated prey on the survival of pink salmon fry during early sea life.

MATERIALS AND METHODS

Fish collection and maintenance

Pink salmon fry in sea water (average length = 31 ± 0.3 mm, wet weight = 190 ± 7 mg) were obtained from the Douglas Pink and Chum, Inc. hatchery at Sheep Creek, near Thane, Alaska, and held in 800-L fiberglass aquaria at the Auke Bay Fisheries Laboratory for 20 d prior to beginning the experiments. The fry were fed Oregon Moist Pellets (Bioproducts, Inc.) for 1 wk then placed on a diet of frozen brine shrimp. When the fry reached 220 mg (± 8 mg) they were transferred to 4-L pvc (polyvinyl chloride) incubators. The sample size was 15 fry/incubator for feeding rate determinations and 30 fry/incubator for growth rate determinations. Sea water flow rate through each incubator was 250 ml/min, water temperature was 8.0 ± 1.0°C, and salinity was 28 ± 2°/$_{oo}$.

Feeding Rate

Live brine shrimp (*Artemia salina*) nauplii were used as the prey organism to simulate the plankton-like organisms that make up an important part of the diet of pink salmon fry (Cooney *et al*., 1981). Oil-contaminated prey (OCP) were prepared by exposing freshly-hatched brine shrimp nauplii to the WSF of Cook Inlet crude oil. The nauplii were placed for 24 h in either 10, 50, or 100% WSF to produce OCP at three dose levels (see Table 1). OCP rations were siphoned from the WSF dosing tank, partitioned with a Bourne plankton splitter (D. Bourne, Marine Research, Inc.), and fed to triplicate fry incubators within each dose level. A fourth group of fry in triplicate incubators was fed uncontaminated prey (controls).

The number of prey introduced to the incubators was estimated by counting three replicate subsamples of prey rations under a dissecting microscope. Uneaten prey were collected on 100 μm plankton netting each day from the incubator effluent and counted to obtain the number of brine shrimp eaten. Sea water entered the incubators from the bottom and exited by a surface drain so that uneaten prey would not settle to the bottom of the incubator. The fish were fed contaminated prey for 10 d, then fed uncontaminated prey for 5 d to measure feeding rates following exposure to OCP. The feeding rates of fish exposed to OCP were compared to controls using a two-way analysis of variance (Sokal & Rohlf, 1969). There was no significant difference in the number of prey fed daily to the fish in each incubator (average = 4420 prey/day/fish, ± 89, n = 60).

Growth Rate

OCP rations used for determining fry growth rates were removed from the WSF dosing tanks, concentrated on plankton netting, and frozen in aluminum weighing dishes. Frozen brine shrimp pellets were punched with a cork borer and placed in incubators containing the fry. Pink fry were fed 3-4% of their wet weight/day based on the wet weight of the control fish. Prey rations were adjusted upwards every 2 wk. The ration size was determined by comparing the proximate analysis of Oregon Moist Pellet (OMP) to that of brine shrmip (Gallagher & Brown, 1975) and increasing the amount of brine shrimp until the caloric value was the same as OMP. This feeding method provided sufficient nutrition to support fry growth. Brine shrimp nauplii were nutritionally inferior to OMP, and live nauplii could not be cultured in the large quantities necessary to feed all the fish.

Fry were fed uncontaminated prey until their wet weight reached 250 mg. Three fry groups were then fed low-, medium-, and high dose OCP, respectively, and a fourth group of fry was fed uncontaminated prey (controls). A fifth group of fish was placed in WSF at 50% of the 96-h LC50 for pink salmon fry, which is approximately 0.7 ppm WSF (Moles & Rice, 1983), and fed uncontaminated prey to compare the effects of WSF on fry growth with the effects of OCP. Fry were sacrificed at 10, 23, 36 and 50 d. Triplicate fry incubators/dose level were sampled on day 50, and average length, wet weight, dry weight, and percent dry weight of fry fed OCP compared to controls by Dunnett's-t (Dunnett, 1964). Fry length (snout to caudal peduncle) and wet weight were measured on all fry/incubator, and 10 of the fry/incubator were dried in an oven (60°C) to determine dry weight and percent dry weight. The remaining fry from each incubator were analyzed for hydrocarbon content.

Hydrocarbon Uptake

In addition to the fish sampled on days 10, 23, 36, and 50 of the growth rate experiment, fry fed the highest dose of OCP were also sampled during the first day of the experiment at 3, 7, and 12 h in order to be certain that fish were acquiring oil from the contaminated prey. Whole fry were processed for hydrocarbons using the methods of Warner (1978), and quantified with a Hewlett-Packard model 5880A gas chromatograph using a fused silica capillary column and methyl silicone fluid as the stationary phase. The hydrocarbon concentrations of OCP (Table 1) were determined by the same methods used to analyze whole fry. Hydrocarbon concentrations in the prey and fish were based on the prevalent mono- and bicyclic aromatic compounds.

Table 1. Average aromatic hydrocarbon concentrations of WSF-exposed brine shrimp prey (n = 5 prey rations). Parentheses contain the standard error of the mean.

Compound	ppm (μg/g) Aromatic Hydrocarbons		
Toluene	0.04 (.01)	0.06 (.02)	0.12 (.02)
Ethyl Benzene	0.03 (.01)	0.11 (.02)	0.21 (.21)
Xylenes	0.05 (.01)	0.26 (.05)	1.26 (.21)
Trimethyl Benzenes	0.21 (.03)	0.24 (.02)	0.37 (.07)
Naphthalene	0.08 (.01)	0.62 (.08)	0.96 (.18)
Methyl Naphthalenes	0.14 (.02)	1.15 (.17)	1.87 (.31)
Dimethyl Naphthalenes	0.07 (.01)	0.80 (.08)	1.75 (.11)
Total	0.62	3.24	6.54

RESULTS

Feeding Rate

There were no mortalities among the fish fed oil-contaminated prey (OCP) but the feeding rates of the frey fed OCP were lower than controls (Fig. 1). All fry fed OCP ate significantly less than controls during the 10-d exposure period (Table 2). The 10-d average daily feeding rates for fish fed control, 0.6, 3.2, and 6.5 ppm OCP were, respectively, 3152 (± 422), 1958 (± 283), 1693 (± 304), and 1428 (± 224) prey/day/fish. Fry fed 0.6, 3.2, and 6.5 ppm OCP ate, respectively, 37, 44, and 55% less than controls. A two-way ANOVA applied to the three fry groups fed OCP indicated that the fish fed 0.6 ppm OCP ate significantly more than fish fed 6.5 ppm OCP (Table 3). The feeding rates of fry fed 3.2 ppm OCP were not significantly different than the feeding rates of fish fed 0.6 and 6.5 ppm OCP.

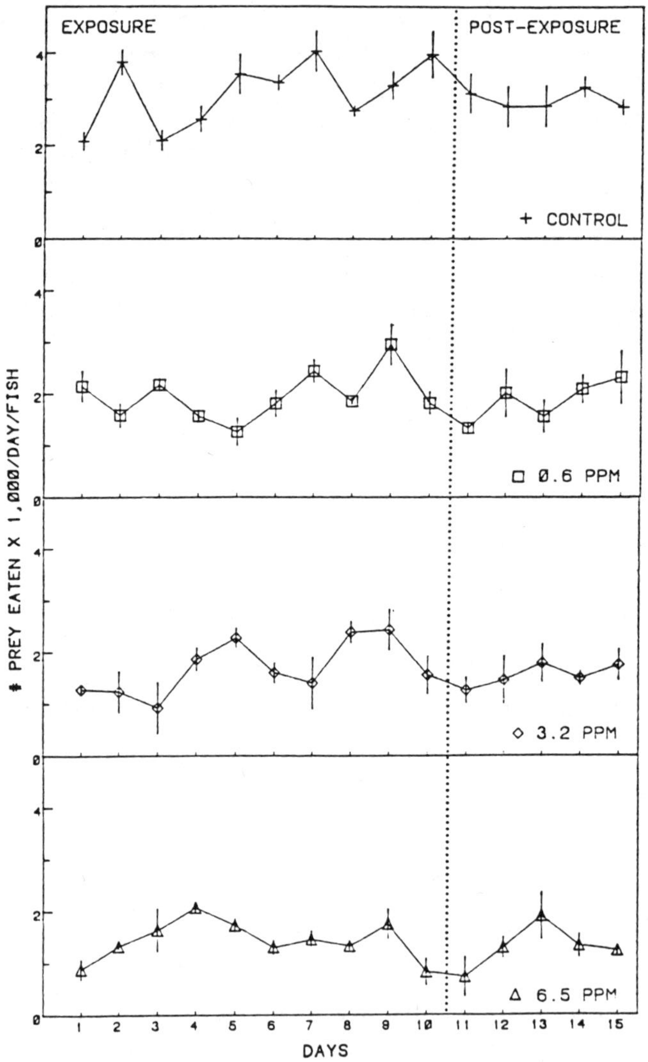

Fig. 1. Average daily feeding rates (± SE) of control pink salmon fry and fry exposed to OCP.

Table 2. Two-way ANOVA tables comparing feeding rates of control pink salmon fry and fry fed oil-contaminated prey during the exposure and post-exposure periods. A = time; B = treatment.

Source of Variation	S.S.	d.f.	M.S.	F-ratio
Exposure				
A	8.96×10^6	9	9.95×10^5	4.325 ***
B	5.21×10^7	3	1.74×10^7	19.892 ***
A x B	2.36×10^7	27	8.72×10^5	3.791 ***
error	1.03×10^8	80	2.30×10^5	
Post-exposure				
A	1.62×10^6	4	4.05×10^5	1.184 ns
B	2.51×10^7	3	8.38×10^6	24.511 ***
A x B	3.53×10^6	12	2.94×10^5	0.860 ns
error	1.37×10^7	40	3.42×10^5	

*** = $p<0.001$; ns = not significant

Table 3. Two-way ANOVA tables comparing feeding rates of pink salmon fry fed oil-contaminated prey during the exposure and post-exposure periods. A = time; B = treatment.

Source of Variation	S.S.	d.f	M.S.	F-ratio
Exposure				
A	7.28×10^6	9	8.09×10^5	3.864 ***
B	4.24×10^6	2	2.12×10^6	3.552 *
A x B	1.08×10^7	18	5.98×10^5	2.856 ***
error	1.26×10^7	60	2.09×10^5	
Post-exposure				
A	2.62×10^6	4	6.55×10^5	2.004 ns
B	2.42×10^6	2	1.21×10^6	3.706 *
A x B	2.03×10^6	8	2.53×10^5	0.775 ns
error	9.80×10^6	30	3.27×10^5	

* = $p<0.05$; *** = $p<0.001$; ns = not significant

Fry fed OCP at each dose level continued to eat significantly less than controls during the 5-d post-exposure period (Table 2). The 5-d average daily feeding rates of fry fed control, 0.6, 3.2, and 6.5 ppm OCP were, respectively, 2992 (± 116), 1861 (± 233), 1551 (± 127), and 1294 (± 242) prey/day/fish during the post-exposure period. The fish fed 0.6 ppm OCP ate significantly more than the fish fed 6.5 ppm OCP (Table 3).

Growth Rate

There were no mortalities among the fish fed OCP during the 50 d growth rate experiment, but the growth of fry fed OCP was reduced at all three OCP dose levels. At the beginning of the experiment, average fry length was 32 mm (± 0.2 mm), wet weight was 255 mg (± 10 mg), and dry weight was 45 mg (± 3 mg). The fry fed uncontaminated prey increased in size during the 50-d experiment. Fry exposed to OCP also increased in size, but were significantly smaller than controls in length (Fig. 2), wet weight (Fig. 3), and dry weight (Fig. 4) by the end of the experiment (Table 4). The wet weight of fish exposed to 3.2 ppm OCP did not increase during the last 14 d of the experiment, and fry fed 6.5 ppm OCP lost weight (Fig. 4). By day 50, the wet weight of fry fed 0.6, 3.2, and 6.5 ppm OCP was, respectively, 16, 28, and 36% less than controls.

Although there was no significant difference in the percent dry weight between fry fed OCP and controls 36 d into the study, the percent dry weight of fry fed 3.2 and 6.5 ppm OCP was significantly less than controls by the last day of the experiment (Table 4). This suggests that during the last 14 d of the experiment the fish fed 3.2 and 6.5 ppm OCP were being depleted of body tissue faster than it could be replaced through the incorporation of food energy.

The reductions in feeding and growth rates with increasing OCP dose levels were very similar, even though feeding and growth rates were obtained from separate experiments. Figure 5 presents the changes in fry growth rate over 50 d (% increase in wet weight/day/fish) and average feeding rates (no. prey eaten x 1000/day/fish) during the exposure period from the previous feeding rate experiment. There was a significant correlation between feeding and growth rates (Pearson's $r = 0.980$, $p<0.05$).

Fry exposed to WSF had greater reductions in growth than the fish fed OCP (Table 4). Fry mortalities from exposure to WSF increased during the experiment, and a sample of the WSF-exposed fish on day 36 was eliminated so that sufficient numbers of fish would be available on day 50. The fish exposed to WSF had more prey than fry fed OCP because prey rations were based on the wet weight of control fish, and the number of fish exposed to WSF continuously decreased. Fry exposed to WSF increased in wet weight by only 27 mg over the 50-d exposure period, and dry weight decreased by 2 mg. Thus, despite an abundant food supply, the growth of these fish was severely reduced.

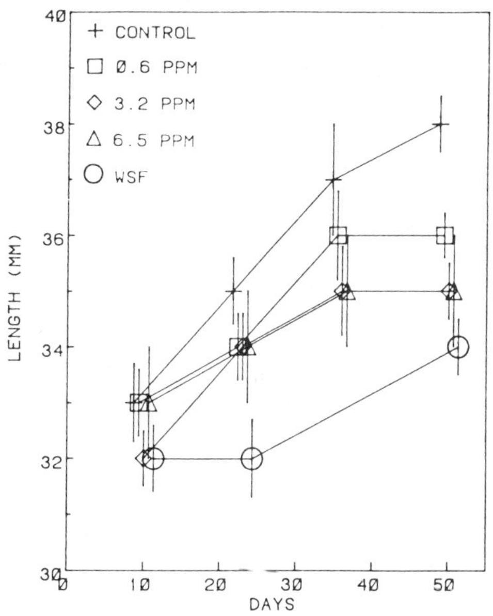

Fig. 2. Average length (± 95% confidence internal) of control pink salmon fry, fry exposed to OCP, and fry exposed to 0.7 ppm WSF in sea water over a 50-d period.

Hydrocarbon Uptake

Naphthalene, methylnaphthalenes, and dimethylnaphthalenes were found in the fry 3 h after the fish were fed 6.5 ppm OCP (Table 5). Mono-aromatic hydrocarbons were not detected in any of the fish fed OCP. After 12 h, aromatic hydrocarbon concentrations were less than half the concentration detected 9 h earlier. In contrast, fry exposed to WSF in sea water rapidly accumulated mono- and bicyclic aromatic hydrocarbons from solution during the first 24 h of exposure. Bicyclic hydrocarbon concentrations in fry exposed to WSF were more than five times the highest concentrations detected in the fry fed 6.5 ppm OCP.

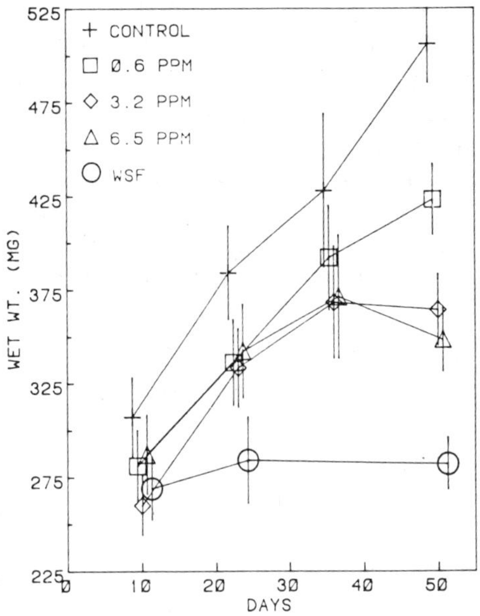

Fig. 3. Average wet weight (± 95% confidence interval) of control pink salmon fry, fry exposed to OCP, and fry exposed to 0.7 ppm WSF in sea water over a 50-d period.

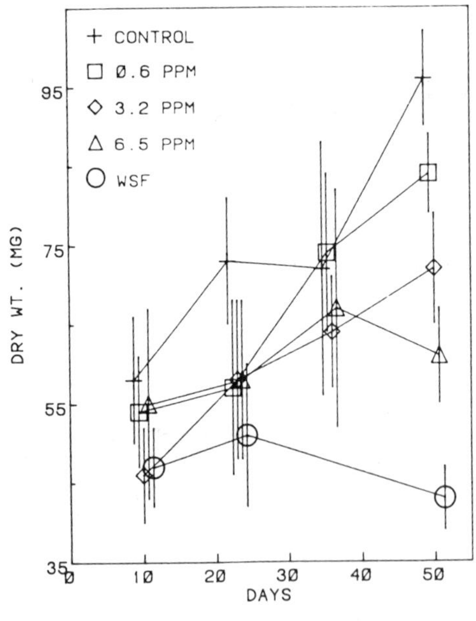

Fig. 4. Average dry weight (± 95% confidence interval) of control pink salmon fry, fry exposed to OCP, and fry exposed to 0.7 ppm WSF in sea water over a 50-day period.

By day 10, small quantities of bicyclic hydrocarbons were detected in fry fed OCP. The fry fed 6.5 ppm OCP still contained naphthalene (0.01 ppm) and methylnaphthalenes (0.02 ppm), fry fed 3.2 ppm OCP contained naphthalene (0.01 ppm), and hydrocarbon concentrations in the fish fed 0.6 ppm OCP were below limits of detection. Aromatic hydrocarbons were below limits of detection on day 23 in fry fed all three OCP dose levels, even though the fish were still ingesting OCP. This indicated that the fry fed OCP were rapidly metabolizing and/or excreting hydrocarbons acquired from the diet, but it is not known if the fish were accumulating metabolites since these were not measured. In contrast, WSF-fry hydrocarbon concentrations at 10 d were the same as those detected after 24 h, and did not change during the experiment.

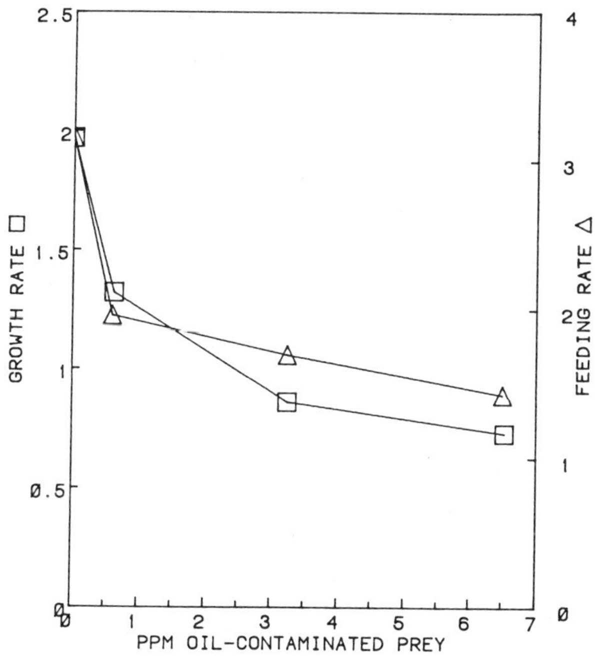

Fig. 5. Pink salmon fry growth rate (% increase in wet wt/day/fish) and corresponding feeding rate (no. prey eaten x 1000/day/fish) for control fry, and fry at each OCP hydrocarbon concentration.

Table 4. Average length, wet weight, dry weight, and percent dry weight of control pink salmon fry, fry fed oil-contaminated prey, and fry exposed 0.7 ppm WSF in sea water for 50 d. Sample size for length and wet weight was 30 fish. Dry weight and percent dry weight sample size was 10 fish. Parentheses contain 95 percent confidence intervals. Levels of significance are based on Dunnett's-t for the three replicates within each fry group.

				ppm Oil-Contaminated Prey						
Replicate No.		Control		0.6		3.2		6.5		WSF	
Length (mm)	1	38	(0.8)	36	(0.9)**	35	(0.8)**	35	(0.9)**	[a]34	(0.9)**
	2	38	(0.9)	36	(0.6)	35	(0.8)	35	(1.0)	[b]33	(0.9)
	3	38	(0.6)	36	(0.8)	35	(0.7)	35	(0.7)	[b]34	(0.7)
Wet wt (mg)	1	491	(36)	423	(39)**	362	(39)**	352	(29)**	[a]288	(32)**
	2	518	(43)	420	(30)	371	(34)	346	(33)	[b]273	(26)
	3	513	(28)	422	(30)	359	(26)	342	(28)	[b]287	(23)
Dry wt (mg)	1	99	(8)	87	(10)*	75	(18)**	66	(8)**	[c]16.3	(0.6)**
	2	100	(17)	80	(11)	73	(13)	60	(14)	[c]15.9	(0.4)
	3	88	(6)	85	(11)	68	(10)	58	(12)	[c]49	(9)
% Dry wt	1	17.8	(0.4)	17.4	(0.2)	16.7	(0.5)*	16.3	(0.5)**	[c]16.3	(0.6)**
	2	17.5	(0.4)	17.1	(0.5)	6.8	(0.3)	15.9	(0.5)	[c]15.9	(0.4)
	3	17.5	(0.4)	17.4	(0.6)	16.6	(0.4)	16.2	(0.5)	[c]16.3	(1.0)

a = 10 fish, b = 13 fish, c = 5 fish
* = $p<0.05$; ** = $p<0.01$

Table 5. Concentrations of aromatic hydrocarbons (ppm) in pink salmon fry exposed to 0.7 ppm WSF in sea water for 24 h, and for fry fed 6.5 ppm oil-contaminated prey for 12 h.

	0.7 ppm WSF	6.5 ppm Oil-contaminated Prey		
Compound	24 h	3 h	7 h	12 h
Toluene	0.82	nd	nd	nd
Ethyl Benzene	0.27	nd	nd	nd
Xylenes	1.23	nd	nd	nd
Trimethyl Benzenes	1.10	nd	nd	nd
Naphthalene	0.59	0.09	0.05	0.03
Methyl Naphthalenes	1.62	0.15	0.08	0.03
Dimethyl Naphthalenes	1.10	0.21	0.18	0.10

nd = not detected; below 0.01 ppm.

DISCUSSION

Pink salmon fry were able to grow when fed oil-contaminated prey (OCP) at hydrocarbon concentrations that would be lethal if present in sea water at the same levels. In contrast, the greatest reduction in fry growth resulted from exposure to WSF in sea water at a hydrocarbon concentration that was nearly an order of magnitude less than the highest hydrocarbon concentration of OCP. Even though OCP hydrocarbon concentrations were higher than the WSF concentration, the actual amount of hydrocarbons exposed to the fish from the diet was considerably less than exposure to WSF. The fry fed 6.5 ppm OCP consumed approximately 4% of their wet weight each day, which amounted to less than 0.15 μg of aromatic hydrocarbons. In contrast, the fry exposed to WSF acquired much greater quantities of aromatic hydrocarbons because the volume of sea water containing hydrocarbons was much greater than the volume of prey consumed by the fish fed OCP. Fishes are capable of metabolizing and excreting aromatic hydrocarbons acquired from their diet (Roubal _et al._, 1977; Thomas & Rice, 1982). This was demonstrated in the fry fed OCP by the decline of aromatic hydrocarbons during the first 12 h of OCP exposure and the absence of aromatic hydrocarbons for the remaining 50 d of exposure, when the rates of metabolism and excretion of aromatic hydrocarbons exceded the rate of hydrocarbon accumulation. The accumulation of metabolites, however, was not measured and they may have been present in the fish. In contrast, the fry exposed to WSF accumulated aromatic hydrocarbons to levels approximately an order of magnitude greater than the highest hydrocarbon concentration in the fry fed OCP, and retained aromatic hydrocarbons during the 50-d study. Hydrocarbons accumulated faster than they could be metabolized and excreted by the fry exposed to WSF because the fry were exposed to a relatively large amount of aromatic hydrocarbons when compared to the amount of hydrocarbons available from OCP. Furthermore, it has been suggested that the accumulation of aromatic hydrocarbons via the gills is a more direct pathway to internal organs (by way of the bloodstream) than from the gut (Dixit & Anderson, 1977). Pink salmon fry exposed to WSF probably accumulated aromatic hydrocarbons because the bloodstream in the gills of the WSF-exposed fish was also in close proximity to aromatic hydrocarbons in sea water, while the fry fed OCP were probably metabolizing and excreting hydrocarbons contained in the digestive tract.

Although the fry fed 0.6 ppm OCP ate more than the fry fed 6.5 ppm OCP, the differences in feeding rates were not as great as the differences in hydrocarbon concentrations. A ten-fold increase in hydrocarbon concentration, for example, produced only a 20% change in feeding rates. The reason for the relatively small difference in feeding rate over a wide range of hydrocarbon concentrations is not clear. Perhaps fry feeding rates were affected at a threshold OCP dose level. Further reductions in feeding rate would therefore not be in proportion to increasing OCP hydrocarbon concentrations, because an impairment in fry feeding behavior was initiated at the lower OCP hydrocarbon concentration.

Pink salmon fry did not avoid OCP. In a preliminary experiment, pink salmon fry did not avoid eating OCP when given a choice between OCP and uncontaminated prey, because petroleum hydrocarbons were later detected in fry tissues (Schwartz, unpubl. data). The ingestion of petroleum-contaminated prey has also been reported with other fish allowed to choose between contaminated and uncontaminated prey (Blackman, 1974). However, fry feeding rates in the present study did not improve when the fry were later fed uncontaminated prey. This suggests that exposure to OCP had a prolonged conditioning effect on the feeding behavior of pink salmon fry. The post-exposure feeding rates of fry previously exposed to each OCP dose level remained at or below the feeding rate of fry exposed to 0.6 ppm OCP, possibly because the effect of OCP at a threshold concentration of 0.6 ppm continued to mask the recognition of uncontaminated prey for the fry previously exposed to 3.2 ppm OCP and 6.5 ppm OCP. It is clear, however, that the effect of OCP on reducing the feeding rate of pink salmon fry is not immediately reversible when oil is removed from the diet.

It is not known if fry metabolic rate is affected by exposure to OCP or if exposure to OCP interferes with the uptake of food energy. Although the similarity between reductions in feeding and growth rates of pink salmon fry at each OCP dose level suggests that reductions in growth rate were mainly the result of reductions in feeding rate, an increase in metabolic rate of fry exposed to OCP may have contributed to reducing fry growth rate. Rice *et al.* (1977) reported that the metabolic rate of pink salmon fry increases with exposure to WSF, and Moles and Rice (1983) concluded that a reduction in feeding rate of pink salmon fry and an additional energy burden, caused by an increase in metabolic rate, were responsible for reducing the growth rate of fry exposed to WSF. Furthermore, previous experiments with petroleum hydrocarbons in the diet of juvenile chinook salmon, *Oncorhynchus tshwaytsha*, indicated that gut tissues were damaged when the fish were exposed to a mixture of aromatic hydrocarbons via the diet, remained damaged when contaminated food was removed from the diet, and suggest that ingested petroleum hydrocarbons could impair the normal uptake of food energy (Hawkes *et al.*, 1980). Hawkes (1977) reported that chronic exposure of rainbow trout, *Salmo gairdneri*, to oil via the diet depleted the trout of energy reserves and reduced growth. Reductions in the percent dry weight of fry fed 3.2 and 6.5 ppm OCP in the present study indicated that after chronic exposure to OCP the fish were beginning to catabolize body tissues. Reductions in percent dry weight could have resulted from reduced food intake, but evidence from previous studies suggests that a reduction in fry growth from chronic exposure to OCP could also result from a reduction in the uptake of food energy and an increase in fry metabolic rate.

Since the survival of juvenile salmonids has been shown to decrease with size because of selective predation by large fishes (Parker, 1971), OCP has the potential for reducing the survival of pink salmon fry by reducing growth rate. Although exposure to WSF had a greater effect on reducing the growth rate of pink salmon fry than OCP in the present study, the fry can detect and avoid petroleum in sea water (Rice, 1973). However, the fry in the present study did not avoid OCP, and it is possible that OCP could have a greater effect than WSF on reducing fry survival when the results of this study are applied to existing conditions in the natural environment. Prey organisms available to salmon fry include pelagic and benthic invertebrate species (Healy, 1979; Cooney *et al.*, 1981). Petroleum hydrocarbons introduced into the marine environment can be absorbed from solution by pelagic prey species, as in the present study using plankton-like brine shrimp nauplii, or taken up by benthic prey species in contact with contaminated bottom sediments and become available to pink salmon fry. Even though OCP were not lethal to pink salmon fry in this study, fry feeding and growth rates were reduced, and fry survival could be inhibited as a result.

ACKNOWLEDGMENTS

Funding for preliminary experiments for this study were provided by the Alaska Sea Grant College Program under grant no. NA79AA-D-00138. Dr. Stanley D. Rice and D. Adam Moles provided helpful advice while conducting this study and with reviewing the manuscript. This research is based on a dissertation submitted in partial fulfillment of the requirements for the Doctor of Philosophy degree with the University of Alaska.

LITERATURE CITED

Blackman, R.A.A. 1974. Effects of sunken oil on the feeding of plaice on brown shrimps and other benthos. Int. Conc. Explor. Sea. CM 1974/E:24, 7 pp.

Conover, R.J. 1971. Some relationships between zooplankton and bunker C oil in Chedabucto Bay following the wreck of tanker *Arrow*. J. Fish. Res. Board Can. 38: 1327-1330.

Cooney, R.T., D. Urquhart, and D. Barnard. 1981. The behavior, feeding biology, and growth of hatchery released pink and chum salmon fry in Prince William Sound, Alaska. Alaska Sea Grant College Program Rpt. No. 81-5.

Dixit, D. and J.W. Anderson. 1977. Distribution of naphthalenes within exposed *Fundulus similis* and correlations with stress behavior. pp. 633-636. *In*: Proceedings of the 1977 Oil Spill Conference, March 8-10, New Orleans, LA. American Petroleum Institute Publ. No 4284.

Dunnett, C.W. 1984. New tables for multiple comparisons with a control. Biometrics. 20: 482-491.

Folmar, L.C., D.R. Craddock, J.W. Blackwell, G. Joyce, and H.O. Hodgins. 1981. Effects of petroleum exposure on predatory behavior of coho salmon (*Oncorhynchus kisutch*). Bull. Environ. Contam. Toxicol. 27: 458-462.

Folmar, L.C. and H.O. Hodgins. 1982. Effects of Aroclor 1254 and no. 2 fuel oil, singly and in combination, on predator-prey interactions in coho salmon (*Oncorhynchus kisutch*). Bull. Environ. Contam. Toxicol. 29: 24-28.

Gallagher, M. and W.D. Brown. 1975. Composition of San Francisco Bay brand brine shrimp (*Artemia salina*). J. Agric. Food Chem. 23(4): 630-632.

Grose, P.L. and J.S. Mattson. 1977. The *Argo Merchant* oil spill, a preliminary scientific report. U.S. Dept. of Commerce, NOAA, Boulder, CO.

Hawkes, J.W. 1977. The effects of petroleum hydrocarbon exposure on the structure of fish tissues. pp. 115-128. *In*: Fate and Effects of Petroleum hydrocarbons in Marine Organisms and Ecosystems. D.A. Wolfe (ed.). Pergamon Press, New York.

Hawkes, J.W., E.H. Gruger, Jr., and O.P. Olson. 1980. Effects of petroleum hydrocarbons and chlorinated biphenyls on the morphology of the intestine of chinook salmon (Oncorhynchus tshwaytsha). Environ. Res. 23: 149-161.

Healy, M.C. 1979. Detritus and juvenile salmon production in the Nanaimo estuary: I. Production and feeding rates of juvenile chum salmon (Oncorhynchus keta). J. Fish. Res. Board Can. 36: 488-496.

Mackie, P.R., R. Hardy, E.I. Butler, P.M. Holligan, and M.F. Spooner. 1978. Early samples of oil in water and some analysis of zooplankton. Mar. Pollut. Bull. 9(11): 296-297.

Moles, A., S. Bates, S.D. Rice, and S. Korn. 1981. Reduced growth of coho salmon fry exposed to two petroleum components, toluene and naphthalene, in freshwater. Trans. Amer. Fish. Soc. 110: 430-436.

Moles, A. and S.D. Rice. 1983. Effects of crude oil and naphthalene on growth, caloric content, and fat content of pink salmon juveniles in sea water. Trans. Amer. Fish. Soc. 112: 205-211.

Parker, R.R. 1971. Size selective predation among juvenile salmonid fishes in a British Columbia inlet. J. Fish. Res. Board Can. 28: 1503-1510.

Rice, S.D. 1973. Toxicity and avoidance tests with Prudhoe Bay oil and pink salmon fry. pp. 667-670. In: Proceedings of the 1973 Oil Spill Conference, 13-15 March, Washington, D.C. American Petroleum Institute publ. no. 4172.

Rice, S.C., R.E. Thomas, and J.W. Short. 1977. Effect of petroleum hydrocarbons on breathing and coughing rates and hydrocarbon uptake-depuration in pink salmon fry. pp. 259-277. In: Physiological Responses of Marine Biota to Pollutants. F.J. Vernberg, A. Calabrese, F.P. Thurberg, and W.B. Vernberg (eds.). Academic Press, Inc., New York.

Roubal, W.T., T.K. Collier, and D.C. Malins. 1977. Accumulation and metabolism of carbon 14 labeled benzene, naphthalene, and anthracene by young coho salmon (Oncorhynchus kisutch). Arch. Environ. Contam. Toxicol. 5: 513-529.

Sokal, R.R. and F.J. Rohlf. 1969. Biometry. The Principles and Practice of Statistics in Biological Research. W.H. Freeman and Co., San Francisco. 776 pp.

Thomas, R.E. and S.D. Rice. 1982. Metabolism and clearance of phenolic and mono-, di-, and polynuclear aromatic hydrocarbons by dolly varden char. pp. 161-176. In: Physiological Mechanisms of Marine Pollutant Toxicity. W.B. Vernberg, A. Calabrese, F.P. Thurberg, and F.J. Vernberg (eds.). Academic Press, Inc., New York.

Warner, J.S. 1978. Chemical characterization of marine samples. American Petroleum Institute Publ. No. 4307. 36 pp.

RESPIRATORY RATE IN THE MYSID *NEOMYSIS AMERICANA*: EFFECTS OF NAPHTHALENE, TEMPERATURE, AND OTHER FACTORS

Roy L. Smith and Bruce R. Hargreaves

SUMMARY

This study was intended to measure sublethal effects of naphthalene (a model of petroleum toxicity) to the mysid *Neomysis americana*. The toxicity assay used was oxygen uptake (VO_2) of individual mysids, measured continuously for 24-h periods using a small continuous-flow respirometer. Mysids were acutely exposed to one of five naphthalene concentrations (0, 0.2, 0.47, 0.73, and 1.0 mg/L) during respirometry. Effects of five acclimation temperatures (5, 10, 15, 20, and 25°C) on respiration, and interactions between temperature and naphthalene concentration were also tested. Other potential predictors of respiratory rate (*e.g.*, size, sex, behavior, and reproductive condition) were tested using a multiple regression model which included quadratic, cubic, and interaction terms.

The mysids from the original March and April collections, which were large individuals from an overwintering population, released their young and died during the course of acclimation to increasing temperatures and day lengths in the laboratory. The offspring matured into smaller individuals whose size was typical of summer populations. Respiratory differences correlated with temperature may therefore have been similar to seasonal differences that occur in natural populations.

Mean VO_2 increased with exposure to naphthalene, suggesting increased maintenance requirements resulting from either naphthalene-induced tissue damage or the additional cost of naphthalene metabolism. Synergistic effects of naphthalene and temperature on mysid respiration were observed. Respiratory compensation for naphthalene toxicity appeared to collapse at high temperature and high exposure combinations sufficient to cause 50% mortality in 96 h.

The time course of respiratory rate increases followed the same pattern as tissue naphthalene accumulation reported in an earlier study with *N. americana*. This pattern suggested that either naphthalene metabolites were not involved in increasing respiration, or that significant metabolism of naphthalene occurred immediately and affected respiration. Respiration in control mysids varied in a daily cycle which may have been endogenous and entrained by feeding.

Naphthalene influenced the respiratory rate of *N. americana* at the lowest concentrations tested both in this study (ca. 0.2 mg/L) and in previous work (ca. 0.08 mg/L). Respiratory effects are thus observable at concentrations of the same magnitude as environmental concentrations of diaromatics (*e.g.*, 0.020 mg/L).

The estimated standard metabolic rate increased linearly with increasing temperature, suggesting acclimation to high temperatures may require more energy for maintenance. Large mysids acclimated to high temperatures showed the greatest respiratory response to naphthalene, suggesting they may be especially vulnerable to toxicants that increase VO_2.

Mass-specific VO_2 declined with increasing length in control mysids, but the rate of decline was reduced by naphthalene exposure, suggesting naphthalene had a greater effect on large mysids. Male mysids had higher respiratory rates than females. The difference between sexes was enhanced by high temperature, large animal size, and naphthalene exposure. The sexual differences appeared gradually during maturation, and were not the result of behavior or reproductive condition of either sex.

These results suggest that all mysids are not equally susceptible to respiratory effects of naphthalene. Large male mysids acclimated to high temperatures have the highest weight-specific respiration, and may therefore be most vulnerable to naphthalene. These observations are consistent with facts already known about *N. americana*: they are more sensitive to handling at high temperatures, the 96-h LC50 for naphthalene decreases with temperature, and males appear to have lower survival than females in wild populations.

INTRODUCTION

Aromatic hydrocarbons ranging in size from naphthalene ($C_{10}H_8$) to coronene ($C_{24}H_{12}$) are important environmental contaminants (Neff, 1979). Diaromatics (naphthalene and its alkyl derivatives) are a prominent fraction of crude and refined oils, and account for much of the toxicity of these oils to marine organisms (Anderson *et al*., 1974; Corner *et al*., 1976; Neff, 1979; Laughlin & Neff, 1981). The mysid crustacean *Neomysis americana*, the test organism in this study, accumulates naphthalene rapidly from a solution of 1.2 mg naphthalene/L sea water, reaching peak tissue concentrations in 2-4 h (Hargreaves *et al*., 1982). The 96-h LC50 of *N. americana* to naphthalene is about 1.4 mg/L at temperatures of 5-20°C, but declines sharply to 0.80 mg/L at 25°C (Hargreaves *et al*., 1982).

Exposure to petroleum and pure aromatic compounds induces increased respiration in crustaceans such as mud crabs (*Rhithropanopeus harrisii*: Laughlin & Neff, 1980, 1981), shrimps (*Palaemonetes pugio*: Dillon, 1983; *Lucifer faxoni*: Lee *et al*., 1978), and mysids (*Neomysis integer*: Laughlin & Linden, 1983). Exposure to pure naphthalene induces increased respiration in *N. americana* acclimated to 15°C (Smith & Hargreaves, 1984). Increased energy metabolism is a significant sublethal toxic effect. It may indicate stress, defined by Dillon (1981) as reduced fitness resulting from overtaxing of the ability to compensate for environmental change.

Reduced growth and reproduction of animals exposed to toxicants may result from changes in energetics, which may include increased energy metabolism or altered feeding or assimilation efficiency. Reduced growth has been observed in the shrimp *Crangon crangon* (Edwards, 1978) and the bivalve molluscs *Modiolus demissus* and *Mytilus edulis* (Gilfillan, 1975) exposed to a water-soluble fraction (WSF) of crude oil. Exposure to the triaromatic compound phenanthrene reduces growth of *R. harrisii* (Laughlin & Neff, 1980). The copepod *Eurytemora affinis* exhibits decreased egg production, brood size, and total fecundity when exposed to WSF of crude oil (Berdugo *et al*., 1977). The apparent increase in energy metabolism of marine animals resulting from exposure to petroleum and pure aromatics suggests that the abundance of sensitive species may be affected in areas chronically polluted by oil.

Marine poikilotherms have considerable ability to adapt to temperature changes by altering membrane viscosity, enzyme structure and conformation, principal metabolic pathways, lipid storage, and internal pH (reviewed by Hochachka & Somero, 1973; White & Somero, 1982). These changes may affect uptake, metabolism, and excretion of naphthalene and other hydrocarbons. Respiratory synergism between temperature and hydrocarbon exposure has been observed in *N. integer* (Laughlin & Linden, 1983) and *R. harrisii* (Laughlin & Neff, 1980, 1981). Acclimation to high temperature increases both naphthalene accumulation and sensitivity to naphthalene (measured by 96-h LC50) in *N. americana* (Hargreaves *et al.*, 1982). Because *N. americana* occurs in environments in which the temperature ranges from 0 to 30°C (Wigley & Burns, 1971; Maurer & Wigley, 1982), temperature may be an important influence on toxicity of hydrocarbons to natural populations of this species.

This study was intended to measure sublethal effects of the aromatic hydrocarbon naphthalene (used in pure form as a model of oil toxicity) on the energetics of *N. americana*. Effects on response to naphthalene of acclimation temperature and other factors (*e.g.*, sex, reproductive condition, duration of exposure, and acclimation time) were also measured.

METHODS AND MATERIALS

Neomysis americana were collected from Delaware Bay on March 16, 1983 (at 5°C) and April 5, 1983 (at 6.2°C) with a modified epibenthic sled (described by Herman, 1963). Mysids were acclimated (by methods described by Hargreaves *et al.*, 1982) in 40-L glass aquaria which contained artificial sea water (Instant Ocean^{tm}) purified by air-lift filters covered by coarse sand. Mysids were fed daily with 24-h *Artemia* nauplii. Initial acclimation conditions were 5°C and 12 h fluorescent light/d. Respiration was measured under the temperature and light conditions to which the mysids were acclimated. When respirometry at 5°C was completed, the remaining mysids were successively acclimated to and tested under the following conditions, which were selected to simulate rising temperature and increasing day length of spring and summer: 10°C, 13 h light/d; 15°C, 13 h light/d; 20°C, 14 h light/d; 25°C, 14 h light/d.

Respiratory rates (VO_2) of individual mysids were measured with a small continuous-flow respirometer. In an earlier study (Smith & Hargreaves, 1984), this respirometer was able to measure brief transient VO_2 during extended tests. The respiratory minima of *N. americana* measured in this study were lower than rates reported in similar mysids measured with sealed-chamber methods, and seemed to more closely approximate standard metabolic rate (Smith & Hargreaves, 1984).

During testing, a single mysid was held in a 1-ml chamber at the end of a glass syringe, and oxygen depletion in the water flowing through the chamber was continuously measured by a Clark electrode. It was hoped the small chamber volume, which allowed rapid response to changes in respiratory rate, would also inhibit swimming, substituting for immobilization (as used by Quetin *et al.*, 1978, with large mysids). The respirometer was isolated from room lights and motion, and provided with an internal light cycle. Flow was adjusted with the intent of maintaining at least 80% oxygen saturation, well above reported tolerance limits for similar mysids (10% for *Holmsiella anomala*, 13% for *Archaeomysis grebnitzkii* and *Neomysis mercedis*; Mauchline, 1980) and critical pO_2 (30-40% for *A. grebnitzkii* and *N. mercedis*; Jawed, 1973). (*N. mercedis* now includes *Neomysis awatschensis* and *N. intermedia* according to Simmons and Knight, 1975). Flow was measured during each test by collecting and weighing the water from both syringes.

Tests were begun at 1000-1100 h and continued for 24 h. Mysids were last fed about 24 h before testing. Mysids were initially placed in the respirometer in clean artificial sea water, and exposure to the toxicant began immediately. Output from the Clark electrode transducer was recorded on a strip chart recorder with a resolution of 0.01 mg/L dissolved oxygen concentration. Barometric pressure was also recorded throughout each test. At the end of the test each mysid was blotted dry and weighed with a Cahn Model M-10 electrobalance having a resolution of 0.002 mg; sex, length, and reproductive condition were also noted. Mysids were individually dried in a vacuum oven at 60°C and 200 torr (absolute) for 24 h and reweighed.

Mysids were acutely exposed during testing to one of five nominal naphthalene concentrations: 1.00 mg/L, 0.73 mg/L, 0.47 mg/L, 0.20 mg/L, or no added naphthalene. Order of testing of naphthalene concentrations was randomized. Tests were repeated three times in a row using the same batch of toxicant mixture but different mysids for each repetition. Additional tests were performed without naphthalene to compare mysids from the two collections. During tests with naphthalene, water samples of effluent were collected every 24 h, placed in amber 10-ml vials (with care taken to exclude air), stored at 5°C, and analyzed for naphthalene by the UV-absorbance method described by Hargreaves et al. (1982).

High bacterial respiration was measured during respirometry with 1.2 mg naphthalene/L sea water in an earlier study (Smith & Hargreaves, 1984). Bacterial respiration was inhibited in this study by additions of 50 mg/L neomycin and 50 mg/L streptomycin (used by Corner et al., 1976, with copepods exposed to naphthalene). Effectiveness of inhibition was confirmed by measuring respiration in empty chambers from which mysids had been removed. To keep bacterial contamination to a minimum, all glass parts were washed with 7X™ cleaner, rinsed with ethanol and deionized water, and baked for 24 h at 200°C; teflon parts were washed the same way and baked for 24 h in a vacuum oven at 60°C and 200 torr (absolute).

Naphthalene is only sparingly soluble in water (30.7 mg/L in deionized water at 25°C: Yalkowski et al., 1983), so it was pre-dissolved in ethanol before dissolution in sea water. Ethanol was chosen as the "carrier solvent" because acetone, frequently used in toxicity tests with lipophilic compounds, has been shown to influence growth and toxicant uptake (Mac & Seelye, 1982). All sea water solutions tested in the respirometer, including controls, contained 40 μL ethanol/L. Additional tests were done without antibiotics or ethanol at 5, 15, and 25°C to confirm that these materials did not affect respiration.

Test mixtures were prepared by mixing antibiotics and artificial sea water, then drawing the mixture through a 0.45 μm nylon filter to remove undissolved particles and microbubbles. The filtered mixture was aerated with 0.45 μm filtered air at the test temperature for 24 h before use. To prevent supersaturation, aeration was done by evacuating the head space, continuously drawing air down a tube to the bottom. Just before use, naphthalene-ethanol stock solution was added to the sea water with a Wiretrol™ positive displacement microsyringe. The solution was mixed by gentle swirling to avoid volatilizing naphthalene or supersaturating the mixture with oxygen.

Behavior of each mysid was noted 3 or 4 times during testing, using the following four digit scale: position 1 (left digit), frantic motion; 2, normal swimming; 3, holding position above the bottom; 4, resting on the bottom. Numbers recorded in these positions indicated the proportion of time, in fifths, the mysid spent in each activity during the observation period (about 30 sec). For example, 0005 indicated the mysid rested on the bottom during the entire observation, and 0320 meant the mysid swam 3/5 of the time and held position 2/5 of the time. These occasional behavior measurements were not intended as a substitute for continuous monitoring, but were recorded in the hope of gaining additional insight into the effects of naphthalene and temperature.

Minimum and maximum oxygen uptake rates (in µl O_2/h) were calculated for each mysid during each 1-h period. Respiratory rates were normalized for animal size by dividing by dry weight. Scaling respiratory rate for size using the allometric equation ($VO_2 = aM^b$; a,b = contants, M = animal mass) was not attempted because only three mysids were measured at each combination of temperature and naphthalene, too few for development of reliable regressions of respiration on size. Temperature and naphthalene exposure were assumed likely to affect the exponent of the allometric respiration curve, so data were not combined. From these hourly dry weight specific minima and maxima, additional respiratory variates (Table 1) were calculated for each animal.

Table 1. *Neomysis americana*. Respiratory variates calculated for each mysid from 24-h chart recordings of respiratory rate in a continuous flow respirometer. All variates are adjusted for mysid size by dividing by dry weight.

Variate	Definition
Hourly VO_2 (µl O_2/h/mg)	lowest rate of O_2 uptake during each 1-h period
Mean VO_2 (µl O_2/h/mg)	mean of 24 hourly VO_2 values
1-h range (µl O_2/h/mg)	highest minus lowest VO_2 during each 1-h period
24-h range (µl O_2/h/mg)	highest minus lowest VO_2 in 24 h
1-h ratio	highest divided by lowest VO_2 during each 1-h period
24-h ratio	highest divided by lowest VO_2 in 24 h.

Statistical analyses of respiratory rate data were performed using the SCSS^{tm} Conversational System (Nie *et al.*, 1980) computer statistics package. Effects of independent variables (fully described in Table 2) on respiration were tested using multiple regression analyses. Behavior was not used as an independent variable because it could not be assumed to be measured without error. Regression analyses were done in two steps: preliminary multiple regressions were used to eliminate independent variables which did not effectively predict respiratory variates, then independent variables which were good predictors of respiration were used in more complex final multiple regressions which included quadratic, cubic, and interaction terms (similar to regressions used by Laughlin & Neff, 1980, 1981). Final regression calculations were iterative: regressions were plotted, compared with actual data, and revised repeatedly with the inclusion of new terms such as logarithms and reciprocals. Terms were retained only if they produced a regression having a higher coefficient of determination and a better visual fit.

Table 2. Independent variables used in preliminary multiple regressions.

Code[1]	Variate	Units
	acclimation time	d
	body water	mg
ct	clock time	h x 2 pi/24
dw	dry weight	mg
ℓ	length	mm
	light	h/d
n	naphthalene concentration	mg/L
r	reproductive condition	1 - 4[2]
s	sex	1 = male; 2 = female
	starting time	h
t	temperature	°C
f	test flow	ml/h
	time of last feeding	h
tss	time since start of test	h

[1]Codes used in multiple regressions (Tables 4-6)
[2]1 = juvenile, 2 = not ripe, 3 = ripe, 4 = brooding

Final multiple regressions were calculated for both 24-h and 1-h respiratory variates. Regressions with 1-h respiratory variates included the same independent variables as those with 24-h variates, with the addition of time since the start of the test and sine of clock time (adjusted such that 24 h = 6.28 radians). Inclusion of the sine of clock time was intended to test the existence of a daily respiratory cycle. An additional final multiple regression was calculated using behavior as the dependent variable and the strong predictors of respiration as independent variables.

Regression equations were plotted using the Autocad-80^{tm} microcomputer graphics program (Autodesk, Inc.), a Scion Microangelo^{tm} S-100 graphics board, and a Houston Instruments DMP-40 plotter. A computer program was written for the purpose of automatically converting data from a microcomputer data base (dBase II^{tm}, Ashton-Tate, Inc.) to high quality line drawings (when interpreted by Autocad-80). This program can be obtained from the Department of Biology, Lehigh University.

RESULTS

Naphthalene concentrations in the respirometer effluent (Table 3) averaged 6.2% above nominal concentrations. Naphthalene declined with time at all temperatures except 5°C, and the rate of decline increased with temperature from 5 to 20°C. The mean rate at which naphthalene concentration decreased over the three day use of each toxicant-sea water batch was 1.9% per day.

Table 3. Mean measured naphthalene concentrations (mg/L), standard errors (SE), and rates of decline of naphthalene with time for each test temperature.

Temp (°C)	Nominal Concentration 0.200	0.467	0.733	1.00	Mean Rate of Decline (%/d)
	Actual Concentration (SE)				
5	0.243 (0.001)	0.481 (0.001)	0.613 (0.001)	0.720 (0.001)	no loss
10	0.304 (0.003)	0.531 (0.008)	0.860 (0.032)	1.089 (0.034)	0.63
15	0.163 (0.001)	0.413 (0.009)	0.722 (0.038)	1.076 (0.001)	2.41
20	0.271 (0.013)	0.488 (0.001)	0.729 (0.024)	1.012 (0.061)	5.33
25	0.233 (0.017)	0.440 (0.005)	0.734 (0.017)	1.086 (0.012)	1.54

Mean VO_2, 24-h range, and 24-h ratio (definitions in Table 1) were not significantly different for animals captured March 16 and April 5, 1983 and tested at 15°C ($p>0.167$, $n = 12$, Student's t-test). Addition of ethanol, streptomycin, and neomycin to the artificial sea water did not significantly affect mean VO_2, 24-h range, or 24-h ratio at 15°C or 25°C ($p>0.499$, $n = 12$, Student's t-test). Mysids acclimated to higher temperatures (20 or 25°C) had significantly reduced mean length ($p<0.002$, ANOVA) and mean dry weight ($p<0.004$, ANOVA). Juveniles were not removed from acclimation tanks, and the size change apparently resulted from death of the original mysids and replacement by their offspring.

Table 4. *Neomysis americana*. Preliminary regressions of respiratory variates on all measured independent variables, without interactions. Importance of variables is ranked (in parentheses) by F test. 1 = best predictor. Variable codes listed in Table 2. Significance: * = $p<0.05$; ** = $p<0.01$; *** = $p<0.001$.

	Regression Coefficients		
Variable	Mean VO_2	24-h Range	24-h Ratio
n	0.842 *** (3)	0.916 *** (2)	
t	0.151 *** (1)	0.0443 ** (3)	-0.119 *** (1)
l		-0.0821 * (4)	
s	-0.806 *** (2)	-0.715 *** (1)	
r	-0.227 ** (4)		
f	0.263 * (5)		
dw			-0.609 ** (2)
constant	1.03 (ns)	3.60 ***	5.25 ***
r^2	0.866	0.519	0.27

Equations:

$$\text{Mean } VO_2 = 1.03 + 0.842(n) + 0.151(t) - 0.806(s) - 0.277(r) + 0.263(f)$$

$$\text{24-h range} = 3.60 + 0.916(n) + 0.0443(t) - 0.0821(l) - 0.715(s)$$

$$\text{24-h ratio} = 5.25 - 0.119(t) - 0.609(dw)$$

In preliminary multiple regressions (Table 4) of mean VO_2, 24-h range, and 24- h ratio on all independent variables (listed in Table 2), temperature, naphthalene concentration, and sex were strong predictors of respiration, and length and test flow were weak predictors (*i.e.*, having $p>0.001$ and significant in only one of the three regressions). All strong predictors were retained in final multiple regressions; length, which provided additional normalization for mysid size, was also retained. Test flow was not included in final regressions because an artificial correlation existed between flow and temperature: VO_2 increased with temperature, and flow consequently had to be increased to maintain minimum dissolved oxygen concentrations. Other independent variables not included in final regressions were reproductive condition, acclimation days, mass of body water, dry weight, hours light per day, starting time, and time of last feeding.

Results of final multiple regressions (Tables 5 & 6) indicated that respiration and behavior were influenced by numerous interactions among the remaining independent variables. Mean VO_2 increased with naphthalene exposure as a curvilinear function, the shape of which was determined by temperature (Fig. 1). Mean VO_2 increased linearly with increasing temperature (Fig. 2). Females had lower respiratory rates than males, and mean VO_2 of males increased more at higher temperatures (Fig. 2). Mean VO_2 declined with mysid length. The rate of decline was less for males (Fig. 3), and decreased with naphthalene exposure (Fig. 4). (Note that figures which do not illustrate the influence of sex show the response of female mysids. Male respiration, which was consistently higher, can be calculated from the regression equations in Tables 5 and 6.)

Naphthalene exposure stimulated 24-h range of males more than of females (Fig. 5). The 24-h ratio declined asymptotically with increasing temperature (Fig. 6), but was not influenced by naphthalene concentration. Mysid behavior increased with increasing temperature, naphthalene exposure, and length, but naphthalene in combination with high temperature and large mysid size reduced activity (behavior regression, Table 5). Activity and oxygen uptake were significantly correlated ($r = 0.431$, $p<0.003$ for activity and hourly minimum VO_2; $r = 0.451$, $p<0.002$ for activity and hourly maximum VO_2).

Table 5. *Neomysis americana*. Final multiple regression coefficients for respiratory variates on independent variables and interactions. Importance of variables is ranked (in parentheses) by F test, 1 = best predictor. Variable/interaction codes listed in Table 2. Significance: * = $p<0.05$; ** = $p<0.01$; *** = $p<0.001$.

Variable/ Inter- action	Regression Coefficients: Mean VO_2	24-h Range	24-h Ratio	Behavior
n		4.85 *** (1)		2.83 ** (1)
n x t	0.500 * (7)			-0.0510 * (3)
n x l	0.0903 *** (4)			-0.123 * (5)
n x s		-1.87 *** (3)		
n^3		-1.25 ** (5)		
n^3 x t	-0.0477 ** (6)			
t	0.280 *** (1)	0.112 *** (2)		0.0396 * (2)
t x s	-0.0780 *** (2)			
t x l		-0.00552 *** (4)		
1/t			46.0 *** (1)	
1/t x s			-4.34 * (3)	
1/t x l			-1.91 ** (2)	
s	1.04 ** (5)			
s x l	-0.0776 *** (3)			
l				0.0721 * (4)
constant	-0.537 ns	0.768 **	0.911 ***	0.0499 ns
r^2	0.898 ***	0.606 ***	0.362 ***	0.251 ***

Equations:

Mean VO_2 = $-0.537 + 0.500(n)(t) + 0.0903(n)(l) - 0.0477(n^3)(t) + 0.280(t) - 0.0780(t)(s) + 1.04(s) - 0.0776(s)(l)$

24-h Range = $0.768 + 4.85(n) - 1.87(n)(s) - 1.25(n^3) + 0.112(t) - 0.00552(t)(l)$

24-h Ratio = $0.911 + 46.0(1/t) - 4.34(1/t)(s) - 1.91(1/t)(l)$

Behavior = $0.0499 + 2.83(n) - 0.0510(n)(t) - 0.123(n)(l) + 0.0396(t) + 0.0721(l)$

Results of final multiple regressions of hourly VO_2, 1-h range, and 1-h ratio (Table 6) indicate that respiration was influenced by time of day and time since the test start. Hourly VO_2 in control mysids varied in a daily cycle (Fig. 7), which was disrupted by exposure to naphthalene. At high naphthalene concentrations, hourly VO_2 increased sharply early in the tests, then slowly declined.

Table 6. *Neomysis americana*. Final multiple regression coefficients for hourly respiratory variates on independent variables and interactions. Importance of variables is ranked (in parentheses) by F test, 1 = best predictor. Variable/interaction codes listed in Table 2. Significance: * = $p<0.05$; ** = $p<0.01$; *** = $p<0.001$.

Variable/ Interaction	Regression Coefficients		
	Hourly VO_2	1-h Range	1-h Ratio
n	2.71 *** (3)	0.408 *** (2)	
n x tss	-0.076 *** (4)		
n x 1/tss	-1.48 *** (8)		
n^3	-0.619 *** (5)	-0.270 *** (4)	-0.0408 *** (3)
t	0.209 *** (1)		-0.00632 *** (2)
t x 1/tss		0.0231 *** (5)	0.00957 ** (5)
1/t	3.23 *** (9)	-1.23 *** (3)	1.20 *** (1)
s	-1.05 *** (2)	-0.182 *** (1)	
1/tss		-0.305 *** (8)	-0.177 *** (4)
sin ct	-0.497 *** (7)	-0.0959 *** (7)	
sin ct x t		0.00573 *** (6)	
sin ct x s	0.319 *** (6)		
constant	0.301 *	0.782 ***	1.25 ***
r^2	0.786 ***	0.169 ***	0.234 ***

Equations:

$$\text{Hourly } VO_2 = 0.301 + 2.71(n) - 0.076(n)(tss) - 1.48(n)(1/tss) - 0.619(n^3) + 0.209(t) + 3.23(1/t) - 1.05(s) - 0.497(\sin ct) + 0.319(\sin ct)(s)$$

$$\text{1-h Range} = 0.782 + 0.408(n) - 0.270(n^3) + 0.0231(t)(1/tss) - 1.23(1/t) - 0.182(s) - 0.305(1/tss) - 0.0959(\sin ct) + 0.00573(\sin ct)(t)$$

$$\text{1-h Ratio} = 1.25 - 0.408(n^3) - 0.00632(t) + 0.00957(t)(1/tss) + 1.20(1/t) - 0.177(1/tss)$$

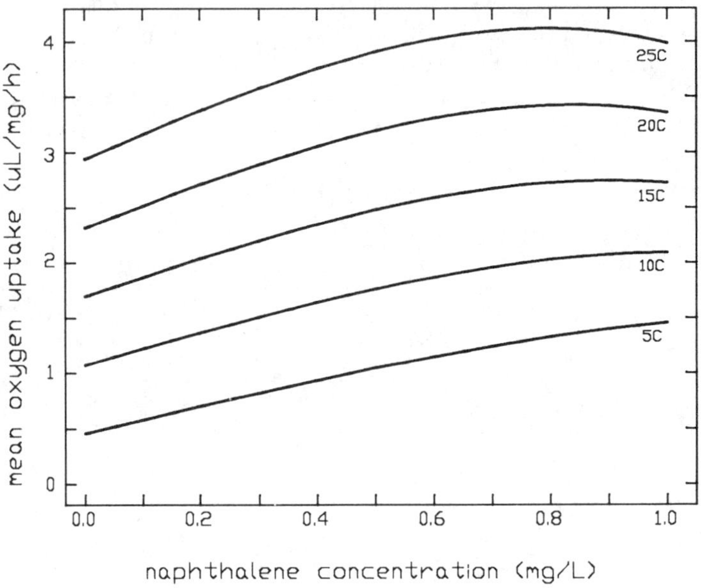

Fig. 1. *Neomysis americana*. Predicted mean VO_2 at different naphthalene concentrations and temperatures (°C). Length = 11 mm, sex = female.

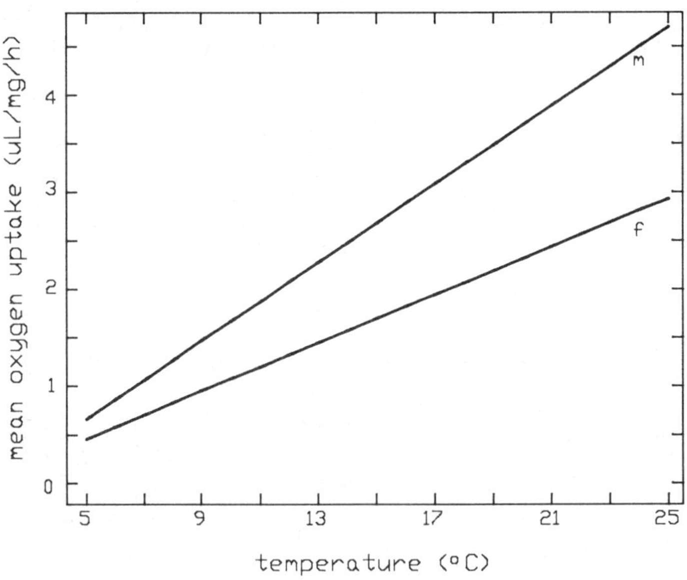

Fig. 2. *Neomysis americana*. Predicted mean VO_2 for male and female mysids at different temperatures. Naphthalene concentration = 0, length = 11 mm, m = male, f = female.

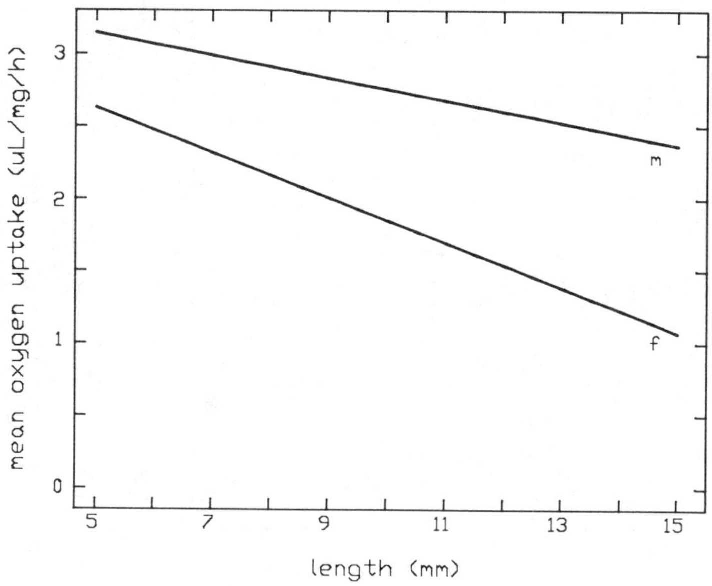

Fig. 3. Neomysis americana. Predicted mean VO_2 for male and female mysids of different total length. Temperature = 15°C, naphthalene concentration = 0, m = male, f = female.

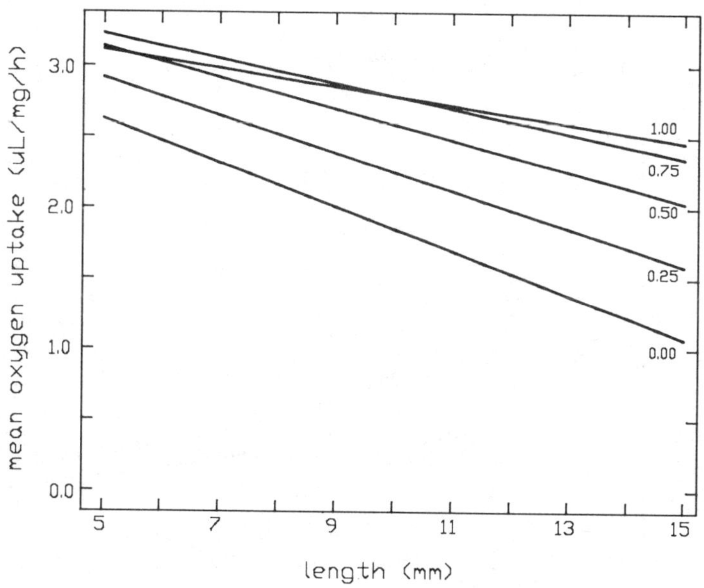

Fig. 4. Neomysis americana. Predicted mean VO_2 for mysids of different total length exposed to naphthalene (mg/L). Temperature = 15°C, sex = female.

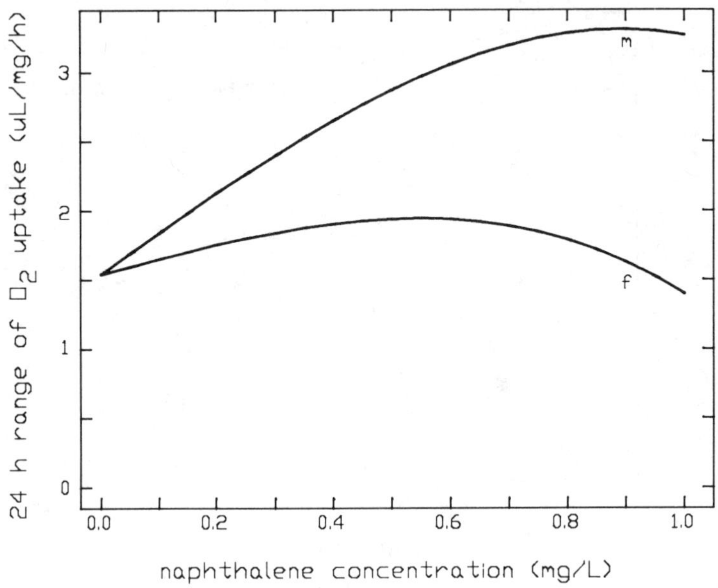

Fig. 5. Neomysis americana. Predicted 24-h range for male and female mysids at different naphthalene concentrations. Temperature = 15°C, length = 11 mm, m = male, f = female.

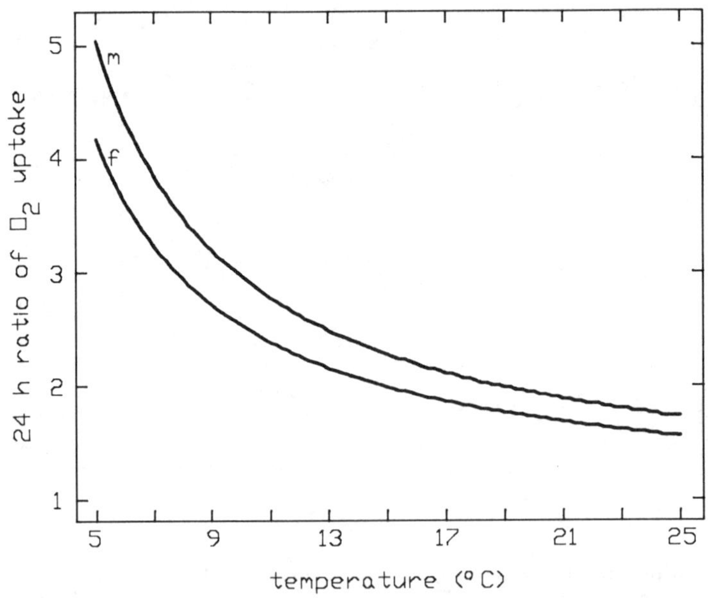

Fig. 6. Neomysis americana. Predicted 24-h ratio at different temperatures for males and females. Length = 11 mm, m = male, f = female.

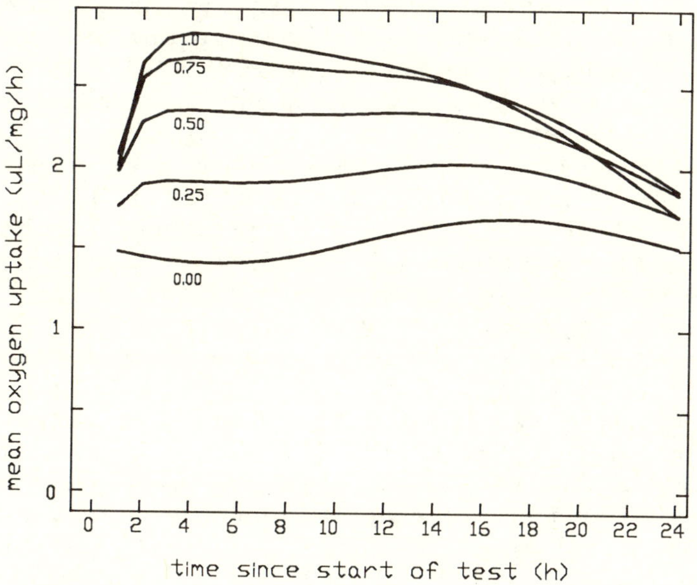

Fig. 7. Neomysis americana. Predicted 1-h minimum VO_2 at different times and naphthalene concentrations (mg/L). Temperature = 15°C, length = 11 mm, sex = female, time 0 = 1100 h.

DISCUSSION

According to Fry (1971), three levels of metabolic rate are distinguished in poikilotherms: standard metabolic rate (SMR; an approximation of the minimum cost of maintenance, observed in a resting, post-absorptive, non-reproductive animal), routine metabolic rate (RMR; observed in an animal exhibiting routine activity), and active metabolic rate (AMR; the maximum rate of aerobic metabolism sustainable during constant activity). Estimation of SMR rather than RMR is particularly useful in toxicity tests with petroleum hydrocarbons, which may simultaneously change normal behavior and SMR. Metabolic scope for activity is defined as AMR minus SMR, and represents the "metabolic headroom" available to perform work. Factorial scope is AMR divided by SMR, representing relative metabolic scope for activity, corrected for size of the animal. If SMR is altered by toxicant exposure, it is possible that metabolic scope for activity and factorial scope may also be influenced.

In this study, respiration of confined, isolated mysids was measured continuously for extended periods by a method capable of detecting transient low rates. The accuracy with which mean VO_2 (the mean of the respiratory minima observed during each hour of the test) estimated the actual SMR depended on patterns of activity spontaneously shown by the mysids. In an earlier study (Smith & Hargreaves, 1984), this respirometer produced respiratory measurements which were generally lower than rates measured by sealed-chamber methods by other investigators, suggesting that mean VO_2 in this study may be similar to SMR. However, the existence of a positive correlation between activity and the 1-h respiratory minima suggests that the 1-h minima were sometimes above resting levels, and their mean may still overestimate SMR.

Because the mysids were not stimulated during testing, the highest rate observed in 24 h should underestimate AMR according to Fry's (1971) definition; although metabolic rates in fish and crustaceans may spontaneously reach 50% or more of AMR without stimulation or overt movement (Fry, 1971; Rutledge & Pritchard, 1981). If the highest 24-h maximum VO_2 underestimated true AMR, 24-h range and 24-h ratio measured in this study must have underestimated metabolic scope for activity and factorial scope, respectively. If the degree of underestimation was consistent, however, examination of changes in 24-h range and ratio with endogenous and exogenous factors may still be useful.

Effect of Temperature

Mysids acclimated to different temperatures in the laboratory had significant differences in respiration, dry weight, and length which did not exist in the original collections. These differences were apparently the result of disappearance of mysids of the original "winter" populations and their replacement by a "summer" population of their offspring, which were significantly smaller at maturity. Temperature acclimation was done in 5°C increments in ascending order from March to July, so the changes in the acclimated mysids may have been analogous to seasonal changes in natural populations. This hypothesis is supported by the observation of similar summer-winter size differences in natural populations of *N. americana* (Pezzack & Corey, 1979; Thompson, 1984) and *N. mercedis* (Toda *et al*., 1982). Seasonal differences in respiration have been observed in *N. mercedis* (Simmons & Knight, 1975), suggesting that the influence of temperature on respiration observed in this study may also have been seasonal.

Temperature was the most important predictor of mean VO_2, 24-h range, and 24-h ratio in multiple regressions, where it produced higher F-values than other independent variables. The log_{10}, the square, and the cube of temperature were weak predictors compared to temperature itself, so increases in mean VO_2 and 24-h range with increasing temperature were apparently linear. Linear increases of respiratory rate with increasing temperature have also been observed in juvenile shrimp (*Peneaus indicus*, Kutty *et al*., 1971), mixed zooplankton (Bishop, 1968), and first zoeae of the shrimp *Pandalus borealis* (Paul & Nunes, 1983). Depending on temperature-related changes in ingestion (which were not measured), increased respiration at high temperature could reduce growth rate. This hypothesis is supported by observations of decreased growth rates of *N. americana* above 20°C (Thompson, 1984).

The rate of decrease of 24-h ratio of *N. americana* with increasing acclimation temperature indicated that 24-h minimum VO_2 increased proportionally more than 24-h maximum VO_2. This decrease in 24-h ratio with increasing temperature conflicts with typical dome-shaped plots of factorial scope vs. temperature (*e.g.*, Brett, 1970). There seem to be two possible causes of this difference: (1) *N. americana* may require a greater proportion of its energy budget for maintenance at high temperatures, or (2) higher levels of activity ordinarily stimulated by high temperature (Brett, 1970) may have been prevented by the small size of the respirometer chamber. If activity levels at high temperatures were inhibited by the respirometer, the accuracy of 24-h maximum VO_2 as an estimate of AMR was decreased at high temperature.

Effect of Naphthalene

Respiratory rate of *N. americana* exposed to naphthalene increased in a pattern determined by acclimation temperature: naphthalene-induced increases in respiration changed from nearly linear at 5°C to curvilinear at 25°C. The increase in VO_2 with naphthalene exposure may indicate compensation for toxic effects, and decreased VO_2 at combinations of high temperature and naphthalene exposure suggests the breakdown of compensation and impending death. This view is supported by the finding of Hargreaves *et al*. (1982) that the 96-h LC50 for *N. americana* exposed to naphthalene decreases from 1.4 to 0.8 mg/L at acclimation temperatures above 20°C. Interacting effects of temperature and hydrocarbon exposure, which have also been found in the mysid

Neomysis integer (Laughlin & Linden, 1983), the mud crab *Rhithropanopeus harrisii* (Laughlin & Neff, 1980, 1981), and the shrimp *Palaemonetes pugio* (Dillon, 1981), appear to be common. The synergism between naphthalene exposure and temperature observed in this study suggests *N. americana* may be most sensitive to aromatic hydrocarbons during the summer when chronic pollution of estuaries by oil, particularly from sewage treatment plants and recreational boating, tends to be greatest.

Although naphthalene exposure influenced mean VO_2 and 24-h range in *N. americana*, naphthalene had no effect on 24-h ratio. The 24-h minimum and maximum VO_2 must therefore have increased proportionally with naphthalene exposure, suggesting naphthalene may have increased the cost of obtaining energy. This implies that higher metabolic rates during naphthalene exposure could have resulted from decreased efficiency of energy conversion rather than increased activity. This view is supported by the finding of Percy (1977) that respiration of a cell-free extract of the marine amphipod *Onisimus affinis* is elevated by exposure to crude oil. Such an effect could arise from uncoupling of oxidative phosphorylation (Neff, 1979), resulting in a decrease in the ratio of ADP phosphorylated to oxygen used (Hanstein, 1976), and requiring increased consumption of oxygen and carbon to maintain life. Naphthalene lacks the acidic sites possessed by typical organic uncouplers (reviewed by Hanstein, 1976), but hydroxylated naphthalene metabolites may be effective uncouplers.

Unless the rise in energy metabolism in *N. americana* exposed to naphthalene is offset by increases in ingestion or assimilation efficiency, it will result in lowered growth or reproduction. Few studies of ingestion of aquatic organisms exposed to oil have been done, but available results show decreased ingestion in exposed crab and slipper shell larvae (Johns & Pechenik, 1980), mussels (Gilfillan, 1975), and Maine lobsters (Capuzzo & Lancaster, 1982). No studies of assimilation efficiency during oil exposure have been reported, but it does not seem likely that increased efficiency could offset the metabolic energy losses observed in this study. The observed respiratory increases therefore could reasonably result in decreased growth or reproduction.

No threshold naphthalene concentration was found for respiratory effects on *N. americana* in this study. The lowest naphthalene concentration tested (0.2 mg/L) resulted in increased mean VO_2 and 24-h range; in an earlier study, concentrations of 0.08 mg/L altered 24-h range in *N.*

americana (Smith & Hargreaves, 1984). The naphthalene threshold for respiratory effects, if any, appears to be less than 0.08 mg/L, considerably lower than naphthalene concentrations supposed to be protective of fish (0.45-0.85 mg/L: DeGraeve et al., 1982; 0.62 mg/L: EPA, 1979). This apparent threshold for effects is of the same order of magnitude as some observed environmental concentrations of diaromatics (e.g., 0.020 mg/L in bottom waters of Raritan Bay, New Jersey; Stainken & Frank, 1979). Toxic effects may be enhanced by the typical associations of naphthalene with more toxic alkylated naphthalenes and triaromatics in oiled environments. It is possible that energy metabolism of natural populations of mysids may be increased by existing levels of chronic oil pollution, and that abundance may be affected. Because of the importance of mysids in estuarine energy exchange (e.g., Johnston & Lasenby, 1982), reduction in mysid populations could affect other species less sensitive to oil.

Effect of Mysid Size

Both dry weight-specific 24-h range and 24-h ratio declined with increasing mysid length, indicating the lowest VO_2 in 24 h increased more rapidly than the highest with increasing length. The narrowing of 24-h range and ratio can be explained either by (1) reduced metabolic scope for activity in large mysids or (2) confinement of large mysids by the respirometer. Confinement appears likely because reduced metabolic scope with increasing size conflicts with typical single-species allometric results (e.g., Brett & Groves, 1979).

Dry weight-specific VO_2 declined with increasing size in control mysids, but the rate of decline was reduced by naphthalene exposure, suggesting naphthalene had a proportionally greater effect on large mysids. The proportionally greater rise in SMR in large mysids exposed to naphthalene suggests they may be especially vulnerable to the effects of naphthalene.

Effect of Sex

Mean VO_2 and 24-h range in N. americana were different for each sex, independent of behavior and reproductive condition, suggesting a permanent energetic dissimilarity between males and females. Simmons and Knight (1975) also found apparently permanent sexual differences in respiration of a mysid, *Neomysis mercedis*. Moshiri *et al*. (1969) reported respiratory variation among male, nonfecund female, and fecund female cladocerans (*Leptodora kindtii*) correlated with swimming speed, and Coull and Vernberg (1970) noted differences in respiration of male and gravid female harpacticoid copepods which they ascribed to depressed metabolism or locomotor activity in females.

Respiratory rate was similar in both sexes for small *N. americana*, but diverged as length increased, suggesting sexual respiratory differences resulted from physiological changes during maturation. Respiratory differences between sexes were slight at 5°C, but became greater with increasing temperature. A similar temperature effect occurs in the harpacticoid copepod *Longipedia helgolandica*, in which respiration of gravid and nongravid animals is the same at 15°C, but respiration of nongravid animals exceeds that of gravid ones at 25°C (Coull & Vernberg, 1970).

Higher energy metabolism in males may result in decreased tolerance to environmental change, particularly rising temperature. In a natural population of *N. americana* in Cape Cod Bay, approximately equal numbers of males and females appear to be produced at the onset of the spring reproductive season, but by October females outnumber males by about 3 to 1 (Maurer & Wigley, 1982). This disappearance of males suggests that males may have lower survival during the summer, and may be related to the respiratory differences found in this study. Naphthalene exposure increased 24-h range to a greater extent in males, although mean VO_2 increased the same amount in both sexes. The greater increase in 24-h range in males probably resulted from increased 24-h maximum VO_2, perhaps caused by short bursts of high respiration. This result suggests that naphthalene exposure affected males more than females, and that males may be more vulnerable to its toxic effects.

Effect of Time

In an earlier respiratory study with *N. americana* (Smith & Hargreaves, 1984), a daily respiratory cycle was observed in which respiratory rate increased from relatively low levels at 0600 h to peak levels at 1800 h. Because animals were not tested in the absence of environmental cues, an endogenous clock was not proved. It was suggested that the increase in hourly VO_2 was in response to lights in the test bath (which were turned on at 0500 h), the most obvious environmental cue. However, it was also possible the mysids increased their metabolic rates in preparation for feeding (which was done at 2000 h in the earlier study). In the present study, feeding time was ca. 1100 h instead of 2000 h. A daily cycle was again observed, but with a maximum at 0600 h and minimum at 1800 h, shifted 12 h from that observed in the earlier study. It now seems more likely the observed respiratory cycle was endogenous, entrained by the predictable cyclic appearance of food.

Mean VO_2 in mysids exposed to higher naphthalene doses rose rapidly, then slowly declined to near control levels. Tissue naphthalene in *N. americana* (acutely exposed to 1.2 mg/L at 15°C) shows a similar pattern: peak tissue concentrations are reached in 3 h, after which tissue naphthalene may decrease (Hargreaves *et al.*, 1982). Accumulation of naphthalene in tissues thus appears to correlate to respiratory effects, suggesting either unmetabolized naphthalene is responsible for increased respiration, or significant metabolism of naphthalene occurs immediately and has a respiratory effect. The decline in respiration observed late in the tests suggests a decline of the mysids' energy stores, although decreased respiration could also result from removal of tissue naphthalene by metabolism or sequestering, or from habituation of the mysids to naphthalene.

ACKNOWLEDGMENTS

We greatly appreciate the assistance and field expertise of Catherine Q. Thompson and Michal Kubik, who collected the mysids used in this study. Special thanks go to Michal, who, after a catastrophic equipment failure wiped out our mysid supply, donated irreplaceable mysids from the same collection and enabled us to finish this work. This study was sponsored by the New Jersey Sea Grant Program under a grant from the Office of Sea Grant, National Oceanic and Atmospheric Administration (NOAA), US Department of Commerce (Publication NJSG-83-124, project R/F-6, grant NA81AA-00065). This paper is based on a dissertation submitted in partial fulfillment of the requirements of a Ph.D. in Biology, Lehigh University, by R.L. Smith.

LITERATURE CITED

Anderson, J.W., J.M. Neff, B.A. Cox, H.E. Tatem, and G.M. Hightower. 1974. Characteristics of dispersions and water-soluble extracts of crude and refined oils and their toxicity to estuarine crustaceans. Mar. Biol. 27: 75-88.

Berdugo, V., R.P. Harris, and S.C.M. O'Hara. 1977. The effect of petroleum hydrocarbons on reproduction of an estuarine planktonic copepod in laboratory cultures. Mar. Pollut. Bull. 8: 138-143.

Bishop, J.W. 1968. Respiratory rates of migrating zooplankton in the natural habitat. Limnol. Oceanogr. 13: 58-62.

Brett, J.R. 1970. Temperature, animals, fishes. pp. 515-560. In: Marine Ecology, Vol. 1. O. Kinne (ed.). Wiley-Interscience, New York.

Brett, J.R. and T.D.D. Groves. 1979. Physiological energetics. pp. 279-352. In: Fish Physiology, Vol. 8. W.S. Hoar, D.J. Randall, and J.R. Brett (eds.). Academic Press, New York.

Capuzzo, J.M. and B.A. Lancaster. 1982. Physiological effects of petroleum hydrocarbons on larval lobsters (*Homarus americanus*): Hydrocarbon accumulation and interference with lipid metabolism. pp. 477-501. In: Physiological Mechanisms of Marine Pollutant Toxicity. W.B. Vernberg, A. Calabrese, F.P. Thurberg, and F.J. Vernberg (eds.). Academic Press, New York.

Corner, E.D.S., R.P. Harris, C.C. Kilvington, and S.C.M. O'Hara. 1976. Petroleum compounds in the marine food web: Short-term experiments on the fate of naphthalene in *Calanus*. J. Mar. Biol. Assoc. U.K. 56: 121-133.

Coull, B.C. and W.B. Vernberg. 1970. Harpacticoid copepod respiration: *Enhydrosoma propinquum* and *Longipedia helgolandica*. Mar. Biol. 5: 341-344.

DeGraeve, G.M., R.G. Elder, D.C. Woods, and H.L. Bergman. 1982. Effects of naphthalene and benzene on fathead minnows and rainbow trout. Arch. Environ. Contam. Toxicol. 11: 487-490.

Dillon, T.M. 1981. Effects of a dimethylnaphthalene-contaminated food source and daily fluctuating temperatures on the survival potential of the estuarine shrimp *Palaemonetes pugio*. Mar. Biol. 63: 227-233.

Dillon, T.M. 1983. Oxygen consumption in the shrimp, *Palaemonetes pugio*, exposed to fluctuating temperatures and food contaminated with the diaromatic hydrocarbon dimethylnaphthalene. Est. Coast. Shelf Sci. 16: 403-413.

Edwards, R.R.C. 1978. Effects of water-soluble oil fractions on metabolism, growth and carbon budget of the shrimp *Crangon crangon*. Mar. Biol. 46: 259-265.

EPA. 1979. Naphthalene. Ambient water quality criteria. Criteria and Standards Division, U.S. Environmental Protection Agency, Washington, DC.

Fry, F.E.J. 1971. The effect of environmental factors on the physiology of fish. pp. 1-98. In: Fish Physiology. W.S. Hoar and D.J. Randall (eds.). Academic Press, New York.

Gilfillan, E.S. 1975. Decrease of net carbon flux in two species of mussels caused by extracts of crude oil. Mar. Biol. 29: 53-57.

Hanstein, W.G. 1976. Uncoupling of oxidative phosphorylation. Biochim. Biophys. Acta 456: 129-148.

Hargreaves, B.R., R.L. Smith, C.Q. Thompson, and S.S. Herman. 1982. Toxicity and accumulation of naphthalene in the mysid *Neomysis americana*: Interactions with environmental temperature. pp. 391-412. In: Physiological Mechanisms of Marine Pollutant Toxicity. W.B. Vernberg, A. Calabrese, F.P. Thurberg, and F.J. Vernberg (eds.). Academic Press, New York.

Herman, S.S. 1963. Vertical migration of the opossum shrimp, *Neomysis americana* Smith. Limnol. Oceanogr. 8: 228-238.

Hochachka, P.W. and G.N. Somero. 1973. Strategies of Biochemical Adaptation. W.B. Saunders Co., Philadelphia. 358 pp.

Jawed, M. 1973. Effects of environmental factors and body size on rates of oxygen consumption in *Archaeomysis grebnitzkii* and *Neomysis awatschensis* (Crustacea: Mysidae). Mar. Biol. 21: 173-179.

Johns, D.M. and J.A. Pechenik. 1980. Influence of the water-accommodated fraction of No. 2 fuel oil on energetics of *Cancer irroratus* larvae. Mar. Biol. 55: 247-254.

Johnston, N.T. and P.C. Lasenby. 1982. Diet and feeding of *Neomysis mercedis* Holmes (Crustacea, Mysidacea) from the Fraser River estuary, British Columbia. Canadian J. Zool. 60: 813-824.

Kutty, M.N., G. Muragapoopathy, and T.S. Krishnan. 1971. Influence of salinity and temperature on the oxygen consumption in young juveniles of the Indian prawn *Peneaus indicus*. Mar. Biol. 11: 125-131.

Laughlin, R.B. and O. Linden. 1983. Oil pollution and Baltic mysids: Acute and chronic effects of the water soluble fractions of light fuel oil on the mysid shrimp *Neomysis integer*. Mar. Ecol. Prog. Ser. 12: 29-41.

Laughlin, R.B. and J.M. Neff. 1980. Influence of temperature, salinity, and phenanthrene (a petroleum derived hydrocarbon) on the respiration of larval mud crabs *Rhithropanopeus harrisii*. Est. Coast. Mar. Sci. 10: 655-669.

Laughlin, R.B. and J.M. Neff. 1981. Ontogeny of respiratory and growth responses of larval mud crabs Rhithropanopeus harrisii exposed to different temperatures, salinities, and naphthalene concentrations. Mar. Ecol. Prog. Ser. 5: 319-332.

Lee, W.Y., K. Winters, and J.A.C. Nicol. 1978. The biological effects of the water-soluble fractions of a No. 2 fuel oil on the planktonic shrimp, Lucifer faxoni. Environ. Pollut. 15: 167-183.

Mac, M.J. and J.G. Seelye. 1981. Potential influence of acetone in aquatic bioassays testing the dynamics and effects of PCBs. Bull. Environ. Contam. Toxicol. 27: 359-367.

Mauchline, J. 1980. Physiology and responses to physical-chemical parameters of the environment. pp. 140-159. In: The Biology of Mysids and Euphausiids. Advances in Marine Biology, Vol. 18. J.H.S. Blaxter, F.S. Russell, and M. Yonge (eds.). Academic Press, New York.

Maurer, D. and R.L. Wigley. 1982. Distribution and ecology of mysids in Cape Cod Bay, Massachusetts. Biol. Bull. 163: 477-491.

Moshiri, G.A., K.W. Cummins, and R.R. Costa. 1969. Respiratory energy expenditure by the predaceous zooplankter Leptodora kindtii (Focke) (Crustacea: Cladocera). Limnol. Oceanogr. 14: 475-484.

Neff, J.M. 1979. Polycyclic aromatic hydrocarbons in the aquatic environment. Applied Science Publishers, London. 261 pp.

Nie, N.H., C.H. Hull, M.N. Franklin, J.G. Jenkins, K.J. Sours, M.J. Norusis, and V. Beadle. 1980. A User's Guide to the SCSS Conversational System. McGraw-Hill, New York. 595 pp.

Paul, A.J. and P. Nunes. 1983. Temperature modification of respiratory metabolism and caloric intake of Pandalus borealis (Kroyer) first zoeae. J. Exp. Mar. Biol. 66: 163-168.

Percy, J.A. 1977. Effects of dispersed crude oil upon the respiratory metabolism of an Arctic marine amphipod, Onisimus (Boekisimus) affinis. pp. 192-200. In: Fate and Effects of Petroleum Hydrocarbons in Marine Organ isms and Ecosystems. D.A. Wolfe (ed.). Pergamon Press, New York.

Pezzak, D.S. and S. Corey. 1979. The life history and distribution of Neomysis americana (Smith) (Crustacea, Mysidacea) in Passamaquoddy Bay. Can. J. Zool. 57: 785-793.

Quetin, L.B., T.J. Mickel, and J.J. Childress. 1978. A method for simultaneously measuring the oxygen consumption and activity of pelagic crustaceans. Comp. Biochem. Physiol. 59A: 263-266.

Rutledge, P.S. and A.W. Pritchard. 1981. Scope for activity in the crayfish *Pacifastacus leniusculus*. Am. J. Physiol. 9(1): R87-92.

Simmons, M.A. and A.W. Knight. 1975. Respiratory response of *Neomysis intermedia* (Crustacea: Mysidacea) to changes in salinity, temperature, and season. Comp. Biochem. Physiol. 50A: 181-193.

Smith, R.L. and B.R. Hargreaves. 1984. Oxygen consumption in *Neomysis americana* (Crustacea: Mysidacea), and the effect of naphthalene exposure. Mar. Biol. 79: 109-116.

Stainken, D. and U. Frank. 1979. Analysis of Raritan Bay bottom waters for polynuclear aromatic hydrocarbons. Bull. Environ. Contam. Toxicol. 22: 480-487.

Thompson, C.Q. 1984. The effects of season and temperature on growth, reproduction, and hardiness of the opossum shrimp *Neomysis americana* (Smith). Ph.D. Dissertation, Lehigh University, Bethlehem, PA. 181 pp.

Toda, H., M. Takahashi, and S. Ichimura. 1982. Abundance and life history of *Neomysis intermedia* Czerniawsky in Lake Kasumigura. Hydrobiologia 93: 31-39.

White, F.N. and G. Somero. 1982. Acid-base regulation and phospholipid adaptations to temperature: Time course and physiological significance of modifying the milieu of protein function. Physiol. Rev. 62: 40-90.

Wigley, R.L. and B.R. Burns. 1971. Distribution and biology of mysids (Crustacea, Mysidacea) from the Atlantic coast of the United States in the NMFS Woods Hole collection. Fish. Bull. 69: 717-746.

Yalkowsky, S.H., S.C. Valvani, and D. MacKay. 1981. Estimation of the aqueous solubility of some aromatic compounds. Residue Reviews 85: 43-55.

EFFECT OF PRETREATMENT EXPOSURE TO TOLUENE AND NAPHTHALENE ON THE SUBSEQUENT METABOLISM OF DIETARY TOLUENE AND NAPHTHALENE BY DOLLY VARDEN, *SALVELINUS MALMA*

Robert E. Thomas and Stanley D. Rice

SUMMARY

Pretreatment exposure of Dolly Varden to naphthalene in sea water resulted in increased metabolism of orally-administered 14C-naphthalene, but pretreatment exposure to toluene did not alter the rate of metabolism of orally-administered 14C-toluene. Naphthalene is less polar than toluene and thus requires significant bioconversion before it is excreted by fish. Toluene is more polar and is readily excreted by fish, therefore inducement of increased bioconversion of toluene in the fish is not needed.

Several pretreatment factors influenced the rate of naphthalene bioconversion, including: 1) length of pretreatment exposure, 2) concentration of pretreatment exposure, and 3) duration of depuration after pretreatment exposure.

Pretreatment exposure of fish to naphthalene (75% of the 96-h LC50) for 48 h doubled or tripled the percentage of naphthalene metabolites in brain, liver, and muscle tissue, and decreased the amount of carbon-14 recovered in those tissues. Pretreatment exposure for only 24 h caused some increases in tissue metabolites, but no significant differences in percent metabolites or tissue burden were measured between the pretreatment fish and controls which were not pretreated.

The concentration of the pretreatment is directly related to the increase in bioconversion rate of naphthalene.

Exposure to 25 and 50% of the 96-h LC50 did not result in increased metabolism of naphthalene or in lower tissue burdens of the hydrocarbons. Exposure at 75% of the LC50 resulted in significant increases in tissue metabolites and decreases in total tissue hydrocarbon burden.

Periods of depuration following pretreatment exposure resulted in a decrease in the induced rate of metabolism of naphthalene. Depuration for as little as 24 h resulted in the return of the naphthalene bioconversion rate to that of control fish which were not pretreated.

Pretreatment exposures of fish can induce increased metabolism of aromatic hydrocarbons, such as naphthalene, and may increase the survival of fish in subsequent exposures. Dinuclear and polynuclear aromatic hydrocarbons are quite lipophilic and must be metabolized before they can be readily excreted. Inducement of increased metabolism of these compounds, resulting from a previous exposure, increases excretion and therefore reduces hydrocarbon accumulation in the tissues of fish.

INTRODUCTION

Fish exposed to petroleum in the environment will accumulate aromatic hydrocarbons in their tissues. Many of the accumulated hydrocarbons will be metabolized or excreted (Thomas & Rice, 1981). Metabolism of toxic aromatic hydrocarbons presumably contributes to the long-term survival of fish by making these hydrocarbons more polar and thus more readily excretable (Thomas & Rice, 1982). A few studies (Lee *et al.*, 1972; Roubal *et al.*, 1977; Varanasi *et al.*, 1978), however, have shown that some hydrocarbon metabolites persist in tissues longer than the parent hydrocarbon. Although most hydrocarbon metabolites are probably less toxic than the parent hydrocarbons, some may still be quite toxic. A metabolite of benzo(a)pyrene, for example, has been identified as the active carcinogen and mutagen in mammals, and Gruger *et al.* (1981) report that a pretreatment exposure of starry flounder (*Platichthys stellatus*) to naphthalene or *p*-cresol shifts the metabolism of 2,6-dimethylnaphthalene toward the production of more toxic metabolites.

Aromatic hydrocarbons are metabolized by many species of fish including salmon (Gruger *et al.*, 1977; Collier *et al.*, 1978), trout and char (Pedersen *et al.*, 1976; Ahokas *et al.*, 1977; Gerhart & Carlson, 1978; Thomas & Rice, 1981, 1982), mullet (Yarbrough & Chambers, 1977), flatfish (Varanasi *et al.*, 1979; Gruger *et al.*, 1981; Varanasi &

Gmur, 1981; Reichert & Varanasi, 1982), killifish (Burns, 1976; Binder & Stegeman, 1980), and cunners (Walton *et al.*, 1978). Most aromatic hydrocarbons are metabolized in the liver by an oxidative biotransformation process catalyzed by an enzyme system known as the mixed function oxidases (MFO). Mixed function oxidases, which are relatively nonspecific and degrade many xenobiotic compounds, including pesticides and aromatic hydrocarbons, have been found at elevated concentrations in several species of fish collected from areas polluted with oil (Payne & Penrose, 1975; Stegeman, 1978). Burns and Teal (1979) reported elevated MFO activity in *Fundulus* collected from an oil-contaminated marsh 5 yr after a Florida oil spill. Because of the effect of xenobiotic exposure on MFO activity, some investigators have suggested that MFO activity be used as a biological indicator of oil contamination (Payne & Penrose, 1975; Payne, 1976).

Although MFOs are inducible in fish exposed to oil, changes in metabolism and disposal of aromatic hydrocarbons by living fish or changes in their survival have not been demonstrated. The *in vitro* measurement of changes in liver enzyme activity is a measure of potential and not a measure of *in vivo* function. In two studies, we have attempted to demonstrate changes in hydrocarbon metabolism and survival of fish after sublethal pretreatment with hydrocarbons. In the first study, we measured the changes in tolerance of pink salmon after pretreatment exposures to toluene or naphthalene in sea water (Rice, unpubl. data), and in this study, we measured the changes in deposition and metabolism of aromatic hydrocarbons given orally to Dolly Varden (*Salvelinus malma*) after pretreatment exposures to toluene or naphthalene in sea water.

The objective of this study was to determine the effects of pretreatment exposure on the metabolism and disposal of these compounds by Dolly Varden. The fish were exposed to either toluene or naphthalene in sea water at various exposure concentrations and durations before labeled toluene or naphthalene was given orally. Twenty-four hours after oral exposure, the fish were killed, tissues sampled for the labeled aromatic hydrocarbon, and the percentage of the label in metabolites determined. We manipulated several exposure variables to accomplish our objective of measuring the effect of a pretreatment exposure to two aromatic hydrocarbons on the tissue accumulation of the hydrocarbons and their metabolites. Specifically, we measured 1) the effect of duration of pretreatment exposure, 2) effect of the concentration of hydrocarbons during the pretreatment exposure, and 3) the effect of depuration after a pretreatment exposure on the accumulation of labeled hydrocarbons and metabolites in the tissues.

MATERIALS AND METHODS

Dolly Varden were obtained by beach seining during late June and early July in the area of Auke Bay, southeastern Alaska. The fish were maintained in running sea water (30 ‰) at ambient sea water temperature (6.5-7.5°C) for at least 3 wk and then acclimated for a minimum of 6 d at 4°C before being used in the study. The fish averaged 36.5 ± 6.1 g and were young adults of both sexes.

Experimental Design

Groups of 10 fish each received different pretreatment exposures to toluene or naphthalene in sea water (different lengths of exposure, concentrations, or length of depuration in clean water after exposure). After the pretreatment, all control and exposed fish received orally administered soluble gelatin capsules containing carbon-14 labeled toluene or naphthalene and were placed in clean sea water for 24 h. Tissues were sampled after 24 h, and accumulated labeled compounds and their metabolites were measured. The 96-h LC50s at 4°C (flow-through) were 0.638 ppm naphthalene and 4.46 ppm toulene.

Tests of Different Hydrocarbons

Groups of 10 fish each were exposed, in flow-through systems, to either naphthalene or toluene in the water column. Concentrations were rigorously maintained at 90% of the 96-h LC50 for toluene and at 75% of the 96-h LC50 for naphthalene according to the technique described by Moles *et al*. (in press). There were no mortalities in the pretreatment exposures. Control fish were handled similarly, except they received no toxicants in their pretreatment exposure. After exposure in the water column, each fish was force-fed a capsule containing either 14C-toluene (4.9 mCi/mM; 3.9 μCi per capsule, fish exposed to toluene) or 14C-naphthalene (4.76 mCi/mM; 4.3 μCi per capsule, fish exposed to naphthalene) dissolved in corn oil.

Tests on Length of Pretreatment Exposures

Groups of fish were exposed to naphthalene in the water column at a concentration which was 75% of the 96-h LC50. Fish were exposed to this concentration for 24 or 48 h. Control fish were handled similarly, except they received no toxicants during pretreatment exposure. After exposure, control and exposed fish were force-fed a capsule containing 14C-naphthalene and were placed in clean sea water.

Tests on Concentration of Pretreatment Exposure

Groups of fish were exposed to naphthalene in the water column at concentrations which were equivalent to 25, 50, or 75% of the 96-h LC50. After 48-h exposure, each fish was force-fed a capsule containing 14C-naphthalene. After receiving the capsule, the fish were placed in clean sea water.

Tests on Length of Depuration After Pretreatment Exposures

Groups of fish were exposed for 48 h to naphthalene in the water column at 75% of the 96-h LC50. Each fish from one group was given a capsule containing 14C-naphthalene immediately after exposure and then placed in clean sea water. A second group was placed in clean sea water for 12 h following exposure but prior to receiving the capsules, and a third group was placed in clean sea water for 24 h before receiving the capsules.

Recovery of Carbon-14 in Tissues

Twenty-four hours after receiving the capsules, all fish in all experimental groups were killed and tissues removed for analysis of carbon-14 content. A modification of the method described by Roubal *et al.* (1977) was used to determine the amount of carbon-14 in the nonpolar (parent hydrocarbon) and polar (metabolite) phases in each tissue after the tissues were digested (for methods, see Thomas & Rice, 1981).

Data are reported as the percentage of administered carbon-14 recovered in each tissue and as the percentage of activity recovered in each tissue in the metabolite phase. Differences of activity recovered in the tissues from all groups of fish were analyzed statistically by one-way analysis of variance and Student's t-test.

RESULTS

Differences in Recovered Metabolites from Tests With Toluene and Naphthalene

Pretreatment exposures to toluene did not cause a significant change in the percent of 14C-metabolites recovered from the tissues of Dolly Varden (Table 1). In contrast, fish that had been treated similarly with naphthalene had increased levels of metabolites in all tissues. The percentage of naphthalene metabolites doubled or tripled in brain, liver, and muscle tissue. The amount of activity in toluene metabolites was generally higher than in naphthalene metabolites in all tissues, and indicated that toluene is more easily metabolized than naphthalene and that tissues other than the liver probably metabolize toluene.

Table 1. Percentage of 14C-metabolites in tissues of Dolly Varden after the fish were pretreated to toluene or naphthalene in sea water then force-fed 14C-toluene and 14C-naphthalene. In the pretreatments the fish were exposed for 48 h to concentrations of the toxicants that were 75-90% of the 96-h LC50. Tissues were sampled for 14C-metabolites 24 h after the fish were force-fed. n = 10; * indicates significant difference between pretreatment groups at $p<0.05$.

Pretreatment	Percentage of metabolites in each tissue			
	Gall Bladder	Liver	Brain	Muscle
		Toluene		
None	87.5	41.9	31.7	58.7
48-h, 90% LC50	82.5	28.0	29.2	54.0
		Naphthalene		
None	66.5	5.1	6.2	17.2
48-h, 75% LC50	75.6	16.7*	15.1*	39.3*

There was also an increase in the percentage of naphthalene metabolites recovered from the gall bladder of fish that received a pretreatment with naphthalene in the water for 48 h (Table 1), but the increases were not as large as those found in the other tissues. The gall bladder is a temporary storage organ that receives metabolites and excretory products from the liver, so the concentration of hydrocarbon metabolites is normally higher in the gall bladder than in other tissues. Neither duration of exposure nor concentration resulted in significant changes in metabolite concentrations in the gall bladder compared to control (no pretreatment) fish.

Factors Affecting Naphthalene Metabolite Levels in Tissues

The recovery of 14C-naphthalene and the percentage of activity in metabolites in tissues are affected by length of pretreatment exposures, concentration of the pretreatment exposures, and length of depuration after pretreatment exposures.

Effect of Length of Pretreatment Exposure to Naphthalene

As the duration of pretreatment with naphthalene to the fish increased, the total amount of carbon-14 recovered in the tissues decreased, and the percentage of the carbon-14 in tissue metabolites increased (Table 2). In this test, we exposed three groups of fish to naphthalene for different periods before oral administration of 14C-naphthalene. Recovery of carbon-14, as a percentage of that administered, was greatest in control fish with no pretreatment exposure and decreased in fish receiving the naphthalene pretreatment exposure. Although the recovery of carbon-14 from tissues at 24 and 48 h was usually less than the recovery from control fish, recovery of carbon-14 from the brain at 48 h was the only tissue significantly less than the recovery of carbon-14 from control fish.

The percentage of metabolites found in the liver, brain, and muscle of fish that received 48 h of pretreatment exposure was two or three times the percentage of metabolites found in the same tissues of the no pretreatment group (Table 2). There were, however, only slight differences in these tissues between the 24-h pretreatment group and the control group. Apparently, at the 4°C temperature used in this study, the inducement of increased naphthalene-metabolizing capability requires exposures greater than 24 h. After 48 h of pretreatment to naphthalene, the rate of bioconversion to metabolites increased and there was generally lower recovery of the carbon-14 from the tissue.

Table 2. Recovery and metabolism of 14C-naphthalene administered orally to Dolly Varden after different lengths of exposure to naphthalene in sea water. Water exposures to naphthalene were for 24 or 48 h at 75% of the 96-h LC50. n = 10; 95% confidence intervals in parentheses; * = significant difference from controls at $p<0.05$.

Pretreatment Time (h)	Gall bladder		Liver		Brain		Muscle	
	% of Adm.	% metab.	% of Adm.	% metab.	% of Adm.	% metab.	% of Adm.	% metab.
0 (control)	0.2 (0.3)	66.5 (9.0)	0.61 (0.16)	5.1 (1.6)	0.12 (0.3)	6.2 (2.8)	6.4 (3.5)	17.2 (5.3)
24	0.04 (0.01)	77.8 (22.6)	0.35 (0.05)	7.9 (4.7)	0.08 (0.03)	7.7 (5.7)	8.3 (1.0)	21.0 (13.7)
48	0.02 (0.01)	75.6 (11.1)	0.44 (0.08)	16.7* (4.8)	0.06* (0.01)	15.1* (4.3)	5.6 (0.05)	39.3* (7.4)

Effect of Naphthalene Concentration During Pretreatment Exposure

The greater the naphthalene concentration in the pretreatment exposure, the lower the total recovery of carbon-14 from the tissues and the greater the percentage of activity in the tissue metabolite fraction (Table 3). In this test, we exposed three groups of fish for 48 h to one of three concentrations of naphthalene (25, 50, or 75% of the 96-h LC50) in sea water before oral administration of 14C-naphthalene. The lower concentrations (25 and 50% of the 96-h LC50) of naphthalene in the pretreatment exposures had little effect on the amount of carbon-14 in the tissues, but the highest preexposure concentration (75% of the 96-h LC50) caused a significant reduction in the amount of carbon-14 in the tissues.

Table 3. Recovery and metabolism of 14C-naphthalene administered orally to Dolly Varden after 48-h exposure to different concentrations of naphthalene in the water. n = 10; 95% confidence intervals in parentheses; * = significant differences from lower exposure dose at $p<0.05$.

Pretreatment % of 96-h LC50	Gall bladder		Liver		Brain		Muscle	
	% of Adm.	% metab.	% of Adm.	% metab.	% of Adm.	% metab.	% of Adm.	% metab.
25	0.06 (0.03)	84.5 (6.5)	0.62 (0.33)	3.2 (1.7)	0.13 (0.04)	0.9 (0.3)	10.5 (1.6)	2.6 (1.5)
50	0.01 (0.04)	86.8 (9.0)	0.75 (0.21)	4.5 (1.6)	0.23 (0.06)	1.7* (0.4)	12.9* (1.0)	1.6 (3.2)
75	0.04 (0.02)	87.5 (10.1)	0.37* (0.05)	10.1* (3.9)	0.08* (0.05)	12.8* (7.5)	3.8* (0.9)	34.3* (9.4)

There was an order of magnitude increase in percentage of carbon-14 found in the tissue metabolite fraction of brain and muscle of the fish receiving pretreatment exposures to 75% of the 96-h LC50 (Table 3). There was also a three-fold increase in metabolite activity in liver tissue. At 4°C and after pretreatment exposures lasting only 48 h, naphthalene concentrations greater than 75% of the 96-h LC50 are above the threshold level required for inducing increased naphthalene bioconversion.

Effect of Depuration After Pretreatment Exposure

The longer the depuration periods after pretreatment exposure to naphthalene, the lower the percentage of carbon-14 recovered in the tissue metabolite fraction (Table 4). In this test, we exposed three groups of fish for 48 h to 75% of the 96-h LC50 followed by three different periods of depuration in clean water (0, 12, or 24 h) before oral administration of 14C-naphthalene. By 12 h of depuration, the percentage of activity in metabolites in liver, brain, and muscle tissue had decreased. By 24 h of depuration, the percentage of activity in metabolites in the same tissues had decreased to about half to one third the percentages of the fish without depuration. These percentages of metabolites were similar to those in control fish that had not received any pretreatment exposures. Although pretreatment exposures at 75% of the 96-h LC50 caused increased bioconversion of naphthalene, the induced rate was rapidly reduced by relatively short periods of depuration.

Table 4. Effect of depuration time on the recovery and metabolism of 14C-naphthalene administered orally to Dolly Varden after 48 h of pretreatment exposure to 75% of the 96-h LC50. Depuration times varied from 0 to 24 h. n = 10; 95% confidence intervals in parentheses; * = significant differences from shorter depuration times at $p<0.05$.

Pretreatment depuration time (h)	Gall bladder		Liver		Brain		Muscle	
	% of Adm.	% metab.	% of Adm.	% metab.	% of Adm.	% metab.	% of Adm.	% metab.
0	0.02 (0.01)	75.6 (11.1)	0.44 (0.08)	16.7 (4.8)	0.06 (0.01)	15.1 (4.3)	5.6 (0.5)	39.6 (7.4)
12	0.04 (0.02)	72.6 (6.4)	0.38 (0.07)	12.7 (3.1)	0.07 (0.03)	11.4* (3.1)	4.2* (0.3)	27.5* (4.3)
24	0.04 (0.02)	87.2 (7.1)	0.34 (0.09)	7.2* (2.0)	0.06 (0.02)	5.6* (1.9)	3.8* (0.3)	19.5* (3.8)
Controls -- No Pretreatment Exposure								
	0.02 (0.01)	66.5 (9.0)	0.61 (0.16)	5.1 (1.6)	0.12 (0.3)	6.2 (2.8)	6.4 (3.5)	17.2 (5.3)

The percentage of metabolites in the gall bladder increased by 24 h of depuration. The gall bladder does not produce metabolites but receives them from other tissues and stores them for a variable time period. Time-course studies on the gall bladder are not very informative.

DISCUSSION

Exposure of Dolly Varden to naphthalene at 75% of the 96-h LC50 for 48 h resulted in a significant increase in the percentage of carbon-14 recovered in the tissue-metabolite fraction. More importantly, the pretreatment also resulted in a significant decrease in the amount of carbon-14 recovered from the tissues. Apparently, increased metabolism of naphthalene results in increased excretion, thereby lowering the tissue burden of hydrocarbons. Increased excretion after pretreatment had been reported by Statham *et al*. (1978) who found that pretreatment of rainbow trout with 2,3-benzanthracene resulted in an increase in biliary excretion of 2-methylnaphthalene metabolites.

The increase in metabolites of aromatic hydrocarbons in fish tissues after pretreatment exposures correlates well with measured increases in enzyme activities (MFOs) of Dolly Varden that were exposed to oil (Collodi, 1983), and is compatible with other observations of induced MFO activity in livers of fish exposed to xenobiotics. The degree to which enzyme activity is increased depends largely on the chemical nature of the compound used in the exposure. Gerhart and Carlson (1978) report that, in general, higher molecular weight compounds (Aroclor 1254, pyrene, chrysene, benzo(a)-pyrene) injected into rainbow trout resulted in increased enzyme activity, whereas lower molecular weight compounds, such as naphthalene, did not. Gruger *et al*. (1981) found that previous exposure of starry founder to naphthalene had a greater effect on increasing the rate of the metabolism of 2,6-dimethylnaphthalene than did exposure to *p*-cresol. They suggest that because *p*-cresol is more polar and thus more soluble in water than naphthalene, *p*-cresol can probably be excreted more effectively. We (Thomas & Rice, 1981) found that Dolly Varden excrete more toluene (as parent compound and metabolite) than naphthalene. Polarity, molecular weight, and partition coefficient all appear to be important in determining the metabolism, elimination, and tissue distribution of aromatic hydrocarbons (Thomas & Rice, 1982).

Because toluene is so readily metabolized and excreted by Dolly Varden, it was not surprising that a previous exposure to toluene did not alter the tissue accumulation of toluene metabolites. Naphthalene, on the other hand, is not as readily excretable: approximately two times as much toluene was excreted during a 24-h period (Thomas & Rice, 1981) and, therefore, naphthalene might require greater biotransformation. Pretreatment exposure to naphthalene resulted in greater biotransformation and, presumably, more excretable products. Rice (unpubl. data) obtained similar data with respect to changes in the tolerance of pink salmon fry after pretreatment with toluene or naphthalene. Pretreatment exposure to toluene did not alter the toluene LC50, whereas pretreatment with naphthalene significantly increased the naphthalene LC50.

Length of the pretreatment exposure has a significant effect on metabolism of naphthalene. Exposure for 24 h at 75% of the 96-h LC50 had very little effect on the tissue metabolite burden. Exposure for 48 h at the same concentration however, doubled or tripled the percentage activity in metabolites. These results again parallel those of Rice (unpubl. data) with pink salmon fry tolerance after pretreatment with naphthalene. He noted that a 24-h pretreatment to naphthalene had little effect on fry sensitivity, whereas a 48-h exposure caused a significant decrease in fry sensitivity to naphthalene.

Our results indicate that a threshold level, in terms of both pretreatment length of exposure and concentration, is necessary to alter the metabolism of naphthalene. Pretreatment exposure for 48 h to 25 and 50% of the 96-h LC50 had little effect on naphthalene metabolism, whereas exposure at 75% had a significant effect. Rice (unpubl. data) found a similar concentration effect on the tolerance of pink salmon fry after pretreatment with naphthalene. Gerhart and Carlson (1978) did not report induction of MFO activity in rainbow trout injected intraperitoneally with naphthalene. Although this might reflect a species difference, it might also indicate the concentration of naphthalene injected was less than the threshold concentration required for induction. Gerhart and Carlson calculated a threshold tissue concentration of benzo(a)pyrene required for MFO induction.

The advantages to fish from pretreatment exposures rapidly disappear when the fish are transferred to clean water. Depuration for as little as 12 h resulted in a significant decrease in metabolite levels, and by 24 h of depuration, metabolite concentrations in liver, brain, and muscle were reduced to the levels in controls which were not pretreated. Similar depuration results were noted in tolerance tests with pink salmon fry by Rice (unpubl. data). The rapid loss of the pretreatment effect on the metabolism of naphthalene in Dolly Varden implies that MFO activity also rapidly returns to preexposure levels after a short depuration period. Therefore, Payne's (1976) suggestion that induction of MFO activity can be used to indicate the presence of certain toxic chemicals in the environment is practical only for chronic continuous exposures.

The induction of increased metabolism of naphthalene by pretreatment of Dolly Varden with naphthalene appears to afford distinct advantages to the exposed fish if they continue to encounter hydrocarbon pollutants. However, there coud be serious consequences for chronic consumers of contaminated fish. The metabolism of some aromatic hydrocarbons results in the production of reactive metabolites with known carcinogenic or mutagenic properties. Varanasi and Gmur (1980) report significant increases in metabolic activation and covalent binding to DNA by fractions of liver homogenates of 3-methylcholanthrene- or benzo(a)pyrene-pretreated starry flounder and coho salmon. Binding to DNA was 10-53 fold greater than in fish that were not pretreated. Gruger *et al*. (1981) report a shift in the metabolism of 2,6-dimethylnaphthalene in naphthalene- or *p*-cresol-pretreated starry flounder toward potentially carcinogenic epoxides. Although there are indications that potentially toxic metabolites could be retained in tissues longer than the parent hydrocarbons (Lee *et al*., 1972; Roubal *et al*., 1977), pretreatment exposure of fish to aromatic hydrocarbons increases the tolerance to subsequent exposures to hydrocarbons.

ACKNOWLEDGMENTS

We thank Jessica Gharrett for long hours of tissue processing. Partial funding support was from the National Oceanic and Atmospheric Administration, Office of Marine Pollution.

LITERATURE CITED

Ahokas, J.T., O. Pelkonen, and N.T. Karki. 1977. Characterization of benzo(a)pyrene hydroxylase of trout liver. Cancer Res. 37: 3737-3743.

Binder, R.L. and J.J. Stegeman. 1980. Induction of aryl hydrocarbon hydroxylase activity in embryos of an estuarine fish. Biochem. Pharmacol. 29: 949-951.

Burns, K.A. 1976. Microsomal mixed function oxidases in an estuarine fish, *Fundulus heteroclitus*, and their induction as a result of environmental contamination. Comp. Biochem. Physiol. 53B: 443-446.

Burns, K.A. and J.M. Teal. 1979. The West Falmouth oil spill: Hydrocarbons in the saltmarsh ecosystem. Estuarine Coast. Mar. Sci. 8: 349-360.

Collier, T.K., L.C. Thomas, and D.C. Malins. 1978. Influence of environmental temperature on disposition of dietary naphthalene in coho salmon (*Oncorhynchus kisutch*): Isolation and identification of individual metabolites. Comp. Biochem. Physiol. 61C: 23-28.

Collodi, P. 1983. Effect of petroleum hydrocarbons on hepatic aryl hydrocarbon hydroxylase activity in coho salmon (*Oncorhynchus kisutch*). M.S. Thesis. Univ. Alaska, Juneau.

Gerhart, E.H. and R.M. Carlson. 1978. Hepatic mixed function oxidase activity in rainbow trout exposed to several polycyclic aromatic compounds. Environ. Res. 17: 284-295.

Gruger, E.H., J.V. Schnell, P.S. Fraser, D.W. Brown, and D.C. Malins. 1981. Metabolism of 2,6-dimethylnaphthalene in starry flounder (*Platichthys stellatus*) exposed to naphthalene and *p*-cresol. Aquat. Toxicol. 1: 37-48.

Gruger, E.H., M.M. Wekell, P.T. Numoto, and D.R. Craddock. 1977. Induction of hepatic aryl hydrocarbon hydroxylase in salmon exposed to petroleum dissolved in sea water and to petroleum and polychlorinated biphenyls, separate and together, in food. Bull. Environ. Contam. Toxicol. 17: 512-520.

Lee, R.F., R. Sauerheber, and G.H. Dobbs. 1972. Uptake, metabolism and discharge of polycyclic aromatic hydrocarbons by marine fish. Mar. Biol. 17: 201-208.

Moles, A., S.A. Andrews, and S.D. Rice. In press. Continuous-flow device(s) for exposure of marine animals to the water-soluble fraction of crude oil and oil components. 10th Aquat. Toxic. Workshop, Halifax, Canada, 1983.

Payne, J.F. 1976. Field evaluation of benzo(a)pyrene hydroxylase induction as a monitor for marine pollution. Science 191: 945-946.

Payne, J.F. 1977. Mixed function oxidases in marine organisms in relation to petroleum hydrocarbon metabolism and detection. Mar. Pollut. Bull. 8: 112-116.

Payne, J.F. and W.R. Penrose. 1975. Induction of aryl hydrocarbon (Benzo[a]pyrene) hydroxylase in fish by petroleum. Bull. Environ. Contam. Toxicol. 14: 112-116.

Pedersen, M.G., W.K. Hershberger, P.K. Zachariah, and M.R. Juchau. 1976. Hepatic biotransformation of environmental xenobiotics in six strains of rainbow trout (*Salmo gairdneri*). J. Fish. Res. Board. Can. 33: 666-675.

Reichert, W.L. and U. Varanasi. 1982. Metabolism of orally administered naphthalene in spawning English sole (*Parophrys vetulus*). Environ. Res. 27: 316-324.

Roubal, W.T., T.K. Collier, and D.C. Malins. 1977. Accumulation and metabolism of carbon-14 labeled benzene, naphthalene, and anthracene by young coho salmon (*Oncorhynchus kisutch*). Arch. Environ. Contam. Toxicol. 5: 513-529.

Statham, C.N., C.R. Elcombe, S.P. Szyjka, and J.J. Lech. 1978. Effect of polycyclic aromatic hydrocarbons on hepatic microsomal enzyme and deposition of methylnaphthalene in rainbow trout *in vivo*. Xenobiotica 8: 65-71.

Stegeman, J.J. 1978. Influence of environmental contamination on cytochrome P-450 mixed function oxidases in fish: Implications for recovery in the Wild Harbor Marsh. J. Fish. Res. Board Can. 35: 668-674.

Thomas, R.E. and S.D. Rice. 1981. Excretion of aromatic hydrocarbons and their metabolites by freshwater and sea water Dolly Varden char. pp. 425-448. In: Biological Monitoring of Marine Pollutants. F.J. Vernberg, A. Calabrese, F.P. Thurberg, and W.B. Vernberg (eds.). Academic Press, New York.

Thomas, R.E. and S.D. Rice. 1982. Metabolism and clearance of phenolic and mono-, di-, and polynuclear aromatic hydrocarbons by Dolly Varden char. pp. 161-176. *In*: *Physiological Mechanisms of Marine Pollutant Toxicity*. W.B. Vernberg, A. Calabrese, F.P. Thurberg, and F.J. Vernberg (eds.). Academic Press, New York.

Varanasi, U. and D.J. Gmur. 1980. Metabolic activation and covalent binding of benzo(a)pyrene to deoxyribonucleic acid catalyzed by liver enzymes of marine fish. Biochem. Pharmacol. 29: 753-761.

Varanasi, U. and D.J. Gmur. 1981. Hydrocarbons and metabolites in English sole (*Parophrys vetulus*) exposed simultaneously to 3H-benzo(a)pyrene and 14C-naphthalene in oil contaminated sediment. Aquat. Toxicol. 1: 49-67.

Varanasi, U., D.J. Gmur, and P.A. Treseler. 1979. Influence of time and mode of exposure on biotransformation of naphthalene by juvenile starry flounder (*Platichthys stellatus*) and rock sole (*Lepidopsetta bilineata*). Arch. Environ. Contam. Toxicol. 8: 673-692.

Varanasi, U., M. Uhler, and S.I. Stranahan. 1978. Uptake and release of naphthalene and its metabolites in skin and epidermal mucus of salmonids. Toxicol. Appl. Pharmacol. 44: 277-289.

Walton, D.G., W.R. Penrose, and J.M. Green. 1978. The petroleum inducible mixed function oxidase of cunner (*Tautogolabrus adspersus*, Walbaum 1792): Some characteristics relevant to hydrocarbon monitoring. J. Fish. Res. Board Can. 35: 1547-1552.

Yarbrough, J.D. and J.E. Chambers. 1977. Crude oil effects on microsomal mixed function oxidase system components in the striped mullet (*Mugil cephalus*). Life Sci. 21: 1095-1100.

LIVER ENERGY METABOLISM AND GILL ATPase ACTIVITY OF *SALMO SALAR* JUVENILES AFTER EXPOSURE TO SUBLETHAL LEVELS OF PHENOLS

K. Haya, B.A. Waiwood, L.E. Burridge and L. Van Eeckhaute

SUMMARY

Gill ATPase activity and liver energy were evaluated as indicators of sublethal effects caused by eight phenols in juvenile Atlantic salmon. Exposure of salmon to 4-methylphenol caused a decrease in gill Na,K-ATPase activity, whereas chloro- and p-nonylphenol caused an increase in residual ATPase activity. None of the other phenols tested had any significant effect on the ATPases. There were no significant differences in energy metabolism (AEC, phosphorylation potential, glucose, glycogen, and creatine phosphate) of the livers of salmon exposed to phenols for 96 h compared to those of controls. There was a trend to decreased levels of glycogen, glucose, and creatine phosphate in livers of treated salmon compared to 0-h controls. However, these decreases are probably the result of starvation as well as exposure to phenols. The results of this study support the view that the determination of sublethal effects caused by xenobiotics will require the development of a series of "clinical tests" rather than a single all encompassing biochemical parameter.

INTRODUCTION

Exposure of animals to sublethal levels of toxicants may inflict stresses on the mechanisms required for maintaining a healthy physiological state. These stresses may result in changes in biochemical, physiological or behavioral processes. Development of biochemical indicators that reflect these changes in aquatic organisms will allow for more accurate prediction of acceptable levels of pollutants in the environment. Such indicators could be also developed as early warning signals of potential pollution problems or for determining the health of an aquatic population. Our research has focused on parameters of energy metabolism as a potential indicator of sublethal effects caused by pollutants.

Adenosine triphosphatases (ATPases) are a group of enzymes which hydrolyze adenosine triphosphate (ATP) to adenosine diphosphate (ADP) and inorganic phosphate (Pi). In the process the energy released becomes available for cation transport (Schuurmans Stekhoven & Bonting, 1981). These enzymes, especially sodium, potassium-activated ATPase (Na,K-ATPase), play a central role in whole body osmoregulation (Towle, 1981). Na,K-ATPase activity is inhibited by ouabain; ATPases not inhibited by ouabain are referred to collectively as residual ATPase. Exposure of aquatic animals to various chemicals alters the activity of ATPases in general, and Na,K-ATPase activity has been evaluated as a potentially useful indicator of sublethal toxicity (Haya & Waiwood, 1983). Gills are the main osmoregulatory surface organ in aquatic animals and are the primary site of uptake of waterborne pollutants. Therefore, gills may be the first site where the sublethal effects of chemicals on Na,K-ATPase activity would be observed.

Adenylate energy charge (AEC) is another potential indicator of sublethal effects in multicellular aquatic organisms (Ivanovici, 1980; Haya & Waiwood, 1983). AEC, (ATP + 0.5ADP)/(ATP + ADP + AMP), is a measure of the amount of energy available from the adenylate pool and is a prime factor in the regulation of flux of energy in catabolic and anabolic processes of cells (Atkinson, 1977). AEC changes in aquatic animals during environmental (salinity, temperature, pH) or xenobiotic stress have been reviewed by Haya and Waiwood (1983).

AEC may not be applicable in systems where phosphagens dominate as energy carriers (*e.g.*, muscle) or where adenylate kinase is not sufficiently active (*e.g.*, mitochondrial matrix) (Reich & Sel'kov, 1981). Creatine phosphate (CP) in vertebrates and arginine phosphate in invertebrates are the main phosphagens that accept or return energy equivalents from the adenylate pool. Cells use this process to maintain a constant AEC until the phosphagen pool is exhausted; provided the inorganic phosphate concentration is well buffered. In those systems where inorganic phoshate levels are limiting or weakly buffered, its concentration is expected to be a factor in mechanisms controlling the energy state of cells (Erecinska et al., 1977). In this situation the phosphorylation potential (PP = ATP/ADP·Pi) may be a better measure of the metabolic energy state (Erecinska et al., 1977; Reich & Sel'kov, 1981; Erecinska & Wilson, 1982). Therefore, phosphagen levels or PP may be altered or may be responsive on exposure to sublethal levels of chemicals in cases where AEC remains constant.

The liver is a very active tissue metabolically, the site of biotransformation of most xenobiotics, an organ in which bioconcentration of chemicals can be large, and the target organ of many toxic compounds. Thus the liver is a likely organ in which sublethal effects on energy metabolism will occur.

The gill Na,K-ATPase activity and liver AEC, creatine phosphate, and PP in Atlantic salmon parr exposed to sublethal levels of eight chlorinated or alkylated phenols were determined. Glucose and glycogen are principal substrates in central pathways of energy-producing carbohydrate metabolism in the liver and their concentrations were also determined. The objective of this study was to evaluate biochemical parameters as indicators of the effects caused by exposure to sublethal levels of phenols.

MATERIALS AND METHODS

Juvenile Atlantic salmon (*Salmo salar*) were obtained from the North Atlantic Salmon Research Center (St. Andrews, New Brunswick, Canada). They were fed commercial fish food (EWOS No. 4) daily. They were held for at least two months in aerated, continuously-flowing freshwater at ambient temperature (3-7°C) prior to the start of the experiments. The fork lengths of the fish used in the experiments had a range of 6-10 cm (mean 7.7 cm). Their weights ranged from 2 to 9.5 g (mean 3.9 g). All the test chemicals except for phenol (Mallinckrodt, St. Louis, MO, USA), 2-chlorophenol (Fisher Scientific, Fair Lawn, NJ, USA) and p-nonylphenol (Eastman Kodak, Rochester, NY, USA) were purchased from Chemical Service (Westchester, PA, USA). All biochemicals were purchased from Sigma (St. Louis, MO, USA) or Behringer-Mannheim (Montreal, Que., Canada).

Determination of LC50

The salmon parr were starved for 24 h prior to and during their exposure to the chemicals. The lethality tests were static with continuous aeration and solution changes at 48 h. Each battery jar (12 L) contained 10 L of fresh water and 10 salmon parr, and was covered with wire mesh. The temperature was maintained between 5 and 7°C by placing the battery jars in water baths cooled with tap water.

Stock solutions of the test compounds were prepared with ethanol. These solutions (0.15 to 2.5 ml) were added directly to the water immediately before addition of fish. Ethanol (2.5 ml) was added to the control tanks. Water samples were taken periodically, but could not be accurately quantified by UV-Vis or HPLC methods. Therefore, 96-h LC50s are for nominal concentrations and were estimated as the geometric mean between the log of the highest concentration at which less than 50% deaths occurred and the lowest concentration at which more than 50% deaths occurred.

Biochemical Studies

Salmon parr were exposed to each of the phenols under the same conditions as above, except that 15 salmon parr were placed in each battery jar. The exposure concentrations were based on, but less than, the nominal LC50 values (Table 1). No deaths occurred in any of the treatment groups except for the 2,4,6-trichlorophenol group where all the fish were dead and the 3,5-dimethylphenol where 10 of 15 fish were dead after 96 h of exposure.

After 96 h of exposure the salmon parr were anaesthetized with t-amyl alcohol. Preliminary studies indicated that anaesthetizing the fish prior to dissection minimized effects caused by handling stress on the levels of adenylates (unpubl. data). The livers and gills were rapidly excised, freeze clamped at -196°C using liquid nitrogen and stored at -80°C. The livers of three fish were pooled prior to homogenization and the extracts were analyzed by enzymatic spectrophotometric techniques using a Beckman Model 25K UV-Vis spectrophotometer. The substrates determined were adenine nucleotides, creatine phosphate, glucose, glycogen, and inorganic phosphate (Bergmeyer, 1974; Heinonen & Lahti, 1981; Haya *et al*., 1983). The gills were analyzed for Na,K- and residual ATPase activity as previously described (Neufeld & Pritchard, 1979; Haya *et al*., 1983) except that the inorganic phosphate was determined by the method of Heinonen and Lahti (1981).

RESULTS AND DISCUSSION

The nominal 96-h LC50 values ranged from 0.04 mg/L for 2,4,6-trichlorophenol to 14.7 mg/L for 2,6-dimethylphenol (Table 1). There was a striking difference between the toxicities of 2,4,6-trichlorophenol and 2,4,5-trichlorophenol (0.04 vs. 3.5 mg/L, respectively). These results are in contrast to other studies where 2,4,5-trichlorophenol was more toxic than 2,4,6-trichlorophenol to intermolt grass shrimp, *Palaemonetes pugio* (96-h LC50 values of 1.12 and 3.95 mg/L, respectively) (Rao *et al.*, 1981) and soft shell clam, *Mya arenaria* (96-h LC50 values of 2.9 and 3.9 mg/L, respectively) (McLeese *et al.*, 1979). Since the LC50 values in the present study were based on nominal concentrations, our LC50 data must be interpreted with caution. No further studies were carried out to define the LC50 values more accurately as our purpose was only to determine sublethal concentrations of the phenols for our biochemical studies.

Table 1. Nominal 96-h LC50 values for Atlantic salmon parr, *Salmo salar*, and exposure concentration of phenols in biochemical studies.

Compound	Exposure conc. (mg/L)	96-h LC50 (mg/L)
2,4,6-Trichlorophenol	0.03	0.04
Pentachlorophenol	0.04	0.07
2-Chlorophenol	0.56	0.75
p-Nonylphenol	0.48	0.90*
4-Methylphenol	1.50	3.00
Phenol	2.71	3.50
2,4,5-Trichlorophenol	1.56	3.50
3,5-Dimethylphenol	11.96	14.30
2,6-Dimethylphenol	10.49	14.70

*McLeese *et al.*, 1980.

In the LC50 experiments, the salmon parr demonstrated a bobbing motion as the first sign of toxicity. This was followed by a period of disorientation, then a period where the parr laid quietly, but not motionless, on their sides at the bottom of the tank. These effects lasted from a few hours prior to death to the full 96 h of the experiment. During exposure of salmon parr to the phenols for the biochemical studies, these signs of toxicity were observed with fish exposed to 3,5-dimethyl-, 2,6-dimethyl- and 2,4,6-trichlorophenol. All the fish exposed to 2,6-dimethylphenol were on their sides after 24 h of exposure, but all survived until sampled after 96 h of exposure.

The Na,K- and residual ATPase activity of gills from 96-h control salmon parr were 0.54 and 2.25 μmole Pi/mg protein·h, respectively (Table 2). These activities are lower than those previously reported for Atlantic salmon parr (7 to 27 μmole Pi/g protein·h for parr with fork lengths of at least 14 cm) (Saunders *et al.*, 1983a,b). The reason for this difference is probably due to the difference in size of the fish as ATPase activity in gills of salmonid juveniles is correlated with fork length (Buckman & Ewing, 1982). The critical length for development of Na,K-ATPase activity is 6-8 cm, therefore the low activities of ATPase found in our studies do not seem unreasonable (Table 2).

Table 2. Mean Na,K- and residual-ATPase activity (± SE) of Atlantic salmon parr, *Salmo salar*, exposed to phenols for 96 h.

Treatment	n	Length (cm)	Weight (g)	ATPase activity (μmole Pi/mg protein·h) Na,K	Residual	Total
0-h control	27	7.7 (0.3)	4.6 (0.7)	0.43 (0.06)	2.03 (0.25)	2.46 (0.28)
96-h control	23	7.5 (0.2)	3.8 (0.4)	0.54 (0.07)	2.25 (0.17)	2.79 (0.20)
Pentachlorophenol	18	7.8 (0.2)	4.4 (0.3)	0.40 (0.07)	2.03 (0.32)	2.43 (0.34)
2-Chlorophenol	15	7.4 (0.2)	4.0 (0.3)	0.49 (0.10)	3.37* (0.40)	3.86* (0.42)
p-Nonylphenol	15	7.6 (0.2)	4.2 (0.3)	0.69 (0.07)	3.09* (0.31)	3.78* (0.39)
4-Methylphenol	15	7.9 (0.3)	4.4 (0.5)	0.29* (0.07)	1.94 (0.20)	2.23 (0.24)
Phenol	21	7.6 (0.2)	3.9 (0.3)	0.40 (0.06)	2.29 (0.21)	2.69 (0.22)
2,4,5-Trichlorophenol	14	7.5 (0.2)	3.7 (0.2)	0.38 (0.07)	2.30 (0.41)	2.68 (0.44)
3,5-Dimethylphenol	15	8.1 (0.2)	5.0 (0.5)	0.45 (0.07)	2.26 (0.17)	2.71 (0.18)
2,6-Dimethylphenol	23	8.0 (0.2)	4.9 (0.4)	0.38 (0.06)	2.75 (0.20)	3.13 (0.23)

*Significantly different ($p<0.05$) from 96-h controls in t-test.

There was no significant difference in Na,K- or residual ATPase activities of the gills of 0-h or 96-h controls. The residual ATPase activity of gills from fish exposed to 2-chloro- or p-nonylphenol was significantly increased (to 3.37 and 3.09 μmole Pi/mg protein·h) compared to those of 96-h controls. On the other hand, Na,K-ATPase activity in gills of parr exposed to 4-methylphenol was significantly decreased (to 0.29 μmole Pi/mg protein·h). In this case, the total ATPase activity was not significantly altered as it was with those showing increased residual ATPase activity above. No other significant differences in Na,K- or residual ATPase activities were observed in the gills of parr exposed to the other phenols.

The effect of xenobiotics on activity of gill ATPases has been variable (see Haya *et al.*, 1983 for a review). For example, oral administration of capsules containing pp'-DDT decreased the activity of Na,K- and residual ATPase in gills of *Salmo gairdneri* (Leadem *et al.*, 1974). Exposure to endosulfan (2-5 μg/L) for 30 d decreased Na,K- and residual ATPase activity of *Channa gachua* gills (Dalela *et al.*, 1978), whereas low concentrations of the related organochlorine pesticides aldrin and dieldrin stimulated the Na,K-ATPase activity of *Labeo rohita* gill homogenates (Verma *et al.*, 1978). Stimulation of Na,K-ATPase activity has also been reported in gills of rainbow trout exposed to zinc (Watson & Beamish, 1980) and in those of lobsters exposed to cadmium (Tucker, 1979). However, gills of lobsters exposed to zinc had decreased Na,K- and residual ATPase activities (Haya *et al.*, 1983) and those of lobsters taken from a Cd-polluted harbor showed no differences in the activities of ATPase to those sampled from a non-Cd-polluted area (Haya *et al.*, 1980).

The levels of adenylates in the livers of 96-h control Atlantic salmon were 1.19, 0.85, and 0.06 μmole/g wet weight for ATP, ADP, and AMP, respectively (Table 3). The levels of ATP and ADP are comparable to those reported for livers of goldfish, *Carassius auratus* (Van Den Thillart *et al.*, 1980). However, the AMP concentration is lower in the salmon parr livers compared to those of goldfish (0.06 vs. 0.52, respectively). Consequently the AEC of the salmon parr liver was higher or equal to the upper range of that for goldfish (0.767 vs. 0.60 to 0.75, respectively). Levels of adenylates in livers of the gulf killifish, *Fundulus grandis*, have been reported (MacFarlane, 1981), but are difficult to compare to these salmon parr values since gulf killifish concentrations were reported on a milligram protein rather than on a gram wet weight basis. The AEC of the gulf killifish liver was 0.712.

The mean creatine phosphate (CP) levels of livers from 96-h control salmon parr was 0.19 μmole/g wet weight and is similar to that found in goldfish (0.21 to 0.39 μmole/g wet weight) (Van Den Thillart *et al.*, 1980). The inorganic phosphate (Pi) concentration of the Atlantic salmon parr livers was 8.4 µmole/g wet weight. Thus, the phosphorylation potential (PP) of livers from 96-h control fish was 0.19/mM. PP has not been reported previously for livers of teleost fish, but the value reported in the present study is in the low end of the range expected for most cells (0.2-0.8/mM; Lehninger, 1975). However, this is about a magnitude lower than that found in mammalian liver cells (Erecinska & Wilson, 1982).

Table 3. Adenylate energy charge (AEC), phosphorylation potential (PP), and levels of adenine nucleotides (ATP,ADP,AMP), creatine phosphate (CP), and inorganic phosphate (Pi) (± SE) in livers of Atlantic salmon parr, *Salmo salar*, exposed to phenols for 96 h. Each n represents a pooled sample of livers from three fish, *i.e.*, 6 to 15 fish per treatment.

		µmole/g wet weight							
Treatment	n	ATP	ADP	AMP	Total	CP	Pi	PP/mM	AEC
0-h control	5	0.99 (0.14)	0.81 (0.10)	0.05 (0.01)	1.85 (0.12)	0.53 (0.33)	6.8 (0.6)	0.23 (0.09)	0.751 (0.029)
96-h control	5	1.19 (0.05)	0.85 (0.04)	0.06 (0.02)	2.10 (0.06)	0.19 (0.02)	8.4 (1.7)	0.19 (0.03)	0.767 (0.009)
Pentachlorophenol	4	1.38 (0.08)	0.74 (0.08)	0.10 (0.01)	2.22 (0.14)	0.17 (0.01)	6.6 (0.4)	0.30 (0.05)	0.789 (0.015)
2-Chlorophenol	5	1.27 (0.08)	0.83 (0.12)	0.01 (0.01)	2.11 (0.08)	0.22 (0.39)	9.0 (2.5)	0.22 (0.04)	0.799 (0.023)
p-Nonylphenol	5	1.00 (0.13)	0.57* (0.06)	0.03 (0.01)	1.60* (0.18)	0.14 (0.01)	6.2 (0.6)	0.31 (0.06)	0.802 (0.013)
4-Methylphenol	5	1.25 (0.12)	0.85 (0.03)	0.04 (0.03)	2.14 (0.11)	0.15 (0.01)	7.8 (0.4)	0.19 (0.03)	0.781 (0.016)
Phenol	4	1.12 (0.03)	0.85 (0.08)	0.09 (0.04)	2.06 (0.08)	0.21 (0.03)	5.6 (0.03)	0.24 (0.03)	0.750 (0.012)
2,4,5-Trichloro-phenol	5	1.34 (0.08)	0.85 (0.07)	0.14 (0.04)	2.33 (0.16)	0.20 (0.02)	8.9 (1.1)	0.19 (0.02)	0.761 (0.018)
3,5-Dimethylphenol	2	1.02 (0.21)	0.72 (0.12)	0.01 (0.01)	1.75 (0.11)	0.01* (0.03)	5.3 (0.5)	0.30 (0.13)	0.785 (0.039)
2,6-Dimethylphenol	5	1.41* (0.04)	0.57* (0.04)	0.05 (0.02)	2.03 (0.06)	0.14 (0.01)	5.6 (0.1)	0.45* (0.05)	0.838* (0.012)

*Significantly different ($p<0.05$) from 96-h controls in t-test.
Total = ATP + ADP + AMP.

No significant differences were found in PP or AEC between livers of Atlantic salmon parr exposed to phenols for 96 h and those of either the 96-h or 0-h controls except for those fish exposed to 2,6-dimethylphenol (Table 3). These fish are in a higher energy state than the control fish as indicated by the increase in AEC and PP. This is contrary to the expected decrease if 2,6-dimethylphenol was interfering with regulatory processes. This effect may or may not be related to the inactive state exhibited by these fish during the exposure (see above). The other isolated significant differences indicated in Table 3 are probably not due to sublethal effects caused by exposure to phenols. The significant decrease observed for creatine phosphate levels in the livers of fish exposed to 3,5-dimethylphenol may be a starvation effect (see below). The primary mechanism of pentachlorophenol toxicity is considered to be uncoupling of mitochondrial oxidative phosphorylation (Rao *et al.*, 1979). Our experiments do not support this hypothesis as pentachlorophenol did not affect any of the parameters of energy metabolism in the livers of the Atlantic salmon parr exposed for 96 h. A number of physiological and xenobiotic stresses have caused a decrease in AEC of many invertebrates and fish muscle (Van Den Thillart, 1982; Vetter & Hodson, 1982; Haya *et al.*, 1983). Decreased liver AEC in teleost fish has only been reported for killifsh exposed to low pH (MacFarlane, 1981) and for goldfish exposed to hypoxic conditions (Van Den Thillart *et al.*, 1980). CP was also depleted in the liver of goldfish during hypoxia.

The levels of glucose and glycogen in livers of the 96-h control Atlantic salmon parr were 4.6 and 250 µmole/g wet weight, respectively (Figs. 1 & 2; glycogen concentrations are reported as µmole of equivalents of glucose/g wet weight of liver). There was no significant difference in liver glycogen or glucose concentration between 0-h and 96-h control fish. The concentrations of liver glycogen and glucose in fish exposed to phenols were all significantly lower than those of the 0-h controls, but not different from those of 96-h controls (t-test; $p<0.05$) in most treatments. The glycogen levels in livers of fish exposed to phenol and 3,5-dimethylphenol were significantly decreased (to 84 and 61 µmole/g wet weight, respectively) compared to those in livers of the 96-h controls. That is, there was a trend for the glycogen and glucose levels in liver of fish exposed to phenols to be lower than that of 96-h control fish which in turn was lower than that of 0-h controls. An exception is the increase in glucose levels for those fish exposed to 2,6-dimethylphenol (to 9.3 µmole/g wet weight). Whether or not this is related to the anaesthetic-like effect (see above) of 2,6-dimethylphenol observed during the exposure will require further studies.

Both increases and decreases in liver glycogen content of teleost fish exposed to chemicals have been reported. For example, chronic exposure of *Mugil cephalus* to pentachlorophenol caused a depletion of liver glycogen reserves (Thomas *et al.*, 1981). Paraquat increased the mobilization of glycogen in the liver of *Cyprinus carpio* L. (Simon *et al.*, 1983). Depending on the length of exposure, endosulfan produced an increase or decrease in the glycogen levels in livers of *Heteropneustes fossilis* (Singh & Srivastava,

1981). An increase in the glycogen content of Platichthys flesus L. liver and a decrease in that of Salmo gairdneri liver was observed on exposure of these fish to cadmium. (250 μg/L Cd) (Larsson & Haux, 1982).

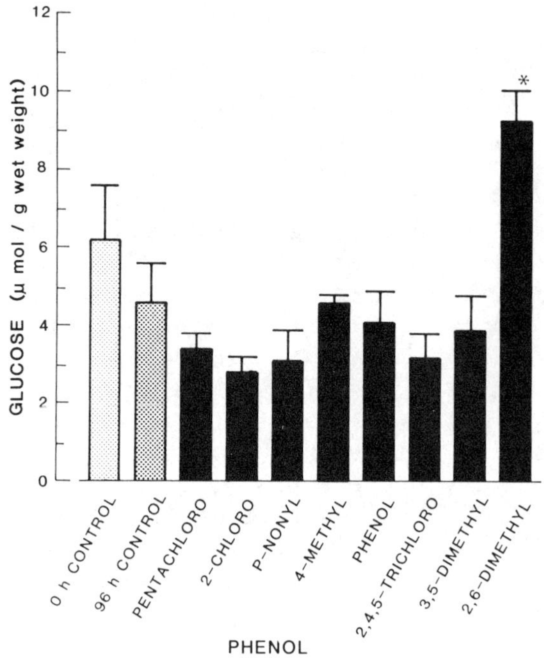

Fig. 1. Levels of glucose (± SE) in livers of Atlantic salmon parr, Salmo salar, exposed to phenols for 96 h. Each represents an n of 2 to 5 pooled samples (see Table 3). *Significantly different ($p<0.05$) from 96-h controls in t-test.

The levels of CP were consistently lower in the livers of 96-h control fish and those exposed to the phenols compared to those of 0-h control fish. As the fish were not fed during the experiment, a probable cause for the decrease in glycogen, glucose and CP levels is starvation. Decreases in liver glycogen and glucose levels occurred in Salmo gairdneri after 5 d without food (Morata et al., 1982). Since the decreases in glucose and glycogen levels were consistently greater in the livers of fish exposed to the phenols for 96 h compared to the 96-h controls, the fish exposed to phenols when not starved may also result in decreased liver concentration of glucose and glycogen. Whether the phenols affect the levels of these substrates directly or increase the effects of starvation will require further studies.

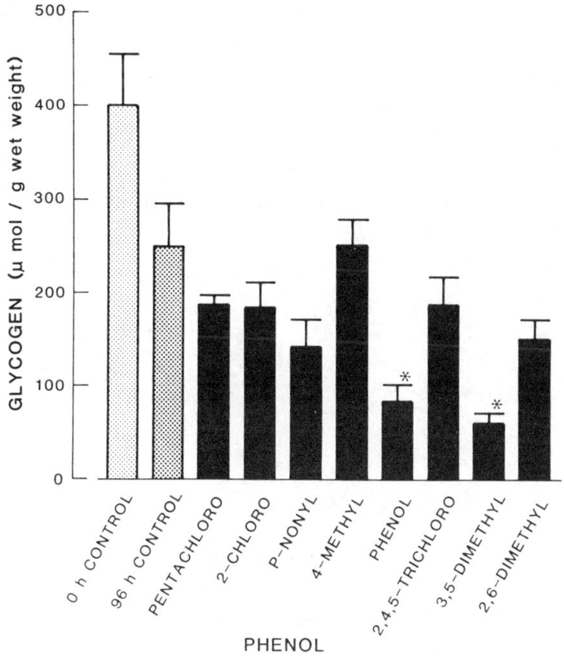

Fig. 2. Levels of glycogen (± SE) in livers of Atlantic salmon parr, Salmo salar, exposed to phenols for 96 h. Each bar represents an n of 2 to 5 pooled samples (see Table 3). *Significantly different ($p<0.05$) from 96-h controls in t-test.

On a theoretical basis, the parameters of energy metabolism considered in the present study have potential as indicators of sublethal effects caused by xenobiotics. However, as more information is obtained, it appears that regulatory processes strive to maintain equilibrium levels of the relative adenylate concentrations. Therefore AEC and PP can remain constant during exposure to toxic concentrations of chemicals. AEC appears to be a useful indicator with invertebrates, but the effects have been variable. Other substrates of energy metabolism (e.g., phosphagens, glucose, glycogen) and energy utilization processes (e.g., ATPases, maintenance of redox potentials) may be affected by sublethal concentrations of chemicals. The biochemical responses of these processes vary with species, organ or tissue sampled, and with many other physiological variables (*e.g.*, sex, stage of life cycle, reproductive status, habitat) as well as with different chemicals. In addition, these parameters can be altered by "natural stressors", for example, salinity, temperature, and tidal effects such as temporary hypoxic or anoxic conditions in the sediment of intertidal zones. It seems that a series of biochemical and physiological tests rather than a single all encompassing test will give a better indication of potential or existing pollution problems, or aid in the assessment of the hazard potential of chemicals in the environment.

ACKNOWLEDGMENTS

We thank Dr. V. Zitko for a critical review of the manuscript, Ms. R. Garnett for editorial comments, Mrs. B. Fawkes for typing and Mr. B. McMullon for preparing the figures.

REFERENCES

Atkinson, D.E. 1977. Cellular Energy Metabolism and Its Regulation. Academic Press, New York, p. 85-107.

Buckman, M. and R.D. Ewing. 1982. Relationship between size and time of entry into the sea and gill (Na+K)-ATPase activity for juvenile spring chinook salmon. Trans. Am. Fish. Soc. 111: 681-687.

Bergmeyer, H.U. (ed). 1974. Methods of Enzymatic Analysis, Academic Press, New York.

Dalela, R.C., M.C. Bhatnagar, A.K. Tyagi, and S.R. Verma. 1978. Adenosine triphosphatase activity in few tissues of a fresh water teleost, *Channa gachua*, following *in vivo* exposure to endosulfan. Toxicology 11: 361-368.

Erecinska, M., M. Stubbs, Y. Miyata, C.M. Ditre, and D.F. Wilson. 1977. Regulation of cellular metabolism by intracellular phosphate. Biochim. Biophys. Acta 462: 20-35.

Erecinska, M. and D.F. Wilson. 1982. Regulation of cellular energy metabolism. J. Membr. Biol. 70: 1-14.

Haya, K., C.E. Johnston, and B.A. Waiwood. 1980. Adenylate energy charge and ATPase activity in American lobster (*Homarus americanus*) from Belledune Harbour. pp. 85-91. *In*: Cadmium Pollution of Belledune Harbour, New Brunswick, Canada. J.F. Uthe and V. Zitko (eds.). Can. Tech. Rep. Fish. Aquat. Sci. 963.

Haya, K. and B.A. Waiwood. 1983. Adenylate energy charge and ATPase activity: Potential biochemical indicators of sublethal effects caused by pollutants in aquatic animals. pp. 307-333. In: Aquatic Toxicology. J.O. Nriagu (ed.). John Wiley and Sons, New York.

Haya, K., B.A. Waiwood, and D.W. Johnston. 1983. Adenylate energy charge and ATPase activity of lobster (*Homarus americanus*) during sublethal exposure to zinc. Aquat. Toxicol. 3: 115-126.

Heinonen, J.K. and R.J. Lahti. 1981. A new and convenient colorimetric determination of inorganic orthophosphate and its application to the assay of inorganic pyrophosphatase. Anal. Biochem. 113: 313-317.

Ivanovici, A.M. 1980. Application of adenylate energy charge to problems of environmental impact assessment in aquatic organisms. Helgol. Wiss. Meeresunters. 33: 556-565.

Larsson, A. and C. Haux. 1982. Altered carbohydrate metabolism in fish exposed to sublethal levels of cadmium. J. Environ. Biol. 2: 71-81.

Leadem, T.P., R.D. Campbell, and D.W. Johnson. 1974. Osmoregulatory responses to DDT and varying salinities in *Salmo gairdneri*. I. Gill Na,K-ATPase. Comp. Biochem. Physiol. 49A: 197-205.

Lehninger, A.L. 1975. Biochemistry. Worth Publishers, Inc., New York. 539 pp.

MacFarlane, R.B. 1981. Alterations in adenine nucleotide metabolism in the gulf killifish (*Fundulus grandis*) induced by low pH water. Comp. Biochem. Physiol. 68B: 193-202.

McLeese, D.W., V. Zitko, C.D. Metcalfe, and D.B. Sergeant. 1980. Lethality of aminocarb and the components of the aminocarb formulation to juvenile Atlantic salmon, marine invertebrates and a fresh water clam. Chemosphere 9: 79-82.

McLeese, D.W., V. Zitko, and M.R. Peterson. 1979. Structure-lethality relationships for phenols, anilines and other aromatic compounds in shrimp and clams. Chemosphere 2: 53-57.

Morata, P., A.M. Vargus, F. Sanchez-Medina, M. Garcia, G. Cardenete, and S. Zamora. 1982. Evolution of gluconeogenic enzyme activities during starvation in liver and kidney of rainbow trout (*Salmo gairdneri*). Comp. Biochem. Physiol. 71B: 65-70.

Neufeld, G.I. and J.B. Pritchard. 1979. Osmoregulation and gill Na,K-ATPase in the rock crab, *Cancer irrotatus*: Response to DDT. Comp. Biochem. Physiol. 62C: 165-172.

Rao, K.R., F.R. Fox, P.J. Conklin, and A.C. Cantelmo. 1981. Comparative toxicology and pharmacology of chlorophenols: Studies on the grass shrimp, *Palaemonetes pugio* pp. 37-72. *In*: Biological Monitoring of Marine Pollutants. F.J. Vernberg, A. Calabrese, F.P. Thurberg and W.B. Vernberg (eds.). Academic Press, New York.

Rao, K.R., F.R. Fox, P.J. Conklin, A.C. Cantelmo, and A.C. Brannon. 1979. Physiological and biochemical investigations of the toxicity of pentachlorophenol to crustaceans. pp. 307-339. *In*: Marine Pollution: Functional Responses. W.B. Vernberg, F.P. Thurberg, A. Calabrese, and F.J. Vernberg (eds.). Academic Press, New York.

Reich, J.G. and E.E. Sel'kov. 1981. Energy Metabolism of the Cell. A Theoretical Treatise. Academic Press Inc. (London) Ltd., pp. 108-198.

Saunders, R.L., E.B. Henderson, P.R. Harmon, C.E. Johnston, and K. Davidson. 1983a. Physiological effects of low pH on smolting Atlantic salmon. pp. 47-49. *In*: Proceedings on the Workshop on Acid Rain. R.H. Peterson and H.H.V. Hord (eds.). Can. Tech. Rept. Fish. Aquat. Sci. 1213.

Saunders, R.L., E.B. Henderson, P.R. Harmon, C.E. Johnston, and J.G. Eales. 1983b. Effects of low environmental pH on smolting of Atlantic salmon (*Salmo salar*). Can. J. Fish. Aquat. Sci. 40: 1203-1211.

Schurrmans Stekhoven, F.M.A.H. and S.L. Bonting. 1981. Sodium-potassium-activated adenosine triphosphatase. pp. 159-182. *In*: Membrane Transport. S.L. Bonting and J.J.H.H.M. Depont (eds.). Elsevier/ North Holland Biomedical Press, Amsterdam.

Simon, L.M., J. Nemcsok, and L. Boross. 1983. Studies on the effect of paraquat on glycogen mobilization in liver of common carp (*Cyprinus carpio* L.). Comp. Biochem. Physiol. 75C: 167-169.

Singh, N. and A.K. Srivastava. 1981. Effects of endosulfan on fish carbohydrate metabolism. Ecotoxicol. Environ. Safety 5: 412-417.

Thomas, P., R.S. Carr, and J.M. Neff. 1981. Biochemical stress responses of mullet *Mugil cephalus* and polychaete worms *Neanthes virens* to pentachlorophenol. pp. 73-103. *In*: Biological Monitoring of Marine Pollutants. F.J. Vernberg, A. Calabrese, F.P. Thurberg and W.B. Vernberg (eds.). Academic Press, New York.

Towle, D.W. 1981. Role of $Na^{++}K^{+}$-ATPase in ionic regulation by marine and estuarine animals. Mar. Biol. Lett. 2: 107-122.

Tucker, R.K. 1979. Effects of *in vivo* cadmium exposure on ATPases in gill of the lobster *Homarus americanus*. Bull. Environm. Contam. Toxicol. 23: 33-35.

Van Den Thillart, G. 1982. Adaptations of fish energy metabolism to hypoxia and anoxia. Mol. Physiol. 2: 49-61.

Van Den Thillart, G., F. Kesbeke, and A. Van Waarde. 1980. Anaerobic energy metabolism of goldfish, *Carassius auratus* (L.). Influence of hypoxia and anoxia on phosphorylated compounds and glycogen. Comp. Biochem. Physiol. 136: 45-52.

Verma, S.R., A.K. Gupta, S.K. Bansal, and R.C. Dalela. 1978. *In vitro* disruption of ATP dependent active transport following treatment with aldrin and its epoxy analog dieldrin in a fresh water teleost, *Labeo rohita*. Toxicology 11: 193-201.

Vetter, R.D. and R.E. Hodson. 1982. Use of adenylate concentrations and adenylate energy charge as indicators of hypoxic stress in estuarine fish. Comp. Biochem. Physiol. 39: 535-541.

Watson, T.A. and F.W.H. Beamish. 1980. Effects of zinc on branchial ATPase activity *in vivo* in rainbow trout, *Salmo gairdneri*. Comp. Biochem. Physiol. 66C: 77-82.

Systematic Index

Subject Index

Liability of lawyers and indemnity insurance